BIM Handbook

A Guide to Building Information Modeling for Owners, Managers, Designers, Engineers, and Contractors

Chuck Eastman
Paul Teicholz
Rafael Sacks
Kathleen Liston

John Wiley & Sons, Inc.

This book is printed on acid-free paper. ♾

Published by John Wiley & Sons, Inc., Hoboken, New Jersey
Published simultaneously in Canada

Library of Congress Cataloging-in-Publication Data:
BIM handbook : a guide to building information modeling for owners, managers, designers, engineers, and contractors / Chuck Eastman . . . [et al.].
p. cm.
Includes bibliographical references and index.
ISBN: 978-0-470-18528-5 (cloth)
1. Building—Data processing—Handbooks, manuals, etc. 2. Building management—Data processing—Handbooks, manuals, etc. 3. Communication in the building trades—Handbooks, manuals, etc. 4. Architectural practice—Handbooks, manuals, etc. 5. Architects and builders—Handbooks, manuals, etc. I. Eastman, Charles M.
TH437.B53 2008
690.0285—dc22

2007029306

Printed in the United States of America

10 9 8 7 6 5 4 3

Contents

Preface

This book is about a new approach to design, construction, and facility management called *building information modeling* (BIM). It provides an in-depth understanding of BIM technologies, the business and organizational issues associated with its implementation, and the profound impacts that effective use of BIM can provide to all members of a project team. The book explains how designing, constructing, and operating buildings with BIM differs from pursuing the same activities in the traditional way using drawings, whether paper or electronic.

BIM is beginning to change the way buildings look, the way they function, and the ways in which they are built. Throughout the book, we have intentionally and consistently used the term 'BIM' to describe an activity (meaning *building information modeling*), rather than an object (*building information model*). This reflects our belief that BIM is not a thing or a type of software but a human activity that ultimately involves broad process changes in construction.

WHY A BIM HANDBOOK?

Our motivation in writing this book was to provide a thorough and consolidated reference to help students and practitioners in the building industry learn about this exciting new approach, in a format independent of the commercial interests that guide vendors' literature on the subject. There are many truths and myths in the generally accepted perceptions of the state of the art of BIM. We hope that the BIM Handbook will help reinforce the truths, dispel the myths, and guide our readers to successful implementations. Many well-meaning decision-makers and practitioners in the building industry at-large have had disappointing experiences after attempting to adopt BIM, because their efforts and expectations were based on misconceptions and inadequate planning. If this book can help readers avoid these frustrations and costs, we will have succeeded.

Collectively, the authors have a wealth of experience with BIM. We believe that BIM represents a paradigm change that will have far-reaching benefits, not only for those in the building industry but for society at-large, as better buildings are built that consume less energy and require less labor and capital resources. We make no claim that the book is objective in terms of our judgment of the necessity for BIM. At the same time, of course, we have made every effort to ensure the accuracy and completeness of the facts and figures presented.

WHO IS THE BIM HANDBOOK FOR, AND WHAT IS IN IT?

The BIM Handbook is addressed to building developers, owners, managers, and inspectors; to architects, engineers of all disciplines, construction contractors, and fabricators; and to students of architecture, civil engineering, and building construction. It reviews Building Information Modeling and its related technologies, its potential benefits, its costs and needed infrastructure. It also discusses the present and future influences of BIM on regulatory agencies; legal practice associated with the building industry; and manufacturers of building products—it is directed at readers in these areas. A rich set of BIM case studies are presented and various BIM tools and technologies are described. Industry and societal impacts are also explored.

The book has four sections:

I. Chapters 1, 2, and 3 provide an introduction to BIM and the technologies that support it. These chapters describe the current state of the building industry, the potential benefits of BIM, parametric modeling of buildings, and interoperability issues.

II. Chapters 4, 5, 6, and 7 provide discipline-specific perspectives of BIM. They are aimed at owners (Chapter 4), designers of all kinds (Chapter 5), general contractors (Chapter 6), and subcontractors and fabricators (Chapter 7).

III. Chapter 8 discusses potential impacts and future trends associated with the advent of BIM-enabled design, construction, and operation of buildings. Current trends are described and extrapolated through the year 2012 as are forecasts of potential long-term developments and the research needed to support them through 2020.

IV. Chapter 9 provides ten detailed cases studies of BIM in the construction industry that demonstrate its use for feasibility studies, conceptual design, estimating, detailing, coordination, construction planning, logistics, operations and many other common construction activities. The case studies include buildings with signature architectural and structural designs (such as the Beijing National Aquatics Center, the 100 11th Avenue apartment building facade in New York City, and the environmentally friendly Federal Building in San Francisco) as well as a wide range of fairly common buildings (a GM production plant, a federal courthouse, a medical office building, a parking structure, a high-rise office building, a mixed commercial and retail development, and a coast-guard training facility).

HOW TO USE THE BIM HANDBOOK

Many readers will find the Handbook a useful resource whenever they are confronted with new terms and ideas related to BIM in the course of their work or study. A thorough first-reading, while not essential, is of course the best way to gain a deeper understanding of the significant changes that BIM is bringing to the building industry.

The first section (Chapters 1–3) is recommended for all readers. It gives a background to the commercial context and the technologies for BIM. Chapter 1 lists many of the potential benefits that can be expected. It first describes the difficulties inherent in current practice within the U.S. building industry and its associated poor productivity and higher costs. It then describes various approaches to procuring construction, such as traditional design-bid-build, design-build, and others, describing the pros and cons for each in terms of realizing benefits from the use of BIM. Chapter 2 details the technological foundations of BIM, in particular parametric and object-oriented modeling. The history of these technologies and their current state of the art are described. The chapter then reviews the leading commercial application platforms for generating building information models. Chapter 3 deals with the intricacies of interoperability, including how building information can be communicated and shared from profession to profession and from application to application. Chapters 2 and 3 can also be used as a reference for the technical aspects of parametric modeling and interoperability.

Readers who desire specific information on how they can adopt and implement BIM in their companies can find the details they need in the relevant chapter for their profession within Chapters 4–7. You may wish to read the chapter closest to your area of interest and then only the executive summaries of each of the other chapters. There is a small degree of overlap within these chapters, where issues are relevant to multiple professions. These chapters make frequent reference to the set of detailed case studies provided in Chapter 9.

Those who wish to learn about the long term economic, organizational, social, and professional implications of BIM and how they may impact your educational or professional life will find an extensive discussion of these issues in Chapter 8.

The case studies in Chapter 9 each tell a story about different professionals' experiences using BIM on their projects. No one case study represents a 'complete' implementation or covers the entire building lifecycle. In most cases, the building was not complete when the study was written. But taken together, they paint a picture of the variety of uses and the benefits and problems that these pioneering firms have already experienced. They illustrate what could be achieved with exsiting BIM technology at the start of the 21st century. There are many lessons learned that can provide assistance to our readers and guide practices in future efforts.

Finally, students and professors are encouraged to make use of the study questions and exercises provided at the conclusion of each chapter.

ACKNOWLEDGMENTS

Naturally, we are indebted first and foremost to our families, who have all borne the brunt of the extensive time we have invested in this book.

Our thanks and appreciation for highly professional work are due to Laurie Manfra, who copyedited the book, and to Jim Harper, our publisher representative at John Wiley and Sons.

Our research for the book was greatly facilitated by numerous builders, designers, and owners, representatives of software companies and government agencies; we thank them all sincerely. Six of the case studies were originally prepared by graduate students in the College of Architecture at Georgia Tech; we thank them, and their efforts are acknowledged personally at the end of each relevant case study. The case studies were made possible through the very generous contributions of the project participants who corresponded with us extensively and shared their understanding and insights.

Finally, we are grateful to Jerry Laiserin, not only for his enlightening foreword, but also for helping to initiate the idea for the BIM Handbook.

Foreword

In recent years, both the concept and nomenclature we now know as BIM— or Building Information Models and Building Information Modeling—have engaged professional and industry awareness sufficiently to justify treatment in handbook format. The result is this book, which amply fulfills the requirements anyone would expect of a handbook, with respect to both breadth of coverage and depth of exposition.

However, neither the concept nor nomenclature of BIM is new—not as of 2007, not as of 2002, nor even 1997. The concepts, approaches and methodologies that we now identify as BIM can be dated back nearly thirty years, while the terminology of the "Building Information Model" has been in circulation for at least fifteen years. Note that my gloss on the history of BIM, below, is necessarily condensed to isolated highlights; my apologies in advance to those whose contributions I may unintentionally have slighted.

The earliest documented example I have found for the concept we know today as BIM was a working prototype "Building Description System" published in the now-defunct *AIA Journal* by Charles M. "Chuck" Eastman, then at Carnegie-Mellon University, in 1975. Chuck's work included such now-routine BIM notions as:

> [designing by] ". . . interactively defining elements . . . deriv[ing] sections, plans, isometrics or perspectives from the same description of elements . . . Any change of arrangement would have to be made only once for all future drawings to be updated. All drawings derived from the same arrangement of elements would automatically be consistent . . . any type of quantitative analysis could be coupled directly to the description . . . cost estimating or material quantities could be easily generated . . . providing a single integrated database for visual and quantitative analyses . . . automated building code checking in city hall or the architect's office. Contractors of large projects may find this representation advantageous for scheduling and materials ordering." (Eastman 1975)

Comparable research and development work was conducted throughout the late 1970s and early 1980s in Europe—especially in the UK—in parallel with early efforts at commercialization of this technology (see below). During the early 1980s this method or approach was most commonly described in the USA as "Building Product Models" and in Europe—especially in Finland—as "Product Information Models" (in both phrases, "product" was used to distinguish this approach from "process" models). The next logical step in this nomenclature evolution was to verbally factor out, so to speak, the duplicated "product" term, so that "Building Product Model" + "Product Information Model" would merge into "Building Information Model." Although the German *BauInformatik* may be translated this way, its usual meaning more closely conforms to the general application of information and computer technology (ICT) to construction. However, the Dutch *Gebouwmodel* was occasionally used in the mid-to-late 1980s in contexts that arguably could be translated to English as "Building Information Model" rather than the literal rendition of "Building Model."

The first documented use of the term "Building Modeling" in English—in the sense that "Building Information Modeling" is used today—appeared in the title of a 1986 paper by Robert Aish, then with GMW Computers Ltd., makers of the legendary RUCAPS software system. Aish, who is today with Bentley Systems, set out in this paper all the arguments for what we now know as BIM and the technology to implement it, including: 3D modeling; automatic drawing extraction; intelligent parametric components; relational databases; temporal phasing of construction processes; and so forth (Aish 1986). Aish illustrated these concepts with a case study applying the RUCAPS building modeling system to the phased refurbishment of Terminal 3 at Heathrow Airport, London (there is, in my opinion, a bit of historical irony in the fact that some twenty years later, the construction of Terminal 5 at Heathrow often is cited as one of the "pioneering" case examples of this technology).

From "Building Model" it was but a short leap to "Building Information Model," for which the first documented use in English appeared in a paper by G.A. van Nederveen and F. Tolman in the December 1992 *Automation in Construction*. (van Nederveen & Tolman 1992).

In parallel with the evolving nomenclature and the R&D efforts centered in academia, commercial products implementing the BIM approach (under whatever commercial moniker at the time) also have a long history. Many of the software functions and behaviors ascribed to today's generation of model-authoring tools, such as AllPlan, ArchiCAD, Autodesk Revit, Bentley Building, DigitalProject or VectorWorks, were also the design goals of earlier commercial software efforts such as: the UK lineage from RUCAPS (cited above) to Sonata and Reflex; another UK lineage from Oxsys to BDS and GDS (the latter still available as MicroGDS); a French lineage that included Cheops and Architrion (the spirit of which lives on in BOA); Brics (a Belgian system that provided the technology core for Bentley's Triforma); the US-based Bausch & Lomb modeling system of 1984; Intergraph's efforts with Master Architect; plus many others, now but dimly remembered (at least, within my recollection).

Thus, by the time of my first attempt to popularize the term (Laiserin, 2002) and craft a multi-vendor consensus around it (Laiserin, 2003) the core nomenclature of Building Information Modeling had been coined at least ten years earlier, and the concept or approach had been established more than a further fifteen years before that (with numerous practical demonstrations along the way, most notably in the Finnish efforts that culminated in the Vera Project by Tekes, the National Technology Agency of Finland).

The present book, as might be expected of any work crafted by multiple authors, contains more than one definition of BIM—from process-oriented to product-oriented and from constructor-practical to defining BIM by what it is not. Allow me to add yet two more of my own: one, broader and more analytic than the subject as covered in this Handbook; the other, more simplistic and functional than the Handbook treatment.

Because my broader definition of BIM-as-process is wholly independent of software for implementation, and therefore outside the scope of this book, I offer only the citations here (Laiserin 2005, 2007). However, the arena of "BIM-ready" or "BIM-worthy" commercial software for design and analysis has so expanded and matured in the decade immediately preceding this writing, that I believe it is now both possible and worthwhile to propose instead some quick and practical principles for qualifying BIM applications. For this purpose, I suggest the following: that IFC certification (per the International Alliance for Interoperability, IAI, and/or the buildingSMART initiative) be deemed a sufficient, but not necessary condition of "BIM-ness" for any design, analysis or collaborative software. IFC certification may be supplemented as needed by local requirements, such as support for the GSA Space Object or conformity with the National BIM Standard in the

USA, or the IFC Code Checking View in Singapore. My point is that while there may be numerous paths to BIM-ness, many of which need not lead through IFC certification, surely it is the case that any design or analysis application that is IFC-certified has thereby attained the requisite BIM-ness to be deemed BIM-ready, BIM-worthy, or indeed just plain BIM.

Notwithstanding the semantic distinctions among various BIM definitions, some people labor under the impression that I—or one or another of the design-software vendors—"coined" the term and/or "originated" or "developed" or "introduced" the concept or approach circa 2002. I have never claimed such distinction for myself, and it is my opinion that the historical record outlined above shows that Building Information Modeling was not an innovation attributable solely to any individual or entity.

Rather than "father of BIM"—as a few well-meaning but over-enthusiastic peers have labeled me—I prefer the unattributed epithet "godfather of BIM," in the sense that a godfather is an adult sponsor of a child not his own. If anyone deserves the title "father of BIM," surely it is Chuck Eastman. From his 1975 pioneering prototype system cited above to his 1999 text, *Building Product Models* (Eastman, 1999—the most authoritative treatment of the subject prior to the present volume), Chuck devoted a quarter century to defining the problems and advancing the solutions, plus a further decade continuing to push forward the frontiers of Building, Information, and Modeling.

In 2005, I had the honor and privilege of co-producing and co-hosting with Chuck (and the PhD program at the College of Architecture, Georgia Institute of Technology) the first industry-academic Conference on BIM (Laiserin, 2005). This venue featured a broad cross-section of design-software and analysis vendors, as well as leading practitioners such as Vladimir Bazjanac of Lawrence Berkeley National Laboratories and Godfried Augenbroe of Georgia Tech. To complement our own opening keynotes, Chuck and I were fortunate to engage for the conference's closing keynote Paul Teicholz, now emeritus at Stanford and the founder of Stanford's Center for Integrated Facility Engineering (CIFE—which has become the leading proponent of Virtual Design and Construction, an approach I see as the ultimate state or end-game of BIM automation). Post-conference correspondence among Paul, Chuck and myself provided the impetus for this Handbook.

In the course of producing the Handbook, Chuck and Paul engaged an additional two formidable collaborators, Rafael Sacks and Kathleen Liston. Rafael's doctoral work at Israel's Technion addressed Computer Integrated Construction and Project Data Models, areas in which he subsequently collaborated with Chuck Eastman (especially regarding structural engineering of steel and precast concrete systems). While pursuing her doctorate at Stanford/CIFE, Kathleen co-authored seminal papers on construction schedule simulation or 4D-CAD, and went on to commercialize that BIM-related technology in the successful software startup company, Common Point, Inc.

All of which brings us back to the present work. It would be difficult to imagine a more accomplished quartet of authors in this field, or any team better suited to undertake a task such as The BIM Handbook. What they have accomplished will stand as the most definitive and authoritative treatment of the subject for many years to come. I now place the reader in the capable hands of my friends and colleagues—Chuck, Paul, Rafael and Kathleen—and their chef d'ouvre—The BIM Handbook.

JERRY LAISERIN
WOODBURY, NEW YORK
NOVEMBER 2007

ACKNOWLEDGMENTS

I wish to thank the following individuals for their generous assistance in documenting the early history of the BIM concept and terminology: Bo-Christer Björk, Hanken University, Finland; Arto Kiviniemi, VTT, Finland; Heikki Kulusjarvi, Solibri, Inc., Finland; Ghang Lee, Yonsei University, Korea; Robert Lipman, National Institute of Standards and Technology, USA; Hannu Penttila, Mittaviiva Öy, Finland.

REFERENCES

Aish, R., 1986, "Building Modelling: The Key to Integrated Construction CAD," CIB 5th International Symposium on the Use of Computers for Environmental Engineering Related to Buildings, 7–9 July.

Eastman, C., 1975, "The Use of Computers Instead of Drawings," *AIA Journal,* March, Volume 63, Number 3, pp 46–50.

Eastman, C., 1999, *Building Product Models: Computer Environments, Supporting Design and Construction,* CRC

Laiserin, J, 2002, "Comparing Pommes and Naranjas," *The LaiserinLetter*™, December 16, Issue 16, http://www.laiserin.com/features/issue15/feature01.php

Laiserin, J, 2003, "The BIM Page," *The LaiserinLetter*™, http://www.laiserin.com/features/bim/index.php

Laiserin, J, 2005, "Conference on Building Information Modeling: Opportunities, Challenges, Processes, Deployment," April 19–20, http://www.laiserin.com/laiserinlive/index.php

Laiserin, J., 2007, "To BIMfinity and Beyond!" *Cadalyst,* November, Volume 24, Number 11, pp 46–48.

van Nederveen, G.A. & Tolman, F., 1992, "Modelling Multiple Views on Buildings," *Automation in Construction,* December, Volume 1, Number 3, pp 215–224.

CHAPTER 1

BIM Handbook Introduction

1.0 EXECUTIVE SUMMARY

Building Information Modeling (BIM) is one of the most promising developments in the architecture, engineering and construction (AEC) industries. With BIM technology, an accurate virtual model of a building is constructed digitally. When completed, the computer-generated model contains precise geometry and relevant data needed to support the construction, fabrication, and procurement activities needed to realize the building.

BIM also accommodates many of the functions needed to model the lifecycle of a building, providing the basis for new construction capabilities and changes in the roles and relationships among a project team. When implemented appropriately, BIM facilitates a more integrated design and construction process that results in better quality buildings at lower cost and reduced project duration.

This chapter begins with a description of existing construction practices, and it documents the inefficiencies inherent in these methods. It then explains both the technology behind BIM and recommends ways to best take advantage of the new business processes it enables for the entire lifecycle of a building. It concludes with an appraisal of various problems one might encounter when converting to BIM technology.

1.1 INTRODUCTION

To better understand the significant changes that BIM introduces, this chapter begins with a description of current paper-based design and construction methods and the predominant business models now in use by the construction industry. It then describes various problems associated with these practices, outlines what BIM is, and explains how it differs from 2D and 3D computer-aided design (CAD). We give a brief description of the kinds of problems that BIM can solve and the new business models that it enables. The chapter concludes with a presentation of the most significant problems that may arise when using the technology, which is now only in its earliest phase of development and use.

1.2 THE CURRENT AEC BUSINESS MODEL

Currently, the facility delivery process remains fragmented, and it depends on paper-based modes of communication. Errors and omissions in paper documents often cause unanticipated field costs, delays, and eventual lawsuits between the various parties in a project team. These problems cause friction, financial expense, and delays. Recent efforts to address such problems have included: alternative organizational structures such as the design-build method; the use of real-time technology, such as project Web sites for sharing plans and documents; and the implementation of 3D CAD tools. Though these methods have improved the timely exchange of information, they have done little to reduce the severity and frequency of conflicts caused by paper documents.

One of the most common problems associated with paper-based communication during the design phase is the considerable time and expense required to generate critical assessment information about a proposed design, including cost estimates, energy-use analysis, structural details, etc. These analyses are normally done last, when it is already too late to make important changes. Because these iterative improvements do not happen during the design phase, *value engineering* must then be undertaken to address inconsistencies, which often results in compromises to the original design.

Regardless of the contractual approach, certain statistics are common to nearly all large-scale projects ($10 M or more), including the number of people involved and the amount of information generated. The following data was compiled by Maged Abdelsayed of Tardif, Murray & Associates, a construction company located in Quebec, Canada (Hendrickson 2003):

- Number of participants (companies): 420 (including all suppliers and sub-sub-contractors)
- Number of participants (individuals): 850
- Number of different types of documents generated: 50

- Number of pages of documents: 56,000
- Number of bankers boxes to hold project documents: 25
- Number of 4-drawer filing cabinets: 6
- Number of 20 inch diameter, 20 year old, 50 feet high, trees used to generate this volume of paper: 6
- Equivalent number of Mega Bytes of electronic data to hold this volume of paper (scanned): 3,000 MB
- Equivalent number of compact discs (CDs): 6

It is not easy to manage an effort involving such a large number of people and documents, regardless of the contractual approach taken. Figure 1-1 illustrates the typical members of a project team and their various organizational boundaries.

There are two dominant contract methods in the U.S, Design-Bid-Build and Design-Build, and many variations of them (Sanvido and Konchar 1999; Warne and Beard 2005).

1.2.1 Design-Bid-Build (DBB)

A significant percentage of buildings are built using the DBB approach (almost 90% of public buildings and about 40% of private buildings in 2002) (DBIA 2007). The two major benefits of this approach are: more competitive bidding to achieve the lowest possible price for an owner; and less political pressure to select a given contractor. (The latter is particularly important for public projects). Figure 1-2 illustrates the typical DBB procurement process as compared to the typical Design-Build (DB) process (see section 1.2.2)

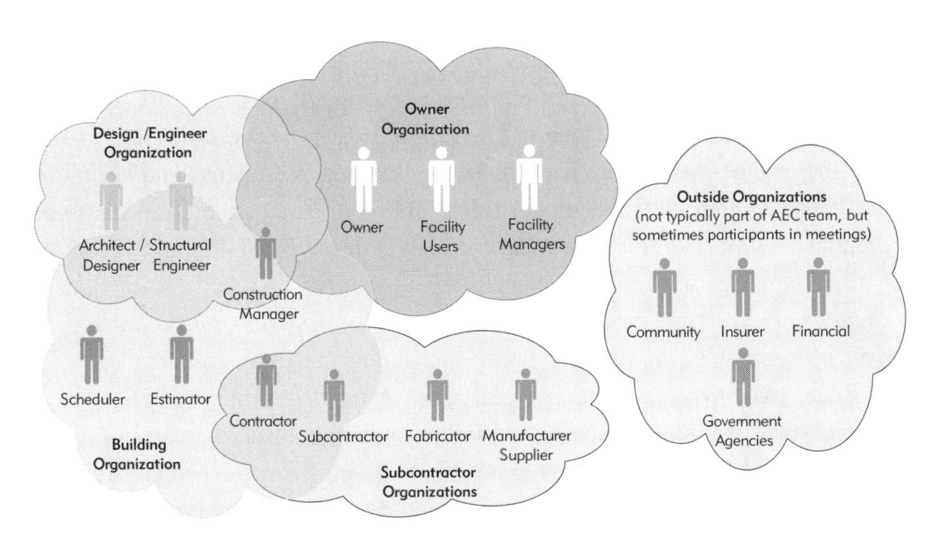

FIGURE 1-1 Conceptual diagram representing an AEC project team and the typical organizational boundaries.

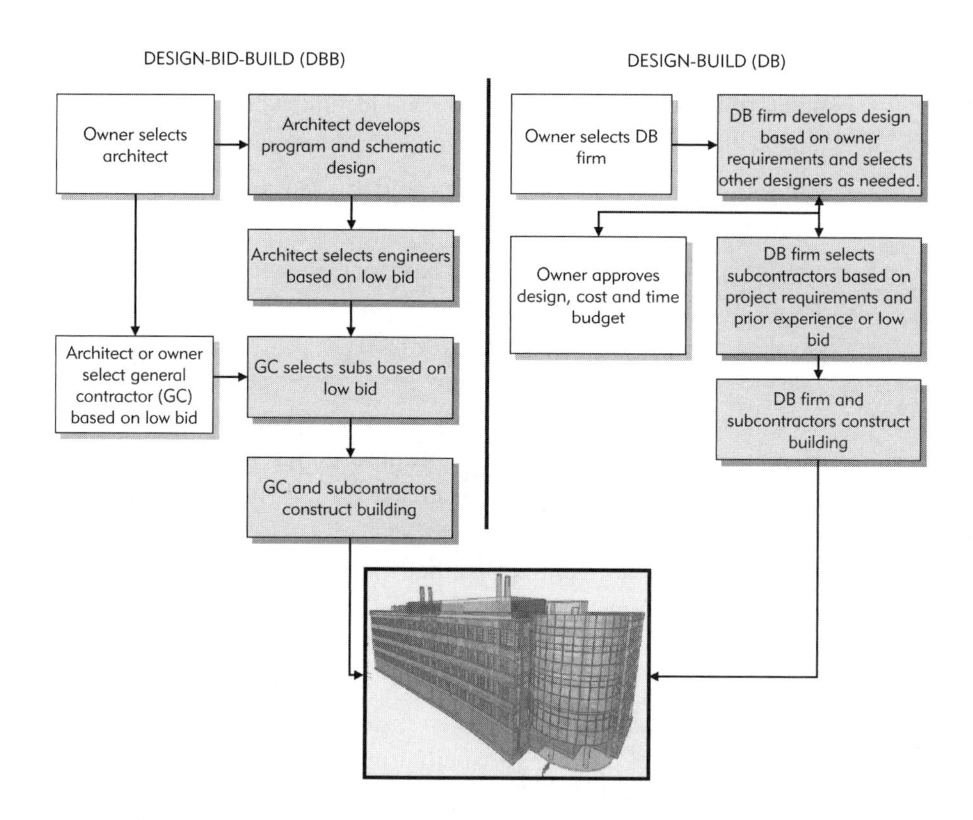

FIGURE 1-2
Schematic diagram of Design-Bid-Build and Design-Build processes.

In the DBB model, the client (owner) hires an architect, who then develops a list of building requirements (a program) and establishes the project's design objectives. The architect proceeds through a series of phases: schematic design, design development, and contract documents. The final documents must fulfill the program and satisfy local building and zoning codes. The architect either hires employees or contracts consultants to assist in designing structural, HVAC, piping, and plumbing components. These designs are recorded on drawings (plans, elevations, 3D visualizations), which must then be coordinated to reflect all of the changes as they are identified. The final set of drawings and specifications must contain sufficient detail to facilitate construction bids. Because of potential liability, an architect may choose to include fewer details in the drawings or insert language indicating that the drawings cannot be relied on for dimensional accuracy. These practices often lead to disputes with the contractor, as errors and omissions are detected and responsibility and extra costs reallocated.

Stage two involves obtaining bids from general contractors. The owner and architect may play a role in determining which contractors can bid. Each contractor must be sent a set of drawings and specifications which are then used to compile an *independent quantity survey*. These quantities, together with the bids from subcontractors are then used to determine their *cost estimate*. Subcontractors selected by the contractors must follow the same process for the part of the project that they are involved with. Because of the effort required, contractors (general and subcontractors) typically spend approximately 1% of their estimated costs in compiling bids.* If a contractor wins approximately one out of every 6 to 10 jobs that they bid on, the cost per successful bid averages from 6% to 10% of the entire project cost. This expense then gets added to the general and subcontractors' overhead costs.

The winning contractor is usually the one with the lowest responsible bid, including work to be done by the general contractor and selected subcontractors. Before work can begin, it is often necessary for the contractor to redraw some of the drawings to reflect the construction process and the phasing of work. These are called *general arrangement drawings*. The subcontractors and fabricators must also produce their own *shop drawings* to reflect accurate details of certain items, such as precast concrete units, steel connections, wall details, piping runs, etc.

The need for accurate and complete drawings extends to the shop drawings, as these are the most detailed representations and are used for actual fabrication. If these drawings are inaccurate or incomplete, or if they are based on drawings that already contain errors, inconsistencies or omissions, then expensive time-consuming conflicts will arise in the field. The costs associated with these conflicts can be significant.

Inconsistency, inaccuracy, and uncertainty in design make it difficult to fabricate materials offsite. As a result, most fabrication and construction must take place onsite and only when exact conditions are known. This is more costly, more time consuming, and prone to produce errors that would not occur if the work were performed in a factory environment where costs are lower and quality control is better.

Often during the construction phase, numerous changes are made to the design as a result of previously unknown errors and omissions, unanticipated site conditions, changes in material availabilities, questions about the design, new client requirements, and new technologies. These need to be resolved by the project team. For each change, a procedure is required to determine the cause, assign responsibility, evaluate time and cost implications, and address

*Based on the second author's personal experience in working with the construction industry. This includes the cost of duplicating the relevant drawings and specs, transporting them to each subcontractor, and the quantity takeoff and cost estimating processes. Electronic plan rooms are sometimes used to reduce the requirement for duplicating and transporting the plans and specs for each bidder.

how the issue will be resolved. This procedure, whether initiated in writing or with the use of a Web-based tool, involves a *Request for Information* (RFI), which must then be answered by the architect or other relevant party. Next a *Change Order* (CO) is issued and all impacted parties are notified about the change, which is communicated together with needed changes in the drawings. These changes and resolutions frequently lead to legal disputes, added costs, and delays. Web site products for managing these transactions do help the project team stay on top of each change, but because they do not address the source of the problem, they are of marginal benefit.

Problems typically arise when a contractor bids below the estimated cost in order to win the job. He will then abuse the change process to recoup losses incurred from the original bid. This, of course, leads to more disputes between the owner and project team.

In addition, the DBB process requires that the procurement of all materials be held until the owner approves the bid, which means that long lead time items cannot be ordered early enough to keep the project on schedule. For this and other reasons (described below), the DBB approach often takes longer than the DB approach.

The final phase is commissioning the building, which takes place after construction is finished. This involves testing the building systems (heating, cooling, electrical, plumbing, fire sprinklers, etc.) to make sure they work properly. Final contracts and drawings are then produced to reflect all *as-built changes,* and these are delivered to the owner along with all manuals for installed equipment. At this point, the DBB process is completed.

Because all of the information provided to the owner is conveyed in 2D (on paper), the owner must put in a considerable amount of effort to relay all relevant information to the facility management team charged with maintaining and operating the building. The process is time consuming, prone to error, costly and remains a significant barrier.

As a result of these problems, the DBB approach is probably not the most expeditious or cost-efficient approach to design and construction. Other approaches have been developed to address these problems.

1.2.2 Design-Build (DB)

The design-build process was developed to consolidate responsibility for design and construction into a single contracting entity and to simplify the administration of tasks for the owner (Beard et al. 2005). Figure 1-2 illustrates this process.

In this model, the owner contracts directly with the design-build team to develop a well-defined building program and a schematic design. The DB contractor then estimates the total cost and time needed to construct the building. After all modifications requested by the owner are implemented, the plan is approved and the final estimate cost for the project is established. It is important to note that because the DB model allows for modifications to be made to the building's design earlier in the process, the amount of money and time needed to incorporate these changes is also reduced. The DB contractor establishes contractual relationships with specialty designers and subcontractors as-needed. After this point, construction begins and any further changes to the design (within predefined limits) become the responsibility of the DB contractor. The same is true for errors and omissions. It is not necessary for detailed construction drawings to be complete for all parts of the building prior to the start of construction on the foundation, etc. As a result of these simplifications, the building is typically completed faster, with far fewer legal complications, and at a somewhat reduced total cost. On the other hand, there is less flexibility for the owner to make changes after the initial design is approved and a contract amount is established.

The DB model is becoming more common in the U.S. and is used widely abroad. Data is not currently available from U.S. government sources, but the Design Build Institute of America (DBIA) estimates that, in 2006, approximately 40% of construction projects in the U.S. relied on a variation of the DB procurement approach. Higher percentages (50%–70%) were measured for some governmental organizations (Navy, Army, Air Force, and GSA). The trend toward increasing use of DB is very strong (Evey 2006).

1.2.3 What Kind of Building Procurement Is Best When BIM Is Used

There are many variations of the design-to-construction business process, including the organization of the project team, how the team members are paid, and who absorbs various risks. There are lump sum contracts, cost plus a fixed or percentage fee, various forms of negotiated contracts, etc. It is beyond the scope of this book to outline each of these and the benefits and problems associated with each of them (see Sanvido and Konchar 1999 and Warne and Beard 2005).

With regard to the use of BIM, the general issues that either enhance or diminish the positive changes that this technology offers depends on how well and at what stage the project team works collaboratively on the digital model. The earlier the model can be developed and shared, the more useful it will be. The DB approach provides an excellent opportunity to exploit BIM technology, because a single entity is responsible for design and construction and both areas participate during the design phase. Other procurement approaches can also benefit from the use of BIM but may achieve only partial benefits, particularly if the BIM technology is not used collaboratively during the design phase.

1.3 DOCUMENTED INEFFICIENCIES OF TRADITIONAL APPROACHES

This section documents how traditional practices contribute unnecessary waste and errors. Evidence of poor field productivity is illustrated in a graph developed by the Center for Integrated Facility Engineering (CIFE) at Stanford University (CIFE, 2007). The impact of poor information flow and redundancy is illustrated using the results of a study performed by the National Institute of Standards and Technology (NIST) (Gallaher et al. 2004).

1.3.1 CIFE Study of Construction Industry Labor Productivity

Extra costs associated with traditional design and construction practices have been documented through various research studies. Figure 1-3 developed by the second author at CIFE, illustrates productivity within the U.S. field construction industry relative to all non-farm industries over a period of forty years, from 1964 through 2004 (the last year for which data is available). The data was calculated by dividing constant contract dollars (from the Department

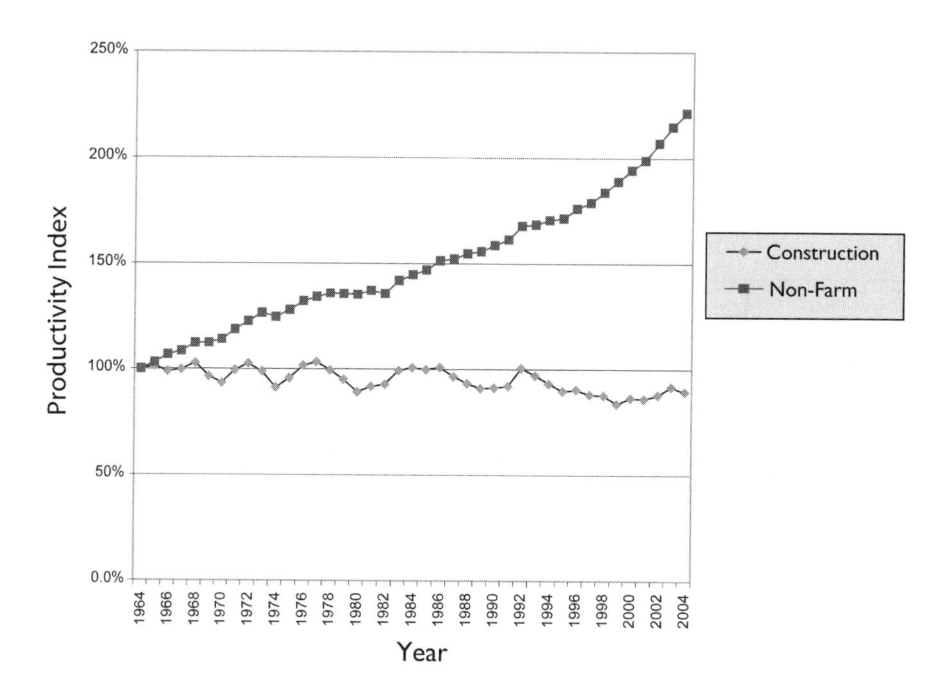

FIGURE 1-3 Indexes of labor productivity for construction and non-farm industries, 1964–2004.
Adapted from research by Paul Teicholz at CIFE.

of Commerce) by field man-hours of labor for those contracts (from the Bureau of Labor Statistics). These contracts include architectural and engineering costs as well as cost for materials and for the delivery of offsite components to the site. Costs associated with the installation of heavy production equipment, such as printing presses, stamping machines, etc., are not included. The amount of man hours required for labor excludes offsite work, such as steel fabrication, precast concrete, etc. During this 40-year-long period, the productivity of non-farm industries (including construction) has more than doubled. Meanwhile, labor productivity within the construction industry alone is estimated to be 10% less than what it was in 1964. Labor represents about 40%–60% of construction's estimated costs. Owners were actually paying approximately 5% more in 2004 than they would have paid for the same building in 1964. Of course, many material and technological improvements have been made to buildings in the last four decades. The results are perhaps better than they appear, because quality has increased substantially. On the other hand, manufactured products are also more complex than they used to be, but they now can be produced at significantly lower cost. The replacement of manual labor with automated equipment has resulted in lower labor costs and increased quality. But the same cannot be said for construction practices.

Contractors have made greater use of offsite components which take advantage of factory conditions and specialized equipment. Clearly, this has allowed for higher quality and lower cost production of components, as compared to onsite work. Although the cost of these components is included in our construction cost data, the labor is not. This tends to make onsite construction productivity appear better than it actually is. The extent of this error, however, is difficult to evaluate because the total cost of offsite production cost is not well-documented.

While the reasons for the apparent decrease in construction productivity are not completely understood, the statistics are dramatic and point at organizational impediments within the construction industry. It is clear that efficiencies achieved in the manufacturing industry through automation, the use of information systems, better supply chain management and improved collaboration tools, have not yet been achieved in field construction. Possible reasons for this include:

- Sixty-five percent of construction firms consist of less than 5 people, making it difficult for them to invest in new technology; even the largest firms account for less than 0.5% of total construction volume and are not able to establish industry leadership (see Figure 6-1 in Chapter 6).
- The real inflation-adjusted wages and the benefit packages of construction workers have stagnated over this time period. Union participation has declined and the use of immigrant workers has increased, discouraging the

> need for labor-saving innovations. While innovations have been introduced, such as nail guns, larger and more effective earth moving equipment and better cranes, the productivity improvements associated with them have not been sufficient to change overall field labor productivity.

The adoption of new and improved business practices within both design and construction has been noticeably slow and limited primarily to large firms. In addition, the introduction of new technologies has been fragmented. Often times, it remains necessary to revert back to paper or 2D CAD drawings so that all members of a project team are able to communicate with each other and to keep the pool of potential contractors and subcontractors bidding on a project sufficiently large.

Whereas manufacturers often have long term agreements and collaborate in agreed upon ways with the same partners, construction projects typically involve different partners working together for a period of time and then dispersing. As a result, there are few or no opportunities to realize improvements over time through applied learning. Rather, each partner acts to protect him or herself from potential disputes that could lead to legal difficulties by relying on antiquated and time-consuming processes that make it difficult or impossible to implement resolutions quickly and efficiently. Of course, this translates to higher cost and time expenditures.

Another possible cause for the construction industry's stagnant productivity is that onsite construction has not benefited significantly from automation. Thus, field productivity relies on qualified training of field labor. Figure 1-4 shows that, since 1974, compensation for hourly workers has steadily declined

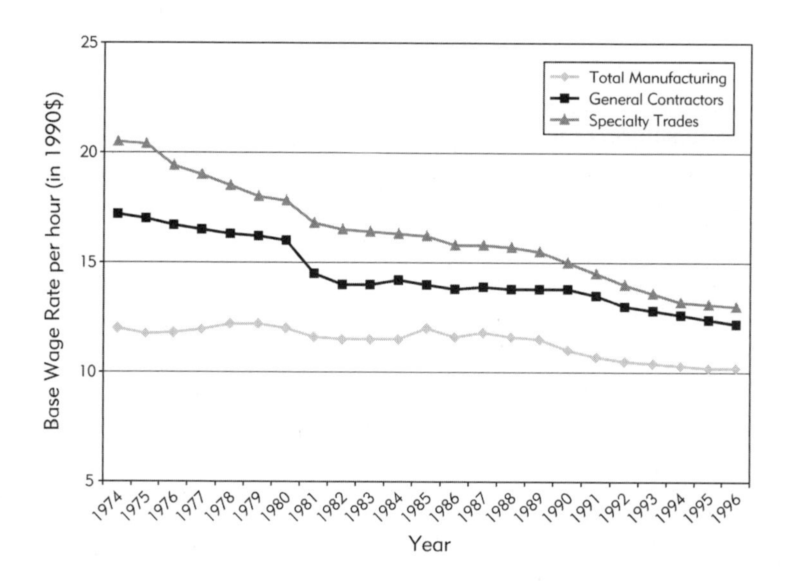

FIGURE 1-4 Trends in real wages (1990 $) for manufacturing and construction hourly workers, 1974–1996. BLS, Series ID : EES00500006.

with the increase in use of non-union immigrant workers with little prior training. The lower cost associated with these workers has discouraged efforts to replace field labor with automated (or offsite) solutions.

Existing problems within the construction industry also involve issues unrelated to the use of advanced technologies. These business practices (described above) limit how quickly new and innovative tools can be adopted. Pushing in the other direction is the growing competitive pressure of globalization which allows overseas firms to provide services and/or materials used in local projects. To cope with this added pressure, leading U.S. firms will need to implement changes that allow them to work faster and more efficiently than they otherwise would.

1.3.2 NIST Study of Cost of Construction Industry Inefficiency

The National Institute of Standards and Technology (NIST) performed a study of the additional cost incurred by building owners as a result of inadequate interoperability (Gallaher et al. 2004). The study involved both the exchange and management of information, in which individual systems were unable to access and use information imported from other systems. In the construction industry, incompatibility between systems often prevents members of the project team from sharing information rapidly and accurately; it is the cause of numerous problems, including added costs, etc. The NIST study included commercial, industrial and institutional buildings and focused on new and "set in place" construction taking place in 2002. The results showed that inefficient interoperability accounted for an increase in construction costs by $6.12 per sf for new construction and an increase in $0.23 per sf for operations and maintenance (O & M), resulting in a total added cost of $15.8 billion. Table 1-1 shows the breakdown of these costs and to which stakeholder they were applied.

In the NIST study, the cost of inadequate interoperability was calculated by comparing current business activities and costs with hypothetical scenarios in which there was seamless information flow and no redundant data entry. NIST determined that the following costs resulted from inadequate interoperability:

- Avoidance (redundant computer systems, inefficient business process management, redundant IT support staffing)
- Mitigation (manual reentry of data, request for information management)
- Delay (costs for idle employees and other resources)

Of these costs, roughly 68% ($10.6 billion) were incurred by building owners and operators. These estimates are speculative, due to the impossibility of providing accurate data. They are, however, significant and worthy of serious consideration and effort to reduce or avoid them as much as possible.

Table 1-1 Additional costs of inadequate interoperability in the construction industry, 2002 (in $millions).

Stakeholder Group	Planning, Engineering, Design Phase	Construction Phase	O&M Phase	Total Added Cost
Architects and Engineers	$1,007.2	$147.0	$15.7	$1,169.8
General Contractors	$485.9	$1,265.3	$50.4	$1,801.6
Special Contractors and Suppliers	$442.4	$1,762.2		$2,204.6
Owners and Operators	$722.8	$898.0	$9,027.2	$1,0648.0
Total	**$2,658.3**	**$4,072.4**	**$9,093.3**	**$15,824.0**
Applicable sf in 2002	1.1 billion	1.1 billion	39 billion	n/a
Added cost/sf	**$2.42/sf**	**$3.70/sf**	**$0.23/sf**	**n/a**

Source: Table 6-1 NIST study (Gallaher et al. 2004).

Widespread adoption of BIM and the use of a comprehensive digital model throughout the lifecycle of a building would be a step in the right direction to eliminate such costs resulting from the inadequate interoperability of data.

1.4 BIM: NEW TOOLS AND NEW PROCESSES

This section gives an overall description of BIM-related terminology, concepts, and functional capabilities; and it addresses how these tools can improve business processes. Specific topics are discussed in further detail in the chapters indicated in parenthesis.

1.4.1 BIM Model Creation Tools (Chapter 2)

All CAD systems generate digital files. Older CAD systems produce plotted drawings. They generate files that consist primarily of vectors, associated line-types, and layer identifications. As these systems were further developed, additional information was added to these files to allow for blocks of data and associated text. With the introduction of 3D modeling, advanced definition and complex surfacing tools were added.

As CAD systems became more intelligent and more users wanted to share data associated with a given design, the focus shifted from drawings and 3D images to the data itself. A building model produced by a BIM tool can support multiple different views of the data contained within a drawing set, including 2D and 3D. A building model can be described by its content (what objects it describes) or its capabilities (what kinds of information requirements it can support). The latter approach is preferable, because it defines what you can do with the model rather than how the database is constructed (which will vary with each implementation).

For the purpose of this book, we define BIM as a modeling technology and associated set of processes to produce, communicate, and analyze *building models*. Building models are characterized by:

- **Building components** that are represented with intelligent digital representations (objects) that 'know' what they are, and can be associated with computable graphic and data attributes and parametric rules.
- **Components that include data that describe how they behave,** as needed for analyses and work processes, e.g., takeoff, specification, and energy analysis.
- **Consistent and non-redundant data** such that changes to component data are represented in all views of the component.
- **Coordinated data** such that all views of a model are represented in a coordinated way.

The following is a definition of BIM technology provided by the M.A. Mortenson Company, a construction contracting firm that has used BIM tools extensively within their practice (Campbell 2006).

Mortenson's Definition of BIM Technology

BIM has its roots in computer-aided design research from decades ago, yet it still has no single, widely-accepted definition. We at the M.A. Mortenson Company think of it as "an intelligent simulation of architecture." To enable us to achieve integrated delivery, this simulation must exhibit six key characteristics. It must be:

- Digital,
- Spatial (3D),
- Measurable (quantifiable, dimension-able, and query-able),
- Comprehensive (encapsulating and communicating design intent, building performance, constructability, and include sequential and financial aspects of means and methods),
- Accessible (to the entire AEC/ owner team through an interoperable and intuitive interface), and
- Durable (usable through all phases of a facility's life).

In light of these definitions, one might argue that few design or construction teams are truly using BIM today. In fact, we may not achieve this high standard for several years. But we believe these characteristics are all essential for reaching the goal of integrated practice.

Furthermore, there are no current implementations of BIM software that meet all of the BIM technology criteria. Over time, the capabilities will grow as will the ability to support better and more extensive practices. The list in the following section is intended to provide a starting point for evaluating specific BIM software tools. See Chapter 2 for more detailed information about BIM technology and an analysis of current BIM tools.

1.4.2 Definition of Parametric Objects (Chapter 2)

The concept of parametric objects is central to understanding BIM and its differentiation from traditional 2D objects. Parametric BIM objects are defined as follows:

- consist of geometric definitions and **associated data and rules**.
- geometry is integrated **non-redundantly**, and allows for no inconsistencies. When an object is shown in 3D, the shape cannot be represented internally redundantly, for example as multiple 2D views. A plan and elevation of a given object must always be consistent. Dimensions cannot be 'fudged'.
- Parametric rules for objects **automatically modify associated geometries** when inserted into a building model or when changes are made to associated objects. For example, a door will fit automatically into a wall, a light switch will automatically locate next to the proper side of the door, a wall will automatically resize itself to automatically butt to a ceiling or roof, etc.
- Objects can be defined at **different levels of aggregation**, so we can define a wall as well as its related components. Objects can be defined and managed at any number of hierarchy levels. For example, if the weight of a wall subcomponent changes, the weight of the wall should also change.
- Objects rules can identify when a particular change violates **object feasibility** regarding size, manufacturability, etc.
- objects have the ability to **link to or receive, broadcast or export sets of attributes**, e.g., structural materials, acoustic data, energy data, etc. to other applications and models.

Technologies that allow users to produce building models that consist of parametric objects are considered BIM authoring tools. In Chapter 2 we elaborate the discussion of parametric technologies and discuss common capabilities in BIM tools including features to automatically extract consistent drawings and to extract reports of geometric parameters. In Chapters 4–7 we discuss these capabilities and others and their potential benefits to various AEC practitioners and building owners.

1.4.3 Support for Project Team Collaboration (Chapter 3)

Open interfaces should allow for the import of relevant data (for creating and editing a design) and export of data in various formats (to support integration with other applications and workflows). There are two primary approaches for such integration: (1) to stay within one software vendor's products or (2) to use software from various vendors that can exchange data using industry supported standards. The first approach allows for tighter integration among products in multiple directions. For example, changes to the architectural model will generate changes to the structural model, and vice versa. This requires, however, that all members of a design team use software provided from the same vendor.

The second approach uses either proprietary or open-source, publicly available, and supported standards created to define building objects (Industry Foundation Classes or IFCs). These standards may provide a mechanism for interoperability among applications with different internal formats. This approach provides more flexibility at the expense of reduced interoperability, especially if the various software programs in-use for a given project do not support the same exchange standards. This allows objects from one BIM application to be exported from or imported into another (see Chapter 3 for an extensive discussion of collaboration technology).

1.5 WHAT IS NOT BIM TECHNOLOGY

The term BIM is a popular buzz word used by software developers to describe the capabilities that their products offer. As such, the definition of what constitutes BIM technology is subject to variation and confusion. To deal with this confusion, it is useful to describe modeling solutions that DO NOT utilize BIM technology. These include tools that create the following kinds of models:

Models that contain 3D data only and no object attributes. These are models that can only be used for graphic visualizations and have no intelligence at the object level. They are fine for visualization but provide no support for data integration and design analysis.

Models with no support of behavior. These are models that define objects but cannot adjust their positioning or proportions because they do not utilize parametric intelligence. This makes changes extremely labor intensive and provides no protection against creating inconsistent or inaccurate views of the model.

Models that are composed of multiple 2D CAD reference files that must be combined to define the building. It is impossible to ensure that the resulting 3D model will be feasible, consistent, countable, and display intelligence with respect to the objects contained within it.

Models that allow changes to dimensions in one view that are not automatically reflected in other views. This allows for errors in the model that are very difficult to detect (similar to overriding a formula with a manual entry in a spreadsheet).

1.6 WHAT ARE THE BENEFITS OF BIM? WHAT PROBLEMS DOES IT ADDRESS?

BIM technology can support and improve many business practices. Although the AEC/FM (Facility Management) industry is in the early days of BIM use, significant improvements have already been realized (compared to traditional 2D CAD or paper-based practices). Though it is unlikely that all of the advantages discussed below are currently in use, we have listed them to show the entire scope of changes that can be expected as BIM technology develops.

1.6.1 Pre-Construction Benefits to Owner (Chapters 4 and 5)

Concept, Feasibility and Design Benefits

Before owners engage an architect, it is necessary to determine whether a building of a given size, quality level, and desired program requirements can be built within a given cost and time budget, i.e. can a given building meet the financial requirements of an owner. If these questions can be answered with relative certainty, owners can then proceed with the expectation that their goals are achievable. Finding out that a particular design is significantly over budget after a considerable amount of time and effort has been expended is wasteful. An approximate (or macro) building model built into and linked to a cost database can be of tremendous value and assistance to an owner. This is described in further detail in Chapter 4 and illustrated in the Hillwood Commercial Project case study in Chapter 9.

Increased Building Performance and Quality

Developing a *schematic model* prior to generating a *detailed building model* allows for a more careful evaluation of the proposed scheme to determine whether it meets the building's functional and sustainable requirements. Early evaluation of design alternatives using analysis/simulation tools increases the overall quality of the building.

1.6.2 Design Benefits (Chapter 5)

Earlier and More Accurate Visualizations of a Design

The 3D model generated by the BIM software is designed directly rather than being generated from multiple 2D views. It can be used to visualize the design at any stage of the process with the expectation that it will be dimensionally consistent in every view.

Automatic Low-Level Corrections When Changes Are Made to Design

If the objects used in the design are controlled by parametric rules that ensure proper alignment, then the 3D model will be constructible. This reduces the user's need to manage design changes (see Chapter 2 for further discussion of parametric rules).

Generate Accurate and Consistent 2D Drawings at Any Stage of the Design

Accurate and consistent drawings can be extracted for any set of objects or specified view of the project. This significantly reduces the amount of time and number of errors associated with generating construction drawings for all design disciplines. When changes to the design are required, fully consistent drawings can be generated as soon as the design modifications are entered.

Earlier Collaboration of Multiple Design Disciplines

BIM technology facilitates simultaneous work by multiple design disciplines. While collaboration with drawings is also possible, it is inherently more difficult and time consuming than working with one or more coordinated 3D models[†] in which change-control can be well managed. This shortens the design time and significantly reduces design errors and omissions. It also gives earlier insight into design problems and presents opportunities for a design to be continuously improved. This is much more cost effective than waiting until a

[†]If a BIM system does not use a single database, which can create problems for very large and/or finely detailed projects, alternative approaches involving automatic coordination of multiple files can also be used. This is an important implementation issue for software vendors. (See Chapter 2 for more discussion of model size issues).

design is nearly complete and then applying value engineering only after the major design decisions have been made.

Easily Check against the Design Intent

BIM provides earlier 3D visualizations and quantifies the area of spaces and other material quantities, allowing for earlier and more accurate cost estimates. For technical buildings (labs, hospitals, etc.), the design intent is often defined quantitatively, and this allows a building model to be used to check for these requirements. For qualitative requirements (this space should be near another, etc.), the 3D model can support automatic evaluations.

Extract Cost Estimates during the Design Stage

At any stage of the design, BIM technology can extract an accurate bill of quantities and spaces that can be used for cost estimation. In the early stages of a design, cost estimates are based primarily on the unit cost per square foot. As the design progresses, more detailed quantities are available and can be used for more accurate and detailed cost estimates. It is possible to keep all parties aware of the cost implications associated with a given design before it progresses to the level of detailing required of construction bids. At the final stage of design, an estimate based on the quantities for all the objects contained within the model allows for the preparation of a more accurate final cost estimate. As a result, it is possible to make better informed design decisions regarding costs using BIM rather than a paper-based system.

Improve Energy Efficiency and Sustainability

Linking the building model to energy analysis tools allows evaluation of energy use during the early design phases. This is not possible using traditional 2D tools which require that a separate energy analysis be performed at the end of the design process thus reducing the opportunities for modifications that could improve the building's energy performance. The capability to link the building model to various types of analysis tools provides many opportunities to improve building quality.

1.6.3 Construction and Fabrication Benefits (Chapters 6 & 7)

Synchronize Design and Construction Planning

Construction planning using 4D CAD requires linking a construction plan to the 3D objects in a design, so that it is possible to simulate the construction process and show what the building and site would look like at any point in time. This graphic simulation provides considerable insight into how the building will be constructed day-by-day and reveals sources of potential problems and opportunities for possible improvements (site, crew and equipment, space conflicts, safety problems, etc.). This type of analysis is not available from paper bid documents. It does, however, provide added benefit if the model includes temporary construction objects such as shoring, scaffolding, cranes, and other major equipment so that these objects can be linked to schedule activities and reflected in the desired construction plan.

Discover Design Errors and Omissions before Construction (Clash Detection)

Because the virtual 3D building model is the source for all 2D and 3D drawings, design errors caused by inconsistent 2D drawings are eliminated. In addition, because systems from all disciplines can be brought together and compared, multi-system interfaces are easily checked both systematically (for hard and soft clashes) and visually (for other kinds of errors). Conflicts are identified before they are detected in the field. Coordination among participating designers and contractors is enhanced and errors of omission are significantly reduced. This speeds the construction process, reduces costs, minimizes the likelihood of legal disputes, and provides a smoother process for the entire project team.

React Quickly to Design or Site Problems

The impact of a suggested design change can be entered into the building model and changes to the other objects in the design will automatically update. Some updates will be made automatically based on the established parametric rules. Additional cross system updates can be checked and updated visually. The consequences of a change can be accurately reflected in the model and all subsequent views of it. In addition, design changes can be resolved more quickly in a BIM system because modifications can be shared, visualized, estimated, and resolved without the use of time-consuming paper transactions. Updating in this manner is extremely error-prone in paper-based systems.

Use Design Model as Basis for Fabricated Components

If the design model is transferred to a BIM fabrication tool and detailed to the level of fabrication objects (shop model), it will contain an accurate representation of the building objects for fabrication and construction. Because components are already defined in 3D, their automated fabrication using numerical control machinery is facilitated. Such automation is standard practice today in steel fabrication and some sheet metal work. It has been used successfully in precast components, fenestration and glass fabrication. This allows vendors world-wide to elaborate on the model, to develop details needed for fabrication and to maintain links that reflect the design intent. This facilitates offsite fabrication and reduces cost and construction time. The accuracy of BIM also allows larger components of the design to be fabricated offsite than would normally be attempted using 2D drawings, due to the likely need for onsite changes (rework) and the inability to predict exact dimensions until other items are constructed in the field.

Better Implementation and Lean Construction Techniques

Lean construction techniques require careful coordination between the general contractor and subs to ensure that work can be performed when the appropriate resources are available onsite. This minimizes wasted effort and reduces the need for onsite material inventories. Because BIM provides an accurate model of the design and the material resources required for each segment of the work, it provides the basis for improved planning and scheduling of subcontractors and helps to ensure just-in-time arrival of people, equipment, and materials. This reduces cost and allows for better collaboration at the job site.

Synchronize Procurement with Design and Construction

The complete building model provides accurate quantities for all (or most, depending upon level of 3D modeling) of the materials and objects contained within a design. These quantities, specifications, and properties can be used to procure materials from product vendors and subcontractors (such as precast concrete subs). At the present time (2007), the object definitions for many manufactured products have not yet been developed to make this capability a complete reality. Where the models have been available (steel members, precast concrete members), however, the results have been very beneficial.

1.6.4 Post Construction Benefits (Chapter 4)

Better Manage and Operate Facilities

The building model provides a source of information (graphics and specifications) for all systems used in a building. Previous analyses used to determine mechanical equipment, control systems, and other purchases can be provided to the owner, as a means for verifying the design decisions once the building is in use. This information can be used to check that all systems work properly after the building is completed.

Integrate with Facility Operation and Management Systems

A building model that has been updated with all changes made during construction provides an accurate source of information about the as-built spaces and systems and provides a useful starting point for managing and operating the building. A building information model supports monitoring of real-time control systems, provides a natural interface for sensors and remote operating management of facilities. Many of these capabilities have not yet been developed, but BIM provides an ideal platform for their deployment. An example of how a building model can serve as a database for facility data is discussed in the Coast Guard Facility Planning case study in Chapter 9.

1.7 WHAT CHALLENGES CAN BE EXPECTED?

Improved processes in each phase of design and construction will reduce the number and severity of problems associated with traditional practices. Intelligent use of BIM, however, will also cause significant changes in the relationships of project participants and the contractual agreements between them. (Traditional contract terms are tailored to paper-based practices.) In addition, earlier collaboration between the architect, contractor, and other design disciplines will be needed, as knowledge provided by specialists is of more use during the design phase. (This is not consistent with the current design-bid-build business model.)

1.7.1 Challenges with Collaboration and Teaming

While BIM offers new methods for collaboration, it introduces other issues with respect to the development of effective teams. Determining the methods that will be used to permit adequate sharing of model information by members of the project team is a significant issue. If the architect uses traditional paper-based drawings, then it will be necessary for the contractor (or a third party) to build the model so that it can be used for construction planning, estimating, and coordination etc. Creating a model after the design is complete adds cost and time to the project, but it may be justified by the advantages of using it for construction planning and detailed design by mechanical, plumbing, other subs and fabricators, design change resolution, procurement, etc. If the members of the project team use different modeling tools, then tools for moving the models from one environment to another or combining these models are needed. This can add complexity and introduce potential errors to the project. Such problems can be reduced by using IFC standards for exchanging data. Another approach is to use a model server that communicates with all BIM applications through IFC or proprietary standards. A number of the case studies presented in Chapter 9 provide background for this issue.

1.7.2 Legal Changes to Documentation Ownership and Production

Legal concerns are presenting challanges, with respect to who owns the multiple design, fabrication, analysis, and construction datasets, who pays for them, and who is responsible for their accuracy. These issues are being addressed by practioners through BIM use on projects. As owners learn more about the advantages of BIM, they will likely require a building model to support operations, maintenance, and subsequent renovations. Professional groups, such as the AIA and AGC, are developing guidelines for contractual language to cover issues raised by the use of BIM technology.

1.7.3 Changes in Practice and Use of Information

The use of BIM will also encourage the integration of construction knowledge earlier in the design process. Integrated design-build firms capable of coordinating all phases of the design and incorporating construction knowledge from the outset will benefit the most. Contracting arrangements that require and facilitate good collaboration will provide greater advantages to owners when BIM is used. The most significant change that companies face when implementing BIM technology is using a shared building model as the basis of all work processes and for collaboration. This transformation will require time and education, as is true of all significant changes in technology and work processes.

1.7.4 Implementation Issues

Replacing a 2D or 3D CAD environment with a BIM system involves far more than acquiring software, training, and upgrading hardware. Effective use of BIM requires that changes be made to almost every aspect of a firm's business (not just doing the same things in a new way). It requires a thorough understanding and a plan for implementation before the conversion can begin. While the specific changes for each firm will depend on their sector(s) of AEC activity, the general steps that need to be considered are similar and include the following:

- Assign top level management responsibility for developing a BIM adoption plan that covers all aspects of the firm's business and how the proposed changes will impact both internal departments and outside partners and clients.
- Create an internal team of key managers responsible for implementing the plan, with cost, time, and performance budgets to guide their performance.

- Start using the BIM system on one or two smaller (perhaps already completed) projects in parallel with existing technology and produce traditional documents from the building model. This will help reveal where there are deficits in the building objects, in output capabilities, in links to analysis programs, etc. It will also provide educational opportunities for leadership staff.
- Use initial results to educate and guide continued adoption of BIM software and additional staff training. Keep senior management apprised of progress, problems, insights, etc.
- Extend the use of BIM to new projects and begin working with outside members of the project teams in new collaborative approaches that allow early integration and sharing of knowledge using the building model.
- Continue to integrate BIM capabilities into all aspects of the firm's functions and reflect these new business processes in contractual documents with clients and business partners.
- Periodically re-plan the BIM implementation process to reflect the benefits and problems observed thus far, and set new goals for performance, time, and cost. Continue to extend BIM-facilitated changes to new locations and functions within the firm.

In Chapters 4 through 7, where specific applications of BIM over the lifecycle of a building are discussed, additional adoption guidelines specific to each party involved in the building process are reviewed.

1.8 FUTURE OF DESIGNING AND BUILDING WITH BIM (CHAPTER 8)

Chapter 8 describes the authors' views of how BIM technology will evolve and what impacts it is likely to have on the future AEC industry and to society at large. There are comments on the near-term future (up to 2012) and the long-term future (up to 2020). We also discuss the kinds of research that will be relevant to support these trends.

It is rather straightforward to anticipate near-term impacts. For the most part, they are extrapolations of current trends. Projections over a longer period are those that to us seem likely, given our knowledge of the AEC industry and BIM technology. Beyond that, it is difficult to make useful projections.

1.9 CASE STUDIES (CHAPTER 9)

Chapter 9 presents ten case studies that illustrate how BIM technology and its associated work processes are being used today. These cover the entire range of the building lifecycle, although most focus on the design and construction phases (with extensive illustration of offsite fabrication building models). For the reader who is anxious to "dive right in" and get a first-hand view of BIM, these case histories are a good place to start.

Chapter One Discussion Questions

1. What is BIM and how does it differ from 3D modeling?
2. What are some of the significant problems associated with the use of 2D CAD, and how do they waste resources and time during both the design and construction phases as compared to BIM-enabled processes?
3. Why has the construction industry not been able to overcome the impact of these problems on field labor productivity, despite the many advances in construction technology?
4. What changes in the design and construction process are needed to enable productive use of BIM technology?
5. How do parametric rules associated with the objects in BIM improve the design and construction process?
6. What are the limitations that can be anticipated with the generic object libraries that come with BIM systems?
7. Why does the design-bid-build business process make it very difficult to achieve the full benefits that BIM provides during design or construction?
8. What kind of legal problems can be anticipated as a result of using BIM with an integrated project team?
9. What techniques are available for integrating design analysis applications with the building model developed by the architect?

CHAPTER 2

BIM Tools and Parametric Modeling

2.0 EXECUTIVE SUMMARY

This chapter provides an overview of the primary technology that distinguishes BIM design applications from other CAD systems. Object-based parametric modeling was originally developed in the 1980s. It does not represent objects with fixed geometry and properties. Rather, it represents objects by parameters and rules that determine the geometry as well as some non-geometric properties and features. The parameters and rules allow the objects to automatically update according to user control or changing contexts. In other industries, companies use parametric modeling to develop their own object representations and to reflect corporate knowledge and best practices. In architecture, BIM software companies have pre-defined a set of base building object families for users, which may be extended, modified, or added to. An object family allows for the creation of any number of object instances, with forms that are dependent on parameters and relationships with other objects. Companies should have the capability of developing user-defined parametric objects and corporate object libraries for customized quality control and to establish their own best practices. Custom parametric objects allow for the modeling of complex geometries, which were previously not possible or simply impractical. Object attributes are needed to interface with analyses, cost estimations, and other applications, but these attributes must first be defined by the firm or user.

Current BIM tools vary in many ways: in the sophistication of their predefined base objects; in the ease with which users can define new object families;

in the methods of updating objects; in ease of use; in the types of surfaces that can be used; in the capabilities for drawing generation; in their ability to handle large numbers of objects and their interfaces with other software.

Most architectural BIM design tools let users mix 3D modeled objects with 2D drawn sections, allowing users to determine the level of 3D detailing while still being able to produce complete drawings. Objects drawn in 2D are not automatically included in bills of material, analyses, and other BIM-enabled applications. Fabrication-level BIM tools, alternatively, typically represent every object fully in 3D. The level of 3D modeling is a major variable within different BIM practices.

This chapter provides an overall review of the major BIM model generation tools and some functional distinctions that can be used for assessing and selecting among them.

2.1 HISTORY OF BUILDING MODELING TECHNOLOGY

2.1.1 Early 3D Modeling of Buildings

The modeling of 3D geometry was a broad research goal that had many potential uses including movies, design, and eventually games. The ability to represent a fixed set of polyhedral forms—shapes defined by a volume enclosing a set of surfaces—for viewing purposes was developed in the late 1960s and later led to the first computer-graphics film, *Tron* (in 1987). These early polyhedral forms could be used for composing an image but not for designing more complex shapes. In 1973, the easy creation and editing of arbitrary 3D solid shapes was developed separately by three groups, Ian Braid at Cambridge University, Bruce Baumgart at Stanford, and Ari Requicha and Herb Voelcker at the University of Rochester (Eastman1999; Chapter 2). Known as *solid modeling,* these efforts produced the first generation of practical 3D modeling design tools.

Two forms of solid modeling were developed and competed for supremacy. The boundary representation approach (B-rep) defined shapes using operations of union, intersection, and subtraction—called Boolean operations—on multiple polyhedral shapes and also utilized refining operations, such as chamfering, slicing, or moving a hole within a single shape. A small set of such operators is shown in Figure 2-1. The sophisticated editing systems developed from combining these primitive shapes and the Boolean operators allowed generation of a set of surfaces that together were guaranteed to enclose a volume.

In contrast, Constructive Solid Geometry (CSG) represented a shape as a tree of operations and initially relied on diverse methods for assessing the final

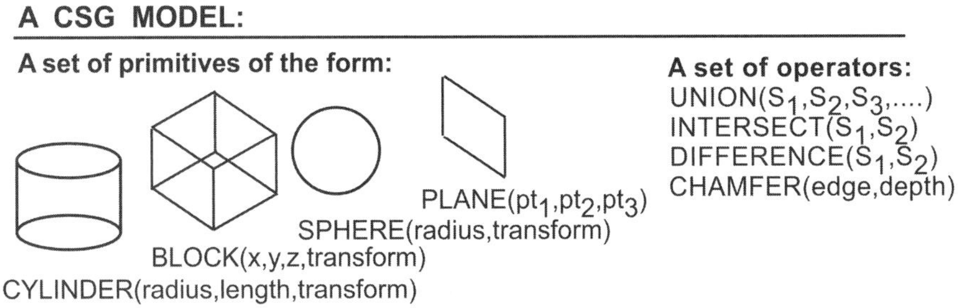

A CSG MODEL:

A set of primitives of the form:

PLANE(pt_1,pt_2,pt_3)
SPHERE(radius,transform)
BLOCK(x,y,z,transform)
CYLINDER(radius,length,transform)

A set of operators:
UNION(S_1,S_2,S_3,....)
INTERSECT(S_1,S_2)
DIFFERENCE(S_1,S_2)
CHAMFER(edge,depth)

FIGURE 2-1 A set of primitive shapes and constructive solid geometry operators.

shape. An example is shown in Figure 2-2. Later, these two methods merged, allowing for editing within the CSG tree (sometimes called the *unevaluated shape*) and also changing the shape through the use of general-purpose B-rep (called the *evaluated shape*). Objects could be edited and regenerated on demand. Figure 2-2 depicts a CSG tree, the unevaluated shapes it references, and the resulting evaluated shape. The result is the simplest of building shapes: a single shape hollowed with a single floor space with a gable roof and door opening. Notice that all locations and shapes can be edited via the shape parameters in the CSG tree, however, shape edits are limited to Boolean or other editing operations shown in Figure 2-1. First generation tools supported 3D facetted and cylindrical object modeling with associated attributes, which allowed objects to be composed into engineering assemblies, such as engines, process plants, or buildings (Eastman 1975; Requicha 1980). This merged approach to modeling was an important precursor to modern parametric modeling.

Building modeling based on 3D solid modeling was first developed in the late 1970s and early 1980s. CAD systems, such as RUCAPS (which evolved into Sonata), TriCad, Calma, GDS (Day 2002), and university research-based systems at Carnegie-Mellon University and the University of Michigan developed their basic capabilities. (For one detailed history of the development of CAD technology see http://mbinfo.mbdesign.net/CAD-History.htm.) This work was carried out in parallel with efforts in mechanical, aerospace, building and electrical product design, where early concepts of product modeling and integrated analysis and simulation were developed. Early conferences in computer-aided design were integrated across all areas of engineering and design, resulting in high levels of synergy. For example: Proceedings 7th–18th Annual Design Automation Conference (ACM 1969–1982); Conferences on Engineering and Scientific Data Management (NASA 1978–1980); Proceedings of CAD76, CAD78, CAD80 (CAD 1976,1978,1980).

Solid modeling CAD systems were functionally powerful but often overwhelmed the available computing power. Some aspects of production, such as

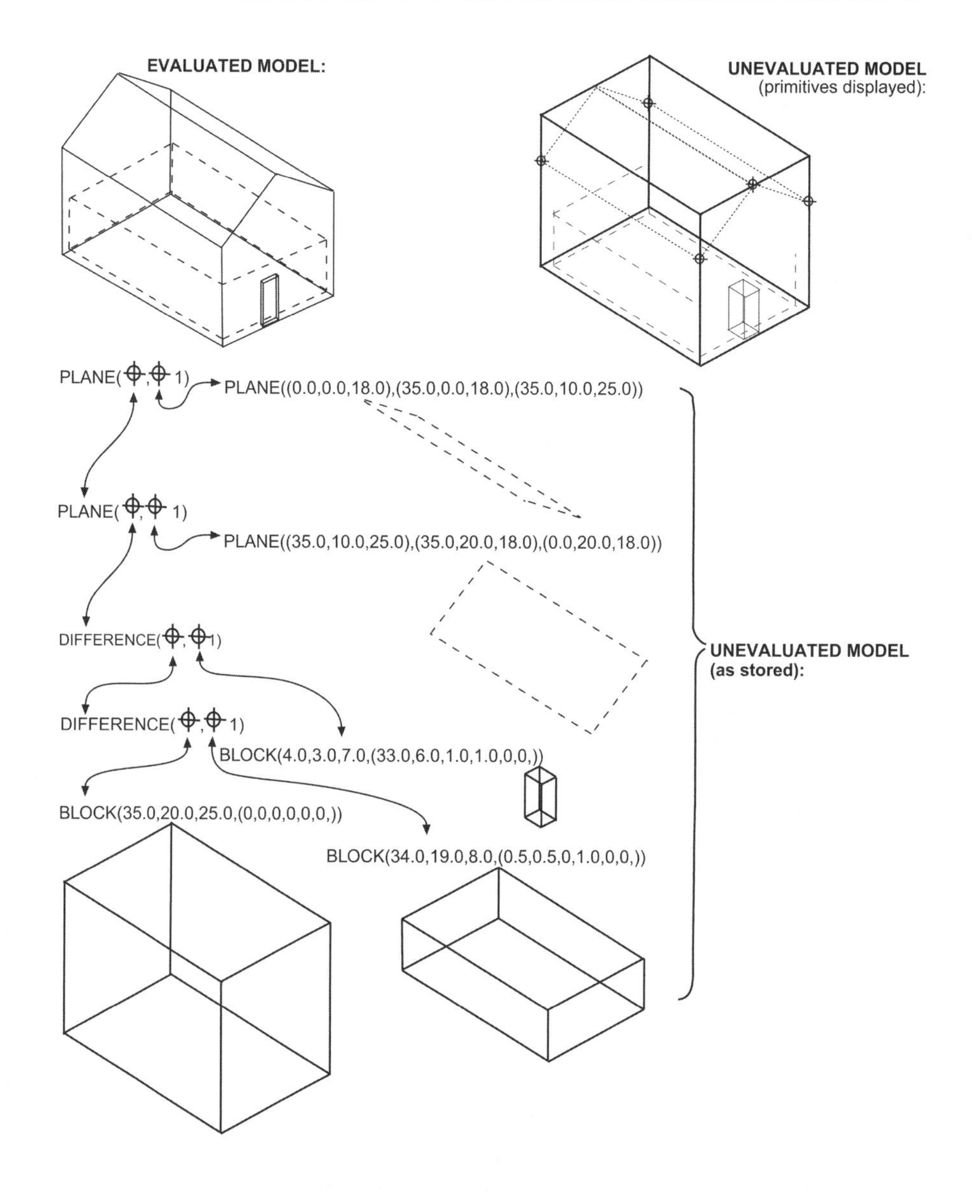

FIGURE 2-2 Example of a CSG model.
From bottom: the unevaluated CSG tree and unevaluated primitive shape that the tree composes. Top right: the unevaluated set of primitives in the parameterized shape and location. Top left: the resulting evaluated B-rep model.

drawing and report generation, were not well developed. Also, designing 3D objects was too conceptually foreign for most designers, who were more comfortable working in 2D. The systems were also expensive, costing upward of $35,000 per seat. The manufacturing and aerospace industries saw the potential benefits in terms of integrated analysis capabilities, reduction of errors, and the move toward factory automation. They worked with CAD companies to resolve the technology's early shortcomings. Most of the building industry did not recognize these benefits. Instead, they adopted architectural drawing editors, such as AutoCAD® and Microstation® that augmented the then-current methods of working and supported the digital generation of conventional 2D construction documents.

2.1.2 Object-Based Parametric Modeling

The current generation of BIM architectural design tools, including Autodesk Revit® Architecture and Structure, Bentley Architecture and its associated set of products, the Graphisoft ArchiCAD® family, and Gehry Technology's Digital Project™ as well as fabrication-level BIM tools, such as Tekla Structures, SDS/2, and Structureworks all grew out of the object-based parametric modeling capabilities developed for mechanical systems design. These concepts emerged as an extension of CSG and B-rep technologies, a mixture of university research and intense industrial development, particularly by Parametric Technologies Corporation® (PTC) in the 1980s. The basic idea is that shape instances and other properties can be defined and controlled according to a hierarchy of parameters at the assembly and sub-assembly levels, as well as at an individual object level. Some of the parameters depend on user-defined values. Others depend on fixed values, and still others are taken from or relative to other shapes. The shapes can be 2D or 3D.

In parametric design, instead of designing an instance of a building element like a wall or door, a designer defines a model family or element class, which is a set of relations and rules to control the parameters by which element instances can be generated but will each vary according to their context. Objects are defined using parameters involving distances, angles, and rules like *attached to, parallel to,* and *distance from*. These relations allow each instance of an element class to vary according to its own parameter settings and contextual relations. Alternatively, the rules can be defined as requirements that the design must satisfy, allowing the designer to make changes while the rules check and update details to keep the design element legal and warning the user if these definitions are not met. Object-based parametric modeling supports both interpretations.

While in traditional 3D CAD every aspect of an element's geometry must be edited manually by users, the shape and assembly geometry in a parametric modeler automatically adjusts to changes in context and to high-level user controls.

One way of understanding how parametric modeling works is by examining the structure of a wall family, including its shape attributes and relations, as shown in Figure 2-3. We call it a wall family, because it is capable of generating many instances of its type in different locations and with varied parameters. While a wall family may focus on straight and vertical walls, varied geometric capabilities are sometimes desired, including those with

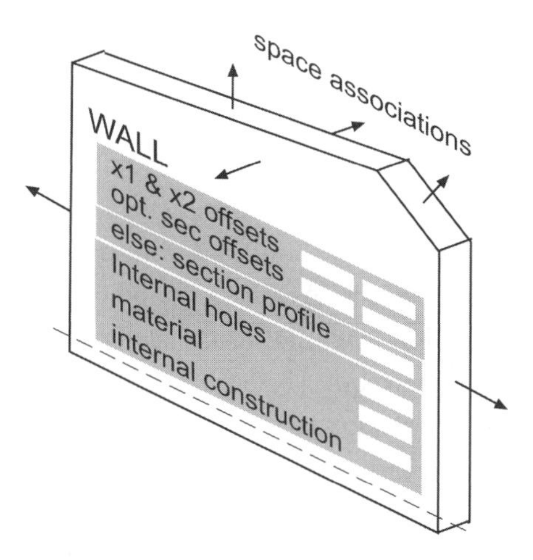

FIGURE 2-3 Conceptual structure of a wall-object family, with various edges associated with bounding surfaces.

curved and non-vertical surfaces. A wall shape is a volume bounded by multiple connected faces, some defined by context and others defined by explicit values. For most walls, the thickness is defined explicitly as two offsets from the wall control line, based on a nominal thickness or the type of construction. Walls with tapered or varying thicknesses have multiple offsets or possibly a vertical profile. The wall's elevation shape is defined by one or more base floor planes; its top face may be an explicit height or possibly defined by a set of adjacent planes (as shown here). The wall ends are defined by the wall's intersection, having either a fixed endpoint (freestanding) or associations with other walls. The control line of the wall (here shown along the bottom) has a start and end point, so the wall does too. A wall is associated with all the object instances that bound it and the multiple spaces it separates.

Door or window openings have placement points defined by a length along the wall from one of its endpoints to a side or to the center of the opening with its required parameters. These openings are located in the coordinate system of the wall, so they move as a unit. A wall will adjust its ends by moving, growing, or shrinking as the floor-plan layout changes, with windows and doors also moving and updating. Any time one or more surfaces of the bounding wall changes, the wall automatically updates to retain the intent of its original layout.

A well-crafted definition of a parametric wall must address a range of special conditions. These include:

- The door and window locations must check that they lie completely within the wall and do not overlap each other or extend beyond the wall boundaries. They typically display a warning if these conditions fail.

- A wall control line may be straight or curved, allowing the wall to take varied shapes.
- A wall may intersect floor, ceiling, or side walls, any of which are made up of multiple surfaces and result in a more complex wall shape.
- Walls may have tapered sections, if they are made of concrete or other malleable materials.
- Walls made up of mixed types of construction and finishes may change within segments of a wall.

As these conditions suggest, significant care must be taken to define even a generic wall. It is common for a parametric building class to have over one hundred low-level rules for its definition. These rules also explain why users may encounter problems with unusual wall layouts—because they are not covered by the built-in rules—and how easy it is to define wall definitions that may be inadvertently limited.

For example, take the clerestory wall and the windows set within it shown in Figure 2-4. In this case, the wall must be placed on a non-horizontal floor plane. Also, the walls that trim the clerestory wall ends are not on the same base-plane as the wall being trimmed. Early BIM modeling tools could not deal with this combination of conditions.

In Figure 2-5, we present a simple sequence of editing operations for the schematic design of a small theater. The upper left view in Figure 2-5 shows the theater with two side walls angling inward toward the stage and a wall separating the back of the theater from the lobby. In terms of boundary associations, the theater side walls are initially attached to the ceiling and floor, and their ends are attached to the rear of the lobby and front wall of the stage. The sloping theater floor is attached to the side walls of the building.

In Figure 2-5, upper right, the lobby side walls are detached from the rear wall and moved part-way open, allowing the lobby to flow around their edges.

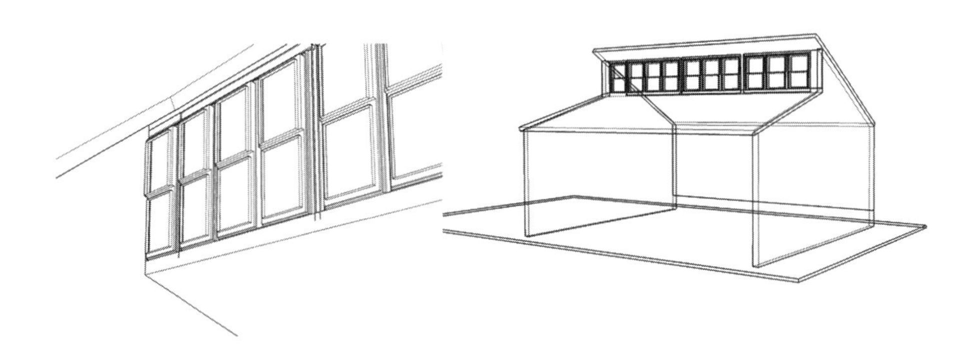

FIGURE 2-4
A clerestory wall in a ceiling that has different parametric modeling requirements than most walls.

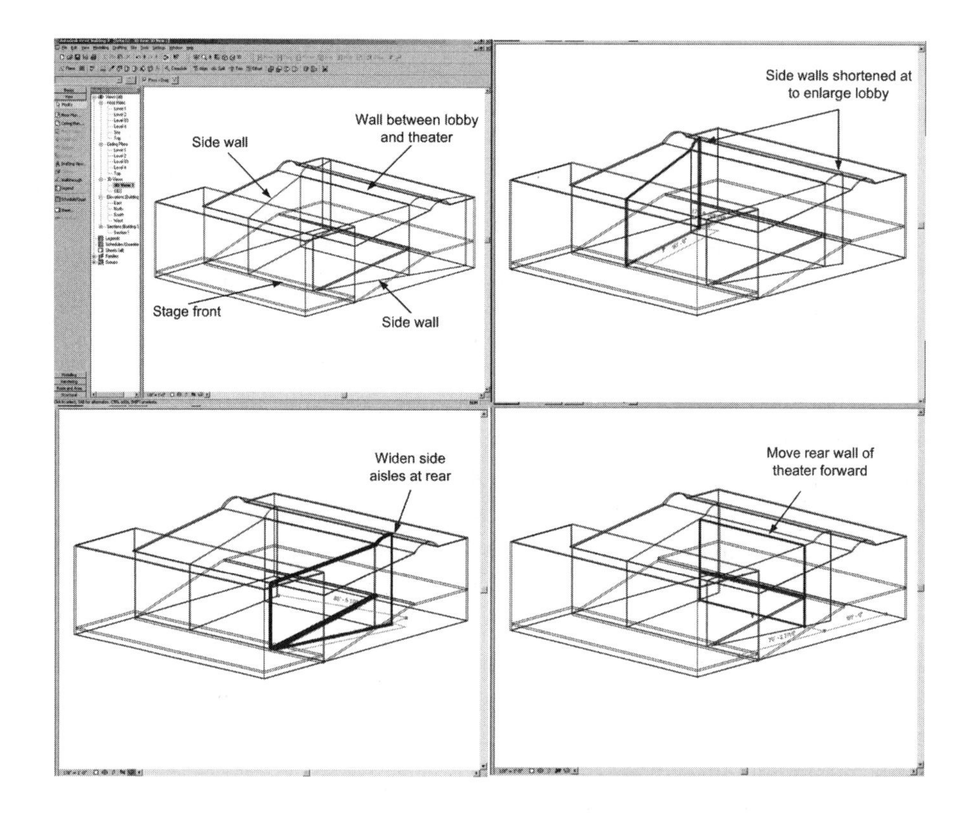

FIGURE 2-5
The conceptual layout of a small theater with lobby at the rear and two side aisles.
The sloped theater house is in the middle. The rear of the side walls is edited, then the angle of the side walls is adjusted. Last the rear of the theater is moved forward.

Notice that the ceiling arc is automatically trimmed. In the lower left figure, the side walls of the theater are re-angled, with the theater rear wall automatically trimmed to abut the walls of the theater. In lower right figure, the back wall is moved forward, making the theater shallower. The back wall automatically shortens to maintain its abutment to the side walls, and the theater and lobby floor automatically adjust to remain abutted to the back wall. The significant point is that the adjustments of the side walls to the roof and slopped theater floor are completely automatic. Once the initial spatial configuration is defined, it is possible to make quick edits and updates. Notice that these parametric modeling capabilities go far beyond those offered in previous CSG-based CAD systems. They support automatic updating of a layout and the preservation of relations set by the designer. These tools can be extremely productive.

2.1.3 Parametric Modeling of Buildings

In manufacturing, parametric modeling has been used by companies to embed design, engineering, and manufacturing rules within the parametric models of their products. For example, when Boeing undertook the design of the 777, they defined the rules by which their airplane interiors were to be defined, for looks, fabrication, and assembly. They fine-tuned the outside shape for aerodynamic performance through many hundreds of airflow simulations—called Computational Fluid Dynamics (CFD)—linked to allow for many alternative shapes and parametric adjustments. They pre-assembled the airplane virtually in order to eliminate more than 6,000 change requests and to achieve a 90% reduction in spatial re-work. It is estimated that Boeing invested more than $1 billion dollars to purchase and set up their parametric modeling system for the 777 family of planes. A good overview of the Boeing effort, its strengths and shortcomings, is available (CalTech 1997).

In a similar way, the John Deere company, working with LMS of Belgium, defined how they wanted their tractors to be constructed. Various models were developed based on John Deere's design-for-manufacturing (DfM) rules. (www.lmsintl.com/virtuallab). Using parametric modeling, companies usually define how their object families are to be designed and structured, how they can be varied parametrically and related into assemblies based on function, and other production criteria. In these cases, the companies are embedding corporate knowledge based on past efforts on design, production, assembly, and maintenance concerning what works and what does not. This is especially worthwhile when a company produces many variations of a product. This is the standard practice in large aerospace, manufacturing, and electronics companies.

Conceptually, building information modeling (BIM) tools are object-based parametric models with a predefined set of object families, each having behaviors programmed within them, as outlined above. A fairly complete listing of the pre-defined object families provided by major BIM architectural design tools is given in Table 2-1 (as of early 2007). These are the sets of pre-defined object families that can be readily applied to building designs in each system.

A building is an assembly object defined within a BIM system. A building model configuration is defined by the user as a dimensionally-controlled parametric structure, using grids, floor levels, and other global reference planes. Alternatively, these can simply be floor planes, wall centerlines, or a combination of them. Together with their embedded object instances and parametric settings, the model configuration defines an instance of the building.

In addition to vendor-provided object families, a number of Web sites make additional object families available for downloading and use. These are the modern equivalent of drafting block libraries that were available for early 2D drafting systems – but of course they are much more useful and powerful. Most of these are generic objects, but a growing capability is the provision of models of specific products. These are discussed in Section 5.4.2 and some of the sites are listed.

There are many detailed differences between the specially-developed parametric modeling tools used in BIM and those used in other industries. Buildings

Table 2-1 The built-in base object families in major BIM tools.

BIM Tool / Base Objects	ArchiCAD v10	Bentley Architecture v8.1	Revit Building v9.1	Digital Project r5.v3
Solid model w/ features	•	•	•	•
Site model	•	• (Contoured model)	• (Toposurface)	
Space definition	Manual	Manual	Room (automatic)	Room (automatic)
Wall	•	•	•	•
Column	•	•	•	•
Door	•	•	•	•
Window	•	•	•	•
Roof	•	•	•	Custom object
Stair	•	•	•	Custom object
Slab	•	•	Floor	•
Wall End	•	•		
Zone	•	Ceiling	Ceiling	
Beam	•			•
Unique Objects for each System	Skylight, Corner window	Shaft	Floor, Curtain System, Railing, Mullion, Brace, Foundation	Opening, Contour opening

are composed of a very large number of simple parts. Its regeneration dependencies are more predictable than for general mechanical design systems; however, the amount of information in even a medium-sized building at construction-level detail can cause performance problems in even the most high-end personal computers. Another difference is that there is a broad set of standard practices and codes that can be readily adapted and embedded to define object behaviors. These differences have resulted in only a few general-purpose parametric modeling tools being adapted and used for Building Information Modeling.

One functional aspect of BIM design tools that is different from those in other industries is their need to explicitly represent the space enclosed by building elements. Environmentally conditioned building space is a primary function of a building. The shape, volume, surfaces, and properties of an interior space are a critical aspect of a building. Previous CAD systems were not good at representing space explicitly. It was generally defined implicitly, as what was left between walls, floor, and the ceiling. We owe thanks to the General Services Administration (GSA) for demanding that BIM design tools be capable of deriving ANSI/BOMA space volumes beginning in 2007. This capability was conveniently overlooked in most BIM systems until GSA mandated it in the GSA BIM Guide, stating that the amount of space produced in government buildings must be accurately assessed. Today, all BIM design tools provide this capability. The GSA BIM Guide is available online at www.gsa.gov/bim. The GSA expands its information requirements annually.

Parametric modeling is a critical productivity capability, allowing low-level changes to update automatically. It is fair to say that 3D modeling would not be productive in building design and production without the automatic update features made possible by parametric capabilities. Each BIM tool varies with regards to the parametric object families it provides, the rules embedded within it, and the resulting design behavior. These important differences are taken up in Section 2.2.

2.1.4 User-Defined Parametric Objects

While each BIM design tool has a growing set of pre-defined parametric object families (see Table 2-1), these are complete only for the most standard types of construction. They are incomplete in two ways:

- Their built-in assumptions about design behavior for the pre-defined object families are normative and do not address some special cases encountered in real world contexts.
- The base object families include the most commonly encountered ones but omit those needed in many special types of construction and building types.

Another perspective is that the base object families in a BIM design tool represent standard practice, as does Ramsey and Sleeper's *Architectural Graphic Standards* (Ramsey et al. 2000). While *standard practice* reflects industry conventions, *best practice* reflects the adjustments to details, the experience a designer or firm has acquired with respect to how elements are to be detailed. Best practices distinguish the quality of design offered by most successful design practices. That is, the predefined objects that come with a

BIM design tool capture design conventions rather than expertise. Any firm that considers itself BIM-capable should have the ability to define its own libraries of custom parametric object families.

All BIM model generation tools support the definition of custom object families. If a needed parametric object family does not exist in the BIM tool, the design and engineering team has the option of either laying out the object instance using fixed B-rep or CSG geometry and remembering to update these details manually or alternatively defining a new parametric object family that incorporates the appropriate design rules and automatic updating behaviors. This embedded knowledge captures, for example, how to frame a particular style of stairway, how to detail the joining of different materials like steel and concrete or synthetic stucco and aluminum extrusions. These objects, once created, can be used in any project in which they are embedded. Clearly, the definition of details is an industry-wide undertaking that defines standard construction practices and a firm-level activity that captures best practice. Detailing is what academics such as Kenneth Frampton have referred to as the tectonics of construction (Frampton et al. 1996). It is an essential aspect of the art and craft of architecture.

If a firm frequently works with some building type involving special object families, the added labor to define these parametrically is easily justified. They provide automatic insertion of company best practice in the various contexts found in different projects. These may be at a high-level for layouts, or for detailing. They may be at various levels of sophistication.

Stadium seating is an example that justifies parametric layouts. It involves constraints that deal with seating capacity and sight lines. An example is shown in Figure 2-6. The two slightly different seating configurations at the top were

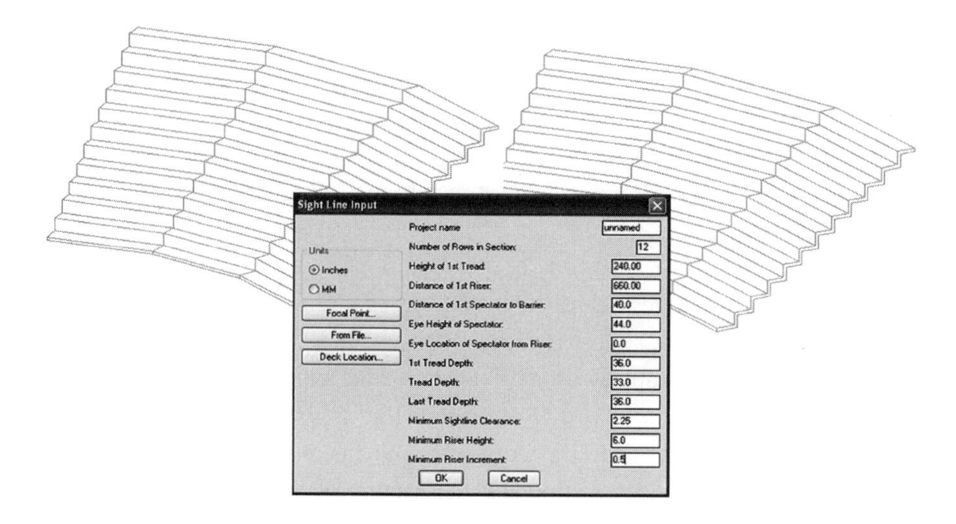

FIGURE 2-6 Custom parametric model of stadium seating.
Courtesy of HOK Sports.

generated from the same object model, which defines a section profile in terms of seating tread width, sightline target, and visual clearance above the tread below. The extruded section is then swept along a three-part path. The seating on the right allows for sightlines at closer points to the playing field. The dialog box for the sightline settings is shown at the bottom. This layout type is straightforward and relies on older solid modeling capabilities. This example was implemented in Architectural Desktop.

A more elaborate custom parametric model for the shell of a stadium is shown in Figure 2-7 and Figure 2-8. The project is for a 50,000-seat soccer and rugby stadium in Dublin, Ireland and was undertaken by HOK Sports. The initial stadium geometry and form was developed through a series of models in Rhino. Defined by a set of site constraints and sightline regulations, a parametric model was constructed in Bentley's Generative Components, using an Excel-based file to store the geometric information. This data was used as a path between Buro Happold and HOK Sports, allowing Buro Happold to develop a roof structure and HOK to develop a facade and roofing system, both in Generative Components. Having established a series of rules for the parametric model, both offices worked to develop the construction information using Generative Components and Excel (as a coordination tool), which is

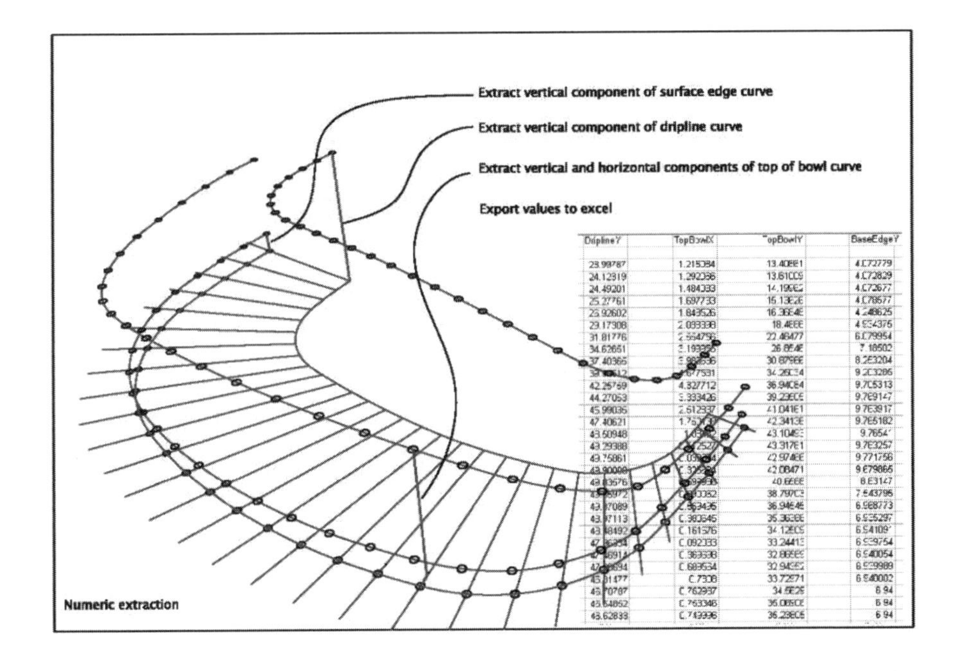

Input of Existing Data into a Parametric Model

FIGURE 2-7
The control curves generated in Generative Components.
Image provided courtesy of HOK sports.

FIGURE 2-8
The detail layout of the stadium shell and its rendering.
Courtesy of HOK Sports.

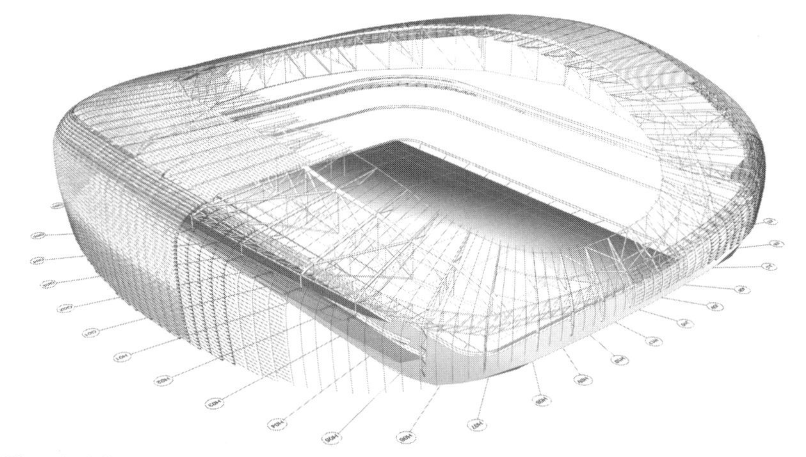

The cladding panel arrangement supporting air flow between louvers.

Rendering of final project in Dublin Ireland.

shown in the top of Figure 2-8. HOK Sports further developed the model in Bentley's Generative Components for the production of a louvered facade system, which allowed ventilation through the stadium shell, as shown. A rendering of the stadium's final design is also shown in the lower part of Figure 2-8. More adventurous architectural firms are using such customized parametric models to lay out and manage complex geometries for single projects, resulting in a new range of building forms that were hardly possible before.

The standard method for defining most custom parametric object families is by using a sketch tool module that is part of all parametric modeling tools. These are primarily used for defining swept shapes. Swept shapes include: extruded profiles, such as steel members; shapes with changing cross sections, such as ductwork fittings; shapes of rotation, such as to sweep a dome section profile around a circle; and other shapes. Sweeps are the most general tool for creating custom shapes. Combined with relations to other objects and Boolean operations, a sketch tools allows constructing almost any shape family.

The sketch tool allows a user to composed of draw a 2D closed section made up of lines, arcs or higher level curves between points, not necessarily to scale, then dimension the sketch and apply other rules to reflect the design intent in terms of the parametric rules. A swept surface may be defined with multiple profiles, either interpolating between them or in some cases with step changes. Four examples from different BIM design tools are shown in Figure 2-9. Each tool has a different vocabulary of rules and constraints that can be applied to a sketch and the operations that can be associated with its parametric behavior.

When developing custom object families using B-rep or parameters, it is important that the objects carry the attributes necessary for the various assessments that the object family's instances must support, such as cost estimation and structural or energy analyses. These attributes are also derived parametrically. These issues are taken up in Section 2.2.2.

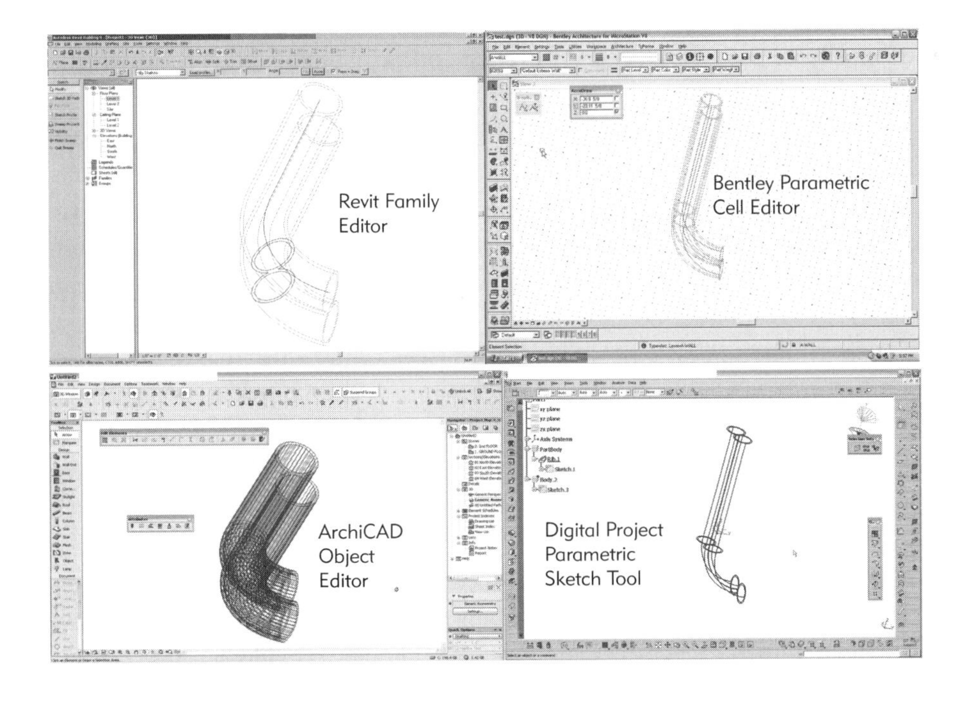

FIGURE 2-9
A figure-eight sketch, swept along a path made up of an arc and a straight segment to create a complex form. Examples are generated using all four of the major BIM design tools.

2.1.5 Design-for-Construction

While all BIM tools allow users to assign layers to a wall section in terms of a 2D section, some architectural BIM authoring tools include parametric layout of nested assemblies of objects, such as stud framing, within generic walls. This allows generation of the detailed framing and derivation of a cut lumber schedule, reducing waste and allowing for faster erection of wood or metal stud framed structures. In large-scale structures, similar framing and structural layout options are necessary extensions for fabrication. In these cases, objects and rules deal with objects as parts and their composition into a system – structural, electrical, piping, etc. In the more complex cases, each of the system's parts are then internally composed of their constituent parts, such as the wood framing, or steel reinforcing in concrete.

BIM tools for building design focus on architectural-level objects, but a different set of authoring tools have been developed for modeling at the fabrication level. These tools provide different object families for embedding different types of expertise. Early examples of such packages were developed for steel fabrication, such as Design Data's SDS/2®, Tekla's X-steel®, and AceCad's StruCad®. Initially, these were simple 3D layout systems with predefined parametric object families for connections, copes that trim members around joining steel sections, and other editing operations. These capabilities were later enhanced to support automatic design based on loads, connections, and members. With associated CNC cutting and drilling machines, these systems have become an integral part of automated steel fabrication. In a similar manner, systems have been developed for precast concrete, reinforced concrete, metal ductwork, piping, and other building systems.

Recent advances have been made in concrete engineering with cast-in-place and precast concrete. Figure 2-10 shows precast reinforcing embedded to meet structural requirements. The layout automatically adjusts to the section size and to the layout of columns and beams. It adjusts for reinforcing around connections, irregular sections, and cut-outs. Parametric modeling operations can include shape subtraction and addition operations that create reveals, notches, bull-noses, and cut-outs defined by the placement of other parts. A precast architectural facade example is shown in Figure 2-11, in terms of the 3D model of the piece and the piece mark (the drawing that describes it). Each building sub-system requires its own set of parametric object families and rules for managing the layout of the system; the rules define the default behavior of each object within the system.

Parametric rules are beginning to encode large amounts of modeling expertise within each building system domain regarding how parts should be laid out and detailed. Current parametric object families provided as base objects

FIGURE 2-10 Automated reinforcing layout and connections for precast concrete in Tekla Structures (See color insert for full color figure).

in BIM tools provide similar information to that provided by *Architectural Graphic Standards* (Ramsey et al. 2000), however, in a form that supports automatic piece definition, layout, connection, and detailing within the computer. More ambitious efforts are now underway among several construction material associations, such as the *American Institute of Steel Construction's Steel Design Guide* (AISC 2007), which now encompasses 21 volumes, and the Precast/Pre-stressed Concrete Institute's *PCI Design Handbook* (PCI 2004). Consortiums of members within these organizations have worked together to draft specifications for defining the layout and behaviors of objects in precast and steel design. Use of these tools by fabricators is discussed in more detail in Chapter 7. It should be noted that despite the fact that fabricators have had a direct hand in defining these base object families and default behaviors, they often need to be further customized so that detailing embedded in the software reflects the company's engineering practices. It should also be noted that architects have not yet taken this road but have relied on BIM design tool developers to define their base object families. Eventually, design handbooks will be delivered in this way, as a set of parametric models and rules.

In fabrication modeling, detailers refine their parametric objects for well understood reasons: such as to minimize labor, to achieve a particular visual appearance, to reduce the mixing of different types of work crews, or to

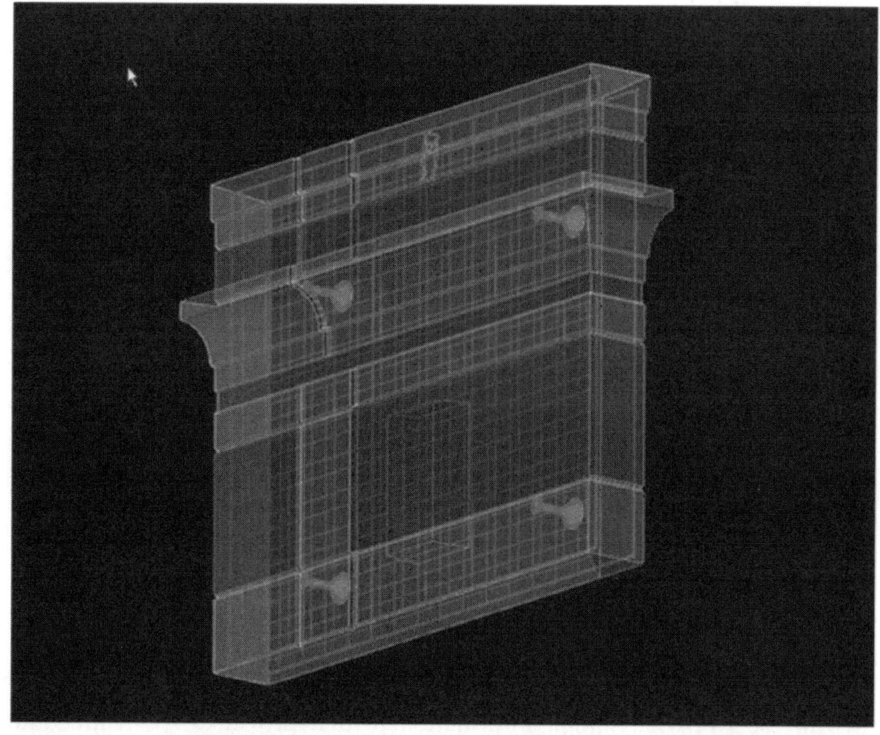

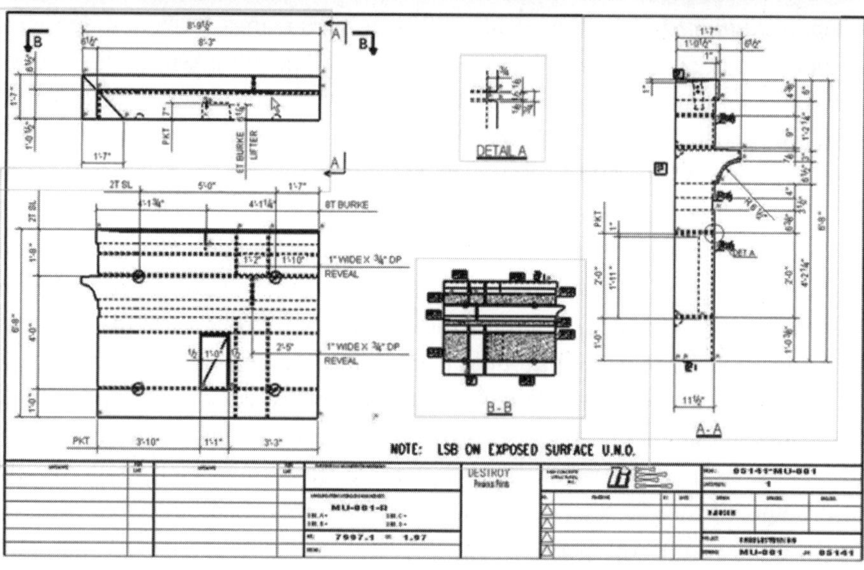

FIGURE 2-11
A parametric model of an architectural precast panel and a piecemark drawing of it.
Image provided courtesy of High Concrete Structures.

minimize the types or sizes of materials. Standard design-guide implementations typically address one of multiple acceptable approaches for detailing. In some cases, various objectives can be realized using standard detailing practices. In other circumstances, these detailing practices can be overridden.

A company's best practices or standard interfacing for a particular piece of fabrication equipment may require further customization.

2.1.6 Object-Based CAD Systems

Several CAD systems in use today are *not* general purpose parametric modeling-based BIM tools such as those that have been reviewed up to this point. Rather, they are traditional B-rep modelers possibly with a CSG-based construction tree and a given library of object classes. AutoCAD®-based construction-level modeling tools, such as CADPipe, CADDUCT, and Architectural Desktop (ADT) are examples of older software technologies. Some Bentley products with fixed vocabularies for object classes are also of this type. Within these CAD system environments, users can select, parametrically size, and lay out 3D objects with associated attributes. These object instances and attributes can be exported and used in other applications, such as for bills of material, fabrication, and other uses.

These systems work well when there is a fixed set of object classes to be composed using fixed rules. Appropriate applications include: piping, ductwork, and cable tray systems for electrical layout. ADT was being developed in this way by Autodesk, incrementally extending the object classes it could model to cover those most commonly encountered in building. ADT also supports custom-defined extrusions and other B-rep shapes but does not support user defined interactions among object instances. New object classes are added to these systems through the ARX or MDL programming language interfaces.

One critical difference with BIM is that users can define much more complex structures of object families and relations among them than is possible with 3D CAD, without undertaking programming-level software development. With BIM, a curtain-wall system attached to columns and floor slabs can be defined from scratch by a knowledgeable non-programmer. Such an endeavor would require the development of a major application extension in 3D CAD.

Another fundamental difference is that in a parametric modeler, users can define custom object families and relate them to existing objects or control grids, also without resorting to computer programming. These new capabilities allow organizations to define object families in their own way and to support their own methods of detailing and layout. Such capabilities were critical for manufacturing applications, such as those that dealt with divergent fabrication processes and product designs. In building, these capabilities are similarly well-utilized by fabrication-level tools, such as those that allow steel fabricators to define connection details and precast and cast-in-place concrete detailers to define connection and reinforcing layouts. Parametric modeling transforms modeling from a geometric design tool to a knowledge embedding tool. The implications of this capability in building design and construction are only beginning to be explored.

2.2 VARIED CAPABILITIES OF PARAMETRIC MODELERS

In general, the internal structure of an object instance as defined within a parametric modeling system is a directed graph, where the nodes are object families with parameters or operations that construct or modify an object and links in the graph reference relations between nodes. At this level, systems vary in how features are pre-defined and embedded in an object (such as a steel connection) and whether parametric objects can be nested into a larger parametric assembly and then further into even larger assemblies, as needed. Some systems offer the option of making the parametric graph visible for editing. Early parametric modeling systems rely on a full re-building of a part-and-assembly model by traversing the complete graph in response to model edits. Modern parametric modeling systems internally mark where edits are made and only regenerate affected parts of the model's graph, minimizing the update sequence. Some systems allow for the optimization of the structure's update graph based on changes made and can vary the sequence of regeneration. Other systems, called *variational systems,* use simultaneous equations to solve equations (Anderl et al. 1996). These capabilities all result in varied performance and scalability for dealing with projects involving a large number of object instances and rules.

The range of rules that can be embedded in a parametric graph determines the generality of the system. Parametric object families are defined using parameters involving distances, angles, and rules, such as *attached to, parallel to,* and *distance from*. Most allow 'if-then' conditions. Their definition is a complex undertaking, embedding knowledge about how they should behave in different contexts. If-then conditions replace one object family or design feature with another, based on the test result of some condition. These are used in structural detailing, for example, to select the desired connection type, depending upon loads and the members being connected. Such rules are also needed to effectively lay out plumbing and duct runs, by automatically inserting the correctly specified elbows and tees.

Some BIM design tools support parametric relations to complex curves and surfaces, such as splines and non-uniform B-splines (NURBS). These tools allow complex curved shapes to be defined and controlled similarly to other types of geometry. Several major BIM tools on the market have not included these capabilities, possibly for performance or reliability reasons.

The definition of parametric objects also provides guidelines for their later dimensioning in drawings. If windows are placed in a wall according to the offset from the wall-end to the center of the window, the default dimensioning will be done this way in later drawings. The *wall control line* and *endpoint intersections* define the placement dimension of the wall. (In some systems, these defaults can be overridden).

A basic capability of 3D parametric modeling systems is conflict detection of objects that spatially interfere with each other. These may be hard interferences, such as a pipe that hits a beam, or soft interferences of objects that are too close together, such as reinforcing in concrete that is too close for aggregate to pass through or a steel beam or pipe that has insufficient clearance for insulation or a concrete cover. Objects that are automatically placed via parametric rules may take interferences into account and automatically update the layout to avoid them. Others do not. This issue varies according to the specific objects being laid out and the rules embedded in them.

Enclosed spaces are the primary functional units realized in most building construction. Their area, volume, surface area composition, and often their shape and interior layout are among the most critical aspects for fulfilling the objectives of the building project. Spaces are the voids within the solid objects of construction and are usually derived from the bounding solids. They are derived in current tools in different ways: automatically defined and updated without intervention by the user; updated on demand; generated in plan defined by a polygon, then extruded to the ceiling height. These methods provide various levels of consistency, management by the user and accuracy. At the same time, space definitions contain the locations of most building functions and many analyses are applied to them, such as for energy, acoustic, and air flow simulations. While the US General Services Administration realized a first-level of capability in this area, this ability will become increasingly important for some uses discussed in Chapter 5. The capabilities of BIM design tools for space definition (as of early 2007) are shown in Table 2-1.

Parametric object modeling provides a powerful way to create and edit geometry. Without it, model generation and design would be extremely cumbersome and error-prone, as was found with great disappointment by the mechanical engineering community after the initial development of solid modeling. Designing a building that contains a million or more objects would be impractical without a platform that allows for effective low-level automatic design editing.

2.2.1 Topological Structures

When we place a wall in a parametric model of a building, we automatically associate the wall to its bounding surfaces, its base floor planes, the walls its ends abut, any walls butting it, and the ceiling surfaces trimming its height. It also bounds the spaces on its two sides. When we put a window or door in the wall, we are defining a *connection* relation between the window and the wall. Similarly, in pipe runs, it is important to define whether connections are threaded, butt welded, or have flanges and bolts. Connections in mathematics are called *topology* and—distinct from geometry—are critical to the representation of a building model and one of the fundamental aspects of parametric modeling.

Connections carry three important kinds of information: what can be connected; what the connection consists of; and how the connection is composed in response to various contexts. Some systems restrict the object types that an object can be connected to. For example, in some systems walls can connect to walls, ceilings, and floors, but a wall-edge may not connect to stairs, windows (perpendicularly), or a cabinet. Some systems encode good practice to exclude such relations. On the other hand, prohibiting them can force users to resort to work-arounds in certain special circumstances. Connections between objects can be handled in different ways in building models. The nailing of sheetrock or wood studs to a base plate are seldom detailed but are covered by a written specification. In other cases, the connection must be defined explicitly with a detail, such as the embedding of windows into precast architectural panels. (Here we use the word *connection* generically to include joints and other attachments between elements.) Topology and connections are critical aspects of a BIM tool that specify what kinds of relations can be defined in rules. They are also important as design objects and often require specification or detailing. In architectural BIM tools, connections are seldom defined as explicit elements. In fabrication-level BIM tools, they are always defined as explicit elements.

2.2.2 Property and Attribute Handling

Object-based parametric modeling addresses geometry and topology, but objects also need to carry a variety of properties if they are to be interpreted, analyzed, priced, and procured by other applications.

Properties include: material specifications needed for fabrication, such as steel or concrete strength and bolt and weld specifications; material properties related to different performance issues such as acoustics, light reflectance, and thermal flows; properties for assemblies like wall and floor-to-ceiling systems or steel and precast concrete assemblies based on weight, structural behavior, etc; and properties for spaces such as occupancy, activities, and equipment needed for energy analysis.

Properties are seldom used singularly. A lighting application requires material color, a reflection coefficient, a specular reflection exponent, and possibly a texture and bump map. For accurate energy analysis, a wall requires a different set. Thus, properties are appropriately organized into sets and associated with a certain function. Libraries of property sets for different objects and materials are an integral part of a well-developed BIM model generation tool and the environment in which the tool resides. The property sets are not always available from the product vendor and often have to be approximated by a user, the user's firm or from the American Society of Testing and Materials data (ASTM). Although organizations such as the Construction Specifications Institute are looking at these issues, the development of property sets for supporting a wide range of simulation and analysis tools has not yet been adequately addressed and is left to users to set up.

Current BIM generation tools default to a minimal set of properties for most objects and provide the capability of adding an extendable set. Several existing BIM tools provide Uniformat™ classes to associate elements for cost estimation. Users or an application must add properties to each relevant object to produce a certain type of simulation, cost estimate, or analysis and also must manage their appropriateness for various tasks. The management of property sets becomes problematic because different applications for the same function may require somewhat different properties and units, such as for energy and lighting.

There are at least three different ways that properties may be managed for a set of applications:

- By pre-defining them in the object libraries so they are added to the design model when an object instance is created.
- By the user adding them as-needed for an application from a stored library of property sets.
- By the properties being assigned automatically, as they are exported to an analysis or simulation application.

The first alternative is good for production work involving a standard set of construction types but requires careful user definition for custom objects. Each object carries extensive property data for all relevant applications, only some of which may actually be used. Extra definitions may slow down an application's performance and enlarge its objects' sizes. The second alternative allows users to select a set of similar objects or property sets to export to an application. This results in a time-consuming export process. Iterated use of simulation tools may require the addition of properties each time the application is run. This would be required, for example, to examine alternative window and wall systems for energy efficiency. The third approach keeps the design application light but requires the development of a comprehensive material tag that can be used by all exporting translators to associate a property set for each object. The authors believe that this third approach is the desired long-term approach for attribute handling. The necessary object classifications and name tagging required of this approach must still be developed. Currently, multiple object tags must be developed, one for each application.

The development of object property sets and appropriate object classification libraries to support different types of applications is a broad issue under consideration by the Construction Specification Institute of North America and by other national specification organizations. It is reviewed in more detail in Section 5.3.3. A comprehensive solution does not yet exist but needs to be developed to support full utilization of BIM technologies.

Object libraries, representing company best practices and specific commercial building products, are an important component of a BIM environment. This important facility is reviewed in Section 5.4.

2.2.3 Drawing Generation

Even though a building model has the full geometric layout of a building and its systems—and the objects have properties and potentially specifications—drawings will continue to be required, as reports extracted from the model, for some time into the future. Existing contractual processes and work culture, while changing, are still centered on drawings, whether paper or electronic. If a BIM tool does not support effective drawing extraction and a user has to do significant manual editing to generate each set of drawings from cut sections, the benefits of BIM are significantly reduced.

With building information modeling, each building object instance—its shape, properties, and placement in the model—is defined only once. From the overall arrangement of building object instances, drawings, reports, and datasets can be extracted. Because of this non-redundant building representation, all drawings, reports, and analysis datasets are consistent if taken from the same version of the building model. This capability alone resolves a significant source of errors and guarantees internal consistency within a drawing set. With normal 2D architectural drawings, any change or edit must be transferred manually to multiple drawings by the designer, resulting in potential human errors from not updating all drawings correctly. In precast concrete construction, this 2D practice has been shown to cause errors costing approximately 1% of construction cost (Sacks et al. 2003).

Architectural drawings do not rely on orthographic projections, as learned in high school drafting classes. Rather, drawings such as plans, sections, and elevations incorporate complex sets of conventions for recording design information graphically on sheets of paper. This includes symbolic depiction of some physical objects, dotted representation of geometry behind the section plane in floor-plans, and very selective dotted-line representation of hidden objects in front of the section plane, in addition to line-weights and annotations. Mechanical, electrical, and plumbing systems (MEP) are often laid out schematically (topologically), leaving the final layout to the contractor, after equipment has already been selected. These conventions require that BIM design tools embed a strong set of representational rules in their drawing extraction capabilities. In addition, drawing conventions of individual firms must be added to the built-in tool conventions. These issues affect both how the model is defined within the tool and how the tool is set up for drawing extraction.

Part of a given drawing definition is derived from how an object is defined, as described earlier. The object has an associated name, annotation, and in some cases line-weights and formats for presentation in different views that are carried in the object library. The placement of the object also has implications. If the object is placed relative to a grid intersection or wall end, that is how its placement will be dimensioned in the drawing. If the object is parametrically defined relative to other objects, such as the length of a beam placed to span between variably placed supports, then the drawing generator will not automatically dimension the length unless the system is told to derive the beam length at drawing generation time.

Most BIM models of buildings do not include 3D and attribute information for all the pieces of a building. Many are shown only in section details. Most BIM design tools provide the means for extracting a drawn section at the level of detail to which they are defined in the 3D model. The location of the drawn section is automatically recorded with a section-cut symbol on a plan or elevation as a cross-reference and the location can be moved if needed. The section is then detailed manually with the needed wood-blocks, extrusions, silicon beading, weather stripping; and associated annotations are provided in the fully detailed drawn section. An example is shown in Figure 2-12, with the figure on the left showing the extracted section and the one on the right showing the detailed section with drafted annotation. In most systems, this detail is associated with the section cut it was based on. When 3D elements in the section change, they update automatically in the section but the hand drawn details must be manually updated.

To produce drawings, each plan, section, and elevation is separately composed based on the above rules from a combination of cut 3D sections and aligned 2D drawn sections. They are then grouped into sheets with normal borders and title sheets. The sheet layouts are maintained across sessions and are part of the overall project data.

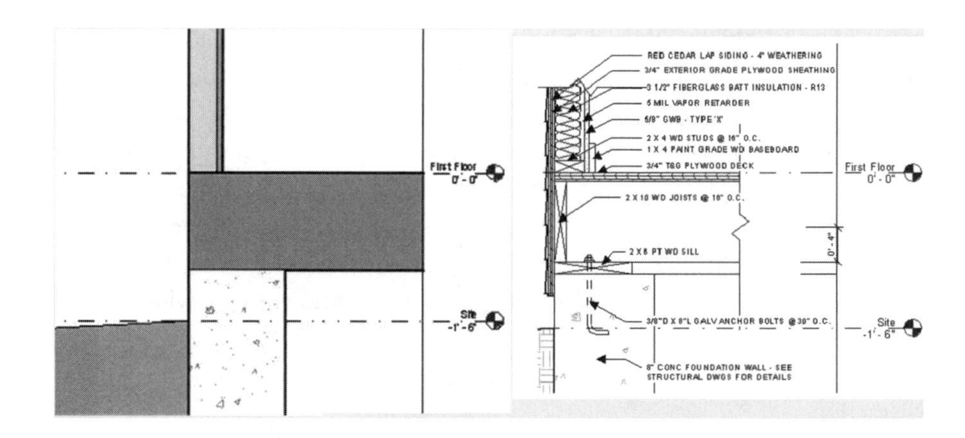

FIGURE 2-12
The initial section extracted from the building model (left) and the manually detailed drawing elaborated from the section (right).
Image provided courtesy of Autodesk.

Drawing production from a detailed 3D model has gone through a series of refinements to make it efficient and easy. Below is an ordered list of the levels of quality that can now be supported technically, though most systems have not realized top-level drawing generation. We start from the weakest level.

1. A weak level of drawing production provides for the generation of orthographic sections cut from a 3D model, and the user manually edits the line formats and adds dimensions, details, and annotations. These details are associative. That is, as long as the section exists in the model, the annotation set-up is maintained across drawing versions. Such association capabilities are essential for effective re-generation of drawings for multiple versions. In this case, the drawing is an elaborated report generated from the model.
2. An improvement upon 1 (above) is the definition and use of drawing templates associated with elements for a type of projection (plan, section, elevation) that automatically generates dimensioning of the element, assigns line weights, and generates annotations from defined attributes. This greatly speeds-up the initial drawing setup and improves productivity, though set-up for each object family is tedious. Only changes to the presentation of data can be made in drawings; edits to the drawing do not change the model. In these first two cases, report management should be provided to inform the user that model changes have been made, but the drawings cannot automatically update to reflect these changes until they are regenerated.
3. Current top-level drawing functionality supports bi-directional editing between models and drawings. If drawings are a specialized view of the model data, then shape-changes made to the drawings should be

permitted in and propagated to the model. In this case, the drawings are updated. If displayed in windows alongside views of the 3D model, updates in any view can be referenced immediately in the other views. Bi-directional views and strong template generation capabilities further reduce the time and effort needed for drawing generation.

Door, window and hardware schedules are defined in a similar way to the three alternatives described above. That is, schedules are also model views and can be updated directly. A static report generator method is weakest, and a strong bi-directional approach is strongest. Such bi-directionality offers important benefits, including the ability to trade hardware used on a set of doors with hardware recommended on the schedule, rather from the model.

In fabrication-level BIM modeling systems, this mixed system of schematic 3D layout and 2D detailing is not used, and the design is assumed to be generated primarily from the 3D object model. In these cases, joists, studs, plates, plywood sills, and other pieces shown in Figure 2-12 would be laid out in 3D. Line-weights and crosshatching are defined for the piece type and applied automatically. Some systems store and place associated annotations with object sections, though these annotations often need shifting to achieve a well-composed layout. Other annotations refer to details as a whole, such as name, scale, and other general notes and these must be associated with the overall detail. Such capabilities come close to automated drawing extraction, but it is unlikely that automation will be complete.

Drawing sheets typically carry more information than plans, sections, and elevations for a building. They include a site plan, which shows the building's placement on the ground plot relative to recorded geo-spatial datum. Some BIM design tools have well developed site-planning capabilities, others do not. Table 2-1 shows which BIM design tools include site objects.

An obvious current goal is to automate the drawing production process as much as possible, since most initial design productivity benefits (and costs) will depend on the extent of automatic generation. At some point, most parties involved in the building delivery process will adapt their practices to BIM technology, such as building inspectors, and financial institutions; we are slowly moving to a paperless world. Drawings will continue to be used, but as throw-away mark-up sheets by construction crews and other users. As these changes take place, the conventions regarding architectural drawings are likely to evolve, allowing them to be customized for the specific task in which they are used. This trend is described in further detail in Section 2.3.3.

It should be clear at this point that BIM technology generally allows designers to use 3D modeling to varying degrees, with 2D drawing sections filling in the missing details. The BIM benefits of data exchange, bills of material, detailed cost estimation, and other actions are lost on those elements defined only in 2D section drawings. Thus, BIM technology allows users to determine the level of 3D modeling detail they wish to use. It can be argued that complete 3D object modeling is not warranted. Few would argue to include nails, flashing, and some forms of vapor barrier as 3D objects in a building model. On the other hand, most projects today are only partially supporting BIM. Fabrication-level models are likely (or should be) full BIM. This mixed technology is also good for firms getting started in BIM, as they can utilize the technology incrementally.

2.2.4 Scalability

A problem that many users encounter is scalability. Problems in scaling are encountered when a project model gets too large for practical use due to its large memory size. Operations become too sluggish, so that even simple operations are unfeasible. Building models are large; even simple 3D shapes take a lot of memory space. Large buildings can contain millions of objects, each with a different shape. Scalability is affected by both the size of the building, say in floor area, and also by the level of detail in the model. Even a simple building can encounter scalability problems if every nail and screw is modeled.

Parametric modeling incorporates design rules that relate geometry or other parameters of one object with those of other objects. Changing one control grid may propagate updates to the whole building. Thus, it is hard to partition a project into parts for separate development. BIM tools developed for architecture do not generally have the means for managing a project spanning multiple object files. Some systems must carry all updated objects in memory simultaneously and are considered *memory-based*. When the model gets too large to be held in memory, virtual memory-swapping occurs, which can result in significant waiting time. Some systems have methods of propagating relations and updates across files and can open, update, and then close multiple files within the span of a single operation. These are called *file-based* systems. File-based systems are generally slower for small projects but their speed decreases only slowly as project size grows.

Based on these definitions, Revit and ArchiCAD are memory-based; Bentley, Digital Project and Tekla Structures are file-based. Tool-specific work processes can mitigate some of the problems associated with scalability. These should be discussed with product vendors.

Memory and processing issues will naturally decrease as computers get faster. Sixty-four-bit processors and operating systems will also provide significant help. There will be the parallel desire, however, for more detailed building models. Issues of scalability will be with us for some time.

2.2.5 Open Questions

Strengths and Limitations of Object-Based Parametric Modeling

One major benefit of parametric modeling is the intelligent design behavior of objects. This intelligence, however, comes at a cost. Each type of system object has its own behavior and associations. As a result, BIM design tools are inherently complex. Each type of building system is composed of objects that are created and edited differently. Effective use of a BIM design tool usually takes months to gain proficiency.

Modeling software that some designers prefer, such as Sketchup, Rhino, and FormZ, are not parametric modeling–based tools. Rather, they have a fixed way of geometrically editing objects, varied only according to the surface types used; and this same functionality is applied to all object types. Thus, an editing operation applied to walls will have the same behavior when it is applied to piping. In these systems, attributes defining the object type and its functional intention, if applied at all, can be added when the user chooses, not when it is created. An argument can be made that for design use, BIM technology with its object-specific behavior is not always warranted. This topic is explored further in Chapter 5.

Why Can't Different Parametric Modelers Exchange Their Models?

It is often asked why firms cannot directly exchange a model from Revit with Bentley Architecture, or exchange ArchiCAD with Digital Project. From the overview discussed previously, it should be apparent that the reason for this lack of interoperability is due to the fact that different BIM design tools rely on different definitions of their base objects. These are the result of different capabilities involving rule types in the BIM tool and also the rules applied in the definition of object families. This problem applies only to parametric objects, not those with fixed properties. These problems could disappear if and when organizations agree on a standard for object definitions. Until then, exchanges for some objects will be limited or will fail completely. Improvements will come about incrementally, as the demand to resolve these issues makes implementation worthwhile, and the multiple associated issues are sorted out. The same issue exists in manufacturing and has not yet been resolved.

Are There Inherent Differences in Construction, Fabrication, and Architectural BIM Design Tools?

Could the same BIM platform support both design and fabrication detailing? Because the base technology for all of these systems has much in common, there is no technological reason why building design and fabrication BIM tools cannot offer products in each other's area. This is happening to some degree with Revit Structures and Bentley Structures. They are developing some of the capabilities offered by fabrication-level BIM tools. Both sides address the engineering market and, to a lesser degree, the contractor market; but the expertise needed to support full production use in these information-rich areas will depend on major front-end embedding of requisite object behaviors, which are distinctly different for each building system. Expert knowledge of specific building system object behaviors is more readily embedded when it is codified, as it is, for example, in structural system design. The interfaces, reports, and other system issues may vary, but we are likely to see skirmishes in the middle-ground for a significant period of time, as each product attempts to broaden its market domains.

Are There Significant Differences between Manufacturing-Oriented Parametric Modeling Tools and BIM Tools?

Could a parametric modeling system for mechanical design be adapted for BIM? Some differences in system architecture are noted in Section 2.3.1. Of course, mechanical parametric modeling tools have already been adapted for the AEC market. Digital Project, based on CATIA, is an obvious example. Also, Structureworks is a precast concrete detailing and fabrication product based on Solidworks. In other areas, such as plumbing, curtain wall fabrication, and ductwork design, we can expect to see both mechanical parametric modeling tools and architectural and fabrication-level BIM tools vying for these markets. The range of functionality offered in each market is still being sorted out. The market is the battleground.

This chapter has provided an overview of the basic capabilities of BIM design tools resulting from their development as object-based parametric design tools. We now turn to reviewing the main BIM design tools and their functional differences.

2.3 OVERVIEW OF THE MAJOR BIM MODEL GENERATING SYSTEMS

Below, we summarize the major functional and performance capabilities that distinguish different BIM design systems, as presented in earlier sections of this chapter. The capabilities apply to both design-oriented systems as well as fabrication BIM tools. These distinguishing capabilities are proposed for those wishing to undertake a first-level review and assessment of alternative systems, so as to make a well-informed decision for a platform at the project, office, or enterprise. The choice affects production practices, interoperability, and to some degree, the functional capabilities of a design organization to do particular types of projects. Current products also have different capabilities regarding interoperability, which may affect their ability to collaborate and can lead to convoluted workflows and replicated data.

We emphasize that no one platform will be ideal for all types of projects. Ideally, an organization would have several platforms that it supports and toggles between for specific projects. At this early date, an effort to adopt any of the available BIM design tools is a significant undertaking and is discussed in later chapters. It involves understanding the new technology, the new organizational skills it presupposes, and then learning and managing those skills. These challenges will recede over time, as the learning curve for one system is scaled. Because the functionality of BIM design tools are changing quickly, it is important to look at reviews of the current versions in *AECBytes, Cadalyst,* or other AEC CAD journals.

2.3.1 Discriminating Capabilities

Within the common framework of providing object-based parametric modeling, BIM authoring tools embody many different kinds of capabilities. Below, we describe them in rough-rank based on our sense of their level of importance.

User Interface: BIM tools are quite complex and have much greater functionality than earlier CAD tools. Some BIM design tools have a relatively intuitive and easy-to-learn user interface, with a modular structure to their functionality, while others place more emphasis on functionality that is not always well-integrated into the overall system. Criteria to be considered here should include: consistency of menus across the system's functionalities following standard conventions; menu-hiding that eliminates irrelevant actions not meaningful to the current context of activities; modular organization of different kinds of functionality and on-line help providing real-time prompts and command-line explanation of operations and inputs. While user interface issues may seem minor, a poor user interface results in longer learning times, more errors, and often not taking full advantage of the functionality built into the application.

Drawing Generation: How easy is it to generate drawings and drawing sets and to maintain them through multiple updates and series of releases? Assessment should include quick visualization of the effects of model changes on drawings, strong associations so that model changes propagate directly to drawings and vice versa, and effective template generation that allows drawing types to carry out as much automatic formatting as possible. A more thorough review of functionality is provided in Section 2.2.3.

Ease of Developing Custom Parametric Objects: This is evaluated with regard to the existence and ease-of-use of a sketching tool for defining parametric objects; determining the extent of the system's constraint or rule-set ('a general constraint rule set should include distance, angle including orthogonality, abutting faces and line tangency rules and 'if-then' conditions) its ability to interface the objects into the user interface for easy embedding in a project') and its ability to support parametric assemblies of objects. These issues are explained further in Section 2.1.4.

Scalability: the ability to handle combinations of a large project scale and modeling at a high level of detail. This involves the ability of the system to remain interactive and responsive regardless of the number of 3D parametric objects in the project. A fundamental issue is the degree that the system is disk-based, in terms of data management, rather then memory-based. Disk-based systems are slower for small projects, but their delay time grows slowly as the project size grows. Memory based systems performance drops quickly once memory space is exhausted. These issues are partially limited by the operating system; Windows XP supports up to 2 GB of working memory for a single process. Sixty-four-bit architectures eliminate the memory use restriction. Graphic card performance is important for some systems. This topic is discussed in Sec. 2.2.4.

Interoperability: Model data is generated, in part, to share with other applications for early project feasibility studies, for collaboration with engineers and other consultants and later for construction. It is supported by the degree that the BIM tool provides direct interfaces with other specific products and, more generally, its import and export support of open data exchange standards, which are reviewed in detail in Chapter 3.

Extensibility: A BIM authoring tool is both for end-use and for use as a platform for customization and extension. Extensibility capabilities are assessed based on whether they provide scripting support—an *interactive language* that adds functionality or automates low-level tasks, similar to AutoLISP® in AutoCAD—an Excel format bi-directional interface, and a broad and well-documented application programming interface (API). Scripting languages and Excel interfaces are generally for end-users, while an API is for software developers. These capabilities are needed depending on the extent to which a firm expects to customize, particularly in the area of interoperability.

Complex Curved Surface Modeling: Support for creating and editing complex surface models based on quadrics, splines, and non-uniform B-splines is important for those firms that do this type of work or that are planning to. These geometric modeling capabilities in a BIM tool are foundational; they cannot be added on later.

Multi-User Environment: Some systems support collaboration among a design team. They allow multiple users to create and edit parts of the same project directly from a single project file and manage user access to these various information parts.

Below, we offer an overview of the current capabilities of the major building model generation platforms. Some reviewed support only architectural design functions, others only various types of fabrication-level building systems, and others both. Each assessment is for the version of the software system noted; later versions may have better or worse capabilities. We review them according to the criteria developed above.

2.3.2 BIM Tools for Architectural Design

Each BIM building design platform is introduced in terms of its heritage, corporate organization, the family of products it is a part of, whether it uses a single file or multiple files per project, support for concurrent usage, interfaces supported, extent of the object library, general price class, building classification system supported, scalability, ease of drawing generation, support for 2D drawn sections, types of objects and derived attributes, and ease of use.

As is broadly understood, the acquisition of a software package is very different from most other purchases we make. Whereas the purchase of a car is based on a very specific product and set of features, a software package involves both its current capabilities and the development path of enhancements that are released regularly, at least annually. A purchaser is buying into both the current product and its future evolutions, as projected by the company. One is also purchasing a support system that at least one person in a firm will be dealing with. The support system is an augmentation of the user-provided documentation and on-line support built into the BIM tool.

Apart from the vendor's support network, a software system owner is also part of a broader user community. Most provide blog communication for peer-to-peer help and open portals for the exchange of object families. These may be free or available at a small cost.

Revit: Revit Architecture is the best known and current market leader for the use of BIM in architectural design. It was introduced by Autodesk in 2002 after the company acquired the program from a start-up. Revit is a completely separate platform from AutoCAD, with a different code base and file structure. The version reviewed here is 9.1. Revit is a family of integrated products that currently includes Revit Architecture, Revit Structure, and Revit MEP. It includes: gbXML interfaces for energy simulation and load analysis; direct interfaces to ROBOT and RISA structural analyses, and the ability to import models from Sketchup, a conceptual design tool, and other systems that export DXF files. Viewing interfaces include: DGN, DWG, DWF™, DXF™, IFC, SAT, SKP, AVI, ODBC, gbXML, BMP, JPG, TGA, and TIF. Revit relies on 2D sections as a way of detailing most types of assemblies.

Revit's strengths: It's easy to learn and its functionality is organized in a well-designed and user-friendly interface. It has a broad set of object libraries developed by third parties. It is the preferred interface for direct link interfaces, because of its market position. Its bi-directional drawing support allows for information generation and management based on updates from drawing and model views; it supports concurrent operation on the same project; and it includes an excellent object library that supports a multi-user interface.

Revit's weaknesses: Revit is an in-memory system that slows down significantly for projects larger than about 220 megabytes. It has limitations on parametric rules dealing with angles. It also does not support complex curved surfaces, which limits its ability to support design with or reference to these types of surfaces.

Bentley Systems: Bentley Systems offers a wide range of related products for architecture, engineering, and construction. Their architectural BIM tool, Bentley Architecture, introduced in 2004, is an evolutionary descendent of Triforma. Integrated with Bentley Architecture are: Bentley Structural, Bentley Building Mechanical Systems, Bentley Building Electrical Systems, Bentley Facilities, Bentley PowerCivil (for site planning), and Bentley Generative Components. These are file-based systems, meaning that all actions are immediately written to a file and result in lower loads on memory. Third parties have developed many different applications on the file system, some incompatible with others within the same platform. Thus a user may have to convert model formats from one Bentley application to another. Currently, Bentley Architecture is in version V8.9.2.42. Interfaces with external applications include: Primavera and other scheduling systems and STAAD and RAM for structural analyses. Its interfaces include: DGN, DWG, DXF™, PDF, STEP, IGES, STL, and IFC. Bentley also provides a multi-project and multi-user model repository called Bentley ProjectWise.

Bentley System's strengths: Bentley offers a very broad range of building modeling tools, dealing with almost all aspects of the AEC industry. Bentley supports modeling with complex curved surfaces, including Bezier and NURBS. It includes multiple levels of support for developing custom parametric objects, including the Parametric Cell Studio and Generative Components. Its parametric modeling plug-in, Generative Components, enables definition of complex parametric geometry assemblies and has been used in many prize-winning building projects. Bentley provides scalable support for large projects with many objects.

Bentley System's weaknesses: It has a large and non-integrated user interface that is hard to learn and navigate; its heterogeneous functional modules include different object behaviors, making it hard to learn. It has less extensive object libraries than similar products. The weaknesses in the integration of its various applications reduce the value and breadth of support that these systems provide individually.

ArchiCAD: ArchiCAD is the oldest continuously marketed BIM architectural design tool availably today. Graphisoft began marketing ArchiCAD in the early 80s. It is the only object-model-oriented architectural CAD system running on the Apple Macintosh. Headquartered in Budapest, Graphisoft was recently acquired by Nemetschek, a German CAD company popular in Europe with strong civil engineering applications. The current version of ArchiCAD is release 11.0. Today, ArchiCAD continues to serve the Mac platform in addition to Windows and has recently released a Mac OS X (UNIX) version. Graphisoft recently introduced a number of construction-oriented applications on the ArchiCAD platform. In early 2007, after Graphisoft was acquired by Nemetschek, the construction applications were spun off to Vico Software, a new company that is actively marketing them. These applications are discussed in Chapter 6.

ArchiCAD supports a range of direct interfaces, with Maxon for curved surface modeling and animation, ArchiFM for facility management and Sketchup. It has interfaces with a suite of interfaces for energy and sustainability (gbXML, Ecotect, Energy+, ARCHiPHISIK and RIUSKA). Custom parametric objects are primarily defined using the GDL (Geometric Description language) scripting language, which relies on CSG-type constructions and a Basic-like syntax. (Basic is a simple programming language often taught to beginners). It contains extensive object libraries for users and also has an OBDC interface.

ArchiCAD's strengths: It has an intuitive interface and is relatively simple to use. It has large object libraries, and a rich suite of supporting applications in construction and facility management. It is the only strong BIM product currently available for Macs.

ArchiCAD's weaknesses: It has some limitations in its parametric modeling capabilities, not supporting update rules between objects in an assembly or automatic application of Boolean operations between objects (Khemlani 2006). While ArchiCAD is an in-memory system and can encounter scaling problems with large projects, it has effective ways to manage large projects; it can partition large projects well into modules in order to manage them.

Digital Project: Developed by Gehry Technologies, Digital Project (DP) is an architectural and building customization of Dassault's CATIA, the world's most widely used parametric modeling platform for large systems in aerospace and automotive industries. DP requires a powerful workstation to run well, but it is able handle even the largest projects. The One Island East case study in Chapter 9 provides an example of DP's ability to model every part of a 70 story office tower. It is able to model any type of surface and can support elaborate custom parametric objects, which is what it was designed to do. The logical structure of CATIA involves modules called *Workbenches*. Until the third release of Version 5, it did not include built-in base objects for buildings. Users could re-use objects developed by others, but these were not supported by DP itself. With the introduction of the Architecture and Structures Workbench, Gehry Technologies has added significant value to the base product. Although not advertised, DP comes with several other workbenches: Knowledge Expert supports rule-based checking of design; the Project Engineering Optimizer allows for easy optimization of parametric designs based on any well-defined objective function; and Project Manager for tracking parts of a model and managing their release. It has interfaces with Ecotect for energy studies.

DP supports VBA scripting and a strong API for developing add-ons. It has the Uniformat© and Masterformat© classifications embedded, which facilitates integration of specifications for cost estimating. It supports the following exchange formats: CIS/2, SDNF, STEP AP203 and AP214, DWG, DXF™, VRML, STL, HOOPS, SAT, 3DXML, IGES, and HCG. In Release 3, it has IFC support.

Digital Project's strengths: It offers very powerful and complete parametric modeling capabilities and is able to directly model large complex assemblies for controlling both surfaces and assemblies. Digital Project relies on 3D parametric modeling for most kinds of detailing.

Digital Project's weaknesses: It requires a steep learning curve, has a complex user interface, and high initial cost. Its predefined object libraries for buildings are still limited. External third party object libraries are limited. Drawing capabilities for architectural use are not well developed; most users output sections to drafting systems for completion.

AutoCAD-based Applications: Autodesk's premier building application on the AutoCAD platform is Architectural Desktop (ADT). ADT was Autodesk's original 3D building modeling tool prior to the acquisition of Revit. It is based on solid and surface modeling extensions for AutoCAD and provides a transition from 2D drafting to BIM. It has a predefined set of architectural objects, and while not fully parametric, it provides much of the functionality offered by parametric tools, including the ability to make custom objects with adaptive behaviors. External Reference Files (XREF) are useful for managing large projects. Drawing files remain separate from the 3D model and must be managed by the user, albeit with a degree of system version control. It relies on AutoCAD's well-known capabilities for drawing production. Interfaces include: DGN, DWG, DWF™, DXF™, and IFC. Its programming extensions include: AutoLISP, Visual Basic, VB Script, and ARX (C++) interfaces.

Additional 3D applications developed on AutoCAD come from a large world-wide developer community. These include Computer Services Consultants (CSC) which offers a number of structural design and analysis packages, AEC Design Group, which offers CADPIPE, COADE Engineering Software, that offers piping and plant design software, SCADA Software AG, that develops control system software, and other groups that produce 3D applications for piping, electrical system design, structural steel, fire sprinkler systems, ductwork, wood framing and others.

AutoCAD-based applications' strengths: Ease of adoption for AutoCAD users because of user interface consistency; easy use because they build upon AutoCAD's well-known 2D drafting functionality and interface.

AutoCAD-based applications' weaknesses: Their fundamental limitations are that they are not parametric modelers that allow non-programmers to define object rules and constraints; limited interfaces to other applications; use of XREFs (with inherent integration limitations) for managing projects; an in-memory system with scaling problems if XREFs are not relied upon; need to propagate changes manually across drawings sets.

Tekla Structures: Tekla Structures is offered by Tekla Corp., a Finnish company founded in 1966 with offices worldwide. Tekla has multiple divisions: Building and Construction, Infrastructure and Energy. Its initial construction product was Xsteel, which was introduced in the mid-1990s and grew to be the most widely used steel detailing application throughout the world.

In response to demand from precast concrete fabricators in Europe and North America (represented by the ad hoc Precast Concrete Software Consortium), the software's functionality was significantly extended to support fabrication-level detailing of precast concrete structures and facades. At the same time, support for structural analysis, direct links to finite-element analysis packages (STAAD-Pro and ETABS), and an open application programming interface were added. In 2004 the expanded software product was renamed Tekla Structures to reflect its generic support for steel, precast concrete, timber, reinforced concrete, and for structural engineering.

Tekla Structures supports interfaces with: IFC, DWG™, CIS/2, DTSV, SDNF, DGN, and DXF™ file formats. It also has export capabilities to CNC fabrication equipment and to fabrication plant automation software, such as Fabtrol (steel) and Eliplan (precast).

Tekla Structures' strengths: Its versatile ability to model structures that incorporate all kinds of structural materials and detailing; its ability to support very large models and concurrent operations on the same project and with multiple simultaneous users. It supports compilation of complex parametric custom component libraries with little or no programming.

Tekla structures' weaknesses: While a powerful tool, its full functionality is quite complex to learn and fully utilize. The power of its parametric component facility, while a strength, requires sophisticated operators who must develop high levels of skill. It is not able to import complex multi-curved surfaces from outside applications, sometimes leading to work-arounds. It is relatively expensive.

DProfiler: DProfiler is a product of Beck Technologies, located in Dallas, Texas. It is based on a parametric modeling platform acquired from Parametric Technologies Corporation (PTC) in the middle 1990s, after PTC decided not to enter the AEC market. DProfiler is an application based on a platform called DESTINI that has evolved from the PTC-acquired software. DProfiler supports very quick definition for the conceptual design of certain building types and then provides feedback regarding construction costs and time. For income generating facilities, such as hotels, apartments, and office buildings, it provides a full economic cash-flow development proforma. It supports planning for: office buildings up to 20 stories; one and two-story medical buildings; apartment buildings and hotels up to 24 stories; elementary, middle, and high schools; and town halls, churches, and movie centers and more. With financial and schedule reporting, the user gains a set of concept design drawings. Users can input their own cost data or use data from RSMeans. Their advertising states 5% accuracy in its business calculations, but these are informal assessments. It currently supports Sketchup and DWG export for further development. The Hillwood case study (Section 9.9) describes its use. Its interfaces include Excel and DWG. Other applications being developed on the DESTINI platform include energy analyses.

DProfiler strengths: DProfiler is being marketed as a closed system, primarily for preliminary feasibility studies before actual design begins. Its ability to generate quick economic assessments on a project plan is unique.

DProfiler weaknesses: DProfiler is not a general purpose BIM tool. Its single purpose (currently) is economic evaluation of a construction project (cost estimating and, where appropriate, income forecasting). Once a model is complete, its interface to support full development in other BIM design tools is limited to 2D DWG files.

2.4 CONCLUSION

Object-based parametric modeling is a major change for the building industry that is greatly facilitating the move from a drawing-based and handcraft technology to one based on digitally readable models that can be exchanged with other applications. Parametric modeling facilitates the design of large and complex models in 3D but imposes a style of modeling and planning that is foreign to many users. Like CADD, it is most directly used as a documentation tool separate from designing. A growing number of firms, however, use it directly for design and for generating exciting results. Some of these uses are taken up in Chapter 5, and the case studies in Chapter 9 provide further examples.

The ability to extract geometric and property information from a building model for use in design, analysis, construction planning, and fabrication, or in operations, will have large impacts on all aspects of the AEC industries. Many of these opportunities are taken up and discussed in the succeeding chapters. The full potential of this enabling capability will not be fully known for at least a decade, because its implications and new uses are discovered gradually. What is currently known is that object-based parametric modeling resolves many of the fundamental representational issues in architecture and construction and allows quick payoffs for those transitioning to it, even with only partial implementation. These payoffs include a reduction in drawing errors due to the built-in consistency of a central building model and the elimination of design errors based on spatial interferences.

While object-based parametric modeling has had a catalytic influence on the emergence and acceptance of BIM, it is not synonymous with BIM tools or the generation of building models. There are many other design, analysis, checking, display, and reporting tools that can play an important role in BIM procedures. Many information components and information types are needed to fully design and construct a building. The authors are of the opinion that many types of software can facilitate the development and maturing of Building Information Modeling. The BIM tools considered here are only the newest in several generations of tools, but in fact they are proving to be revolutionary in their impact.

Chapter 2 Discussion Questions

1. Summarize the major functionalities that distinguish the capabilities of a BIM design tool from 3D CAD modeling tools.
2. Most BIM design tools support both 3D object models as well as 2D drawn sections. What considerations should be made when determining the changeover level of detail, such as when to stop modeling in 3D and complete the drawings in 2D?
3. Why is it unlikely that a single integrated system will incorporate a unified parametric model of all of a building's systems? On the other hand, what would be the advantages if it could be achieved?
4. In what ways are some of the current popular design tools not BIM tools? Sketchup? 3D Max Viz? FormZ? Rhino?
5. What are the essential differences between a manufacturing parametric modeling tool, such as Autodesk Inventor, and a BIM design tool, such as Revit?
6. Do you think there may be additional manufacturing oriented parametric modeling tools used as a platform to develop BIM applications? What are the marketing costs and benefits? What are the technical issues?
7. Suppose you are a Chief Information Officer for a medium-sized architectural firm (with fewer than 25 employees). The firm specializes in school builings. Propose an outline structure for the firm's custom object library. Relate to the list of built-in objects in Table 2-1 when considering your answer.
8. You are part of a small team of friends that has decided to start an integrated design-build firm comprised of both a small commercial contractor and two architects. Lay out a plan for selecting one or more BIM-model creation tools. Define the general criteria for the overall system environment.

CHAPTER 3

Interoperability

3.0 EXECUTIVE SUMMARY

No single computer application can support all of the tasks associated with building design and production. *Interoperability* depicts the need to pass data between applications, allowing multiple types of experts and applications to contribute to the work at hand. Interoperability has traditionally relied on file-based exchange formats, such as DXF (Drawing eXchange Format) and IGES that exchange only geometry.

Starting in the late 1980s, *data models* were developed to support *product* and *object model* exchanges within different industries, led by the ISO-STEP international standards effort. Data model standards are developed both through the ISO organization and by industry-led efforts, using the same technology, specifically the *EXPRESS* data modeling language. EXPRESS is machine-readable and has multiple implementations, including a compact text file format, SQL and object database implementations and XML implementations. All are in use.

The two main building product data models are the *Industry Foundation Classes* (IFC) – for building planning, design, construction and management and *CIMsteel Integration Standard Version 2,* (CIS/2) – for structural steel engineering and fabrication. Both IFC and CIS/2 represent geometry, relations, processes and material, performance, fabrication and other properties, needed for design and production, using the EXPRESS language. Both are frequently extended, based on user needs.

Because EXPRESS supports applications with multiple redundant types of attributes and geometry, two applications can export or import different information for describing the same object. Efforts are being made to standardize the data required for particular workflow exchanges. In the US, the main effort is called the *National BIM Standards* (NBIMS) project. Interoperability imposes a new level of modeling rigor that firms are still learning to manage. Other formats for model viewing – 3D PDF and DWF – provide capabilities that resolve some types of interoperability problems.

While files support exchange between two applications, there is a growing need to coordinate data in multiple applications through a *building model repository*. Only in this way can the consistency, data and change management be realized for large projects. However, there are still some unresolved issues in the general use of building model repositories.

3.1 INTRODUCTION

The design and construction of a building is a team activity and increasingly, each activity and each type of specialty is supported and augmented by its own computer applications. Beside the capability to support geometry and material layout, there are structural and energy analyses, cost estimation and scheduling the construction, fabrication issues for each subsystem, plus much more. *Interoperability* identifies the need to pass data between applications, and for multiple applications to jointly contribute to the work at hand. Interoperability eliminates the need to replicate data input that has already been generated, and facilitates smooth workflows and automation. In the same way that architecture and construction are collaborative activities, so too are the tools that support them.

Even in the earliest days of 2D CAD in the late 1970s and early 1980s, the need to exchange data between different applications was apparent. The most widely used AEC CAD system at that time was Intergraph. A set of businesses arose to write software to translate Intergraph project files to other systems, especially for process plant design, for example exchanging data between the piping design software and the piping bills of material or analysis applications.

Later, in the post-Sputnik era, NASA found that they were expending significant amounts of money paying for translators among all their CAD developers. The NASA representative, Robert Fulton, brought all the CAD software companies together and demanded that they agree on a public domain exchange format. Two NASA-funded companies, Boeing and General Electric, offered to adapt some initial efforts they had undertaken separately. The resulting exchange standard was christened IGES (Initial Graphics Exchange Specification). Using IGES, each software company need only develop two translators (it was thought), for exporting from and importing to their application, instead of developing a translator for every pair-wise exchange. IGES was an early success that is still widely used throughout all design and engineering communities.

One of the impetuses for the development of BIM design tools was the already existing development of object-based parametric design being used in many construction support activities. Anyone visiting the kitchen department of a Home Depot or Lowes has been able to select, configure and review a kitchen design while making a purchase. What you might not have seen, however, is how these tools plan the cutting of the hardwood, plywood or other construction materials, how the software automatically defines the joinery and even the production plan. Similar tools have existed in structural steel for 3D design, analysis and fabrication since the mid-1990s. Sheet metal and ductwork fabrication has been available since that time also (these technologies are reviewed in Chapter 7).

In effect, most of the downstream fabrication of building systems has been moving to parametric modeling and computer-aided fabrication. The gap was at the front-end involving the building design itself (Eastman et al. 2002). In addition, of course, are the various analysis applications for structures, energy use, lighting, acoustics, air flow, etc. that have the potential to inform design (as well as review it at the end, which is where these tools are mostly used today). Because the BIM design tools have been developed in an industry where these diverse applications already exist, the need to interface or more intimately interoperate with these tools is a basic requirement.

3.2 DIFFERENT KINDS OF EXCHANGE FORMATS

Data exchanges between two applications are typically carried out in one of the four main ways listed below:

1. Direct, proprietary links between specific BIM tools
2. Proprietary file exchange formats, primarily dealing with geometry
3. Public product data model exchange formats
4. XML-based exchange formats

Direct links provide an integrated connection between two applications, usually called from one or both application user interfaces. Direct links rely on middleware software interfacing capabilities such as ODBC or COM or proprietary interfaces, such as ArchiCad's GDL or Bentley's MDL. These are all programming level interfaces, relying on C, C++ or now C# languages. The interfaces make portions of the application's building model accessible for creation, export, modification or deletion.

A proprietary exchange file format is one developed by a commercial organization for interfacing with that company's application. While a direct linking of applications is a runtime and binary interface, an exchange format is implemented as a file in a human readable text format. A well known proprietary exchange format in the AEC area is DXF (Data eXchange Format) defined by Autodesk. Other proprietary exchange formats include SAT (defined by Spatial Technology, the implementer of the ACIS geometric modeling software kernel), STL for stereo-lithography and 3DS for 3D-Studio. Because each of these has their own purpose, they address functionally specific capabilities.

The public level exchange formats involve using an open-standard building model, of which the IFC (Industry Foundation Class) (IAI 2007), or CIS/2 (CIS/2 2007) for steel, are the principle options. Notice that the product model formats carry object and material properties and also relations between objects, in addition to geometry. These are essential for interfacing to analysis and construction management applications.

Software companies quite reasonably prefer to provide exchanges to specific companies using a direct link, because they can support them better, and it keeps customers from using competitor's applications. The functionality supported is determined by the two companies (or divisions within the same company). However, because they have been developed, debugged and maintained by the two companies involved, they are typically robust for the versions of the software designed for, and the functionality intended. The resulting interface usually reflects a joint business agreement regarding marketing and sales. The interfaces are maintained as long as their business relationship holds.

On the other hand, there is a natural desire to 'mix-and-match' applications to provide functionality beyond what can be offered by any single software company. The method of integration becomes critical for projects involving large teams, because gaining interoperability of different systems used by the team is easier than moving all team firms to a single platform. The public sector also wishes to avoid a proprietary solution that gives any one software platform a monopoly. Only IFC and CIS/2 (for steel) are public and internationally recognized standards today. Thus the IFC data model is likely to become the international standard for data exchange and integration within the building construction industries.

XML is eXtensible Markup Language, an extension to HTML, the base language of the Web. XML allows definition of the structure and meaning of some data of interest; that structure is called a *schema*. The different XML schemas support exchange of many types of data between applications. XML is especially good in exchanging small amounts of business data between two applications set up for such exchanges.

A summary of the most common exchange formats in the AEC area is listed in Table 3-1. Table 3-1 groups file exchange formats with regard to their main usage. These include 2D raster image formats for pixel-based images, 2D vector formats for line drawings, 3D surface and solid shape formats for 3D forms. 3D object-based formats are especially important for BIM uses and have been grouped according to their field of application. These include the ISO-STEP based formats that include 3D shape information along with

Table 3-1 Common exchange formats in AEC applications.

Image (raster) formats	
JPG, GIF, TIF, BMP, PIC, PNG, RAW, TGA, RLE	Raster formats vary in terms of compactness, number of possible colors per pixel, some compress with some data loss.
2D Vector formats	
DXF, DWG, AI, CGM, EMF, IGS, WMF, DGN	Vector formats vary regarding compactness, line widths and pattern control, color, layering and types of curves supported.
3D Surface and Shape formats	
3DS, WRL, STL, IGS, SAT, DXF, DWG, OBJ, DGN, PDF(3D), XGL, DWF, U3D, IPT, PTS	3D surface and shape formats vary according to the types of surfaces and edges represented, whether they represent surfaces and/or solids, any material properties of the shape (color, image bitmap, texture map) or viewpoint information.
3D Object Exchange formats	
STP, EXP, CIS/2	Product data model formats represent geometry according to the 2D or 3D types represented. They also carry object properties and relations between objects.
Game formats	
RWQ, X, GOF, FACT	Game file formats vary according to the types of surfaces, whether they carry hierarchical structure, types of material properties, texture and bump map parameters, animation and skinning.
GIS formats	
SHP, SHX, DBF, DEM, NED	Geographical information system formats
XML formats	
AecXML, Obix, AEX, bcXML, AGCxml	XML schemas developed for the exchange of building data. They vary according to the information exchanged and the workflows supported.

connectivity relations and attributes, of which the IFC building data model is of highest importance. Also listed are various game formats, which support fixed geometry, lighting, textures along with actors, and dynamic, moving geometry, and geographical information system (GIS) public exchange formats for 3D terrain, land uses and infrastructure.

All methods of interoperability must deal with the issue of versions. When an application is updated with new capabilities, it may make the exchange mechanism faulty, if it is not maintained and versions of the standard are not well-managed.

3.3 BACKGROUND OF PRODUCT DATA MODELS

Until the mid-1980s, almost all data exchange in all design and engineering fields was based on various file formats. DXF and IGES are well-known examples. These provided effective exchange formats for shapes and other geometry, which is what they were designed to do. However, object models of piping, mechanical, electrical and other systems were being developed at this time. If data exchange was to deal with models of complex objects with their geometry, attributes and relations, any fixed file exchange format quickly became so large and complex as to be useless. These issues arose in both Europe and the US about the same time. After some back and forth, the International Standards Organization (ISO) in Geneva, Switzerland, initiated a Technical Committee, TC184, to initiate a subcommittee, SC4, to develop a standard called STEP (STandard for the Exchange of Product Model Data), ISO-10303, to address these issues.

The ISO-STEP organization developed a new set of technologies, based on:

- Use of a machine readable modeling language instead of a file format
- The language emphasizes data declarations but includes procedural capabilities for rules and constraints
- The language has mappings to different implementations, including a text file format, database schema definitions, and recently, XML schemas
- Reference sub-models that are shared and re-used subsets of larger standard models for geometry, measurements, representation classification and other generic needs

One of the main products of ISO-STEP was the EXPRESS language, developed by Douglas Schenck and later contributed to by Peter Wilson (Schenck and Wilson 1994). EXPRESS adopts many object-oriented concepts, including multiple inheritance. Here, object refers to a computer language concept that is broader than just representing physical objects. Thus objects can be used to represent conceptual or abstracted objects, materials, geometry, assemblies, processes and relations, among other things.

EXPRESS has become the central tool to support the modeling of products across a broad range of industries: mechanical and electrical systems, process plants, shipbuilding, processes, furniture, finite element models, plus others, as well as the AEC. It also includes a large number of libraries of features, geometry, classifications, measurements and others to use as common foundations for product data models. Both metric and imperial measurements are supported. As a machine-readable language, it is excellent for computational use, but difficult for human users; thus a graphical display version of the language was developed and is commonly used, called EXPRESS-G. The product data model for an application domain is called an Application Protocol, or AP. All the ISO-STEP information is in the public domain.

Because EXPRESS is machine readable, it can have multiple implementations. These include a compact text file format (called a Part-21 or P-21 file), SQL and object-based database implementations, and XML implementations (Part 28 format).

Surrounding the STEP standard is a group of software companies providing toolkits for implementing and testing software based on EXPRESS. These facilitate the various implementations, testing and deployment of STEP based exchange capabilities. These include graphical viewers and model navigators, testing software, and other implementation tools.

3.3.1 IFC Relation to STEP

AEC organizations may participate in TC-184 meetings and initiate STEP AP development projects. Also, non-TC184 organizations may use the STEP technologies to develop industry-based product data models. There are examples of both these approaches in the AEC, all based on the ISO-STEP technology:

- **AP 225** – Building Elements Using Explicit Shape Representation – the only completed building oriented product data model developed and approved by TC184. It deals with the exchange of building geometry.
- **IFC** – Industry Foundation Classes – an industry-developed product data model for the design and full life cycle of buildings, supported by the IAI.
- **CIS/2** – CimSteel Integration Standard, Version 2 – is an industry-developed standard for structural steel design, analysis and fabrication, supported by the American Institute of Steel Construction and the Construction Steel Institute of the UK. CIS/2 is widely used and deployed.

- **AP 241** – Generic Model for Life Cycle Support of AEC Facilities – addresses industrial facilities, and overlaps with IFC functionality – proposed in 2006 by the German National Committee; and is under review as a new AP.

AP 225 is used in Europe, mostly in Germany, as an alternative to DXF. Not many CAD applications support it. The IFC has thus far limited but growing worldwide use. Most of the BIM design tools support it, to varying degrees. It is described in more detail in Section 3.3.3. CIS/2 is broadly used in the North American structural steel fabrication industry. AP 241 is a recent proposal by the German STEP Committee to develop a product data model for factories and their components in a fully ISO-STEP compatible format. A debate is ongoing whether such an effort, parallel to the IFC, should proceed.

We see that there are multiple building product data models with overlapping functionality. All these data models are defined in the EXPRESS language. They vary in the AEC information they represent, and how they depict it. IFC can represent building geometry, as can AP225. There is overlap between CIS/2 and IFC in the schematic design of structural steel. Because of this overlap, but with significant unique capabilities on both sides, harmonization of IFC and CIS/2 has been undertaken by Georgia Institute of Technology, funded by AISC. As a result, for steel layout and design, the two product data models have had their definitions adjusted to be compatible. Also a translator supporting common exchange workflows has been developed. The website for this endeavor is: http://www.arch.gatech.edu/~aisc/cisifc/.

3.3.2 Organization of IAI

The IFC has a fairly long history. In late 1994, Autodesk initiated an industry consortium to advise the company on the development of a set of C++ classes that could support integrated application development. Twelve US companies joined the consortium. Initially defined as the Industry Alliance for Interoperability, the Alliance opened membership to all interested parties in September, 1995 and changed its name in 1997 to the International Alliance for Interoperability. The new Alliance was reconstituted as a non-profit industry-led international organization, with the goal of publishing the Industry Foundation Class (IFC) as a neutral AEC product data model responding to the AEC building lifecycle. A good historical overview of the IFCs is available on the IAI website: http://www.iai-international.org/About/History.html.

As of 2006, the IAI has eleven chapters in 19 countries worldwide, with about 450 corporate members. It is truly an international effort.

All chapters may participate in Domain Committees, each of which addresses one area of the AEC. Currently, the Domains include:

- AR – Architecture
- BS – Building Services
- CM – Construction:
  - CM1 – Procurement Logistics,
  - CM2 – Temporary Construction
- CS – Codes and Standards
- ES – Cost Estimating
- PM – Project Management
- FM – Facility Management
- SI – Simulation
- ST – Structural Engineering
- XM – Cross Domain

By participating in a Domain Committee, all members have input to the portion of the IFC that corresponds to their interests. Different national chapters are focusing on different domains.

The International Council Executive Committee is the overall lead organization in the IAI, made up of eight members. The North American chapter is administered by NIBS, the National Institute of Building Science, in Washington D.C.

3.3.3 What Are the IFCs?

The Industry Foundation Classes (IFC) was developed to create a large set of consistent data representations of building information for exchange between AEC software applications. It relies on the ISO-STEP EXPRESS language and concepts for its definition, with a few minor restrictions on the language. While most of the other ISO-STEP efforts focused on detailed software exchanges within specific engineering domains, it was thought that in the building industry this would lead to piecemeal results and a set of incompatible standards. Instead, IFC was designed as an extensible 'framework model.' That is, its initial developers intended it to provide broad general definitions of objects and data from which more detailed and task-specific models supporting particular workflow exchanges could be defined. This is characterized in Figure 3-1. In this regard, IFC has been designed to address all building information, over the whole building lifecycle, from feasibility and planning, through design (including analysis and simulation), construction, to occupancy and operation (Khemlani 2004).

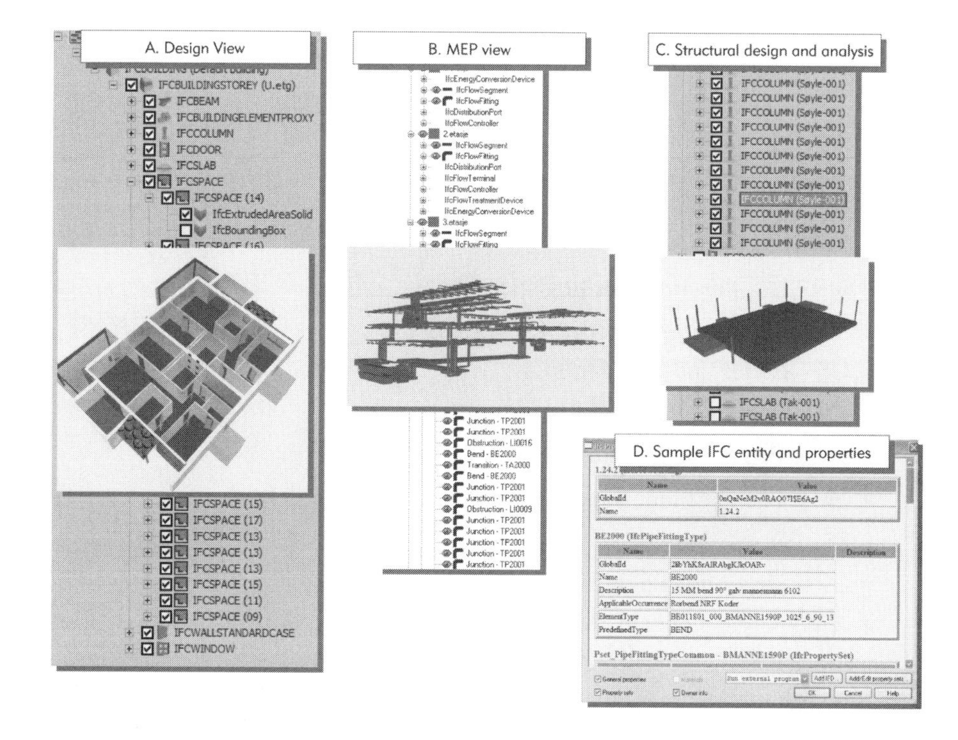

FIGURE 3-1
IFCs consists of a library of object and property definitions that can be used to represent a building project and support use of that building information for particular use. The figure shows three examples of specific domain uses from a single IFC project: (A) An architectural view, (B) a mechanical system view, and (C) a structural view. Also shown are (D) a sample IFC object or entity and sample properties and attributes.

As of 2007, the current release of the IFC is Version 2x3. It is available for review at: http://www.iai-international.org/Model/IFC(ifcXML)Specs.html.

All objects in EXPRESS are called *entities.* The conceptual organization of IFC *entities* is diagrammed in Figure 3-2. At the bottom are twenty-six sets of base entities, defining the base reusable constructs, such as Geometry, Topology, Materials, Measurements, Actors, Roles, Presentations and Properties. These are generic for all types of products and are largely consistent with ISO-STEP Resources, but with minor extensions.

The base entities are then composed to define commonly used objects in AEC, termed Shared Objects in IFC model. These include building elements, such as generic walls, floors, structural elements, building service elements, process elements, management elements, and generic features. Because IFC is defined as an extensible data model and is object-oriented, the base entities can be elaborated and specialized by subtyping* to make any number of sub-entities.

The top-level of the IFC data model are the domain-specific extensions. These deal with different specific entities needed for a particular use. Thus

*Subtyping provides for defining a new class of building object that 'inherits' the properties of its 'parent' class and adds new properties that make it distinct from its parent and any possible 'sibling' classes. IFC superclasses, subclasses and inheritance behavior conform to accepted principles of object-oriented analysis. For more detail, see (Booch 1993).

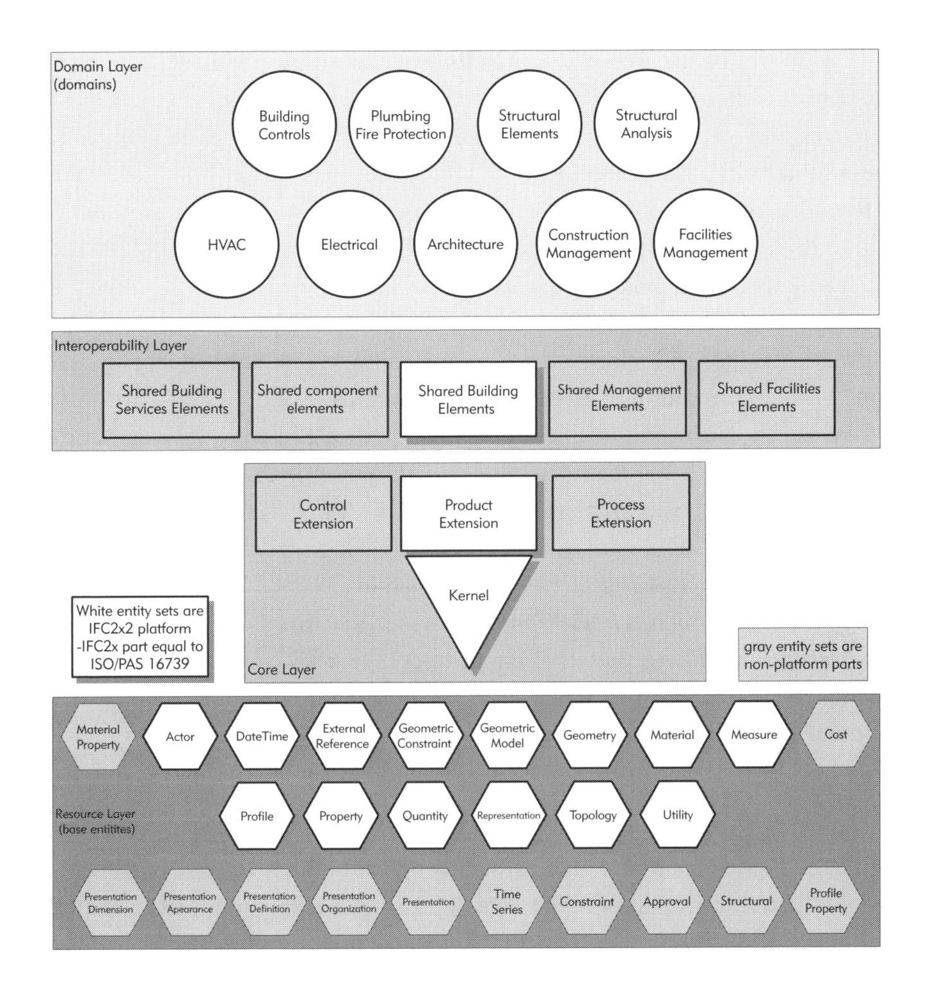

FIGURE 3-2
The system architecture of IFC subschemas.
Each Resource and Core subschema has a structure of entities for defining models, specified at the Interoperability and Domain Layers. Adapted from IAI international IFC/ifcXML online specifications for IFC2x Edition 3 at http://www.iai-international.org/Model/R2x3_final/index.htm.

there are Structural Elements and Structural Analysis extensions, Architectural, Electrical, HVAC, Building Control element extensions.

Each of the geometric shapes in the system architecture diagram in Figure 3-2 identifies a set of EXPRESS language *entities, enumerations,* and *types.* The architecture thus provides a type of indexing system into the IFC model, which is also defined in EXPRESS. The IFC model is quite large and still growing. As of the current release 2x3, there are 383 Kernel-level entities, 150 shared entities in the middle level, and 114 domain-specific entities in the top level.

Given the IFC hierarchical object subtyping structure, the objects used in exchanges are nested within a deep sub-entity tree. For example, a wall entity has a trace down the tree:

IfcRoot → IfcObjectDefinition → IfcProduct → IfcElement → IfcBuildingElement → IfcWall

Each level of the tree introduces different attributes and relations to the wall entity. *IfcRoot* assigns a Global ID and other identifier information. *IfcObjectDefinition* optionally places the wall as part of a more aggregate assembly, and also identifies the components of the wall, if these are defined. *IfcProduct* defines the location of the wall and its shape. *IfcElement* carries the relationship of this element with others, such as wall bounding relationships, and also the spaces (including exterior space) that the wall separates. It also carries any openings within the wall and optionally their filling by doors or windows. Many of these attributes and relations are optional, allowing implementers to exclude some of the information from their export routines.

Products, including walls, may have multiple shape representations, depending upon their intended uses. Within the IFC, almost all objects are within a composition hierarchy defined by *IfcObjectDefinition;* that is, they are both part of a composition and have their own components. IFC also has a general purpose *IfcRelation,* which has different kinds of relations as subtypes, one of which is *IfcRelConnects,* which in turn has the subclass *IfcRelConnectswithRealizing* that is used to reference wall connections. This one example indicates the extensiveness of the IFC model. This type of approach is followed for all IFC modeled objects.

3.3.4 IFC Coverage

While the IFC is able to represent a wide range of building design, engineering, and production information, the range of possible information to be exchanged in the AEC industry is huge. The IFC coverage increases with every release and addresses limitations, in response to user and developer needs. Here we summarize the major coverage and limitations as of early 2007.

All application-defined objects, when translated to an IFC model, are composed of the relevant object type and associated geometry, relations and properties. It is in the area of geometry, relations and properties that most limitations are encountered. In addition to objects that make up a building, IFC also includes process objects for representing the activities used to construct a building, and analysis properties, which are the input and results of running various analyses.

Geometry: The IFC has means to represent a fairly wide range of geometry, including extrusions, solids defined by a closed connected set of faces (B-Reps), and shapes defined by a tree of shapes and union-intersection operations (Constructive Solid Geometry). Surfaces may be those defined by extruded shapes (including those extruded along a curve) and Bezier surfaces. These cover most construction needs. IFC omits shapes constructed of multi-curved surfaces, such as B-splines and non-uniform B-splines (NURBS) that can be defined in design applications such as Rhino®, Form-Z®, Maya®, Digital Project and some Bentley applications. In these cases, a shape with these surfaces will be translated with missing surfaces and possibly other errors. These errors must be recognized, and then the geometry managed in some other manner – by changing the geometry representation in the exporting application, for example. In most applications, the conversion from NURBS surfaces to meshes is automatic, but one way. The IFC geometry was designed to support exchange of simple parametric models between systems, such as wall systems and extruded shapes. However, few translators have made use of these capabilities and their power is just beginning to be explored.

Relations: Care has been taken in the IFC data model to represent a rich set of relations between objects in some BIM design tool for translation into IFC. They are defined according to abstract classification as:

- *Assigns* – deals with relations between heterogeneous objects and a group, or selection of parts of assemblies for particular uses; for example, all entity instances installed by a particular trade may be referenced by the assigns relation;
- *Decomposes* – is the general relation dealing with composition and decomposition, of assemblies and their parts;
- *Associates* – relates shared project information, such as external equipment specifications, with model instances; an example may be a modeled piece of mechanical equipment which 'associates' with its specification in the supplier's catalog;
- *Defines* – deals with the relation between a shared description of an object and the various instances of that object, for example a description of a window type and the various instances of the window;
- *Connects* – defines a general topological relationship between two objects, which is defined functionally by subclasses. For example walls have 'connects' with their bounding walls, floors or ceilings.

There are many subclasses of *IfcRelations* covering almost any desired relation. No omissions are known.

Properties: IFC places emphasis on *property sets,* or P-sets. These are sets of properties that are used together to define material, performance, and contextual properties, e.g., wind, geological or weather data. There are collected P-sets for many types of building objects, such as common roof, wall, window glazing, window, beam, reinforcement. In addition, many properties are associated with different material behaviors, such as for thermal material, products of combustion, mechanical properties, fuels, concrete, and others. IFC includes properties for costs, time, quantity takeoffs, spaces, fire safety, building use, site use and others.

Several omissions can be identified. There are very limited properties for specialized space functions, such as required for security in public buildings or functional zoning, as in a theater. Measurements lack tolerance properties; there is no explicit way to represent uncertainty. In such cases, options are available to define and depict user-defined property sets. These must be managed by user agreement, as they are not yet built into the specification.

Meta-properties: IFC designers have thought about the use of information over time and the meta-data needed to deal with information management. IFC is strong in addressing information ownership, identification, change management and tracking of changes, controls and approvals. IFC also has capabilities to define constraints and objectives for describing intent. However, we are not aware of these capabilities being used.

The IFC has well-developed object classes for buildings at the contracting level of detail. In general, it currently is weak in representing the details needed for fabrication and manufacturing. It only partially addresses reinforcing in concrete, metal welds and their specification, concrete mix and finish definition, or fabrication details for window wall systems, for example. This level of detail may either be defined in more detailed product data models, such as CIS/2, or they may be added to IFC later.

These different descriptions are brought together to describe the information represented in some design application, or to be received by a building application from some other application or repository. The current limitations are in no way intrinsic, but reflect the priority needs of users up to now. If extensions are needed to deal with the limitations noted, these can be added, through a regularly scheduled extension process.

3.3.5 IFC in Use

A typical data exchange scenario is shown in Figure 3-3. A Source Application has modeled information to be used by a Receiving Application. The source application has a translator written for it that extracts information instances from the application's native data structure and assigns them to appropriate IFC entity classes. The entity instance data are then mapped from the IFC objects into (in this case) a text file format defined by the ISO-STEP Part-21. This file is then received by the other application and interpreted by the Receiving Application's translator in terms of the IFC object instances it

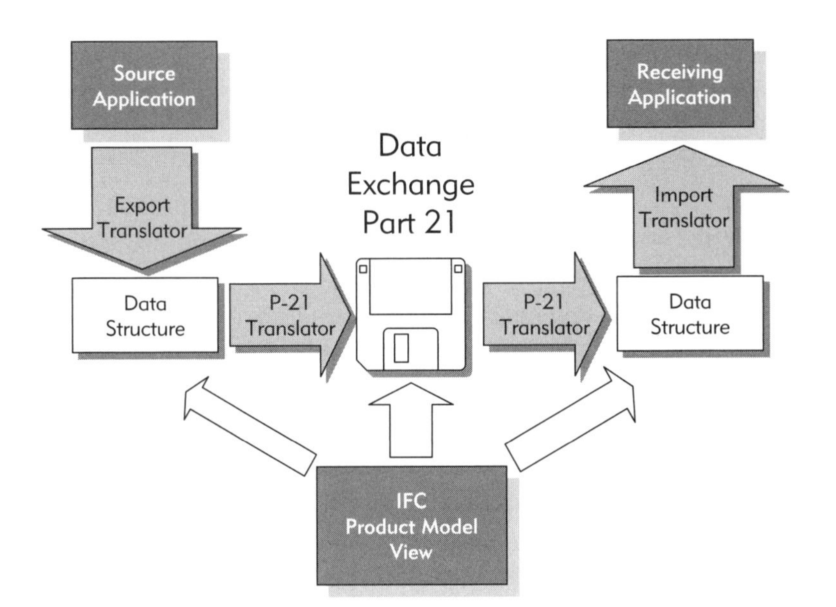

FIGURE 3-3
A scenario of the most common type of IFC exchange between two applications.

represents. The translator in the Receiving Application writes the relevant IFC objects into its native data structure for use.

Different BIM tools have their own proprietary data structures for representing a building and other design information. Some explicitly store properties and relations, while others compute them on demand. They internally use different geometric representations. Thus, two building modeling tools can both have perfectly good IFC translators to export and import data, but may still be able to exchange very little useful data. The translator's capabilities and object coverage are supposed to be defined in the translator's documentation. For these reasons, IFC model exchange currently needs careful initial testing to determine what information is carried by the exchanging applications.

IFC Viewers

A number of companies have developed geometry and geometry-and-properties viewers for IFC models. Most of these are freely available for downloading.

Available IFC Geometry and Property Viewers:

DDS IfcViewer, at: http://www.dds.no (free)

IfcStoreyView, at: http://www.iai.fzk.de/ifc (free)

IFC Engine Viewer, at: http://www.ifcviewer.com (free)

ISPRAS IFC/VRML Converter, at: http://www.ispras.ru/~step (free)

Octaga Modeler, at: http://www.octaga.com (commercial)

Solibri Model Viewer™, at: http://www.solibri.com/ (free)

Some viewers display the attributes of selected objects and provide means to turn on and off sets of entities. IFC viewers are useful for debugging IFC translators, and to verify what data has been translated.

IFC Views

Because of the variable representation of objects in IFC, two parallel and related efforts are being undertaken to define IFC subsets more precisely. In both cases, data exchanges are prescribed in terms of particular tasks and workflows. The American effort is the National BIM Standard (NBIMS), initiated by the Facilities Information Council, an organization of US government procurement and facility management groups—within the Department of Defense—and administered through the National Institute of Building Sciences (NIBS). The European effort is being led by Norway, developing an Information Delivery Manual (IDM). Both of these efforts are directed toward specifying IFC Views—specific subsets of the IFC—to be used for specific workflow exchanges. Both groups rely on the buildingSMART™ name, recently adopted by the IAI for promoting the IFC (IAI 2007).

In the USA, the plan for development and implementation of IFC Views is to encourage different business domains within the construction industry to identify data exchanges that, if automated, would provide high value payoffs. These would be specified at a functional level by the AEC business domain. The assumption is that building associations, such as the American Institute of Architects, Associated General Contractors, Precast Concrete Institute, Portland Cement Association, American Institute of Steel Construction and other institutions would represent the business domains. Once specified, the business domain, working with the NIBS-buildingSMART® organization, would fund information technology specialists to specify the IFC Views to be exchanged, and to develop the functional specifications. Last, NIBS-buildingSMART® and the AEC domain will work with the BIM software tool developers to implement the View translators. There is also discussion of certification.

When these specific workflow-based translators are implemented, they will be explicitly incorporated into translators, based on either P-21 files or database queries. Some views are being defined for one-way passing of a dataset, while others are anticipating multiple iterated exchanges, such as might be desired between a design and an analysis tool to allow interactive performance improvement of the design. These Views, when certified, will add significantly to the robustness of IFC exchanges and eliminate the need for pre-testing and exchanges, as is required today.

A possible example of a fine-grain workflow exchange is shown in Figure 3-4, showing exchanges between building designers and structural engineers. Six

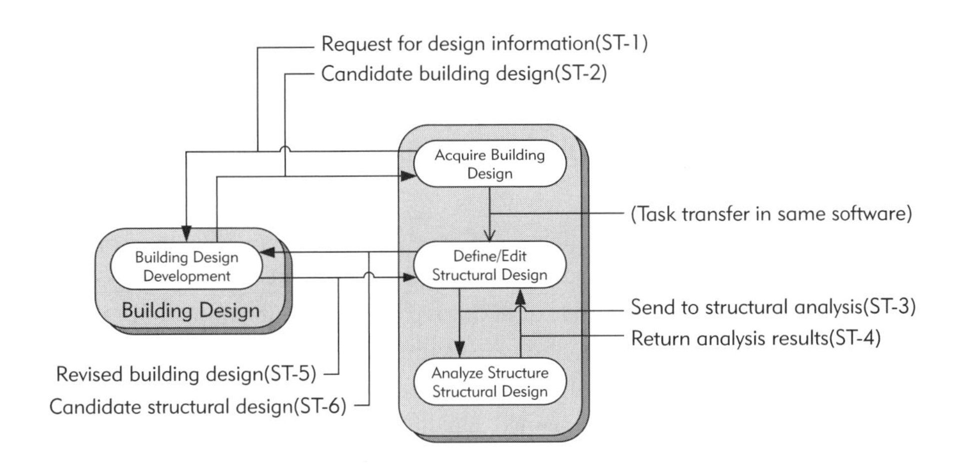

FIGURE 3-4
Example workflows between building designer and structural engineer.

information exchanges are detailed, denoted in the figure as (ST-1) to (ST-6). These include three iterated exchanges: (1) first to lay out the structural system conceptually (ST-1 and ST-2); (2) the structural engineer undertaking iterated analyses to develop a good design based on the engineer's knowledge of the project, (ST-3 and ST-4), and (3) exchanges between the designer and the structural engineer, coordinating details with the rest of the building systems and reflecting design intent (ST-5 and ST-6). The structural engineer optionally has two applications, a structural design application, such as Revit® Structures, Bentley Structures or Tekla Structures, and a structural analysis application. In many cases (ST-3) and (ST-4) will be directly integrated through the applications' APIs. Of course, coverage of all relevant AEC domain exchanges will require definition of hundreds of workflows, each with different intent and data.

IFC Initiatives

At this point in time (2007) there are significant efforts to apply the IFC in various parts of the world:

- CORENET is driven by the Singaporean Building and Construction Authority in collaboration with other public and private organizations. It is a major initiative to re-engineer the business processes of the construction industry to integrate the major processes of a building project life cycle: supported by key infrastructures, supporting electronic submission and recording, checking and approval processes, communication methods for dealing with submittal, paperwork, and recording of the review, as well as documentation and training (CORENET 2007).

- Australia is undertaking a similar effort as Singapore, under the trademark of DesignCheck™ (Ding et al. 2006).
- The International Code Council in the US has developed a plan that goes down a different path from the CORENET efforts, called SMARTcodes (ICC 2007).
- The Norwegian government and construction industry are working together to initiate changes in their construction industry, including building control (automatic code checking), planning (e-submission of building plans), and integration in design, procure, build, and facility management. Their initiative is also called BuildingSmart® and there are efforts to coordinate the two similarly named but different parallel efforts. The Norwegian one is expected to produce a significant impact on the efficiency, productivity and quality of the construction industry. See "Industry Initiatives and Norwegian Solution" at http://www.iai-international.org/.
- The General Services Administration of the US government has undertaken a series of BIM demonstration projects, addressing various applications, many relying on exchanges based on the IFC. These are described on the same IAI website above, under Industry Solutions, GSA Pilots.
- Based on these demonstrations, all GSA building projects starting in 2007 are to utilize BIM design tools and use of an exported model in IFC format to support checking of the preliminary concept design against the specific project's programmatic spatial requirements. This is the initial functional application of BIM being mandated in the US. Other mandated applications are in development. Further requirements will be introduced in succeeding years. These activities have led to the draft development of GSA BIM guidelines to be followed for all new GSA projects (GSA 2007).

Additional initiatives are being undertaken in Finland, Denmark, Germany, Korea, Japan, China and other countries.

3.3.6 Implications of IFC Use

As the IFC data model becomes adopted by various governmental organizations for code checking and design review (as being undertaken by GSA and in Singapore), it will have an increasingly strong impact on aspects of architectural and contractor practice. This impact simultaneously affects users and BIM design tool developers.

The completion of a set of contract drawings in traditional practice imposes a level of rigor and discipline in the final generation of those drawings. This discipline and rigor will increase even more significantly in the creation and definition of building models used for code checking and design review.

An early example of this required rigor are the GSA submittal requirements for checking building area calculations against the building's program (GSA 2006), calculated according to the ANSI-BOMA space calculation methods (ANSI 1996). In order to carry out this type of check, the building model must have the following information available:

- All rooms and spaces must be labeled in a manner consistent with the space program
- The 3D boundaries of all spaces must be defined by their bounding surfaces and closed in a manner allowing area and volume calculation
- If the BIM design tool does not automatically calculate and maintain the consistency of the space volumes and areas, these must be created and maintained by the user, in the manner consistent with the ANSI-BOMA area calculation method. The final area calculation is made by the reviewing application, so that this is guaranteed to be done in a consistent manner.

These requirements indicate that future firms will have to carefully prepare and possibly pre-check their model structure to make sure it is modeled appropriately for automatic review. Programs already exist to do this checking for the GSA BIM uses. For example, a check can be run that all of the spaces are tagged as closed spaces with appropriate codes defining their intended function.

Currently design reviews are being performed on conceptual design phase models that do not include materials, hardware or detailing, or the review ignores such building model information. Later more elaborate types of testing for other programmatic issues will apply to more detailed design, and finally full construction level models. Most building models today do not carry design in 3D to the level of interior finishes and use 2D sections to represent such details (see Figure 2-13). This practice will have to be coordinated with the expected reviews and other uses of the models and the accuracy of results expected.

This rigor also applies to interfaces with analysis applications and to the hand-off between a design BIM tool and fabrication-level ones. The broader implication is that special care will be required in the definition of a building information model, because the information will be used by other applications. There will be a great difference between the sketch models often developed today for design review or rendering, and the rigorously crafted building models used for analysis and design review in the future.

3.3.7 The Future of IFC

The IFC is the only public, non-proprietary and well-developed data model for buildings and architecture existing today. It is a de facto standard worldwide and is being formally adopted by different governments and agencies in various parts of the world. It is being picked up and used for a growing number of uses, in both the public and private sectors. In the 2007 AIA BIM Awards program, administered by the Technology in Architectural Practice Knowledge Community, six of 32 project submissions included use of the IFC. While these are meant to provide examples of best use, it does indicate a broad take-up of IFC usage.

The IFC data model standard is continuously evolving. A new version with extensions is released every two years. The extensions are undertaken in two phases. First, various domain-specific teams are assembled around specific issues or targeted use – say structural analysis or reinforced concrete construction. A group carefully generates a set of requirements. Then a candidate extension to the IFC is generated and proposed by modeling experts. The extension is voted on and approved by its participants. Later, the different extensions that have been made over a two year cycle are collected and then integrated in a logical way by the Model Support Group (MSG) to provide consistent extensions for the next release. The IFC release is documented and distributed and reviewed with the AEC software firms, who then develop translator implementations of the IFC model extensions, which are then tested for certification. Certification until recently has been at the IFC Release level, where some subset of IFC entities are correctly read and/or written, and the capabilities documented.

Because the IFC Views supporting various specific exchanges are just now being developed in the National BIM Standard and in Europe by building SMART®, the current IFC translators are often lacking in needed reliability without extensive pre-testing. These limitations are expected to recede as IFC Views are defined and implemented. The IFC and NBIMS efforts are currently being undertaken with minimal funding and much volunteer effort. The level of support may become an Achilles' heel of the widespread use of this important technology.

3.4 XML SCHEMAS

An alternative way to exchange data is through XML. XML is an extension to HTML, the language used to send information over the Web. HTML has a fixed set of tags (a tag tells what kind of data follows) and has focused on presentation, different kinds of media and other types of fixed format Web data. XML expands upon HTML by providing user-defined tags to specify an intended meaning for data transmitted, allowing user-defined schemas. XML has become very popular for exchange of information between Web applications, e.g., to support ecommerce transactions or collect data.

There are multiple methods for defining custom tags, including Document Type Declarations (DTDs). DTDs have been developed for mathematical formulas, vector graphics and business processes, among many others. There are other ways to define XML schemas, including XML Schema (http://www.w3.org/XML/Schema), RDF (Resource Description Framework) (http://www.w3.org/RDF/) and OWL Web Ontology Language (http://www.w3.org/TR/2004/REC-owl-features-20040210/). Research is proceeding to develop even more powerful tools around XML and more powerful schemas, based on precise semantic definitions called ontologies, but practical results for these more advanced approaches has thus far been limited.

Using current readily available schema definition languages, some effective XML schemas and processing methods have been developed in AEC areas. Five of them are described in the box that follows:

XML Schemas in AEC Areas

OGC (Open Geospatial Consortium) has developed the OpenGIS® Geographic Objects (GO) Implementation Specification. It defines an open set of common, language-independent abstractions for describing, managing, rendering, and manipulating geometric and geographic objects within an application programming environment (OGC 2007).

gbXML (Green Building XML) is a schema developed to transfer information needed for preliminary energy analysis of building envelopes, zones and mechanical equipment simulation (gbXML 2007).

aecXML is administered by FIATECH, a major construction industry consortium supporting AEC research, and the IAI. It can represent resources such as contract and project documents (Request for Proposal (RFP), Request for Quotation (RFQ), Request for Information (RFI), specifications, addenda, change orders, contracts, purchase orders), attributes, materials and parts, products, equipment; meta data such as organizations, professionals, participants; or

(Continued)

XML Schemas in AEC Areas *(Continued)*

activities such as proposals, projects, design, estimating, scheduling and construction. It carries descriptions and specifications of buildings and their components, but does not geometrically or analytically model them (FIATECH 2007).

IFCXML is a subset of the IFC schema mapped to XML, supported by IAI. It also relies on XML Schema for its mapping. It currently supports the following use cases: Material Catalogs, Bill of Quantities and adding User Design Quantities. Support for additional use cases are planned (IAI 2007a).

BLIS-XML is a subset of the IFC Release 2.0, developed to support a small number of use cases. It was developed in 2001–2002 in an effort to get into use a practical and productive version of the IFC. BLIS-XML uses the BLIS schema with a schema converter developed by Secom Co. Ltd. (BLIS 2002).

The Associated General Contractors (AGC) has announced that it will develop agcXML, a schema for its construction business processes. It was under development at the time this book was written.

Each of these different XML schemas defines its' own entities, attributes and relations. They work well to support work among a group of collaborating firms that implement a schema and develop applications around it. However, each of the XML schemas is different and incompatible. IFCXML provides a global mapping to the IFC building data model, for cross-referencing. The OGC GIS schema is being harmonized with IFC efforts. XML formatting takes more space than, say, IFC clear text files (between 2 and 6 times more space). The longer term issue is to harmonize the other XML schemas with equivalence mappings between them and with data model representations. The analogy is when the railroads in the US all rapidly built tracks over the country, each with their own gage; they worked fine within their own community, but could not link up.

3.5 PORTABLE, WEB-BASED FORMATS: DWF AND PDF

Two widely available formats are 3D PDF (Portable Document Format) developed by Adobe® and DWF (Design Web Format) developed by Autodesk®. These two formats support a 'publishing' information workflow and do not address the interoperability issues supported by IFC and the XML schemas discussed in the previous section. That is, these Web formats provide design and engineering professionals with a way to publish the building information model for review and viewing, with markup and query capabilities; but not to enable modification of the model information. The widespread availability of these formats is likely to lead to their playing a significant role in the exchange and viewing of project information. Here is a brief overview of some of the features of these formats (see Chapter 4 for a more comparative and detailed discussion of their role in BIM implementation):

- **Generic, non-domain specific, and extensible schema:** These formats do not have domain-specific schemas, rather they have schemas with general classes of entities, from geometric polygonal entities and solid entities to markup objects and sheet objects. These formats are designed to meet the broad needs of engineering and design disciplines including manufacturing and the AEC industry. PDF was originally designed for exchange of textual-based documents and extended the format to include support for U3D (Universal 3D) elements. The DWF schema was designed specifically for exchange of intelligent design data and is based upon Microsoft's XML-based XPS (XML paper specification) format and extensions, allowing anyone to add objects, classes, views, and behaviors. Although PDF is an ISO standard, neither DWF nor the 3D PDF extensions are ISO standards.
- **Embedded views of the project information:** Both formats represent the model data and views of that data. Data views include 2D plot views, 3D model views, or raster image views. The 2D and 3D model representations are fully navigable, selectable and support queries. They include object meta-data, but object parameters are not editable.
- **Widely available viewing tools:** Both formats are distributed with free, publicly available viewers.
- **High fidelity, accuracy, and precision:** Both formats were designed for plot-capable printing with high degrees of accuracy and precision.
- **Highly compressible:** Both formats are optimized for portability and are highly compressible. IFC and many of the other XML or 3D formats are not.

3.6 FILE EXCHANGE VERSUS BUILDING MODEL REPOSITORIES

Production use of IFC-based data exchange and XML-based e-business exchanges have begun with file exchanges. Quickly, however, people are learning that management of the versions, updating, and change management of the data associated with an increasingly complex number of heterogeneous applications, leads to major data management challenges. Issues that arise that are best resolved by repositories:

- Supporting exchanges between multiple concurrent applications that both read and write project data; that is, the workflows are not linear
- Propagating and managing changes that impact multiple application datasets
- When there are multiple authoring applications that must be merged for later use
- Supporting very frequent or realtime coordination between multiple application users.

The technology associated with the resolution of these types of issues, and also the smoothing of exchange of data between combinations of applications, is a *building model repository*. A building model repository is a database system whose schema is based on a published object based format. It is different from existing project data management (PDM) systems and web-based project management systems in that the PDM systems are file based, and carry CAD and analysis package project files. Building model repositories are object-based, allowing query, transfer, updating and management of individual project objects from a potentially heterogeneous set of applications. The only broad building level object schema is the IFC, (but also CIS/2 and AP225 could be used for limited applications – see Section 3.3.1). A number of companies have developed building model repositories using the IFCs.

An IFC repository can support integration of the data generated by multiple applications for use in other applications, support iterations on parts of the design, and track changes at the object level. They provide access control, version management, and various levels of design history, relating the various geometric, material, and performance data that represent a building (Eurostep 2007; Jotne 2007).

While basic capabilities exist today, development is ongoing to provide a range

Current building model repositories available:

- Jotne EDM Model Server
- LKSoft IDA STEP Database
- EuroSTEP Model Server
- EuroSTEP SABLE Server
- Oracle Collaborative Building Information Management

of effective services for building model repositories. Some early supporting software includes:

- viewers for inspecting the geometrical and property data in building model repositories (see Section 3.3.5);
- checking product data model repositories, for logical correctness ('building spell- and grammar-checking'), for building program checking, or building code checking (CORENET 2007; Jotne 2007; Solibri 2007).

Future areas where repositories are expected to provide important services include: dataset preparation for multiple analyses, such as energy analyses of building shells, of interior energy distribution and mechanical equipment simulation; bills of material and procurement tracking; construction management; building commissioning; facility management and operations. Some of these issues are explored in Section 4.5.5.

In practice, each design participant and application is not involved with the complete representation of the building's design and construction. Each participant is interested in only a subset of the building information model, defined as particular views of the building model. Similarly, coordination does not apply universally; only a few users need to know reinforcing layouts inside concrete or weld specifications. It is still an open question, then, as to whether there is a need for a single integrated database, or instead, multiple federated databases that can provide limited specific consistency checking between dispersed models.

This issue is further complicated by the challenges with storing the required data in the appropriate format to archive and recreate the native project files required by the various BIM authoring tools. The neutral format that repositories carry data in is inadequate to recreate the native data formats used by applications, except in a few limited cases. Today, these can only be re-created from the native application datasets themselves. This is due to the basic heterogeneity of the built-in behavior in the parametric modeling design tools (described in Section 2.2.7). Thus any neutral format exchange information, such as IFC model data, must be augmented by or associated with the native project files produced by the BIM authoring tools. While the future for managing project data, especially for large projects, seems to belong to building model repositories, many issues remain to be sorted out for their effective use.

Other industries have recognized the need for product model servers. Their implementation in the largest industries – electronics, manufacturing, and aerospace — has led to a major industry involving Product Lifecycle Management (PLM). These systems are custom engineered for a single company and typically involve system integration of a set of tools including product model management, inventory management, material and resource tracking and scheduling, among others. They rely on supporting model data in one of a few proprietary native formats, possibly augmented by ISO-STEP based exchanges. These have penetrated only the largest businesses, because the current business model of PLM is based on system integration services. What is not available is a ready-to-use product that can support medium or small scale organizations that dominate the make-up of construction industry firms. Thus the medium and small industries – in both construction and manufacturing - are waiting for PLM systems that can be easily tailored for various kinds of use.

3.7 SUMMARY

Some different popular exchange formats and their capabilities in terms of geometry, on one axis, and modeling power on the other, are diagrammed in Figure 3-5. It indicates both our current status and frustrations. The overall issues of

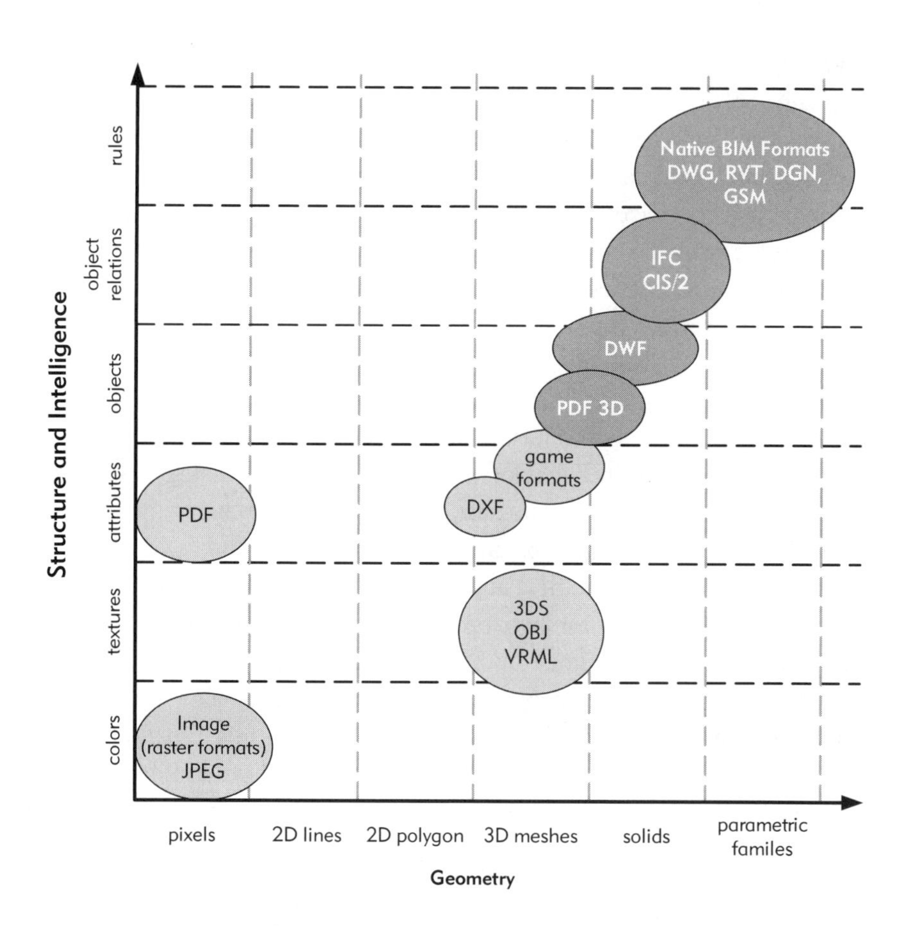

FIGURE 3-5 Comparison of different project data exchange formats according to the geometry supported and their attributes and associativity.

interoperability are not yet resolved. Some people claim that IFC and public standards are the only resolution, while others say that the move by public standards to resolve outstanding issues is too slow and proprietary solutions are preferable.

In the last three years, significant movement has been made in both directions. On the one hand, all BIM design tools now support IFC fairly well, allowing basic exchanges to be made with adequate completeness and accuracy. Only a few exchanges allow editing; most only static viewing. There is a small movement for using IFC for analysis interfaces, but these too are just becoming available. Various XML schemas are being used for different business exchanges. On the other side, formats such as DWF and PDF hold the potential to become richer and support exchange as well as viewing. (Both have XML capabilities.) We expect to see both approaches co-exist, side by side. But we are all actors in this decision, both through purchase decisions and advocacy. Users' and owners' wishes will prevail.

Chapter 3 Discussion Questions

1. What are the major differences between DXF as an exchange format and an object-based one like IFC?
2. Choose a design or engineering application that has no effective interface with a BIM design tool you use. Identify the types of information the BIM design tool needs to send to this application.
3. Extend this to think what might be returned to the BIM design tool, as a result of running this application?
4. Take a simple design of some simple object, such as a Lego sculpture. Using IFC, define the IFC entities needed to represent the design. Check the description using an EXPRESS parser, such as the free EXPRESS-O checker available from the Sourceforge open software website.
5. For one or more of the coordination activities below, identify the information that needs to be exchanged in both directions:
   a. building design that is informed by energy analysis of the building shell.

(Continued)

Chapter 3 Discussion Questions *(Continued)*

   b. building design that is informed by a structural analysis.
   c. steel fabrication level model that coordinates with a shop scheduling and materials tracking application.
   d. cast-in-place concrete design that is informed by a modular formwork system.
6. What are the distinguishing functional capabilities provided by a Building Model Repository and database as compared to a file-based system?
7. Explain why file exchange between design systems using IFC can result in errors. How would these errors be detected?

CHAPTER 4

BIM for Owners and Facility Managers

4.0 EXECUTIVE SUMMARY

Owners can realize significant benefits on projects by using BIM processes and tools to streamline the delivery of higher quality and better performing buildings. BIM facilitates collaboration between project participants, reducing errors and field changes and leading to a more efficient and reliable delivery process that reduces project time and cost. There are many potential areas for BIM contributions. Owners can use a building information model to:

- **Increase building value** through BIM-based energy design and analysis to improve overall building performance
- **Shorten project schedule** from approval to completion by using building models to coordinate and prefabricate design with reduced field labor time
- **Obtain reliable and accurate cost estimates** through automatic quantity take-off from the building model, providing feedback earlier in a project when decisions will have the greatest impact
- **Assure program compliance** through ongoing analysis of the building model against owner and local code requirements
- **Produce market-ready facilities** by reducing time between procurement decisions and actual construction, allowing for the selection of the latest technologies or trend finishes

- **Optimize facility management and maintenance** by using the as-built building information model as the database for rooms, spaces, and equipment.

These benefits are available to all types of owners: small and large, serial or one-time builders, private or institutional. Owners have yet to realize all of the benefits associated with BIM or employ all of the tools and processes discussed in this book. Significant changes in the delivery process, selection of service providers, and approach to projects are necessary to fully realize BIM's benefits. Today, owners are rewriting contract language, specifications, and project requirements to incorporate the use of BIM processes and technologies into their projects as much as possible. Owners investing in BIM efforts are reaping advantages in the marketplace through the delivery of higher value facilities and reduced operational costs. In concert with these changes, some owners are actively leading efforts to implement BIM tools on their projects by facilitating and supporting BIM education and research.

4.1 INTRODUCTION: WHY OWNERS SHOULD CARE ABOUT BIM

Lean processes and digital modeling have revolutionized the manufacturing and aerospace industries. Early adopters of these production processes and tools, such as Toyota and Boeing, have achieved manufacturing efficiencies and commercial successes (Laurenzo 2005). Late adopters were forced to catch-up in order to compete; and although they may not have encountered the technical hurdles experienced by early adopters, they still faced significant changes to their work processes.

The AEC industry is facing a similar revolution, requiring both process changes and a paradigm shift from 2D-based documentation and staged delivery processes to a digital prototype and collaborative workflow. The foundation of BIM is a coordinated and information-rich building model with capabilities for virtual prototyping, analysis, and virtual construction of a project. These tools broadly enhance today's CAD capabilities with an improved ability to link design information with business processes, such as estimating, sales forecasts, and operations. With drawing-based processes, analyses must be done independently of the building design information, often requiring duplicate, tedious, and error-prone data entry. The result is loss of value in information assets across phases and increased effort to produce project information, as the conceptual diagram in Figure 4-1 shows. Consequently, such analyses can be out of sync with design information and lead to errors. With

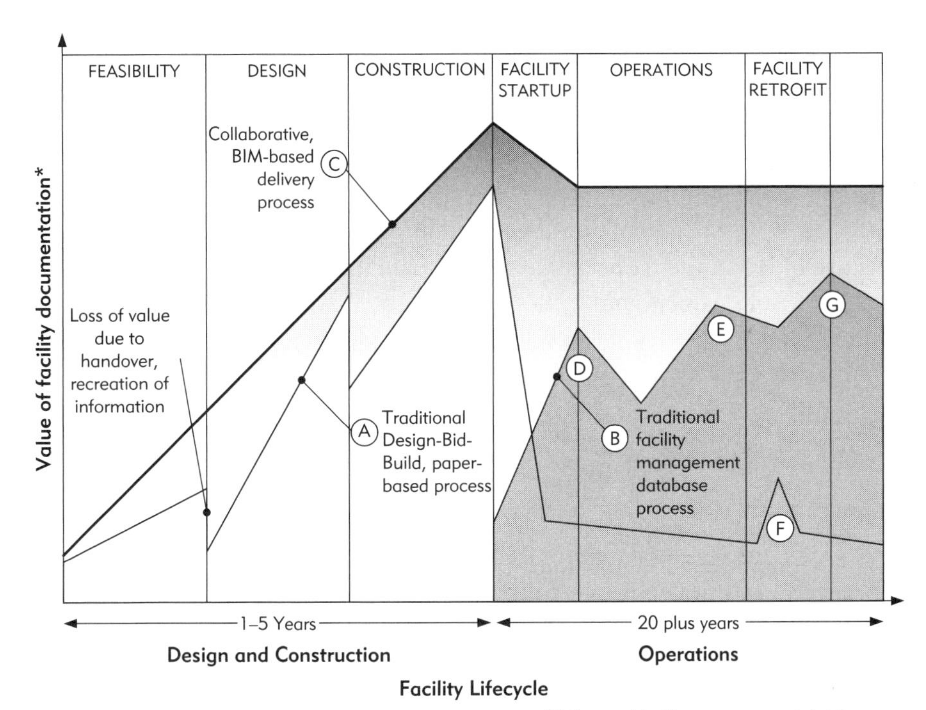

FIGURE 4-1
A) Traditional single-stage drawing-based deliverables,
B) traditional facility management database system, and C) BIM-based deliverables throughout the project delivery and operation process.

BIM-based processes, the owner can potentially realize a greater return on his or her investment as a result of the improved design process, which increases the value of project information in each phase and decreases the effort required to produce that information. Simultaneously, owners can reap dividends in project quality, cost, and future retrofits to the facility.

This chapter discusses how owners can use BIM to manage project risk, improve project quality, and deliver value to their businesses. It also shows how facility managers can use BIM to better manage their facilities. Owners here are the organizations that initiate and finance building projects. They make strategic decisions in the facility delivery process through the selection of service providers and the type of delivery processes they use. These decisions ultimately control the scope and effectiveness of BIM on a project.

This chapter begins with a discussion of BIM applications for all types of building owners and facility managers. This is followed by a discussion of how these different applications address issues for different types of owners: owner operators vs. owner developers; serial vs. one-time builders; private vs. government; and local vs. global. Section 4.4 discusses how different project delivery methods impact the implementation of various BIM applications. It shows why a collaborative delivery process is the best approach for successful application of BIM on a project.

Section 4.5 provides a guide to BIM tools that are suitable or better oriented for owners. Most of the BIM tools available today are targeted toward service providers, such as architects, engineers, contractors, and fabricators; they are not specifically targeted for owners. Other tools are discussed in Chapters 5, 6, and 7, and references are provided for those sections. Section 4.6 discusses the owner's building information model and how the owner's perspective of it and the scope and level of detail may differ from those discussed in subsequent chapters.

Owners play a significant education and leadership role in the building industry. Section 4.7 discusses different ways for owners to implement BIM applications on their projects, including pre-qualification of service providers, education and training seminars, contractual requirements, and changing their internal processes. Section 4.8 follows with a discussion of the risks and the process and technology barriers associated with BIM implementation. The chapter concludes with guidelines for successful implementation.

4.2 BIM APPLICATION AREAS FOR OWNERS

Traditionally, owners have not been agents of change within the building industry. They have long been resigned to typical construction project problems, such as cost overruns, schedule delays, and quality issues (Jackson 2002). Many owners view construction as a relatively small capital expenditure compared to the lifecycle costs or other operational costs that accrue over time. Changing marketplace conditions, however, are forcing owners to rethink their views and place greater emphasis on the building delivery process and its impact on their business (Geertsema et al. 2003; Gaddie 2003).

The firms that provide services to owners (AEC professionals) often point to the short-sightedness of owners and the frequent owner requested changes that ultimately impact design quality, construction cost, and time.

Because of the considerable potential impact that BIM can have on these problems, the owner is in the position to benefit most from its use. Thus, it is critical that owners of all types understand how BIM applications can enable competitive advantages and allow their organizations to better respond to market demands and yield a better return on their capital investments. In those instances in which service providers are leading the BIM implementation—seeking their own competitive advantage—educated owners can better leverage the expertise and know-how of their design and construction team.

In the following sections, we provide an overview of drivers that are motivating all types of owners to adopt BIM technologies, and we describe the different types of BIM applications available today. These drivers are:

- Cost reliability and management
- Time to market
- Increasing complexity in infrastructure and marketplace
- Sustainability
- Labor shortages
- Language barriers
- Asset management

Table 4-1 summarizes the BIM applications reviewed in this chapter from the owner's perspective and the respective benefits associated with those applications. Many of the applications referenced in this chapter are elaborated on in greater detail in Chapters 5, 6 and 7, and in the case studies in Chapter 9.

4.2.1 Cost Reliability and Management

Owners are often faced with cost overruns or unexpected costs that force them to either "value engineer," go over budget, or cancel the project. Surveys of owners indicate that up to two-thirds of construction clients report cost overruns (Construction Clients Forum 1997; FMI/CMAA 2005, 2006). To mitigate the risk of overruns and unreliable estimates, owners and service providers add contingencies to estimates or a "budget set aside to cope with uncertainties during construction" (Touran 2003). Figure 4-2 shows a typical range of contingencies that owners and their service providers apply to estimates, which vary from 50% to 5% depending on the project phase. Unreliable estimates expose owners to significant risk and artificially increase all project costs.

The reliability of cost estimates is impacted by a number of factors, including market conditions that change over time, the time between estimate and execution, design changes, and quality issues (Jackson 2002). The accurate and computable nature of building information models provides a more reliable source for owners to perform quantity take-off and estimating and provides faster cost feedback on design changes. This is important because the ability to influence cost is highest early in the process at the conceptual and feasibility phase, as shown in Figure 4-3. Estimators cite insufficient time, poor documentation, and communication breakdowns between project participants, specifically between owner and estimator, as the primary causes of poor estimates (Akintoye and Fitzgerald 2000).

Table 4-1 Summary of BIM application areas and potential benefits to all owners, owner-operators, and owner-developers; and a cross-reference to case studies presented in Chapter 9.

Book section	Specific BIM application areas for Owner (referenced in this Chapter)	Market Driver	Benefits to all Owners	Relevant case study (CS) or reference
Chapter 5: designers and engineers	Space planning and program compliance	Cost management marketplace complexity	Ensure project requirements are met	Coast Guard Facility Planning Federal Courthouse
	Energy (environmental) analysis	Sustainability	Improve sustainability and energy efficiencies	Federal Office Building
	Design configuration/ scenario planning	Cost management complexity	Design quality communication	Coast Guard Facility Planning
	Building system analysis/simulation	Sustainability	Building performance and quality	National Aquatics Center
	Design communication/ review	Marketplace complexity and language barriers	Communication	All case studies
Chapters 5 and 6: designers, engineers, contractors	Quantity take-off and cost estimation	Cost management	More reliable and accurate estimates	Hillwood Commercial Project Penn National Parking Structure 100 11th Ave. Apartments
	Design coordination (clash detection)	Cost management and infrastructure complexity	Reduce field errors and reduce construction costs	Camino Medical Building
Chapters 6 and 7: contractors and fabricators	Schedule simulation/ 4D	Time to market, labor shortages, and language barriers	Communicate schedule visually	One Island East Office Tower
	Project controls	Time to market	Track project activities	GM Flint project
	Pre-fabrication	Time to market	Reduce onsite labor and improve design quality	Camino Medical Building 100 11th Ave. Apartments
Chapter 4: owners	Pro forma analysis	Cost management	Improve cost reliability	Hillwood Commercial Project
	Operation simulation	Sustainability/Cost management	Building performance and maintainability	
	Asset management	Asset management	Facility and asset management	Coast Guard Facility Planning

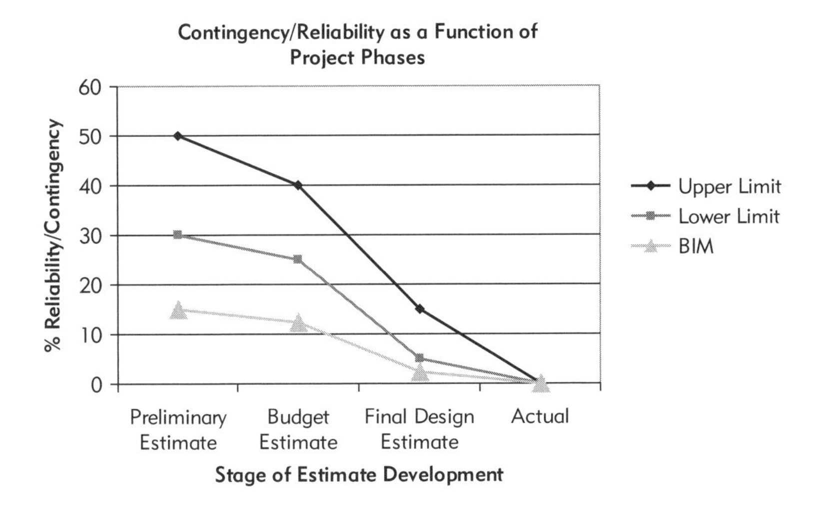

FIGURE 4-2 Chart showing the upper and lower limits that an owner typically adds to the contingency and reliability of an estimate over different phases of a project (data adapted from United States 1997; Munroe 2007; Oberlander and Trost 2001) and the potential targeted reliability improvements associated with BIM-based estimating.

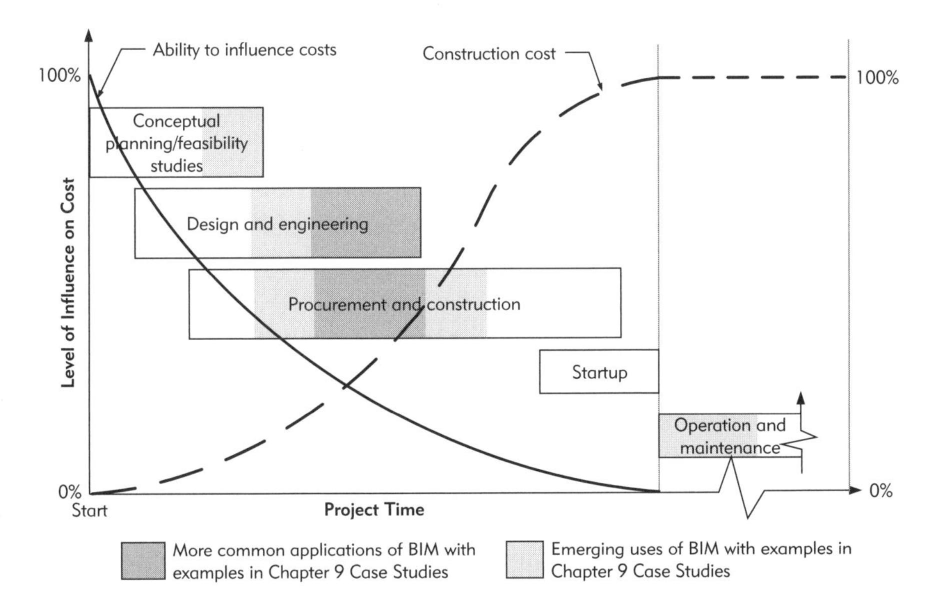

FIGURE 4-3 Influence of overall project cost over the project lifecycle. Today's use of BIM is typically limited to the late phase of design and engineering or early phases of construction. Use of BIM earlier in design process will have greater influence on cost. Improving overall cost reliability is a key motivator for employing BIM-based cost estimating methods.

Owners can manage cost with BIM applications to provide:

More reliable estimates early in the process with conceptual BIM estimating. Estimates that use conceptual building information models consisting of components with historical cost information, productivity information, and other estimating information can provide owners with quick feedback on various design scenarios. Accurate estimates can be very valuable early in the project, particularly for assessing a project's predicted cash flow and procuring finance. The Hillwood Commercial project case study demonstrates how owners working with a service provider employing a conceptual BIM-based estimating tool called DProfiler are able to reduce overall contingency and reliability and ultimately save money by borrowing less.

Faster, better detailed, and more accurate estimates with BIM quantity take-off tools. Both owners and estimators struggle with the ability to respond to design and requirement changes and understand the impact of those changes on the overall project budget and estimate. By linking the design model with the estimating processes, the project team can speed up the quantity take-off and overall estimating process and get faster feedback on proposed design changes (see Chapters 5 and 6). For example, owners can automatically derive accurate quantities and in turn streamline and verify estimates of designers and subcontractors (Rundell 2006). The Hillwood Commercial project case study cites evidence that estimating with BIM early in design can result in a 92% time reduction to produce the estimate with only a 1% variance between the manual and BIM-based processes. In the One Island East Office Tower case study, the owner was able to set a lower contingency in their budget as a result of the reliability and accuracy of the BIM-based estimate.

Owners, however, must realize that BIM-based takeoff and estimating is only a first step in the whole estimating process; it does not thoroughly address the issue of omissions. Additionally, the more accurate derivation of components that BIM provides does not deal with specific site conditions or the complexity of the facility, which depend on the expertise of an estimator to quantify.

4.2.2 Time to Market: Schedule Management

Time-to-market impacts all industries, and facility construction is often a bottleneck. Manufacturing organizations have well-defined time to market requirements, and companies like General Motors, as discussed in the GM Flint case study, must explore methods and technologies that enable them to deliver facilities faster, better, and cheaper. The GM Flint team completed the project five weeks ahead of schedule by using innovative BIM processes, which provide owners and their project teams with tools to automate design, simulate operations, and employ offsite fabrication. These innovations—initially targeted towards manufacturing or process facilities—are now available to the general commercial facility industry and its service providers. The innovations provide owners with a variety of BIM applications to respond to the following time to market needs:

Reduce time to market through the use of parametric models. Long building cycles increase market risk. Projects that are financed in good economic times may reach the market in a downturn, greatly impacting the project's ROI (Return on Investment). BIM processes, such as BIM-based design and prefabrication, can greatly reduce the project duration, from project approval to facility completion. The component parametric nature of the BIM model makes design changes easier and the resulting updates of documentation automatic. The Coast Guard Facility Planning case study is an excellent example of a parametric-based design to support rapid scenario planning early in a project. This type of BIM application allows owners to better respond to market trends or business missions closer to construction and adjust project requirements in collaboration with the design team.

Reduce schedule duration with 3D coordination and prefabrication. All owners pay a cost for construction delays or lengthy projects, either in interest payments on loans, delayed rental income, or other income from sales of goods or products. In the Flint and Camino Medical Building case studies in Chapter 9, the application of BIM to support coordination and prefabrication led to improved field productivity, reduced field effort, and reductions in the overall construction schedule, which resulted in on-time delivery for the owner.

Reduce schedule-related risk with BIM-based planning. Schedules are often impacted by activities involving high risk, dependencies, multiple organizations, or complex sequences of activities. These often occur in projects such as renovations of existing facilities, where construction must be coordinated with ongoing operations. For example, a construction manager representing the owner used 4D models (see Chapter 6 and Figure 4-4) to communicate a

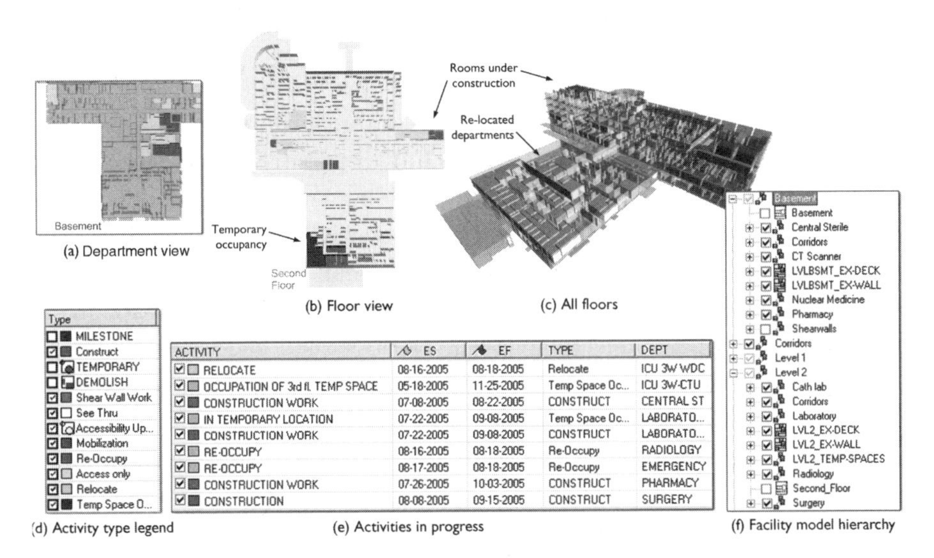

FIGURE 4-4 A) Views of a 4D model for a nine-floor hospital facility showing concurrent retrofit activities across departments and floors: a) 4D view of a department; b) 4D view of a floor; c) 4D view of all floors; d) activity type legend showing the types of activities the construction management team and owner communicated in the 4D model; e) the activities in progress; and f) the 4D hierarchy showing the organization by floor and department.
Image provided courtesy of URS.

schedule to hospital staff and mitigate the impact of activities on their operations (Roe 2002).

Quickly respond to unforeseen field conditions with 4D-coordinated BIM models. Owners and their service providers often encounter unforeseen conditions that even the best digital models cannot foresee. Teams using digital models are often in a better position to respond to unforeseen conditions and get back on schedule. For example, a retail project was slated to open before Thanksgiving for the holiday shopping season. Three months into the project, unforeseen conditions forced the project to stop for three months. The contractor used a 4D model (see Chapter 6) to help plan for the recovery and open the facility on time (Roe 2002).

4.2.3 Complexity of Building Infrastructure and Building Environment

Modern buildings and facilities are complex in terms of the physical infrastructure and the organizational, financial, and legal structures used to deliver them. Complicated building codes, statutory issues, and liability issues are now common in all building markets and are often a bottleneck or a significant hurdle for project teams. Often, owners must coordinate the design and approval efforts simultaneously. Meanwhile, facility infrastructures have grown increasingly complex. Traditional MEP systems are being integrated with data/telecom, building sensors or meters, and in some cases sophisticated manufacturing or electrical equipment.

BIM tools and processes can support owners' efforts to coordinate the increasingly complex building infrastructure and regulatory process by:

Coordinating infrastructure through fully-integrated 3D models of MEP, architectural, and structural systems. A building information model enables virtual coordination of a building's infrastructure across all disciplines. The owner of a facility can include its own representatives from its maintenance and operations staff to provide input and review of the model. Rework due to design flaws can potentially be avoided. The GM Flint and Camino Medical Building projects demonstrate how an owner can work with a construction team to coordinate the MEP systems using digital 3D models.

Producing higher quality and maintainable infrastructure through interactive review of coordinated models. Many owners need to go beyond typical MEP coordination to ensure that the MEP, data/telecom, and equipment are accessible and maintainable. This is particularly crucial for companies that depend heavily on these systems, such as biotech and technology companies, which demand reliable 24/7 service. Interactive review of the model allows owners to virtually access and simulate maintenance procedures.

Conforming to codes and requirements through BIM-based automatic code-checking. Often, owners and their design teams must work with a variety of jurisdictions to ensure their facilities meet design, performance, and workplace safety codes. Regulatory personnel also face challenges for ensuring compliance and conformance during design and construction. A potential benefit of a building information model is the ability to automatically analyze and check the model for program and code compliance. In Singapore, the government and building industry have implemented an electronic approval process, called CORENET (CORENET 2007). The use of building models can break the review bottleneck by giving regulatory agencies better tools to analyze facility designs. BIM functions for anything beyond basic code-checking, however, still require research before they will become available for adoption by statutory agencies.

Preventing litigation through collaborative creation and sign-off of building information models. Today, many projects invoke litigation to resolve payment issues due to changes. These issues include: designers citing owner-initiated changes; owners arguing that designers did not meet contractual requirements; and contractors arguing about scope of work and lack of information or inaccurate project documentation. Processes that center on a building model can mitigate such situations simply due to the level of accuracy and resolution necessary for creating a model; the collaborative effort of creating the model often leads to better accountability amongst project participants.

4.2.4 Sustainability

The green building trend is leading many owners to consider the energy efficiency of their facilities and the overall environmental impact of their projects. Sustainable building is good business practice and can lead to greater marketability of a facility. Building models provide several advantages over traditional 2D models due to the richness of object information needed to perform energy or other environmental analyses. Specific BIM analysis tools are discussed in detail in Chapter 5. From the owner's perspective, BIM processes can help:

Reduce energy consumption through energy analysis. On average, energy accounts for $1.50–$2.00 per square foot of operational costs (Hodges and Elvey 2005). For a 50,000 square foot facility, this amounts to $75–$100K annually. Investment in an energy-saving building system, such as enhanced insulation, reduces energy consumption by 10% and translates to $8–$10K annual savings. The break-even point for an up-front investment of $50K would occur by the 6th year of operation. The challenge when making such assessments is to compute the actual reduction in energy consumption achievable by any specific design. There are many tools for owners to evaluate the payoff and return on energy-saving investments, including life-cycle analysis, and these are discussed in Chapter 5. While these analysis tools do not absolutely require the use of a building information model for input, a model greatly facilitates their use. The Federal Office Building case study demonstrates the kinds of energy conservation analyses that can be integrated using BIM tools.

Improve operational productivity with model creation and simulation tools. Sustainable design can greatly impact overall workplace productivity. 92% of operating costs are spent on the people who work in the facility (Romm 1994). Studies suggest that day-lighting in retail and offices improves productivity and reduces absenteeism (Roodman and Lenssen 1995). BIM technologies provide owners with tools needed for assessing the appropriate trade-offs when considering the use of daylighting and the mitigation of glare and solar heat gain, as compare with project cost and overall project requirements. The Federal Office Building case study compared different scenarios to maximize the potential benefits of daylighting.

4.2.5 Overcoming Labor Shortage, Education, and Language Barriers

Owners face global and local market issues. All projects must meet local approval and typically rely on local resources to perform onsite activities. Even on local projects, it is common for multiple languages to be spoken. Many projects integrate global resources, particularly for international work; and many industrialized countries are experiencing a growing shortage of skilled workers and an aging workforce in the construction and facility industry (McNair and Flynn 2006). BIM-based delivery of a project can potentially mitigate the impact of these trends by:

Maximizing labor efficiency through BIM design linked with prefabrication and field planning. Combined with management approaches like lean construction, BIM can help project teams improve labor productivity and decrease field labor demands. The Camino Group Medical Building case study is an excellent example of using a 3D information model to support both prefabrication and optimization of field activities through virtual coordination and planning, reducing demands on field labor. In fact, because lean construction requires process changes that transcend the boundaries of individual companies in construction projects, the most successful applications have been those promoted by *owners,* such as Sutter Health, the HMO behind the Camino Medical Building. BIM has a central role to play in facilitating the application of lean techniques, such as pull flow control of labor and material flows.

Overcoming language barriers through BIM simulation and communication. Domestically and abroad, most projects involve personnel who speak multiple languages or dialects in the office and field. Efforts that embed translation tools into CAD or even BIM fall short due to the jargon and trade-specific information commonly notated on drawings. BIM can be used to communicate daily field activities to foreign field workers (Sawyer 2006). This interactive Game Boy-like view provides workers with a highly interactive way to navigate and query project information.

Educating the project team through interactive BIM reviews. Projects often continue over long periods and involve numerous service providers. Project teams must continually educate new project participants during each phase of the project. The computable nature of the 3D building information models makes them an excellent tool for quickly bringing new team members up to speed, so they can understand the scope, requirements, and status of a project. This communication is vital, particularly as a project ramps up or new participants join the team. The fourth author participated on a project where the owner (Disney) used 3D, 4D, and simulated operation models to educate over 400 interested parties, ranging from local agencies and executives to prospective bidders and new employees (Schwegler et al. 2000). The rich interactive nature of the information greatly enhanced understanding of the project. Unlike today's 2D disparate project documentation or even 3D models, the BIM design and review tools provide highly interactive features that not only

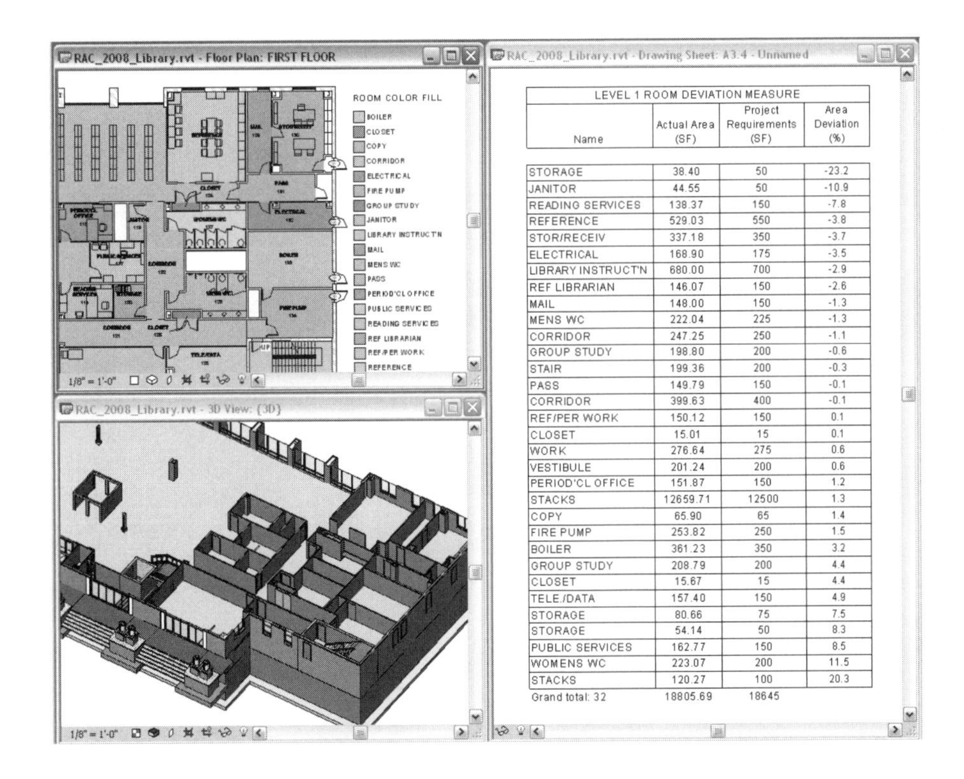

LEVEL 1 ROOM DEVIATION MEASURE			
Name	Actual Area (SF)	Project Requirements (SF)	Area Deviation (%)
STORAGE	38.40	50	-23.2
JANITOR	44.55	50	-10.9
READING SERVICES	138.37	150	-7.8
REFERENCE	529.03	550	-3.8
STOR/RECEIV	337.18	350	-3.7
ELECTRICAL	168.90	175	-3.5
LIBRARY INSTRUCT'N	680.00	700	-2.9
REF LIBRARIAN	146.07	150	-2.6
MAIL	148.00	150	-1.3
MENS WC	222.04	225	-1.3
CORRIDOR	247.25	250	-1.1
GROUP STUDY	198.80	200	-0.6
STAIR	199.36	200	-0.3
PASS	149.79	150	-0.1
CORRIDOR	399.63	400	-0.1
REF/PER WORK	150.12	150	0.1
CLOSET	15.01	15	0.1
WORK	276.64	275	0.6
VESTIBULE	201.24	200	0.6
PERIOD'CL OFFICE	151.87	150	1.2
STACKS	12659.71	12500	1.3
COPY	65.90	65	1.4
FIRE PUMP	253.82	250	1.5
BOILER	361.23	350	3.2
GROUP STUDY	208.79	200	4.4
CLOSET	15.67	15	4.4
TELE./DATA	157.40	150	4.9
STORAGE	80.66	75	7.5
STORAGE	54.14	50	8.3
PUBLIC SERVICES	162.77	150	8.5
WOMENS WC	223.07	200	11.5
STACKS	120.27	100	20.3
Grand total: 32	18805.69	18645	

FIGURE 4-5
A screenshot showing the spatial analysis information in a schedule and in the shaded 2D view.
Image provided courtesy of Autodesk.

allow you to view the project but query the model and probe a variety of information about the project's components.

4.2.6 Design Assessment

Owners must be able to manage and evaluate the scope of the design against their own requirements at every phase of a project. During conceptual design, this often involves spatial analysis. Later on, this involves analyses for evaluating whether the design will meet its functional needs. Today, this is a manual process, and owners rely on designers to walkthrough the project with drawings, images, or rendered animations. Requirements often change, however, and even with clear requirements, it can be difficult for an owner to ensure that all requirements have been met.

Additionally, an ever increasing proportion of projects involve either the retrofit of existing facilities or building in an urban setting. These projects often impact the surrounding community or users of the current facility. Seeking input from all project stakeholders is difficult when they cannot adequately interpret and understand the project drawings and schedule. Owners can work with their design team to use a building information model to:

Improve program compliance through BIM spatial analyses. Owners such as the United States Coast Guard are able to do rapid spatial analyses with BIM authoring tools (See Coast Guard Facility Planning case study). Figure 4-5 shows how a building model can communicate in real-time both spatially and in data form, to check compliance with requirements. Different

FIGURE 4-6 Snapshots showing the owner (GSA) and judges in a Virtual Reality Cave environment while interactively reviewing the design. *Image provided courtesy of Walt Disney Imagineering.*

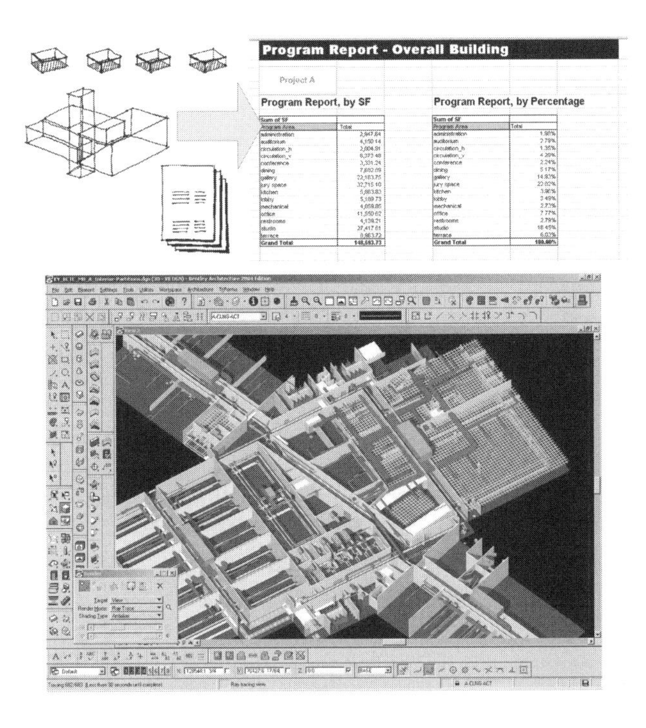

FIGURE 4-7 **Example of BIM space modeling by Jacobs Facilities, where they used spatial information to check the design against program requirements and to evaluate such things as natural lighting and energy efficiencies during the conceptual design process.**
Image provided courtesy of Jacobs.

colors are automatically assigned to rooms based on their dimensions and function. In some cases, the color-coding can alert designers or owners of rooms that exceed or don't meet existing requirements. This visual feedback is invaluable during conceptual and schematic design. Thus, the owner can better ensure that the requirements of their organization are met and that operational efficiencies of the program are realized.

Receive more valuable input from project stakeholders through visual simulation. Owners often need adequate feedback from project stakeholders, who either have little time or struggle with understanding the information provided about a project. Figure 4-6 is a snapshot of judges reviewing their planned courtroom. Figure 4-4 shows a 4D snapshot of all floors of a hospital to communicate the sequence of construction for each department and get feedback on how it will impact hospital operations. In both projects, the building information model and rapid comparison of scenarios greatly enhanced the review process. The traditional use of real-time and highly-rendered walkthrough technologies are one-time events, whereas the BIM and 4D tools make what-if design explorations far easier and more viable economically.

Rapidly reconfigure and explore design scenarios. Real-time configuration, however, is possible either in the model generation tool or a specialized configuration tool. Figure 4-7 shows an example from the Jacobs Facilities project, where BIM was used to quickly evaluate scenarios and to analyze requirements, needs, budget, and owner feedback (McDuffie 2007). Figure 4-8

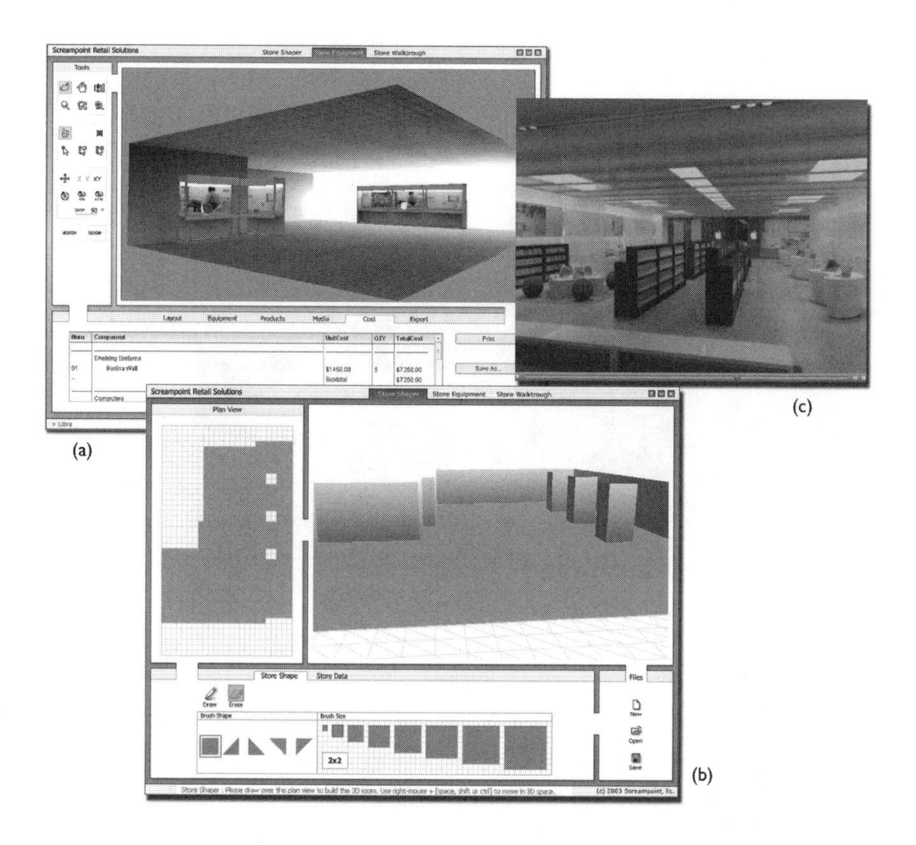

FIGURE 4-8 (A) A tool for quickly exploring retail space design and cost from (B) a set of retail shapes. (C) A rendering of a retail store for a specific design scenario.
Images provided courtesy of Screampoint.

shows a store configuration tool that allows a retail owner to quickly configure, visualize, and price different layouts for a retail space. These objects have rules and behaviors that constrain placement based on clearances and adjacencies and allow design teams to quickly layout a room and virtually walk around it.

Simulate facility operations. Owners may need additional types of simulations to assess the design quality beyond walkthroughs or visual simulations. These may include crowd behavior or emergency evacuation scenarios. Figure 4-9 shows an example crowd simulation for a typical day at a metro station with related analysis. The simulations used the building information model as a starting point for generating these scenarios. Such simulations are labor intensive and involve the use of specialized tools and services. For facilities where such performance requirements are critical, however, the initial investment in a building information model can pay off due to the more accurate 3D input that these specialized tools require.

4.2.7 Facility and Information Asset Management

Every industry is now faced with understanding how to leverage information as an asset; and facility owners are no exception. Today, information is generated

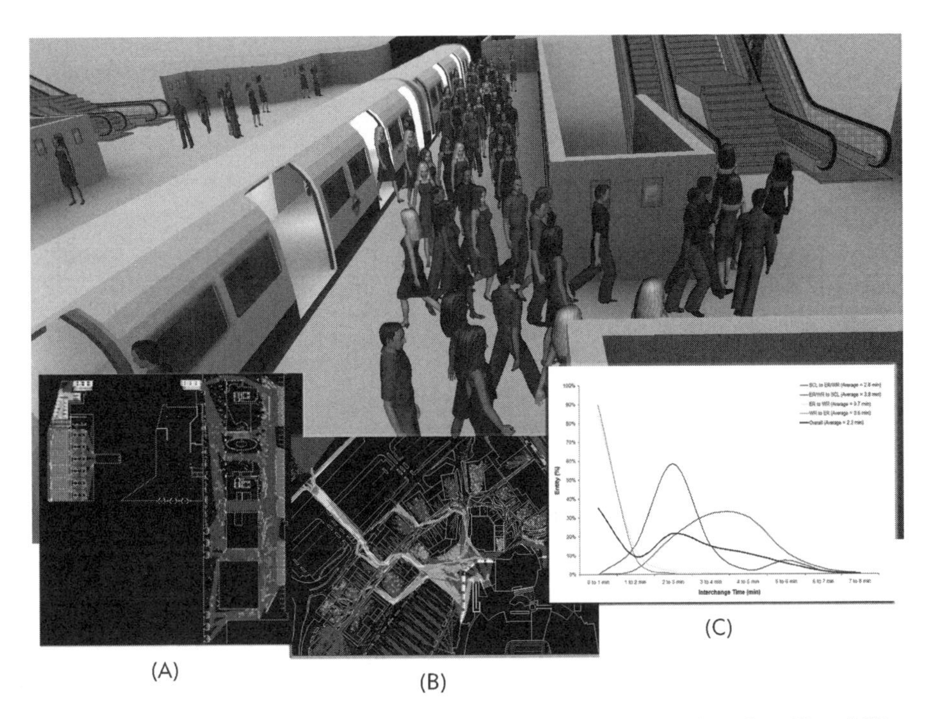

FIGURE 4-9 Examples of Legion Studio's visual and analytical outputs based on 2D and 3D building information data.
The main 3D rendering shows a simulation of a metro station during a weekday morning peak.
(A) A map of an airport uses color to show average speed, with red indicating slow movement and blue indicating free-flowing movement;
(B) a map of a stadium with access routes and adjacent retail facilities showing mean density, with red and yellow indicating the locations of highest density; and (C) a graph comparing passenger interchange times between several origin-destination pairs. (See color insert for full color figure.) *Images provided courtesy of Legion Limited.*

during each project phase and often re-entered or produced during hand-offs between phases and organizations, as shown in Figure 4-1. At the end of most projects, the value of this information drops precipitously, because it is typically not updated to reflect as-built conditions or in a form that is readily accessible or manageable. Figure 4-1 shows that a project involving collaborative creation and updating of a building model potentially will see fewer periods of duplicate information entry or information loss. Owners who view the total lifecycle ownership of their projects can use a building model strategically and effectively to:

Quickly populate a facility management database. In the Coast Guard Facility Planning case study, the team realized a 98% time savings by using building information models to populate and edit the facility management database. These savings are attributed to a reduction in labor needed to enter the spatial information.

FIGURE 4-10 The United State Coast Guard Web-based facility asset and portfolio system. *Image provided courtesy of Onuma and Associates, Inc.*

Manage facility assets with BIM asset management tools. The United States Coast Guard is integrating BIM into its portfolio and asset management, as discussed in Coast Guard Facility Planning case study. Figure 4-10 shows several views of their Web-based asset management tool, which integrates GIS data and building models for various facilities. Building components and assemblies are associated with facility information and used to support critical analyses, such as mission readiness. Another example is a 4D financial model shown in Figure 4-11 that associates each building object or objects with a

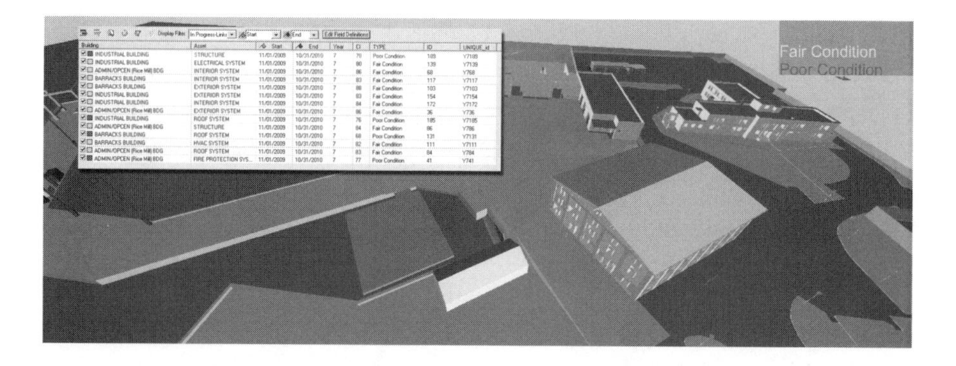

FIGURE 4-11 A 4D financial model showing how the "assessed" condition of facilities, ranging from good (green) to fair to poor (red)as indicated by different colors, changes over time. (See color insert for full color figure.) *Image provided courtesy of PBS&J, Common Point, Inc., AEC Infosystems, Inc., and MACTEC, Inc.*

condition assessment over time. The owner can view the facility or facilities periodically to get a "big picture" view of its condition assessment.

Rapidly evaluate the impact of retrofit or maintenance work on the facility. Another example is the use of visual and intelligent models to help facility managers assess the impact of retrofit or maintenance work. For example, a BIM - based FM system was applied during maintenance work on the Sydney Opera House (Mitchell and Schevers 2005). The maintenance team used the model to visually assess which areas would be affected when power was cut to a specific room.

4.3 TYPES OF OWNERS: WHY, HOW OFTEN, AND WHERE THEY BUILD

The many varied BIM applications discussed in the previous sections highlight the broad set of uses and benefits that BIM technologies offer. For many owners, these benefits must directly tie in to their business models. The following sections discuss the high-level and key differences between owner types and how different BIM applications support their specific needs.

4.3.1 The BIM Business Case

When initiating a project, every owner considers the overall economics of building a facility. Thus, the decision to apply BIM technology is often supported by an economic justification, either in terms of decreasing first or recurring costs or potential gains in income. Table 4-2 summarizes the typical costs and prospective income for a facility by category. To justify a BIM application, it must either decrease the first or recurring costs or increase income. Recurring costs can exceed first costs by as much as 2:1(Dolan 2006). Thus, when considering a BIM application, owners should not simply look at its impact on first costs but on recurring costs. In the following sections, we consider how different owners and delivery processes factor into this analysis.

4.3.2 Operating Owners vs. Developer Owners

Owners who build to operate have a strong incentive to consider long-term costs of owning and operating a facility. Examples of owners who build to occupy include corporations, retail, healthcare facilities, schools and universities, sports and recreation facilities, and the federal government. In the U.S. this type of building comprises nearly 50% of total construction and totaled $350 billion in 2006, with general commercial and educational comprising

Table 4-2 Summary of typical costs and income for a facility and factors that increase or decrease the cost/income and the type of BIM application that can positively impact the costs/income.

	Expense or Income Category	Description	Factors that increase/decrease cost (relative)	BIM Application
First costs (all owners)	Land acquisition	Cost to purchase land or existing facility	Economy and market demand	None
	Design	Services related to designing the building to meet owner requirements	Scope (extent of requirements)	Program compliance and automated-drawing production
			Level of information development, and complexity	Coordination
			Regulatory compliance	Code-checking
	Construction	Services related to physically constructing the facility based on design documentation	Labor market	Prefabrication and 4D
			Material costs and accurate quantities	Prefabrication and coordinated design
			Rework due to errors and quality of construction	Clash detection and coordination
			Site constraints	4D
			Complexity	Coordination and 4D
	Project management	Costs associated with managing the project	Coordination costs and number of organizations	
	Furnishings	Furniture, fixtures, and equipment	Configuration, availability, and coordination	Design configuration
Recurring costs (primarily operating owners)	Carrying charges	Interest on financing	Project duration	4D and prefabrication
			Over-estimating the project	Takeoff conceptual estimating
	Insurance	Costs associated with insuring facility	Operational risks	Operation simulation
	Replacement	Costs associated with replacing furnishings, finishes (carpets and paints), equipment, including roofs or other building components	Selecting low-end and poorly performing materials	Asset management and analysis

(Continued)

	Expense or Income Category	Description	Factors that increase/decrease cost (relative)	BIM Application
	Maintenance	Costs associated with maintaining the facility including landscape and grounds	Poor design and construction and the maintainability of the design	Operation simulation
	Operations and utilities	Energy and voice data	Energy efficiency and building performance	Energy analysis and performance monitoring
	Personnel	Salary and benefits for non-operation staff	Poorly functioning space or non-ideal work environment	Energy analysis and scenario design
Income	Sale/lease	Income from selling or renting a property	Time to sale/market	4D, prefabrication, and coordination
			Marketability of facility	Design configuration
				Operation simulation (pre-sales visualization)
				Energy analysis
			Market demand and the economy	None
	Production	Goods manufactured or sold in a facility (retail)	Time to market	4D, prefabrication, and coordination
	Services	Income from services performed in facility	Workplace productivity of services	Energy and lighting analysis, operation simulation
	Indirect	Reduction in operational costs	Sustainability	Energy analysis and design options

the largest portion of this market (Tulacz 2006). The case studies at the end of this book demonstrate various BIM applications and benefits for the following types of operating owners:

- Healthcare (Camino Group Medical Building)
- Federal facilities (Federal Courthouse, Federal Office Building, and Coast Guard Facility Planning)
- General commercial space (One Island East Office Tower and Hillwood Commercial Project)
- Manufacturing (General Motors Production Plant)

These types of owners—owners who operate—benefit from any of the BIM applications areas listed in Table 4-2 that are associated with first and recurring costs.

Owner developers, on the other hand, build to sell and are motivated to maximize their rate of return on their investment in the land and the facility. Developers who sell have a shorter ownership lifecycle compared to the owner who operates. Owners who sell are typically real estate developers or multi or single family home builders. * Their financial models are primarily driven by the cost of the land and permitting and the risks associated with market downturns and changing market trends. The changes in construction cost have less impact than these factors in their business models. The indirect costs that stem from design and construction services can impact their business models in the following ways:

1. Unreliable or poor estimates result in higher (than necessary) carrying costs.
2. Lengthy permitting phases result in higher carrying costs and missed market opportunitites.
3. Rework and poor planning increase schedule duration and increase carrying costs, time to market/sale
4. Design commitments early in a length design a and schedule process reduce opportunities to respond to market changes and to maximize sale price.

Owner developers can avoid these hidden costs and increase ROI by applying BIM on their project to improve estimate reliability (e.g., conceptual estimating), better communicate project to the community, reduce time to market, and improve marketability of the project.

4.3.3 When Owners Build: One-Time or Serial

Another factor related to the owner's business model is how often they build. Owners that build once are initially less likely to invest in process changes either due to lack of time for research and education or perceived value to invest in change. Whereas owners building serially, such as universities, developers, or retail owners, often recognize the inefficiencies in current delivery methods and their impact on overall project cost. These repeat builders are more likely to consider building delivery processes that are not based on minimizing

*The scope of this discussion does not address the unique demands of the home builder market and its own network of service providers. While many of the benefits we cite certainly apply to home builders, the technologies and tools we discuss in this book focus on commercial construction.

initial first-costs to delivery models that optimize the overall building delivery and yield quality facilities based on reliable costs and schedules. Consequently, serial owners tend to represent a high proportion of early BIM adopters.

One-time owners, though, can benefit just as much as serial owners and can do so easily by collaborating with design and construction providers familiar with BIM processes. Many of the case studies discuss projects where a service provider initiated the use of BIM.

4.4 HOW OWNERS BUILD

There are a number of procurement methods available to owners and the type of method chosen will impact the owner's ability to manage the BIM process and realize its benefits. The most important differentiator between owners is the degree to which they participate in the design and construction or facility development process. For example, owners who employ project management staff, including architects and planners, will use BIM within their own organizations. Owners who employ service-providers will either instruct others to use BIM through contractual imperatives or rely on service providers that choose to use BIM. These methods often depend on how the owner procures the project. Although there are numerous ways to procure a project, the three main facility delivery processes are: single stage or traditional (design-bid-build), design-build, and collaborative. The following sections discuss these three different delivery processes and how they impact the potential use of BIM on a project.

4.4.1 Design-Bid-Build

In the U.S., 60% of facility design and construction is still performed under the design-bid-build process (American Institute of Architects 2006). As described in Chapter 1, the single stage or traditional method involves an owner maintaining contractual relationships with the designer and builder and managing the relationship and flow of information between the two groups, as shown in Figures 4-12 and 4-13. The architect and builder manage the relationships and information flow between their sub-consultants and subcontractors. This contractual divide results in a sequential process with specific deliverables at specified time periods, creating an information barrier between organizations and often prohibiting the electronic exchange of timely project information. In some cases, the owner hires a construction manager, either external or internal, to oversee and manage communication within the design and construction organizations. This process does not lend itself well to supporting the adoption

FIGURE 4-12 The traditional organization of a project team involves contracts between the owner and the primary architect and builder, who in turn maintain contracts between these organizations and sub-consultants. This type of contracting often prevents the flow of information, responsibility, and ultimately the ability to effectively use BIM tools and processes.

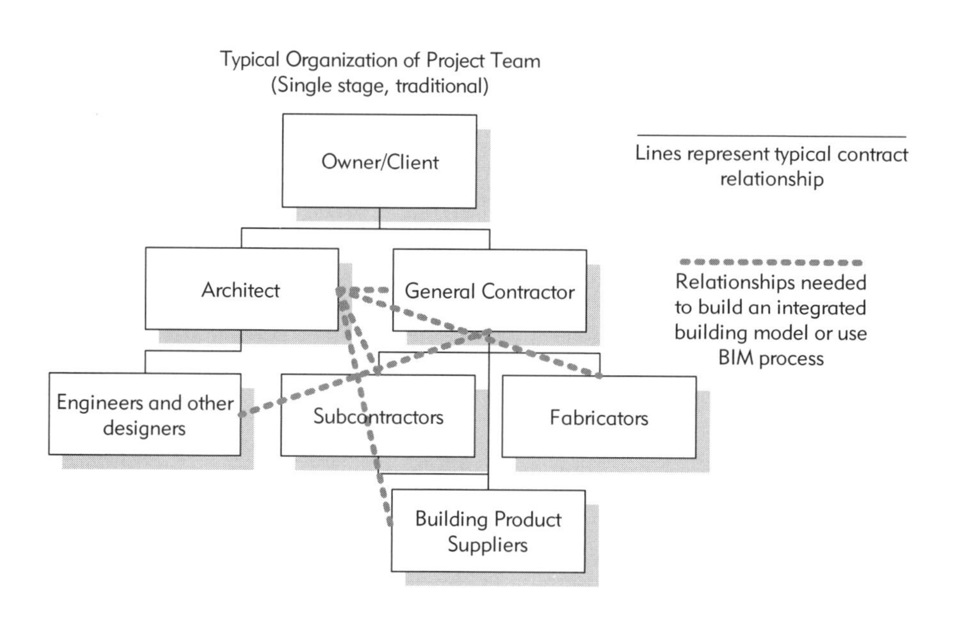

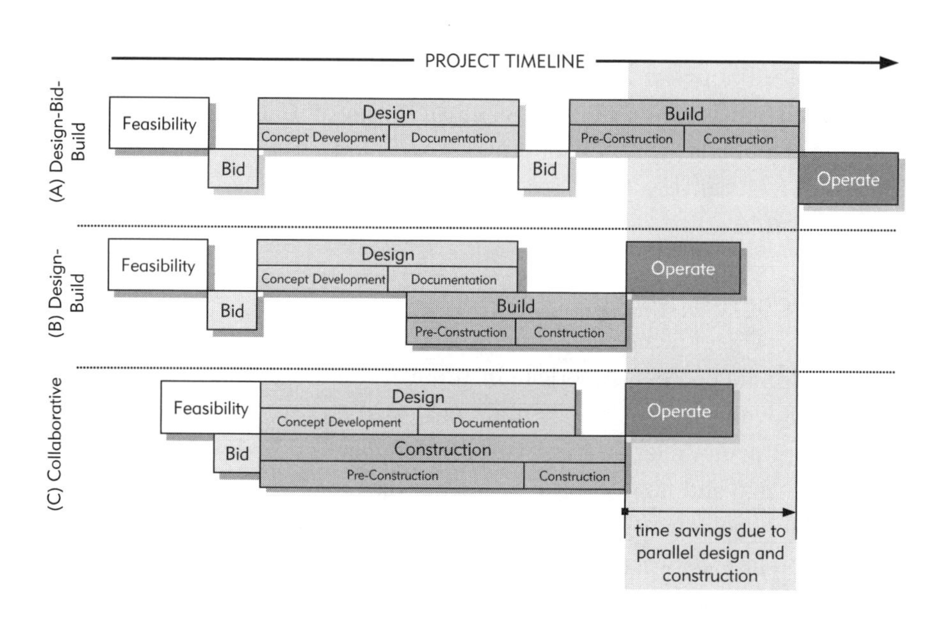

FIGURE 4-13 Diagram comparing three different delivery processes.
A) The traditional single-stage involves the completion of each phase prior to the start of the next phase, often involving a different organization performing each phase in a non-integrative process; B) the design-build process involves an overlap of development phases leading to a shortened overall schedule and requires integration between designers and builders;
c) a collaborative process involves participation by all key participants as early in the process as possible and ongoing collaboration.

of technologies or processes across the project team, due to the many contractual divides. Typically, this process is defined by the "wall" of deliverables whereby at the end of each phase the deliverables are handed "over the wall" with little or no integration or collaboration between the participants in each phase. The deliverables are typically paper-based. Often at the start of each phase, information is re-created or duplicated, reducing the value of project documentation and resulting in a loss of project knowledge (See Figure 4-1). Contractors or their subcontractors and fabricators often re-draw architect's construction documentation in the form of configuration drawings, because the architectural and engineering drawings do not dimension or lay out spaces or detail the connections needed for subcontractor coordination.

This deliverable-based approach makes it difficult to successfully implement those BIM tools and processes that require a model-based handover between organizations; in these scenarios, the organizations agree a priori on the format and content of the model itself. For this reason, BIM application in a DBB project is often limited to single phase BIM applications, such as 4D or energy analysis. An owner, however, can expand the potential BIM application on DBB projects to include:

1) Contractual requirements specifying the format and scope of the building model and other information at each deliverable phase
2) The cooperation of individual organizations to support or promote the BIM process within their own phase

4.4.2 Design-Build

As discussed in Chapter 1, an increasingly common model of facility delivery is design-build (DB), where a contractual relationship is established between the owner and a single organization and often represents the architect and builder either as a single company or as a partnership. In design-build, the individual phases are not necessarily shortened, but the overlap of phases reduces the overall project duration, as shown in Figure 4-13(B). In some cases, the integration of construction details earlier in the process can inform the design and lead to improved coordination and constructability and reduced construction time.

Studies comparing these approaches conclude that the DB process does have benefits in terms of project timeliness and schedule reliability (Debella and Ries 2006; Konchar and Sanvido 1998; USDOT - Federal Highway Administration 2006; Ibbs et al. 2003). The Konchar and Sanvido study found benefits of DB in terms of cost and quality, but other studies do not show significant cost or quality differences between the two approaches. The DB delivery process creates a potentially smoother flow of information between design and construction organizations. It still requires that organizations (or groups) spend time at the beginning of the project to define the work process, build and maintain the building model, outline its specific uses including scope and responsibility of each participating organization, and test the process. Unlike the DBB method, where such collaboration is often limited to specific handover periods or deliverables, the DB team can collaboratively identify BIM applications across project phases and focus less on specific deliverables between organizations and more on overall deliverables to the owner. Section 4.9 provides some guidelines and tips for owners wanting to promote a successful BIM effort.

4.4.3 Collaborative Process

A third and new approach to procuring a project is a variant of the design-build, has yet to be named, and emphasizes a collaborative, alliance-based relationship between owner and AEC service providers. A study by the University of Texas (Geertsema et al. 2003) documents a significant shift in non-collaborative and collaborative relationships between owner and contractors from "winning bid first use" to "alliance preferred provider." That is, owners are moving away from the traditional selection of providers based on 'low-bid,' which is common in DBB scenarios, to preferred providers. The same study notes that the projects with collaborative relationships are more successful from both the owner and contractor's perspectives.

This collaborative approach involves the selection and participation of all key project participants as early in the process as possible, either with single prime (representing multiple organizations) or multiple prime contractual relationships. Figure 4-13(C) shows how this process may occur during the feasibility stage, when the participation of service providers potentially has the greatest impact and can help define project requirements. Where the design-build process typically involves a partial overlap of designers and builders, the collaborative process requires overlap beginning with the project's initial phase and by more participants, requiring more intimate involvement of the owner or owner's representative. The collaborative process may not necessarily yield a shorter overall project duration or earlier start of construction, as shown in Figure 4-13, but it does achieve participation of the construction team, including fabricators and suppliers who are involved early and often. This approach is often combined with incentives for the entire project team to meet specific project goals and targets and also involves risk sharing. The primary difference between this process and the design-build approach is the collective sharing of risk, incentives, and the method of creation of facility documentation.

The collaborative approach is an ideal procurement and delivery method for reaping the benefits of BIM applications on a project. The participation of all project participants in the creation, revision, and updating of the building information model forces participants to work together and virtually build the project.

The technical challenges discussed in the previous section will persist, and owners along with the entire project team should follow the guidelines in Section 4.9 to maximize the team's effort.

4.4.4 Internal or External Modeling

These different procurement methods do not address situations where owners perform some or all of the design, engineering, or construction services. Outsourcing is a common trend for many owners (Geertsema et al. 2003). There are some owner organizations that have construction management and construction superintendents on staff. In such cases, as discussed in Section 4.9, the owner must first assess their internal capabilities and work processes. The "wall" of deliverables can exist internally, and defining model handover requirements between internal groups is just as critical. The owner must ensure that all participants, internal or external, can contribute to the creation, modification, and review of the building model. This may involve the owner requiring the use of specific software or data formats to exchange data.

Outsourcing, however, does have an impact on the overall BIM effort, and owners who choose to hire a third party to produce the building information model independent of the project's internal and external team of service providers should carefully consider full outsourcing of the model. Typically, the outsourcing effort leads to a building information model that is under-utilized, outdated, and of poor quality. This occurs for several reasons. First, the internal or external team has to reach a specified point in the project to hand over the traditional documentation. Second, the outsource team must spend significant time, often with little contact since the team is now busy working towards the next deliverable, to understand and model the project. Finally, the outsource team does not typically have highly-skilled or experienced staff with building knowledge. Thus, outsourcing should be done with considerable attention and management oversight or used as an effort to support the BIM effort, not replace it. The One Island East Office Tower case study is an excellent example of working with external resources to develop the building model while integrating its resources into the project team both physically and virtually. Another example is the Letterman Digital Arts project in San Francisco, where the owner hired an outside firm to build and maintain the building model (Sullivan 2007). In both cases, the critical success factor was attributed to bringing the resources onsite and mandating participation by all project participants.

4.5 BIM TOOL GUIDE FOR OWNERS

In the previous sections, we reference several BIM technologies that owners and their service providers are employing. In this section, we provide an overview of BIM tools or features of those tools intended to fulfill owners' needs and other owner-specific BIM applications. Chapters 5–7 discuss the specific BIM design and construction technologies, such as model generation tools, energy analysis, 4D, and design coordination.

4.5.1 BIM Estimating Tools

Owners use estimates to baseline their project cost and perform financial forecasting or pro forma analyses. Often, these estimates are created using square foot or unit cost methods, by an owner representative or estimating consultant. Some estimating software packages, such as U.S. Cost Success Estimator (U.S. Cost 2007), are designed specifically for owners. Microsoft© Excel, however, is the software most commonly used for estimating. In 2007, U.S. Cost provided their customers with functionality to extract quantity takeoff information from a building model created in Autodesk Revit®. Another product targeted to owners is Exactal's CostX® product (Exactal 2007), which imports building models and allows users to perform automatic and manual takeoffs. Chapter 6 provides a more detailed overview of BIM-based estimating. Owners should evaluate and consider the following features of such applications:

- **Level of estimating detail.** Most owners rely on unit-cost or square footage estimating methods, using values for specific facility types. Some estimating tools are designed for this basic estimating. Others use "assemblies" or estimating items that include detailed line-item costs. Owners need to make sure that the estimating tool extracts BIM component information at the level of detail required: quantities, per component, or square footage.
- **Organization formats.** Most owners will organize their estimate into a Work Breakdown Structure (WBS). In the U.S., most owners use either MasterFormat or Uniformat WBS. For example, the GSA is working with the Construction Specification Institute (CSI) to expand this format and requires estimating tools that support extensions to these formats.

- **Integration with custom cost/component databases.** BIM-based estimating often requires establishing links between BIM components and estimating items and assemblies via a specific component attribute or through a visual interface. This may require significant setup and standardization.
- **Manual intervention.** The ability to manually modify, adjust, or enter quantity take-off information is critical.
- **Model aggregation support.** The ability to import multiple models and combine take-off and estimate information from different models. Tools like U.S. Cost allow owners to aggregate and re-use estimating items across projects. With BIM-based estimating, an owner may need to perform the quantity take-off from multiple models, multiple facilities, or multiple domain models, i.e., architectural, structural, etc.
- **Versioning and comparison.** One of the most important potential features of a BIM quantity take-off and estimating tool is the ability to compare versions and track the differences between any two or more design scenarios or versions. For example, Exactal CostX® provides a visual comparison between two versions of a design to show where changes have been made and their impact on quantity take-off information. This feature is potentially invaluable for owners seeking to understand the cost impact of design changes.
- **Reporting features.** Most estimating tools provide reporting features to print hard copy reports of the take-off and estimate. Some estimating software tools provide a viewer, as shown in Figure 4-14, to enable an owner to interactively view the cost information and review it in 2D, 3D, or spreadsheet format.

4.5.2 Model Validation, Program, and Code Compliance

An emerging group of BIM software tools is called *model checkers*. These tools are discussed in greater detail in Chapter 5, as many of the features are related to services provided by the designer. From the owner or construction manager's perspectives, these tools perform a variety of important functions:

Check against program requirements. This type of feature compares owner requirements with the current design (See Chapter 5 for further discussion of these tools). These may include spatial, energy, and distance and height requirements for specific spaces or between spaces as well as adjacency requirements. Owners may have their own staff do these checks, or they may require that the design team or a 3rd party do them.

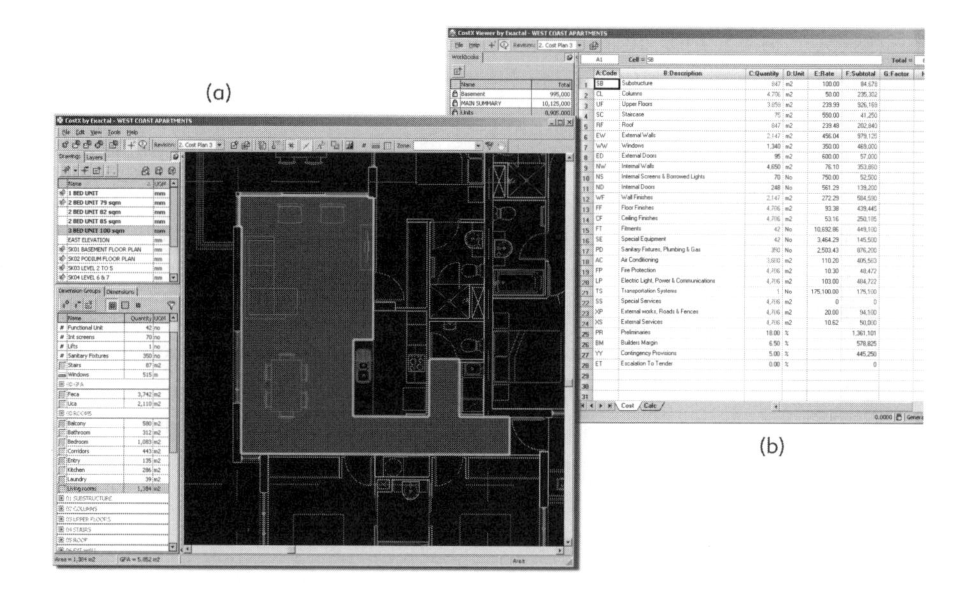

FIGURE 4-14 Screenshot of the Exactal CostX® Viewer that (a) allows owners to view the project in 2D or 3D along with its related cost items and (b) view detailed information for the cost estimate.

Validate building information model. Today, owners are able to quickly and thoroughly assess the quality of a building information model or, if an owner requires specific types of information input, determine whether that information exists and is in the specified format. Tools such as Solibri Model Checker™ (Solibri 2007) (Figure 4-15) provide both types of validation. At a generic level of validation, the owner can test for duplicate components, components within components, or components missing critical attributes or that do not comply with standards established by the team. The Solibri Model Checker also allows users to provide more sophisticated queries and tests; for example, whether the model contains specified information types. These features are relevant to owners who are considering using the model during post-construction phases or require operation-specific information.

4.5.3 Project Communication and Model Review Tools

Project communication occurs formally and informally on a variety of levels. Through contractual requirements, owners dictate the format, timing, and method of communication related to project deliverables. In this way, they often establish: the baseline project documentation format, modes of project information exchange, and expected project workflows. Within this formal communication is the daily exchange of information between project participants; and this is often impacted by the contractual relationship between those participants and the overall procurement method. Traditional

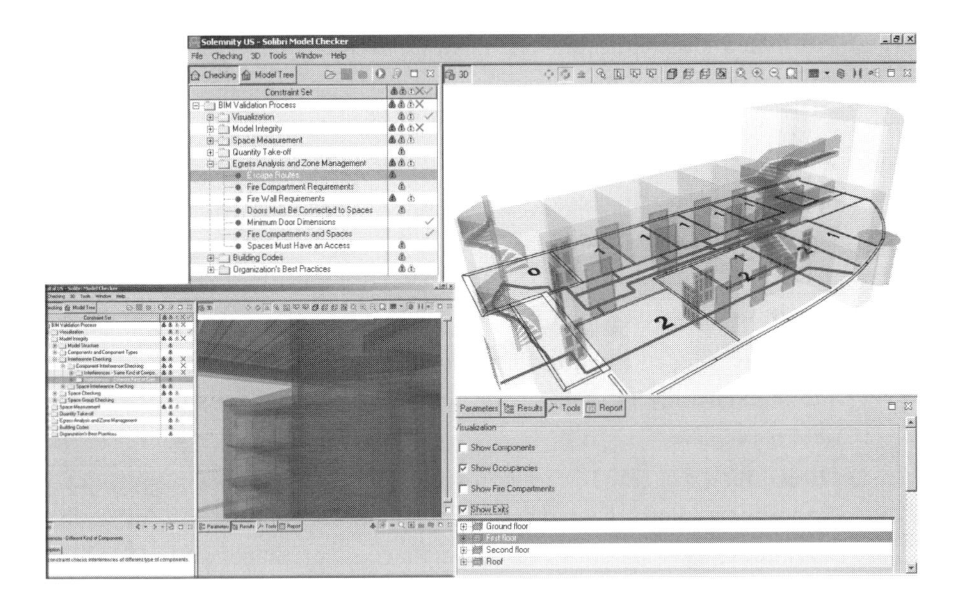

FIGURE 4-15 Snapshots of Solibri Model Checker™ showing: A) the validation of escape routes for a design; and B) the validation of model integrity.
Image provided courtesy of Solibri.

single-stage delivery methods tend to limit exchange between disciplines, and each individual discipline develops their own internal methods of project communication.

Thus, BIM-based project communication as it relates to the exchange of building information model data is quite different compared to traditional paper-based or 'publish' modes of information exchange. Consequently, the team (including the owner) needs to establish the protocols and tools that support the ongoing exchange, modification, and review of the building information model. In some cases, the owner may opt to control this communication, as was done by Swire Properties in the One Island East Office Tower case study. In other cases, such as the General Motors Production Plant case study, the design-build team jointly develops and maintains the communication and model management tools.

Within these different structures of communication are detailed technical issues and protocols for managing that communication. Our discussion of these issues is limited to the exchange of building model information and does not address the challenges of managing communication and storing a wide variety of project information, such as contracts, specifications, RFIs, change orders, etc.

There are four different types of communication exchange related to the building information model (these are different in nature and format than the content in the exchange discussed in Chapter 3):

a) **Published snapshots** are one-directional static views of the building information model that provide the receiving party with access only to the visual or filtered meta-data, such as bitmap images.

b) **Published building information model views and meta-data** provide the receiving party with viewing access to the model and related data with limited capabilities to edit or modify data. PDF or DWF are such examples that allow users to view 2D or 3D and to mark-up, comment, and change certain view parameters or perform query functions on the model.

c) **Published files of the building information model** include proprietary and standard file formats (.dwg, .rvt, .ifc), where access to the native data is possible.

d) **Direct database (DB) access** provides users access to the project database either on a dedicated project server or distributed project server. The model data is controlled through access privileges and may allow only those users that create the content to edit it, or alternately the DB may provide more sophisticated edit and change capabilities.

The "publish" methods of BIM exchange are common in BIM practice today. The use of a project database accessible by project participants is less common, but the General Motors Production Plant case study describes such a scenario. Figure 4-16 shows a sample information workflow for a BIM-based project that includes all four of these information exchanges. Most projects will continue to use a variety of exchange methods to support various roles within a project team.

While service providers may select the tools and perform many of the functions related to communication and management of the building information model, they must do so in compliance with the needs and requirements of the owner and the project. As a rule of thumb, for every designer or engineer that uses a BIM tool, there will be another ten individuals seeking to view and review the building information model. Communication methods (c) and (d), which require proprietary tools and steep learning-curves to implement, may be impractical. Likewise, limiting methods of exchange to methods (a) and (b) will hinder any ability to collaboratively create and modify the building information model.

In the following section, a review of the tools that support these communication methods is organized into two categories: model viewing and review tools and model management tools.

4.5.4 Model Viewing and Review

Two types of digital model viewing technologies are increasing in use. The first are specialized model viewing tools that run on the desktop and can import and

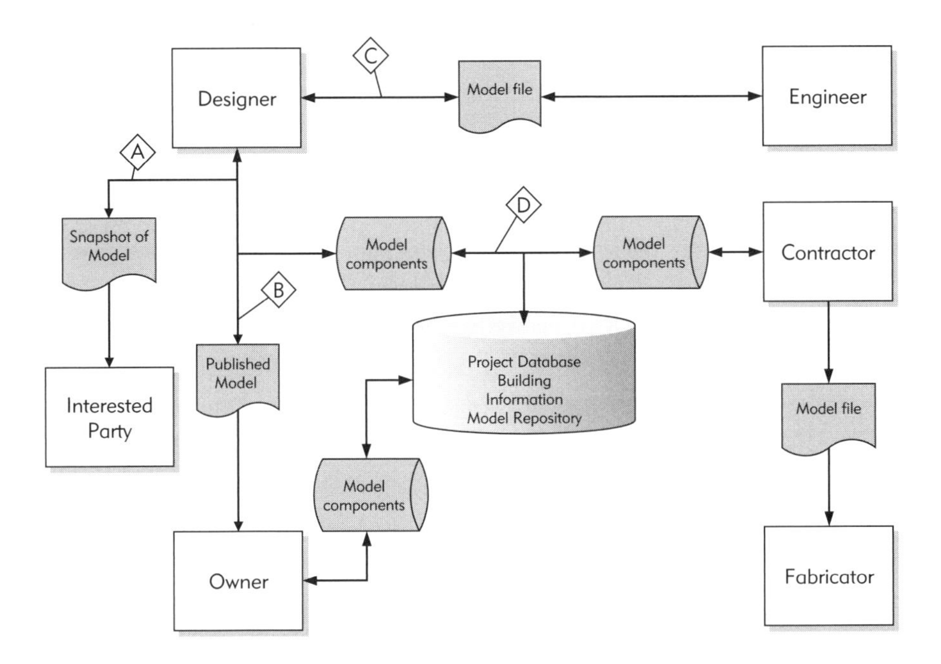

FIGURE 4-16 Conceptual diagram of the different types of communication exchange that might occur on a typical BIM-based project.
This scenario shows a project using a central repository for the project, with the architect, owner, and contractor having access to that model. The architect publishes (A) snapshots and (B) models for interested parties and owners to review. The engineer exchanges model files (C) with the designer such as a .dwg, .rvt. or .dgn file. The contractor exchanges published views of the model, (D) such as PDF or DWF for their subcontractors.

integrate a variety of model formats. These tools allow users to navigate through the model and query it interactively as well as to see sections or to view parts of the model; these support exchange methods B, C, and D. The second type includes Web-based viewing tools that import a standard format, such as Adobe® PDF or DWF™ and supports methods A and B. These tools also offer interactive navigation but require the model creator to publish in these formats. Owners will most likely encounter both types of tools on their projects and should consider what types of features the project team will use, including: comments/markup, take-off (dimensional queries), viewing, presentation, round-trip markup, and who will have access to the model. Examples of model viewers are Adobe® Acrobat® Professional, Autodesk® Design Review, and NavisWorks™ Roamer. Examples of Web-based model viewers are NavisWorks™ Freedom and Actify Spinfire™ Reader.

When evaluating these tools an owner should consider the features listed in Table 4-3.

Table 4-3 Model review features to consider.

File import features	What formats does the tool import? (See Chapter 2 for a list of file formats.)
Integrate models	Can the tool merge and integrate different types of file formats into one view and model?
Data import types	What types of non-geometric data does the tool import, and how can the user view the meta-data or model properties?
Multi-user support	Does the tool support multi-user access to a file or model or "shared" viewing of a model over the network?
Mark-up and comment tools	Can users mark up and comment in the tools? Are these mark-ups time stamped and tracked for review?
Model view support	Can the user view multiple views simultaneously? For example: plan, section, and 3D views?
Document view	Does the tool support viewing of related documents, such as text files or images or spreadsheets?
Dimension queries	Can a user easily measure in 2D and 3D?
Property queries	Can the user select a building object and view the object properties or perform a query to find all objects with a specific property or property value?
Clash-detection	Does the model review tool support clash-detection? If so, can you track the status of the clashes or classify the clashes?
4D	Does the model review tool include features to link the model objects to schedule activities or support other types of time-based simulations?
Re-organization of the model	Can the user re-organize the model into functional or user-defined groups and control viewing or other functions with these custom-defined groups?

4.5.5 Model Servers

Model servers are designed to store and manage access to the model and its data either at a host site or on an internal server and network. Today, most organizations use file-based servers that utilize standard file transfer protocols to exchange data between the server and client. From the BIM perspective, file-based servers do not yet directly link to or work with building information models or building objects within those models, as they store data and provide access only at the file level.

Today, most collaboration management for BIM projects involves a model manager manually integrating the models and creating project model files. The case studies in Chapter 9 illustrate the challenges many companies face in setting up work processes to manage models. In most cases, the organizations work with their internal IT groups to set up and maintain a server to host model files. In the General Motors Production Plant case study, the company used Bentley's ProjectWise®; the One Island East Office Tower used Digital Project's™ built-in features to support model management.

Off the shelf commercial model server solutions are not yet widely implemented and often require custom installation and trained IT staff to operate and maintain. Examples of model servers are:

- Bentley ProjectWise® (www.bentley.com)
- Enterprixe Model Server (www.enterprixe.com)
- EPM Technology EDMserver (www.epmtech.jotne.com)
- Eurostep modelserver for IFC (www.eurostep.com)
- SABLE developed by EuroSTEP (Hobaux 2005)

4.5.6 Facility and Asset Management Tools

Most existing facility management tools either rely on polygonal 2D information to represent spaces or numerical data entered in a spreadsheet. From most facility manager's perspectives, managing spaces and their related equipment and facility assets does not require 3D information; but 3D, component-based models can add value to facility management functions.

Building models provide significant benefits in the initial phase of entering facility information and interacting with that information. With BIM, owners can utilize "space" components that define space boundaries in 3D, thus greatly reducing the time needed to create the facility's database, since the traditional method involves manual space creation once the project is complete. The Coast Guard Facility Planning case study recorded a 98% reduction in time and effort to produce and update the facility management database by using a building information model.

Today, few tools exist that accept the input of BIM space components or other facility components representing fixed assets. Some of the tools that are currently available are:

- ActiveFacility (www.activefacility.com)
- ArchiFM (www.graphisoft.co.uk/products/archifm)
- Autodesk® FM Desktop™ (www.autodesk.com) (see Figure 4-17A)
- ONUMA Planning System™ (www.onuma.com/products/OnumaPlanningSystem.php)
- Vizelia suite of FACILITY management products (www.vizelia.com) (see Figure 4-17B)

In addition to the general features that any FM system should support, owners should consider the following issues with respect to the use of such tools with building models:

- **Space object support.** Does the tool import "space" objects from BIM authoring tools, either natively or via IFC? If so, what properties does the tool import?
- **Merging capabilities.** Can data be updated or merged from multiple sources? For example, MEP systems from one system and spaces from another system?
- **Updating.** If retrofit or reconfiguration of the facility takes place, can the system easily update the facility model? Can it track changes?

Leveraging a building information model for facility management may require moving to purpose-built BIM facility tools, such as Autodesk® FM Desktop™ or to third-party BIM add-on tools.

The use of BIM to support facility management is in its infancy and the tools have only recently become available in the marketplace. Owners should work with their facility management organizations to identify whether current facility management tools can support BIM spaces or whether a transition plan to migrate to BIM-capable facility management tools is required.

4.5.7 Operation Simulation Tools

Operation simulation tools are another emerging category of software tools for owners that use data from a building information model. These include crowd

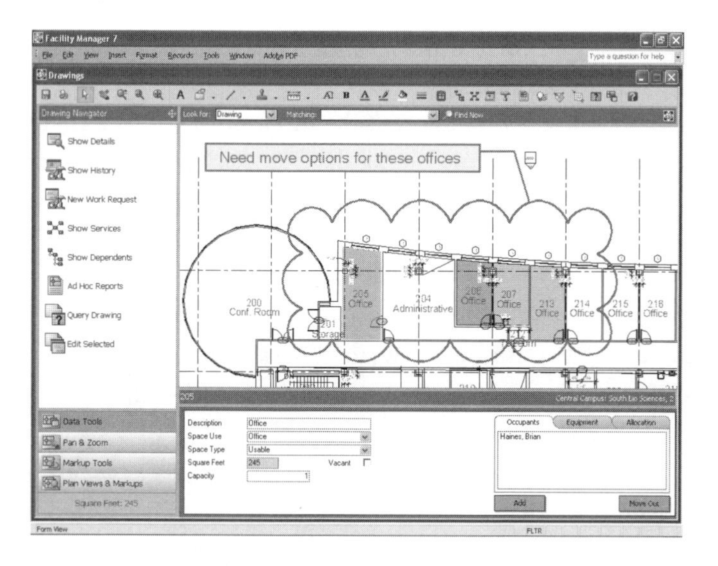

FIGURE 4-17A Autodesk®FM Desktop™ showing a visual interface of the facility spaces and data views. *Image provided courtesy of Autodesk, Inc.*

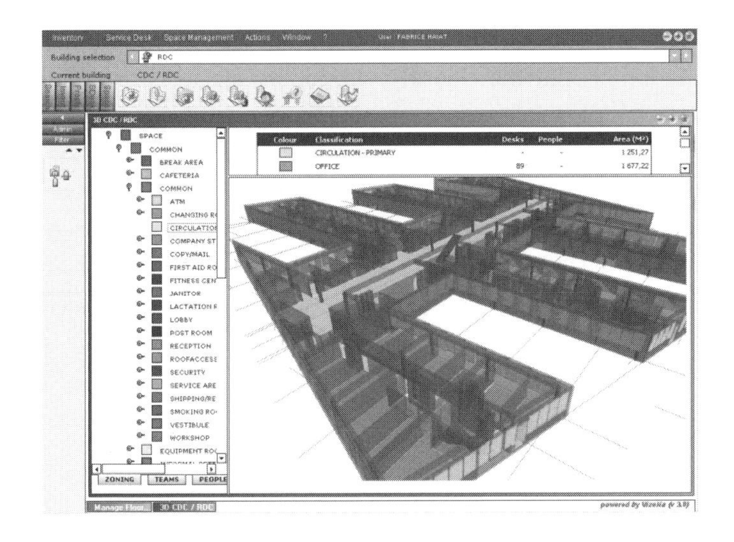

FIGURE 4-17B Screenshot Vizelia FACILITY Space showing the 3D, color (shaded) coded view of spaces by type. *Image provided courtesy of Vizelia, Inc.*

behavior (Still 2000), manufacturing, hospital procedure simulation, and emergency evacuation or response simulations. Many of them are provided by firms that also offer the services to perform the simulations and add necessary information. In all cases, the tools require additional input of information to perform the simulations; and in some cases, they only extract the geometric properties from the building information model. Table 4-4 summarizes some of the commercially available simulation tools.

Table 4-4 Summary of commercially available operation simulation tools.

Simulation Type	Company and Software Name	Input BIM?
Crowd behavior	Legion Studio (www.legion.com)	No
	eRENA ViCROWD (www.erena.kth.se/crowds.html)	No
	Crowd Dynamics (www.crowddynamics.com)	No
Evacuation	IES Simulex (www.iesve.com)	Yes (via gbXML ModelBuilder environment or DXF file)
	buildingExodus (http://fseg.gre.ac.uk/exodus/)	No (DXF file)
Operation	Common Point OpSim (www.commonpointinc.com)	Yes (geometry)
	buildingExodus (fseg.gre.ac.uk/exodus/)	No (DXF file)
	SIMSuite (www.medsimulation.com) Flex-Sim: (www.hospital-simulation.com)	

More typical examples of operation simulation tools do not involve specialized simulations but the use of real-time visualization or rendering tools that take the building information model as input. For example, the fourth author participated in the development of a 3D/4D model for Disney California Adventure. With specialized tools and services, the same model was used to simulate emergency scenarios and the ride rollercoaster (Schwegler et al. 2000). Likewise, the Letterman Lucas Digital Arts center team used their model to evaluate evacuation and emergency response scenarios (Boryslawski 2006; Sullivan 2007).

4.6 AN OWNER AND FACILITY MANAGER'S BUILDING MODEL

Owners need not only to be conversant in the kinds of BIM tools but also the scope and level of detail they desire for a building model of their project. In Chapters 5, 6, and 7, we discuss the types of information that designers, engineers, contractors, and fabricators create and add to building information models to support many of the BIM applications. To take advantage of post-construction BIM applications, as discussed in Section 4.2 and listed in Figure 4-18, owners need to work closely with their service providers to ensure that the building model provides adequate scope, level of detail, and information for the purposes intended. Figure 4-18 provides a framework for owners to understand the relationship between the level of detail in a model—masses, spaces, and construction-level detail (see vertical direction)—and the scope of a model, including spatial and domain-specific elements such as architectural and detailed MEP elements.

Often, each service provider defines the scope and level of detail required for their work. The owner can mandate the scope and level of detail required for post-construction use of the model. For example, at the feasibility stage, masses and spaces are sufficient to support most BIM applications for conceptual design. If the owner requires more integrative BIM applications, then both the level of integration in the model (horizontal) and level of detail (vertical) are increased in the effort to produce the model.

Table 4-5 provides a partial list of some key types of information that the building model needs to support for post-construction use. Some of this information is represented in the IFC schema, as discussed in Chapter 2, and there is a working group within the IAI, the "Facility Management Domain" (www.iai-na.org/technical/fmdomain_report.php) that addresses facility-specific

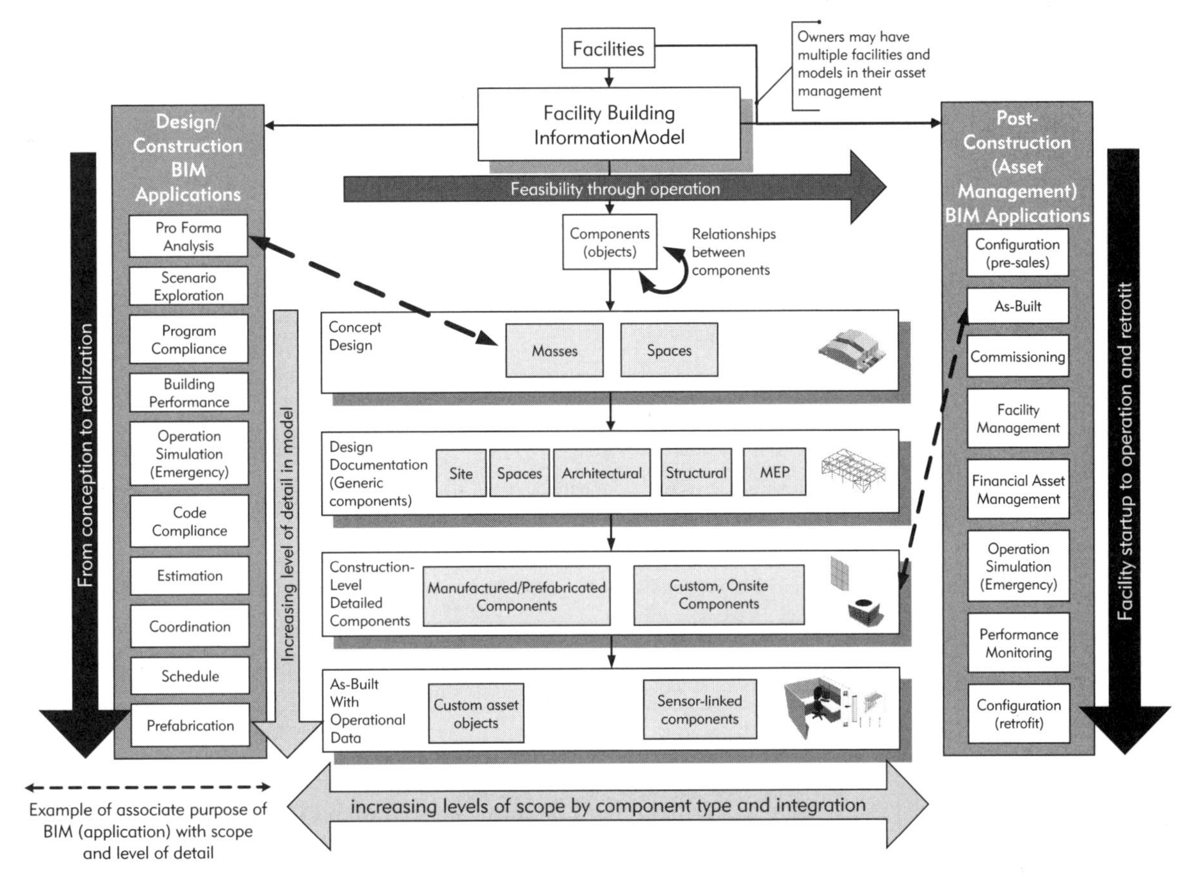

FIGURE 4-18 Conceptual diagram showing the relationship between various BIM applications during the facility delivery process; post-construction and their relationship to the level of scope and detail in the model.

scenarios, such as move management, work order flows, costs, accounts, and financial elements in facility management. The IAI focuses on the representation of this information within the building model.

Other resources for owners with respect to understanding and defining building information requirements are:

- **OSCRE®** (Open Standards Consortium for Real Estate, www.oscre.org). This non-profit organization is defining information requirements and standards for transaction-based scenarios, including appraisal, commercial property information exchange, and facilities management work orders.

Table 4-5 Owner's building information model.

Purpose	Type of Model Information
To support program compliance and facility management. In a typical design process, the spatial information is defined to meet program compliance and support code-checking analysis. These are critical for program compliance and use of the BIM for facility management.	Spaces and functions
To support commissioning activities such as performance specifications.	Performance specifications for HVAC and other facility operation equipment
For post-construction analysis and tracking as well as data for future forecasting.	As-built schedule and cost information
To budget and schedule maintenance.	Manufactured product information
For replacement costs and time periods and assessment information (See Coast Guard Facility Planning case study)	Financial asset management data
To plan and prepare for evacuation and other emergency crises.	Emergency information
To monitor and track progress of design, construction, or maintenance activities.	Activity status

- **Capital Facilities Information Handover Guide** (NIST and FIATECH 2006). This document defines information handover guidelines for each phase of facility delivery and the building's lifecycle and elaborates many of the information issues discussed in this section.
- **OGC** (Open Geospatial Consortium, www.opengeospatial.org). This non-profit standards organization is developing standards for geospatial data and has a specific working group looking at the integration of GIS and building model data.

4.7 LEADING THE BIM IMPLEMENTATION ON A PROJECT

Owners control the selection of design service providers, the type of procurement and delivery processes, and the overall specifications and requirements of a facility. Unfortunately, many owners accept the current status quo and may not perceive their ability to change or control how a building is delivered. They may even be unaware of the benefits that can be derived from a BIM process.

Owners cite challenges with changing standard design or construction contracts produced by governing associations such as the American Institute of Architects (AIA) or the Association of General Contractors (AGC). The federal government, for example, faces many barriers to changing contracts since these are governed by agencies and legislatures. These challenges are real and the AIA, AGC, and federal agencies such as the GSA and Army Corps of Engineers are working toward instituting the contracting methods necessary to support more collaborative and integrated methods of procurement (see Chapters 5 and 6 for a discussion of these efforts). Yet, the case studies and the various projects cited in this book demonstrate a variety of ways in which owners can work within current contractual arrangements and overcome the barriers presented in Section 4.8. Owner leadership and involvement is a prerequisite for optimal use of BIM on a project.

Owners can deliver maximum value to their organization by building internal leadership and knowledge, by selecting service providers with BIM project experience and know-how, and by educating the network of service providers and changing contractual requirements.

4.7.1 Build Internal Leadership and Knowledge

The owner-led BIM efforts in Chapter 9 (Camino Group Medical Building, One Island East Office Tower, and Coast Guard Facility Planning) share two key processes: (1) the owner first developed internal knowledge about BIM technologies; and (2) the owner dedicated key personnel to lead the effort. For example, in the Camino Medical Building project, the owner examined internal work processes intensively and identified the tools and lean methods that could deliver the facilities more efficiently. In these projects, the owners did not develop the full knowledge of how to implement various BIM applications but created a project environment where service providers could constructively apply appropriate BIM applications.

The One Island East Office Tower case study shows a slightly different approach to building that knowledge. The owner, Swire Properties Inc., had done extensive research to improve the company's ability to better deliver and manage their facilities and properties. They identified barriers related to the management of 2D information and the wide variety of project information. When they were presented with the concepts of building information modeling, they had the internal knowledge to know where to apply and leverage available BIM technologies.

The U.S. Coast Guard is building its internal knowledge and defining a roadmap for implementing BIM, as discussed in the Coast Guard Facility Planning case study (Brucker et al. 2006). This roadmap is a phased approach to implementing BIM across their organization and various facility projects. The knowledge necessary to build such a roadmap was the result of pilot projects and a significant investigation and research effort led by various groups within the U.S. Coast Guard. The roadmap includes both milestones related to specific BIM technology applications for managing project information and facility assets as well as milestones for procuring and delivering facilities using various BIM applications.

All of these cases demonstrate owners that developed knowledge through an exploration of their own internal business models and work processes related to delivering and operating facilities. They understood the inefficiencies inherent in their current work processes and how they impacted the bottom line. In so doing, key members of the staff were equipped with the knowledge and skills to lead the BIM effort.

4.7.2 Service Provider Selection

Unlike the case in global manufacturing industries, such as that of automobiles or semiconductors, no single owner organization dominates the building market. Even the largest owner organizations, which are typically government agencies, represent only a small fraction of the overall domestic and global facility markets. Consequently, efforts to standardize processes, technologies, and industry standards are far more challenging within the AEC industry than in industries with clear market leaders. With no market leaders, owners often look at what their competition is doing or to industry organizations as guides for best practice or latest technology trends. In addition, many owners build or initiate only one project and lack expertise to take a leadership position. What all owners share, though, is the control over how they select service providers and the format of project deliverables.

Owners can use a number of methods to ensure that the service providers working on their project are conversant in BIM and its related processes:

Modifying job skill requirements to include BIM related skills and expertise. For internal hires, owners can require prospective employees to have specific skills, such as 3D and knowledge of BIM or component-based design. Many organizations are now hiring employees with BIM-specific job titles such as *BIM Specialist, BIM Champion, BIM Administrator, 4D Specialist,* and *Manager, Virtual Design and Construction*. Owners may hire employees with these titles or find service providers that bear similar ones. Some examples of job skill requirements are detailed in the box titled "Examples of Job Skill Requirements." (J.E. Dunn 2007).

Including BIM-specific pre-qualification criteria. Many Requests for Proposals (RFPs) by owners include a set of pre-qualification criteria for prospective bidders. For public works projects, these are typically standard forms that all potential bidders must fill out. Commercial owners can formulate their own pre-qualification criteria. An excellent example is the qualification requirements formulated by hospital owner Sutter Health that are described in the Camino Group Medical Building case study. These include explicit requirements for experience and the ability to use 3D modeling technologies. Similarly, GM and the design-build team of Ghafari/Ideal (General Motors Production Plant case study) selected consultants and subcontractors with 3D experience and a willingness to participate in the use of 3D building models on the project.

Interviewing prospective service providers. Owners should take the time to meet designers face-to-face in the pre-qualification process, since any potential service provider can fill out a qualification form and note experience with specific tools without having project experience. One owner even prefers meeting at the designer's office to see the work environment and the types of tools and processes available in the workplace. The interview might include the following types of questions:

- **What BIM technologies does your organization use and how did you use them on previous projects?** Perhaps use a modified list of BIM application areas from Table 4-1 as a guide.
- **What organizations collaborated with you in the creation, modification, and updating of the building model?** If the question is asked to

Examples of Job Skill Requirements

- Minimum 3-4 years experience in the design and/or construction of commercial buildings structures
- BS Degree (or equivalent) in construction management, engineering, or architecture
- Demonstrated knowledge of building information modeling
- Demonstrated proficiency in one of the major BIM applications and familiarity with review tools
- Working knowledge and proficiency with any of the following: Navisworks, SketchUp, Autodesk® Architectural Desktop, and Building Systems (or specific BIM applications that your organization uses)
- Solid understanding of the design, documentation, and construction processes and the ability to communicate with field personnel

an architect, then find out if the structural engineer, contractor, or prefabricator contributed to the model and how the different organizations worked together.

- **What were the lessons learned and metrics measured on these projects with respect to the use of the model and BIM tools? And how were these incorporated into your organization?** This helps to identify evidence of learning and change within an organization.
- **How many people are familiar with BIM tools in your organization and how do you educate and train your staff?**
- **Does your organization have specific job titles and functions related to BIM (such as those listed previously)?** This indicates a clear commitment and recognition of the use of BIM in their organization.

4.7.3 Build and Educate a Qualified Network of BIM Service Providers

One of the challenges for owners is finding service providers proficient with BIM technologies within their existing network. This has led several owners to lead proactive efforts to educate potential service providers, internal and external, through workshops, conferences, seminars, and guides. Here are three examples:

Formal education. The United States General Services Administration has established a National 3D/4D BIM Program (General Services Administration 2006). Part of this effort includes educating the public and potential service providers and changing how they procure work (see next section). The educational efforts include working with BIM vendors, professional associations such as the AIA and AGC, as well as standards organizations and universities, by sponsoring seminars and workshops. Each of the ten GSA regions has a designated BIM "champion" to push adoption and application to projects in their respective regions. For example, the authors have each been invited to present BIM concepts to various owners' groups, both in the U.S. and other parts of the world. Unlike some commercial organizations, the GSA does not view its BIM expertise and knowledge as proprietary and recognizes that for the GSA to ultimately benefit from the potential of BIM, all project participants need to be conversant with BIM technologies and processes.

Informal education. Sutter Health's educational efforts are largely centered around implementing lean processes and BIM technologies on their projects (Sutter Health et al. 2006). Sutter invited service providers to attend informal workshops with presentations on lean concepts, 3D, and 4D. Sutter also supports project teams using BIM technologies to conduct similar workshops open to industry professionals. These informal workshops provide ways for professionals to share experiences and learn from others and ultimately to widen the number of service providers available to bid on future Sutter projects.

Training support. A critical part of education, beyond teaching BIM concepts and applications, is related to technical training for specific BIM tools. This often requires both technical education of BIM concepts and features for transitioning from 2D- to 3D-component parametric modeling as well as software training to learn the specific features of the BIM tools. For many service providers, the transition is costly, and it is difficult to justify initial training costs. Swire Properties (One Island East Office Tower) recognized this as a potential barrier and paid for the training of the design team to use specific BIM tools on their project.

4.7.4 Change Deliverable Requirements: Modify Contracts and Contract Language

Owners can control which BIM applications are implemented on their projects through the type of project delivery process they select and with BIM-specific contractual or RFP requirements. Changing the delivery process is often more difficult than changing the requirements. Many owners first start with changes in the RFP and contracts in three areas:

1. **Scope and detail of the model information.** This includes defining the format of project documentation and changing from 2D paper to a 3D digital model. Owners may choose to forego specific requirements pertaining to the 3D format and the types of information service providers include in the model (see Figure 4-18 and Section 4.6); or owners can provide detailed language for those requirements (see the Camino Group Medical Building case study). As owners gain experience, the nature of these requirements will better reflect the types of BIM applications an owner desires and the information that the owner team demands throughout the delivery process and subsequent operation of the facility. Table 4-6 provides a reference for the types of information an owner should consider relative to desired BIM applications.

2. **Uses of model information.** This includes specifying services more readily performed with BIM tools, such as 3D coordination, real-time review of design, frequent value engineering using cost estimating software, or energy analysis. All of these services could be performed with traditional 2D and 3D technologies; but providers using BIM tools would most likely be more competitive and capable of providing such services. For example, 3D coordination is greatly facilitated through BIM tools. Tables 4-1, 4-2, and 4-6 provide a

summary of the BIM applications owners can use as a basis to describe the services relevant to their specific projects.

3. Organization of model information. This includes project work breakdown structure and is discussed in Section 4.5.1. Many owners overlook this type of requirement. Today, CAD layer standards or Primavera activity fields are templates for how designers organize the project documentation and the building information. Similarly, owners or the project team need to establish an initial information organization structure. This may be based on the geometry of the project site (Northeast section) or the building structure (East wing, Building X). The One Island East and Camino Medical Building case studies both discuss the project work breakdown structure that the teams employed to facilitate the exchange of building information model information and project documentation. Efforts are underway to establish building model standards, such as the National Building Information Model Standard. The NBIMS should provide much-needed definition and a useful resource for owners to define the project work breakdown structure. The U.S. Coast Guard, for example, references these within their milestones.

These requirements, however, are often difficult to meet without some modifications to the fee structure and relationships between project participants or without the use of incentive plans that define the workflow and digital hand-offs between disciplines. Often, these are more difficult to define in a workflow centered on a digital model, as opposed to files and documents. Additionally, approval agencies still require 2D project documentation as do standard professional contracts (as was the case in the Penn National Parking Structure case study). Consequently, many owners maintain the traditional document and file-based deliverables (see Figure 4-19); and they insert digital 3D workflows and deliverables into the same process. That is, each discipline works independently on their scope and BIM applications and hands-off the 3D digital model at specified times.

The case studies contained in Chapter 9 provide excellent examples of how owners have modified their delivery processes to support a more collaborative real-time workflow. These modifications include:

Modified design-build delivery. The GM Production Plant project demonstrates a collaborative process achieved through modifications to the design-build delivery process. GM hired the design-build team and then participated in the selection of subcontractors and additional design consultants. The goal was to form the team as early as possible and engage them from the outset.

Table 4-6 Relationship between the BIM application area and the required scope and level of detail in the building model.

	BIM Scope									Component Property Types					
	Concept			Generic				Construction Level							
BIM Application Area	Masses	Spaces	Site	Architectural	Structural	MEP	Utilities	Prefabricated Components	Custom Components	Geometric Properties	Material Properties	Functional	Cost	Constructability (Method)	Manufacturer
Design/Construction															
Pro Forma Analysis	•	•	•							•		•	•		
Scenario Exploration	•	•	•	•						•					
Program Compliance		•		•						•		•			
Building Performance				•	•	•				•	•	•			•
Operation Simulation				•	•	•				•		•			•
Code Compliance				•	•	•				•					
Cost				•	•	•		•	•	•	•		•	•	•
Coordination				•	•	•		•	•	•					
Schedule				•	•	•				•	•			•	•
Prefabrication				•	•	•		•	•	•	•				
Post Construction/Operations															
Configuration				•	•			•	•	•		•		•	•
Commissioning				•	•	•		•	•	•					•
Facility Management		•								•		•			
Financial Asset management		•		•	•	•				•			•		
Operation Simulation				•	•	•				•		•			
Performance monitoring				•	•	•				•		•			
Built				•	•	•		•	•	•	•			•	•
Configuration (retrofit)				•	•	•		•	•	•	•	•		•	•

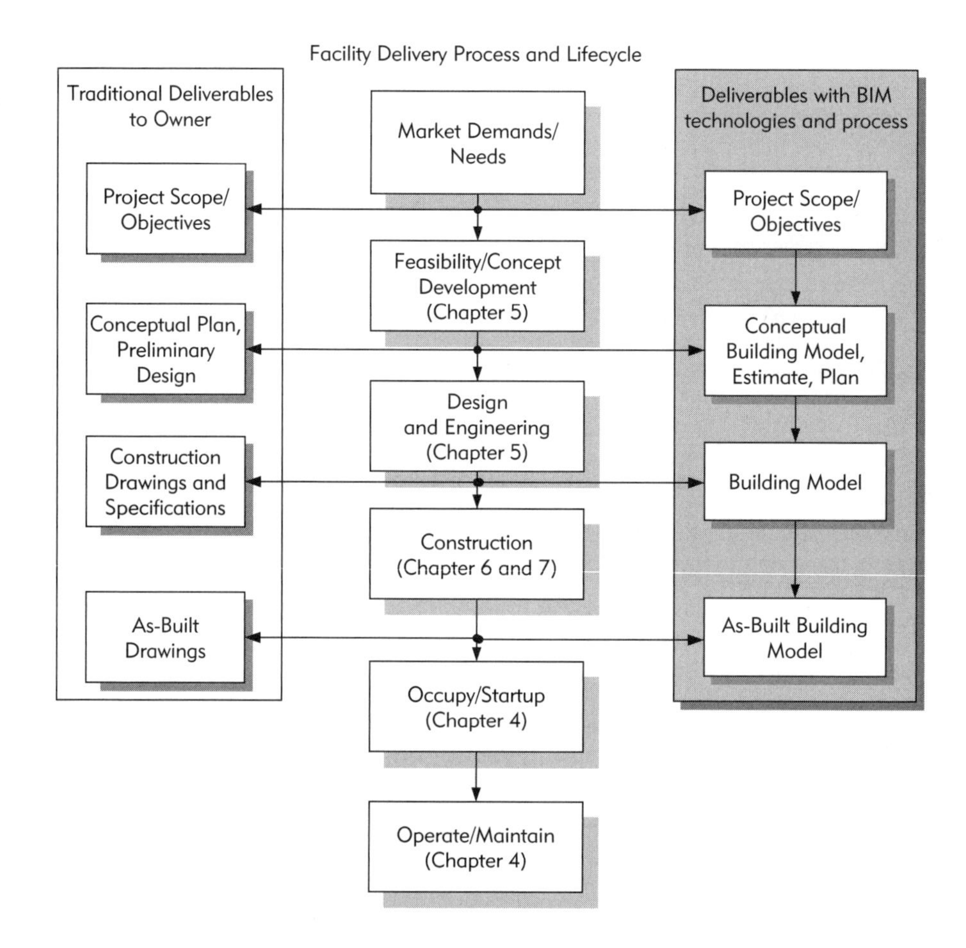

FIGURE 4-19
Typical contract deliverables as a function of the traditional design-bid-build process compared to the types of deliverables that result from a collaborative BIM-based process. Owners will need to change contracts and language to promote the use of BIM.

Performance-based contracts. Performance-based contracts or performance-based acquisition (PBA) focus on results, are typically fixed-fee, and allow service providers to deliver a facility or their services using their own best practices (Department of Defense 2000). This emphasizes the outcome, as defined by the owner, rather than intermediate milestones or deliverables. Many government agencies are moving to this approach, targeting 40%–50% of new work using this approach (General Services Administration 2007). This type of contract typically requires that the owner spend more time early in the project to define the facility requirements and structure the contracts to accommodate such an approach. This approach may seem a contradiction to the previous recommendations; but service providers utilizing BIM will most likely be more competitive and requirements can be BIM-based.

Shared Incentive plans. Performance-based contracts are often implemented with shared incentive plans. When all members collaborate on most phases of building, there is no clear partitioning of organization contributions. The Camino Group Medical Building case study provides an example of a shared incentive plan designed to distribute cost savings to the project team. It provides financial incentives based on the overall project performance and not solely on individual organizational performance. These plans are often difficult to define and implement, as the case study demonstrates. Nonetheless, shared incentive plans reward teams for collaborative performance rather than local optimization of discipline-specific performance.

4.8 BARRIERS TO IMPLEMENTING BIM: RISKS AND COMMON MYTHS

There are risks associated with any changes to work processes. Realistic and perceived barriers and changes related to implementing BIM applications on projects are no exception. These barriers fall into two categories: process barriers to the business, including legal and organizational issues that prevent BIM implementation; and technology barriers related to readiness and implementation. These are summarized below.

4.8.1 Process Barriers

The market is not ready—it's still in the innovator phase. Many owners believe that if they change the contracts to require new types of deliverables, specifically 3D or building information models, they will not receive competitive bids, limiting their potential pool of bidders and ultimately increasing the price of the project. Recent surveys do indicate that a majority of service providers are not actively using BIM technologies on their projects. These same surveys, however, indicate increasing rates of adoption and awareness of BIM concepts and applications:

- 75% of architects are using some level of 3D, with one-third of those using BIM as a construction resource and one-third using it for "intelligent modelling" (Gonchar 2007). These users preferred BIM to traditional 2D CAD, because of the visualization and value to the customer (owner), greater accuracy, coordination, and efficiency. This indicates early success and potential ongoing growth. These firms are continuing to invest and expand their use of BIM.

- Service providers would use 3D and BIM approaches more if owners (clients) asked for it (Green Building Studio 2004).

The case studies and many additional references in this book also indicate a transition from innovator to early adopter phase for design-related BIM applications. As the use of BIM increases, owners will find increasing numbers of service providers capable of using BIM.

The project is already financed and design is complete—it's not worth it to implement BIM. As a project nears construction, it's true that owners and the project team will miss valuable opportunities available through the use of BIM applications, such as conceptual estimating and program compliance. There is still ample time and opportunity, however, to implement BIM in the latter stages of design and through the early phases of construction. For example, the BIM implementation in the One Island East Office Tower case study began after construction documents were started. The BIM implementation on the Letterman Digital Arts Center, driven by the owner, began post-design and resulted in significant identification of design discrepancies and estimated cost savings of $10 million (Boryslawski 2006). The team, however, recognized that had the effort started earlier even more cost savings and benefits would have been realized.

Training costs and the learning curve are too high. Implementing new technologies such as BIM technologies is costly in terms of training and changing work processes and workflows. The dollar investment in software and hardware is typically exceeded by the training costs and initial productivity losses. This can be seen clearly in the adoption cash flow example in Chapter 7. Often, most service providers are not willing to make such an investment unless they perceive the long-term benefit to their own organization and/or if the owner subsidizes the training costs. In the One Island East Office Tower case study, the owner understood that the potential gains in productivity, quality, and asset management outweighed the initial costs and paid for the training.

Everyone must be on board to make the BIM effort worthwhile. It is often difficult to ensure that all project participants have the know-how and willingness to participate in the creation or use of the building information model. Many of the case studies in Chapter 9 demonstrate the benefits of BIM implementation without full participation but also highlight challenges with re-creating information from organizations not participating in the modeling effort. The Letterman Digital Arts Center (Boryslawski 2006) demonstrates how an owner can mitigate the risk of non-participation. The owner hired a third-party to manage the BIM effort and recreate information as required. As the project progressed, the third-party was able to identify organizations with BIM modeling capabilities and have them participate in the process. The owner supported the effort but did not mandate full participation in the BIM effort.

Too many legal barriers exist and they are too costly to overcome. Contractual and legal changes are required on several fronts to facilitate the use of BIM and more collaborative project teams. Even the digital exchange of project information is sometimes difficult today, and teams are often forced to exchange only paper drawings and rely on old-fashioned contracts. Public institutions face even greater challenges, since they are often governed by laws that take considerable time to change. Nonetheless, several government agencies and private companies have overcome these barriers and are working toward contract language that not only changes the nature of how information is exchanged within the project team but the liability and risks associated with a more collaborative effort. The Camino Medical Building, GM Production Plant, and Federal Courthouse case studies are examples.

The primary challenge is the assignment of responsibility and risk. BIM implementation centralizes information that is "broadly accessible," depends on constant updating, and subjects designers to increased potential liability (Ashcraft 2006). The legal profession recognizes these barriers and the necessary risk-allocation changes that need to take place. This is a real barrier, one that will continue to persist and will depend on professional organizations such as the AIA and AGC to revise standard contracts and/or owners to revise their own contract terms.

Issues of model ownership and management will be too demanding on owner resources. BIM potentially requires insight across multiple organizations and aspects of the project. Typically, a construction manager (CM) provides the oversight by managing communication and reviewing project documentation. The CM also oversees that the process is aligned with specific deliverables and milestones. With BIM, issue discovery and problem identification occur early and more frequently, enabling teams to resolve issues early; but this often requires owner input, which should be seen as a benefit and not a drawback. The current slack in the delivery process is significantly reduced, demanding more direct owner involvement. The process is more fluid and interactive. Owner requested changes will become less transparent and the impacts of these changes will demand ongoing participation. Managing this process and the related management of the model will become critical to the project. Owners need to establish clear roles and responsibilities and methods to communicate with the project team and ensure that an owner representative is available as-needed.

4.8.2 Technology Risks and Barriers

Technology is ready for single-discipline design but not integrated design. It is true that two–five years ago the creation of an integrated model required extensive effort on the part of a project team and dedicated technical expertise to support that integration. Today, many of the BIM design tools reviewed in Chapter 2 have matured and provide integration capabilities between several disciplines at the generic object level (see Figure 4-18). As the scope of the model and number and types of building components increase, however, performance issues also increase. Thus, most project teams choose to use model review tools to support integration tasks, such as coordination, schedule simulation, and operation simulation. The Camino Medical Building and GM Production Plant projects, for example, used the Navisworks model review tool to perform clash detection and design coordination. Currently, BIM design environments are typically good for one or two-discipline integration. The integration of construction-level detail is more difficult, and model review tools are the best solution to achieve this.

A greater barrier is related to work process and model management. Integrating multiple disciplines requires multi-user access to the building information model. This does require technical expertise, establishment of protocols to manage updates and edits of the model, and establishing a network and server to store and access the model. It also provides an excellent context for new users to learn from more experienced ones.

Owners should perform audits with their project teams to determine the type of integration and analysis capabilities that are desired and currently available and prioritize accordingly. Full integration is possible but does require expertise, planning, and proper selection of BIM tools.

Standards are not yet defined or widely adopted—so we should wait. Chapter 3 discusses the various standards efforts, such as IFCs and the National BIM Standards, which will greatly enhance interoperability and widespread BIM implementation. In practice, however, these formats are rarely used, and most organizations use proprietary formats for model exchange. For many owners, this poses a risk to the short and long term investments in any building information modelling effort. There are owner-specific standardization efforts related to real estate transactions and facility management, as discussed previously; however, the case studies in this book, with the exception of the Jackson Courthouse and the GM Flint plant, demonstrate that a variety of successful BIM implementations have been achieved without reliance on these standards; and it is not a barrier to implementation.

4.9 GUIDELINES AND ISSUES FOR OWNERS TO CONSIDER WHEN ADOPTING BIM

Adopting BIM alone will not necessarily lead to project success. BIM is a set of technologies and evolving work processes that must be supported by the team, the management, and a cooperative owner. BIM will not replace excellent management, a good project team, or a respectful work culture. Here are some key factors an owner should consider when adopting BIM.

Perform a pilot project with a short time frame, small qualified team, and a clear goal. The initial effort should use either internal resources or trusted service providers that your organization has worked with. The more knowledge an owner builds with respect to the implementation and application of BIM, the more likely future efforts will succeed, as the owner develops core competencies to identify and select qualified service providers and forge cooperative teams.

Do a prototype dry run. When doing a pilot project, it's always best to do a dry run and make sure the tools and processes are in place to succeed. This may be as simple as giving the designer a small design task that showcases the desired BIM applications. For example, the owner can ask the design team to design a conference room for twenty people, with specific targets for budget and energy consumption. The deliverable should include a building information model (or models to reflect two or three options) and the related energy and cost analysis. This is an example of a design task that is achievable in one or two days. The architect can build the model and work with an MEP engineer and estimator to produce a set of prototype results. This requires that the project participants work out the kinks in the process, so to speak, and also allows the owner to provide guidance regarding the types of information and formats of presentation that provide clear, valuable, and rapid feedback.

Focus on clear business goals. While this chapter cites many different benefits, no single project has yet achieved all of these benefits. In many cases, the owner started with a specific problem or goal and succeeded. The GSA's pilot project efforts (Dakan 2006), for example, each involved one type of BIM application for nine different projects. The application areas included energy analysis, space planning, Lidar scanning to collect accurate as-built data, and 4D simulation. The success in meeting focused and manageable goals led to expanded use of multiple BIM applications on projects such as the Federal Courthouse case study.

Establish metrics to assess progress. Metrics are critical to assessing the implementation of new processes and technologies. Many of the case studies include project metrics, such as reduced change orders or rework, variance from baseline schedule or baseline cost, reduction in typical square footage cost. There are several excellent sources for metrics or goals relevant to specific owner organizations or projects, including:

- **Construction Users Roundtable** (CURT). This owner-led group holds workshops and conferences and issues several publications on their Web site (www.curt.org) for identifying key project and performance metrics.
- **CIFE Working Paper on Virtual Design and Construction (Kunz and Fischer, 2007).** This paper documents specific types of metrics and goals along with case study examples.

Also, see Section 5.4.1 for the development of assessment metrics related to design.

Participate in the BIM effort. An owner's participation is a key factor of project success, because the owner is in the best position to lead a project team to collaborate in ways that exploit BIM to its fullest benefit. All of the case studies in which owners took leadership roles—General Motors Production Plant, Federal Courthouse, Camino Group Medical Building, One Island East Office Tower, and Coast Guard Facility Planning—demonstrate the value of the owner's participation in proactively leading the BIM implementation. They also highlight the benefits of ongoing involvement in that process. BIM applications, such as those for BIM design review, enable owners to better participate and more easily provide the necessary feedback. The participation and leadership of owners is critical to the success of the collaborative project teams that exploit BIM.

Chapter 4 Discussion Questions

1. List three types of procurement methods and how these methods do or do not support the use of BIM technologies and processes.
2. Imagine you are an owner embarking on a new project and have attended several workshops discussing the benefits of BIM. What steps would you take to identify whether you should support and promote the use of BIM on your project?
3. If the owner did decide to adopt BIM, what types of decisions would be needed to ensure the project team's success in using BIM at each stage of the building life-cycle?
4. With respect to the application and benefits of BIM technologies and processes, what are the key differences between an owner who builds to sell the facility vs. an owner who builds to operate?
5. Imagine you are an owner developing a contract to procure a project using a collaborative approach through the use of BIM. What are some of the key provisions that the contract should include to promote team collaboration, the use of BIM, and project success?
6. List and discuss three risks associated with using BIM and how they can be mitigated.
7. List two or three processes or project factors that influence the success of BIM implementation.
8. Imagine you are an owner building your first project and plan to own and occupy the facility for the next 15–20 years. You do not plan to build another facility and will outsource its design and construction. Should you consider BIM? If so, list two or three reasons why BIM would benefit your organization, and describe what steps you might take to achieve the benefits you cite. If you believe that BIM would not benefit your project, explain why.
9. List three market trends that are influencing the adoption and use of BIM and how BIM enables owners to respond to those market trends.

CHAPTER 5

BIM for Architects and Engineers

5.0 EXECUTIVE SUMMARY

Building Information Modeling can be considered an epochal transition in design practice. Unlike CADD, which primarily automates aspects of traditional drawing production, BIM is a paradigm change. By partially automating the detailing of construction level building models, BIM redistributes the distribution of effort, placing more emphasis on conceptual design. Other direct benefits include easy methods guaranteeing consistency across all drawings and reports, automating spatial interference checking, providing a strong base for interfacing analysis/simulation/cost applications and enhancing visualization at all scales and phases of the project.

This chapter examines the impact of BIM on design from four viewpoints:

- *Conceptual design* typically includes resolution of siting, building orientation and massing, satisfaction of the building program, addressing sustainability and energy issues, construction and possibly operating costs and sometimes issues requiring design innovation. BIM potentially supports much greater integration and feedback for early design decisions.
- *The integration of engineering services*; BIM supports new information workflows and integrates them more closely with existing simulation and analysis tools used by consultants.
- *Construction level modeling* includes detailing, specifications and cost estimation. This is the base strength of BIM.

- *Design-construction integration* addressing the scope of innovation that can potentially be achieved through a collaborative design-construction process, such as with the design-build procurement model.

Different design projects can be categorized according to the level of information development required for realizing them, ranging from predictable franchise-type buildings to experimental architecture. The information development concept facilitates distinguishing the varied processes and tools required for designing and constructing all varieties of buildings.

This chapter also addresses issues of adoption of BIM into practice, such as: the evolutionary steps to replace 2D drawings with 3D digital models; automated drawing and document preparation; managing the level of detail within building models; the development and management of libraries of components and assemblies; and new means for integrating specifications and cost estimation. The chapter concludes with a review of the practical concerns that design firms face when attempting to implement BIM, including: the selection and evaluation of BIM authoring tools; training; office preparation; initiating a BIM project; and planning ahead for the new roles and services that a BIM-based design firm will evolve toward.

Building Information Modeling provides major challenges and opportunities to design firms. While there are clear economic benefits for fabricators to employ 3D parametric modeling and interoperability for use in automation and other productivity benefits, the direct benefits to design are more difficult to quantify. However, BIM integration with analysis and simulation, and improving the quality of design, especially in its early stages, provides value to owners equivalent or even greater than any savings in construction costs; design quality is long lasting, offering benefits over the life of a building. Developing these new capabilities and services, then employing them to gain recognition and serve as a basis for new and repeat work, is a path encouraged for all design firms, large and small. BIM increases the value that designers can offer to clients and the public.

5.1 INTRODUCTION

In 1452, early Renaissance architect Leon Battista Alberti distinguished architectural design from construction by proposing that the essence of design lay in the thought processes associated with conveying lines on paper. His goal was to differentiate the intellectual task of design from the craft of construction. Prior to Alberti, in the first century BC, Vitruvius discussed the value inherent in using plans, elevations, and perspectives to convey design intent in what is considered to be the first treatise on architecture (Morgan 1960; Alberti 1987). Throughout architectural history, drawing has remained the dominant mode of representation. Even now, contemporary authors often critique how different architects use drawings and sketches to enhance their thinking and creative work (Robbins 1994). The extent of this time-honored tradition is further apparent in the way that computers were first adopted as an aid for automating certain aspects of the design-to-construction process, which was succinctly expressed in the original meaning of the term CADD – computer-aided design and drafting.

Because of this history, what is being called Building Information Modeling (BIM) can be considered revolutionary in the way it transforms architectural thinking by replacing drawings with a new foundation for representing design and for aiding communication, construction, and archiving based on 3D digital models.

BIM can also be thought of as the conceptual equivalent of crafting scale models of buildings within a virtual digital 3D computer-based environment. Unlike physical models, virtual models can be accurate at any scale; they are digitally readable and writable; and they can be automatically detailed and analyzed in ways that are not possible with physical scale models. They may contain information that is not expressible in a physical model, such as analyses for structures and energy and in-place cost codes for interfacing with a variety of other software tools.

BIM facilitates interactions between different design tools by improving the availability of information for those tools. It supports feedback from analyses and simulations into the design development process, and these changes will, in turn, affect the way designers think and the processes they undertake. BIM also facilitates the integration of construction and fabrication thinking into the building model, encouraging collaboration beyond what is involved with drawings. As a result, BIM will likely redistribute the time and effort designers spend in different phases of design.

Outline of traditional architectural services

Feasibility studies

Non-spatial quantitative and textual project specification, dealing primarily with cash flows, function or income generation; associates areas and required equipment; includes initial cost estimation; may overlap and iterate with pre-design; may overlap and iterate with production or economic planning.

Pre-design

Fixes space and functionality requirements, phasing and possible expansion requirements; site and context issues; building code and zoning constraints; may also include updated cost estimation based on added information.

(Continued)

Outline of traditional architectural services *(Continued)*

Schematic Design (SD)

Preliminary project design with building plans, showing how the pre-design program is realized; massing model of building shape and early rendering of concept; identifies candidate materials and finishes; identifies all building subsystems by system type.

Design Development (DD)

Detailed floor plans including all major construction systems (walls, facades, floor and all systems: structural, foundation, lighting, mechanical, electrical, communication and safety, acoustic, etc.) with general details; materials and their finishes; site drainage, site systems and landscaping.

Construction Detailing (CD)

Detailed plans for demolition, site preparation, grading, specification of systems and materials; member and component sizing and connection specifications for various systems; test and acceptance criteria for major systems; all chaises, block-outs, and connections required for intersystem integration.

Construction Review

Coordination of details, reviews of layouts, material selection and review; changes as required when built conditions are not as expected, or due to errors.

This chapter addresses how BIM influences the entire range of design activities, from the initial stages of project development, dealing with feasibility and concept design, to design development and construction detailing. In a narrow sense, it addresses building design services however this role is realized: carried out by autonomous architectural or engineering firms; as either part of a large integrated architecture/engineering (AE) firm or through a development corporation with internal design services. Within these varied organizational structures, a wide variety of contractual and organizational arrangements may be found. This chapter also introduces some of the new roles that will arise with this technology and considers the new needs and practices that BIM supports.

5.2 SCOPE OF DESIGN SERVICES

Design is the activity where a major part of the information about a project is initially defined and the documentation structure is laid out for information added in later phases. A summary of the services provided within the traditional phases of design is shown in Figure 5-1. The traditional contract for architectural services suggests a payment schedule (and thus the distribution of effort) to be 15% for schematic design, 30% for design development, and 55% for construction documents (AIA 1994). This distribution reflects the weight traditionally required for the production of construction drawings.

Due to its ability to automate standard forms of detailing, BIM significantly reduces the amount of time required for producing construction documents. Figure 5-1 illustrates the relationship between design effort and time, indicating how effort is traditionally distributed (line 3) and how it can be re-distributed as a result of BIM (line 4). This revision aligns effort more closely with the value of decisions made during the design and build process (line 1) and the growth in the cost of making changes within the project lifetime (line 2). The chart emphasizes the impact of early design decisions on the overall functionality,

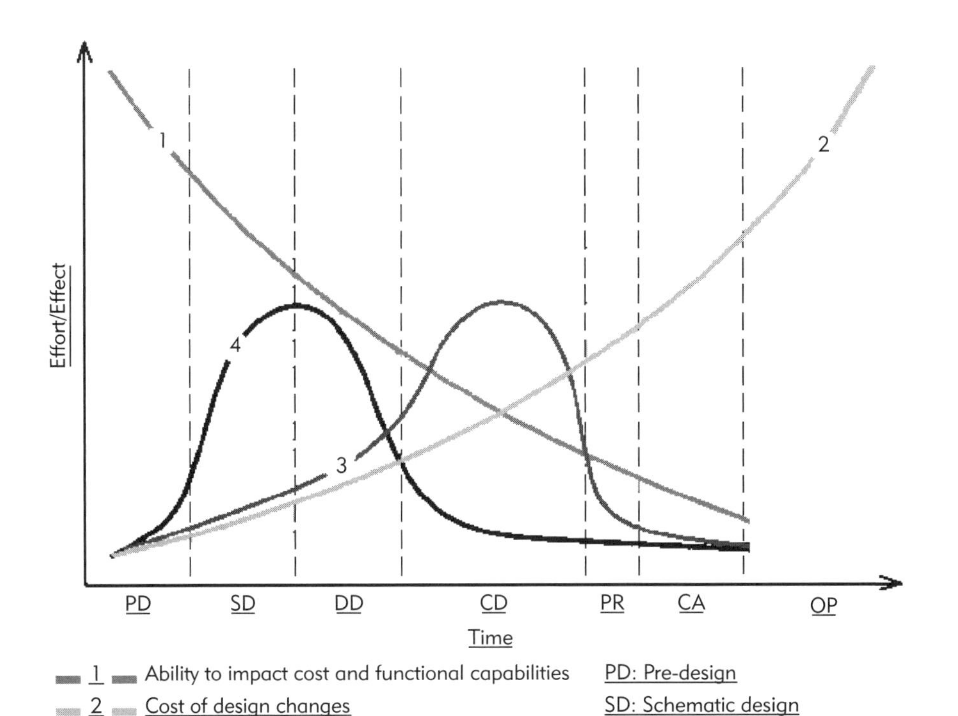

FIGURE 5-1
Value added, cost of changes and current compensation distribution for design services (CURT 2007).

costs and benefits of a building project. The fee structure in some projects is already changing to reflect the value of decisions made during schematic design and the decreased effort required for producing construction documents.

5.2.1 The Concept of Information Development

Various building projects begin at different levels of information development, including definition of the building's function, style, and method of construction. At the low end of the information development spectrum are franchise buildings, including warehouses and roadside service stations, often called "big boxes," and other buildings with well defined functional properties and fixed building character. With these, minimal information development is required, and the client often knows ahead of time what is going to be delivered. Knowledge of the expected outcome is prescribed, including design detailing, construction methods, and environmental performance analyses.

At the other end of the spectrum—involving the highest level of information development—are owners interested in developing facilities for new social functions or attempting to re-think existing functions, such as combining an airport with a seaport, an undersea hotel, or a theater for experimental multimedia performances. Other instances of high information development involve agreements between the owner and designer to explore the application of non-standard materials, structural systems or environmental controls. Two of the case studies in Chapter 9—the Beijing Aquatic Center and the San Francisco Federal Building—are examples of high information development projects. Their respective functions led to the development of new and untried systems that were generated from first principle analyses. For some time, progressive architecture firms and students have expressed an interest in fabricating buildings using non-standard materials and forms, following the inspiration of Frank Gehry, Sir Norman Foster, and others. These projects involve higher levels of information development in the short term, until such cladding or construction practices become part of the arsenal of standard and conventional practices.

In practice, most buildings are functionally and stylistically a composition of well-understood social functions, with some variations in detail practices and procedures, styles and image. On the construction side, most architecture conforms to well-understood construction practices, with only occasional innovations regarding materials, fabrication and onsite assembly.

The scope of design services, considered from the level of information development, can be simple or elaborate, depending on the needs and intention of the client as well as the level of sophistication within the project delivery team, as outlined above. Traditionally, the level of information development is conveyed in the scope of contracts that define architectural services, as outlined

Range of Technical Services Used During Design

- Financial and cash flow analyses
- Analysis of primary functions including services in hospitals, rest homes, airports, restaurants, convention centers, parking garages, theater complexes, etc.
- Site planning, including parking, drainage, roadways
- Design and analysis/simulation of all building systems, including:
    - structure
    - mechanical and air handling systems
    - emergency alarm/control systems
    - lighting
    - acoustics
    - curtainwall systems
    - energy conservation and air quality
    - vertical circulation
    - security
- Cost estimation
- Accessibility assessment
- Landscaping, fountains and planting
- External building cleaning and maintenance
- External lighting and signage

in the box on pages 151 and 152 ("Outline of traditional architectural services".) In projects with well-defined data for function and construction, the initial phase may be abbreviated or omitted, with design development (DD) and construction detailing (CD) being the main tasks. In other instances, feasibility, pre-design, and schematic design (SD) may be of critical importance, where the major costs and functional benefits are determined. With such projects, steeper fee structures are justified.

5.2.2 Technical Collaborations

Design services potentially involve a great range of technical issues, involving various building systems, different building types, and the specialty services required of them, such as equipment for laboratories or artificial materials used on the playing fields of stadiums. A sample of these typical services is summarized in the box titled "Range of Technical Services Used During Design" listed on this page. While some of the services listed in this box are carried out by the primary design firm, more often, they are undertaken by external consultants. For example, in a study of collaborative architectural services (Eastman et al. 1998) with the firm of John Portman and Associates in Atlanta, a large building project in Shanghai was found to include over twenty-eight different types of consultants.

From this overview, we can appreciate what is well-known by most large architectural firms but less understood by many clients, developers, contractors, and even small design firms—that building design is a broad and collaborative undertaking, involving a wide range of issues that require technical detailing and focused expertise. It is in this broad context that BIM must operate, by both enhancing quality and coordination. We can also see from the diversity of contributors for this book that the main challenge in adopting BIM technology is getting all parties of a design project to agree on new methods of working, and for documenting and communicating their work. In the end, everyone must adapt to the practices associated with this new way of doing business. This point is emphasized—implicitly and explicitly—in the case studies in Chapter 9.

Today, an additional set of collaborators is frequently becoming involved in the early design stages. These are contractors and fabricators who are candidates for executing the project downstream, who form the basis of a design-build effort or other type of teaming arrangements. These experts address constructibility, procurement, scheduling and similar issues.

5.3 BIM USE IN DESIGN PROCESSES

The two technological foundations of Building Information Modeling reviewed in Chapters 2 and 3—parametric design tools and interoperability—offer many process improvements and information enhancements within traditional design practices. These benefits span all phases of design. Some of the potential uses and benefits of BIM have yet to be conceived, but several tracks of development have evolved far enough to demonstrate significant payoffs. Here, we consider the role and process of design from four of those viewpoints which apply in varying degrees to different projects, depending on their level of information development.

The first viewpoint addresses **conceptual design**, as it is commonly conceived. This involves generating the basic building plan, its massing and general appearance, determining the building's placement and orientation on the site, its structure and how the project will realize the basic building program. These are the typical and traditional unknowns of most building projects. BIM can have a huge impact in strengthening the quality of decisions made in this phase, based on quick feedback. These initial decisions can be much better informed regarding such aspects as the building program, construction and operating cost constraints, and increasingly, environmental considerations.

A second viewpoint addresses the **use of BIM for design and analysis** of building systems. Analysis in this respect may be thought of as operations to measure the fluctuations of physical parameters that can be expected in the real building. Analysis covers many functional aspects of a building's performance, such as structural integrity, temperature control, ventilation, lighting, circulation, acoustics, energy distribution and consumption, water supply and waste disposal, all under varied use or external loads. This viewpoint concerns collaboration with the various professions involved supported by integration of the analysis software those professions utilize. They in turn produce the design layouts that are used to plan and coordinate the various systems. In exceptional cases involving high-level information development, the early design process can involve experimental analyses of structure, environmental controls, construction methods, use of new materials or systems, detailed analyses of user processes, or other technical aspects of a building project. There isn't one set of problems in need of analysis. This last viewpoint demands intense collaboration among a team of specialists, requiring a mix of configurable tools that, when combined, form a design workbench.

The third viewpoint is the conventional BIM viewpoint of its **use in developing construction-level information**. Building modeling software includes placement and composition rules that can expedite the generation of standard or pre-defined construction documentation. This provides the option of both speeding up the process and enhancing quality. Construction modeling is a basic strength of current BIM authoring tools. Today, the primary product of this phase is construction documents. But this is changing. In the future, the building model itself will serve as the legal basis for construction documentation.

The fourth and last viewpoint involves **design and construction integration**. At the more obvious level, this view applies to well-integrated design-build processes in conventional construction, facilitating fast, efficient construction of the building after design, or indeed in parallel with it. It emphasizes the growing use of a building model for direct use in construction and generating initial input for fabrication-level modeling. In its more ambitious aspect, this view involves working out non-standard fabrication procedures, working from carefully developed detailed design models supporting 'design for fabrication.'

In the sections that follow, these viewpoints are described in greater detail. In lieu of the milestones in traditional design contracts, we consider these four broad areas with an understanding of the fluidity of changes inherent in current design development sequences. We also address a number of practical issues: model-based drawing and document preparation; development and management of libraries; integration of specifications and details for cost estimation. The chapter concludes with some practical issues of design practice: selecting a BIM authoring tool, training and introduction into projects, and issues of staffing.

5.3.1 Concept Design and Preliminary Analyses

As the previous sections indicated, major decisions regarding the value, performance and costs of a building are made during concept design. Thus the potential benefits that design firms can offer to clients will increasingly focus on the differentiating services they can offer within the concept design phase. Concept design based on early analysis feedback is especially important for projects involving medium or high levels of information development. We expect this to become an increasingly important area of design firm differentiation.

Today, a growing number of easy-to-use tools are available that were conceived not for heavy-duty production design but as light-weight, intuitive tools that are relatively easy to use, so much so, that they become invisible aspects of the designer's thinking process. Each provides functionality that is important for preliminary design. Some tools focus on quick 3D sketching and form generation, such as Google SketchUp® (Google 2007), or for larger and more geometrically complex projects, form·Z. Other software programs support layout according to a building program, such as Facility Composer and Trelligence (Trelligence 2007) or simple layouts and interfaces for energy, lighting, and other forms of analysis relevant to conceptual designs, such as EcoTect (Square One Research 2007), IES and Green Building Studio. Another important area for conceptual design is cost assessment, which is offered by Dprofiler (Beck Technology 2007) plus others. Unfortunately, no one of these programs provides the broad spectrum of functionality needed for general concept design, and smooth interoperability between these tools is not yet a reality. In practice, most users rely on one of the aforementioned software tools. Of these, few are able to interface easily and efficiently with existing BIM authoring tools. A quick review of these various secondary tools follows.

3D Sketching Tools

The earliest available conceptual design tools were those for 3D sketching. form·Z can be considered the grandparent of this category, having served as a 3D design tool since 1989. It supports powerful 3D solid and surface modeling capabilities that can represent any form imaginable, for architects and product designers. In its early days, it was considered the most intuitive solid modeler available, but with growth in functionality and new competitors, form·Z is a trade-off between powerful freeform surface modeling and intuitiveness. An example project developed using form·Z, for a high speed train station in Pusan, Korea, is shown in Figure 5-2. Other products with similar functionality include Rhino and Maxon, which also emphasize freeform capabilities.

FIGURE 5-2 Competition models using form·Z. *Courtesy of View by View, San Francisco.*

Other applications focus on the quick sketch 3D layout of building spaces and envelopes. Products of this type include SketchUp® and the now defunct Architectural Studio® from Autodesk. These tools support quick generation of schematic designs and rendering in a manner conveying the character of the proposed space and building shell. An example project and multiple views generated in SketchUp are shown in Figure 5-3.

These applications support easy sketch definitions of the geometric forms used in architecture but typically do not carry object types or properties other than material color and/or transparency, which are only useful for rendering. As a result, they do not interoperate well with other concept design tools, such as those described below.

Space Planning

A building's requirements often center on a set of spatial needs defined by the program, describing the number and types of spaces that the client expects, their respective square footages, the environmental services they require, and in some cases the materials and surfaces desired. Critical relations between these spaces are further detailed according to organizational practices; for example, access required between different wards and treatment facilities within a hospital.

Space planning involves organizing the spatial needs defined by the client and expanding them to include storage, support, mechanical, and other ancillary spaces. Typically, these applications depict the space program in two forms, as a series of line items on a spreadsheet and as a block diagram layout of the proposed floor plan. Software products that incorporate this functionality

FIGURE 5-3
A massing design study done in SketchUp.

include Visio Space Planner®, Vectorworks® Space Planning Tool, Trelligence®, and the Army Corps of Engineers' Facility Composer. Autodesk Revit® and ArchiCAD®'s links with Trelligence provide similar spreadsheet capabilities within these BIM design tools. A composite of three screen shots from Facility Composer is shown in Figure 5-4. The figure shows the program spreadsheet of space types, indicating what has been laid out (actual) and what remains to be allocated (available) by building and by story. A plan of one story and the schematic massing of the entire building are also shown in Figure 5-4.

These applications explicitly represent the spaces within a building, with or without development of space enclosures. A spreadsheet shows the current layout allocation in relation to the space program requirements. All current space planning programs support massing based on a blank sheet of paper, without envelope constraints; thus none appear to support generation of layouts within the confines of a given building shell, or within a form that captures a target image. These tools provide another set of important but incomplete schematic design capabilities.

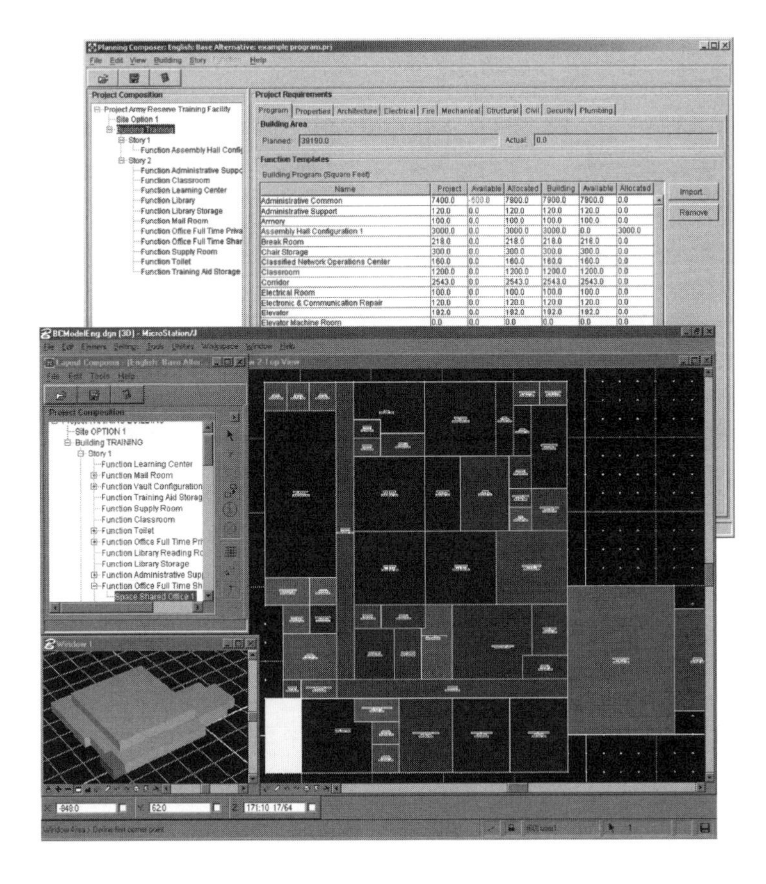

FIGURE 5-4 **Like most space planning systems, Facility Composer supports development of massing diagrams based on a program spreadsheet and compares the current layout in relation to the given program.**

Environmental Analysis

The third type of application and interface focuses on energy and environmental aspects of a candidate design. IES Virtual Building®, Ecotect® and Green Building Studio are three products in this area. Some images from Ecotect, providing performance feedback to the building model, are shown in Figure 5-5. These products operate through a mixture of a simplified building model and also with direct translators to existing analysis/simulation applications. Ecotect and IES have their own building model with form generation and editing capabilities and have some of the functionality of sketching applications.

With drawing systems, the preparation of datasets to run an application was so cumbersome, that if applied at all, they were relegated to the later stages of design. With BIM, the interfaces to applications can be automated, allowing almost real-time feedback on design actions. These environmental analysis applications incorporate interfaces to a set of energy, artificial and natural lighting analyses, fire egress and other assessment applications, allowing quick analysis of schematic-level designs. gbXML provides an interface from existing BIM design tools to its set of analysis applications. The different interfaces supported by Ecotect, IES and gbXML are shown in Table 5-1.

These environmental analysis tools offer insight into the behaviors associated with a given design, and provide an early assessment of gross energy, lighting use, as well as estimated operating costs. Until now, such performances relied mainly on designer experience and rules of thumb. These application suites offer

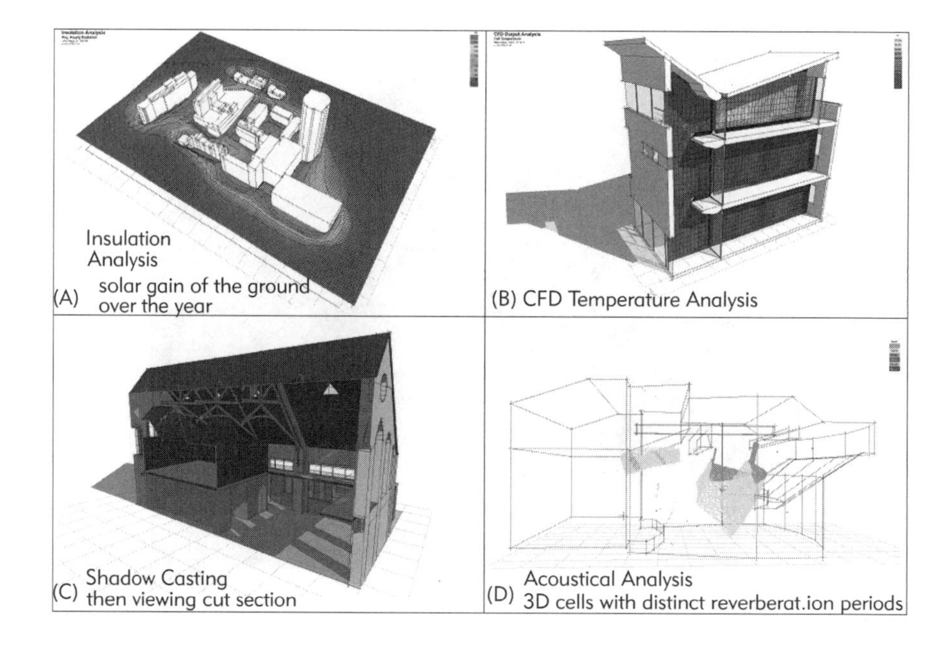

FIGURE 5-5 Images from Ecotect, showing A) solar heat gain between buildings, B) computational fluid dynamics of thermal flows within a building, C) sunlight inside a building and D) acoustics analysis. (See color insert for full color picture).

Table 5-1 Analyses supported from application suite's building model.

Ecotect - own building model plus IFC input	
DAYSIM	lighting simulator
Radiance	lighting simulator
CIBSE	energy analysis
Energy+	energy analysis
	solar radiation analysis
	reverberation time acoustic analysis
NIST-FDS, Fluent and WinAir4	general interface for multiple computational fluid dynamic analyses
IES – own building model plus direct link with Autodesk Revit®	
ApacheCalc	heat loss and gain
ApacheLoads	heating and cooling loads
ApacheSim	dynamic thermal simulation
ApacheHVAC	HVAC plant simulation
SunCast	sun shading
MacroFlo	simulates natural ventilation and mixed mode systems
MicroFlo	interior computational fluid dynamics application
Deft	value engineering
CostPlan	capital cost estimates
LifeCycle	estimates lifetime operating costs
IndusPro	ductwork layout and sizing
PiscesPro	pipework systems
Simulex	building evacuation
Lisi	elevator simulation
gbXML – XML link from Autodesk Revit®, Bentley Architecture and ArchiCAD®	
DOE-2	energy simulation
Energy+	energy simulation
Trane2000	equipment simulation
	building product information

only limited compatibility with existing BIM design tools (as reviewed in Chapter 2). In this regard, gbXML export interfaces are available within ArchiCAD®, Bentley Architect and Revit®. Ecotect has IFC interfaces with ArchiCAD® and Digital Project. IES has a direct interface with Revit®.

Environmental analysis tools also require significant amounts of non-project specific information, including details that may affect incident sunlight and any objects or effects that may restrict sunlight or views of existing structures, such as geographic location, climatic conditions, structures, or topography. This information is not typically carried within BIM design tools but by secondary analysis tools. These distributed datasets often introduce management-level problems, such as determining which analysis run gave which results and based on which version of the design. In this respect, repositories can play an important role (see Chapter 3).

More broadly, the exchange formats supported by all existing conceptual design tools are shown in Table 5-2. At the moment, most of the information generated by these tools must be regenerated in the transfer to a BIM authoring tool.

Table 5-2 Concept design applications and the exchange formats they support.

	Import formats							Export formats						
Application	3DS	DWG	DXF	VRML	IGES	SAT	**Others**	3DS	DWG	DXF	VRML	IGES	SAT	**Others**
Massing and Sketching														
SketchUp	•	•	•					•	•	•	•			obj, Epix
form.Z	•	•	•	•	•		act, obj, rib, w3d, stp, stl, 3dmf	•	•	•	•	•	•	act, Epix, ArtL, obj, rib, w3d, stp, stl, 3dm
Rhino	•	•	•		•		obj, stp, stl, 3dm	•	•	•	•	•	•	3dm, stp, obj, x_t, stl, xgl
Maxon	•		•	•			direct3d, stl, cct	•		•	•			direct 3d, stl, cct
Space Planning														
MS Visio		•	•				wmf, xls							xls
Vectorworks	•	•	•	•	•	•						•	•	stl, epix
Facility Composer					•		IFC, stp					•		IFC, stp
Trelligence									•	•				gml, csv
Environmental Analysis														
IES		•	•				dgn, gbXML, Revit®							Revit®
Ecotect			•				IFC			•	•			
GreenBuilding Studio		•					rvt, dgn, XML				•			

Other Issues of Conceptual Design

For completing what has traditionally been schematic design, two other aspects of a design must also be defined: site development (including existing conditions) and typological identification of all building systems. Site development involves building placement, elevation, definition of all major site and ground contour changes and enhancements, and the general scope of site development. Use of the site is often an important aspect of an overall conceptual design. Some BIM design tools support site planning, as listed in Table 2-1, and some environmental analysis tools support site as well as exterior solar and wind studies. To our knowledge, outside space functions can currently be dealt with in space planning tools using block spaces to make site layout allocations.

Conceptual design usually involves identifying the 'type' for each of the building systems, including structural, exterior envelope, energy and HVAC, lighting, and vertical circulation. This information is needed for generating initial cost estimates at an early stage; verifying that the project is within economic scope and satisfying the program. Most existing environmental analysis applications identify HVAC system types but not structural or other systems.

Traditionally, cost estimation for small projects at the conceptual stage involved 'back-of-envelope' calculations based on a single gross unit measure, such as square footage or the number of rooms. Validation of the proposed design concept for larger projects usually involves a detailed cost estimate prepared by a consultant. With the proper setup, BIM supports rapid generation of cost estimates, so it is now practical to undertake cost estimates throughout the concept exploration and development process. While at this level of design development only rough estimates of construction cost can be generated, the information provided can inform the designer of potential issues early, or alternatively provide confidence that the proposed design can be developed within range of the project's budget. The issues associated with generating almost real-time cost estimates are reviewed in Sections 5.3.2 and 6.6.

The only software currently available for representing all building systems and supporting concept-level cost estimation is DProfiler, which enables rapid composition of a concept model and generation of a cost estimate. It can only be used for certain predefined building types. Like the energy-related concept tools, the DProfiler building model is distinct and supports only DXF/DWG export to other applications, such as the concept design tools listed above or the BIM design tools. DProfiler is described more fully in the Hillwood Commercial Project case study in Chapter 9.

Another aspect of understanding the building context is in capturing as-built conditions. This is a critical issue for retrofit work and remodeling.

New surveying techniques, based on laser scanning with point clouds, offer a valuable new technique to capture as-built conditions. These are discussed in Chapter 8.

Concept Design Summary

Concept design tools must balance the need to support the intuitive and creative thinking process when a basic design scheme is first being defined and explored with the ability to provide fast assessment and feedback based on a variety of simulation and analysis tools, allowing more informed design. Unfortunately, each of these tools only does part of the overall task, requiring translation between them and later with the major BIM tools discussed in Chapter 2.

None of the tools available today support the full scope of conceptual design services. They require users to either gain and maintain competency in a number of different software programs, each with different user interfaces, or fill the gaps by relying on manual paper-based modes of assessment (or more likely, intuition). Data exchange and workflows between these applications are also limited, as shown in Table 5-2. In large firms, the various tasks associated with schematic design (and their supporting applications) may be split among multiple people using custom API-based interfaces. Small firms are likely to select one of these tools and forgo the benefits of using multiple, due to the costs of developing custom workflows between them.

Toward the end of the schematic design phase, output from these applications must be transferred to a general BIM design tool. Ideally, such an exchange would be easy and two-way, allowing for simple analyses or complex form-generations to be revised and re-evaluated. But as noted earlier, even in the primary direction this transfer is not yet well-supported by existing BIM tools.

How could these diverse conceptual design applications be integrated? There are at least four ways to integrate the different functionality needed for schematic design, as reviewed here: (1) **a single application** is developed that covers all the functionality; this has not yet happened; (2) **a suite of integrated applications** could be developed based on a business plan that is mutually beneficial to various companies, using a set of direct translators or plug-ins; this also has not happened, but some such arrangements exist; (3) the applications support **a neutral public standard exchange interface** (such as IFC) and rely on it to support integration; this has begun, as in the case of Ecotect; (4) the easier-to-use **BIM authoring tools expand their capabilities**, such as Revit® or ArchiCAD®, to include the functionality reviewed here. All but the first of these integration efforts are being applied in varying degrees.

To summarize, while the front-end services associated with conceptual design are likely to become increasingly important as BIM is adopted, the current technology base is not yet in place to support such a change. Existing conceptual design tools provide only very limited solutions. On the other hand, BIM model creation tools are generally too complex to be used for sketching and form-generation. Paper and pencil remain the dominant tools for such work. In the near future, the authors anticipate evolutionary progress in this area, with effective concept design systems solutions emerging soon.

5.3.2 Building System Design and Analysis/Simulation

As design proceeds past the conceptual stage, systems require detailed specification. Mechanical systems need sizing, and structural systems must be engineered. These tasks are usually undertaken through collaboration with engineering specialists, internal or external to the design organization. Like concept design, effective collaboration among these activities provides an area of market differentiation.

In this section, we review the general issues associated with applying analysis and simulation methods to design. First, we focus on the use of such applications as part of the normal performance assessment process during the detailing of building systems in the later stages of design. In contrast to the earlier applications, the applications in this phase are complex and usually operated by technical domain specialists. We consider areas of application and existing software alternatives: some of the issues concerning their use and exchange; and integration methods and general concerns relating to collaboration. We conclude by examining the special use of analysis and simulation models that explore innovative applications of new technologies, materials, controls, or other systems to buildings. It is important to note that such experimental architecture generally requires specialized tools and configurations.

Analysis/Simulation Software

As design development proceeds, details concerning the building's various systems must be determined in order to validate earlier estimates and to specify the systems for bidding, fabrication, and installation. This detailing involves a wide range of technical information.

All buildings must satisfy structural, environmental conditioning, fresh water distribution and waste water removal, fire retardance, electrical or other power distribution, communications and other basic functions. While each of these capabilities and the systems required to support them may have been identified earlier, their specification for conformance to codes, certifications and client objectives require more detailed definition. In addition, the spaces in a building are also systems circulation and access, systems of organizational functions supported by the spatial configuration. Tools for analyses of these systems are also coming into use.

In simple projects, the need for specialized knowledge with respect to these systems may be addressed by the lead members of a design team, but in more complex facilities, they are usually handled by specialists who are located either within the firm or hired as consultants on a per-project basis.

Over the past two decades, a great many computerized analysis capabilities were developed, long before the emergence of BIM. One large set of these is based on building physics. With drafting (electronic or manual), significant effort was required to prepare a dataset needed to run these analyses. With automated interfaces, a more collaborative workflow is possible, allowing multiple experts from different domains to work together to generate the final design.

An effective interface between a BIM authoring tool and an analysis/simulation application involves at least three aspects:

(1) Assignment of specific attributes and relations in the BIM authoring tool consistent with those required for the analysis.

(2) Methods for compiling an analytical data model that contains appropriate abstractions of building geometry for it to function as a valid and accurate representation of the building for the specified analysis software. The analytical model that is abstracted from the physical BIM model will be different for each type of analysis.

(3) A mutually supported exchange format for data transfers. Such transfers must maintain associations between the abstracted analysis model and the physical BIM model and include ID information to support incremental updating on both sides of the exchange.

These aspects are at the core of BIM's fundamental promise to do away with the need for multiple data entry for different analysis applications, allowing the model to be analyzed directly and within very short cycle times. Almost all existing building analysis software tools require extensive preprocessing of the model geometry, defining material properties and applying loads. Where BIM tools incorporate these three capabilities, the geometry can be derived directly from the common model, material properties can be assigned automatically for each analysis and the loading conditions for an analysis can be stored, edited and applied.

The way in which structural analyses are handled illustrates these aspects well. Because architectural design applications do not generate or represent structural members in a way that is suitable for performing structural analyses, some software companies offer separate versions of their BIM software to provide these capabilities. Revit® Structures and Bentley Structures are two examples that provide the basic objects and relationships commonly used by structural engineers—such as columns, beams, walls, slabs, etc.—in forms that are fully interoperable with the same objects in their sibling architectural BIM applications. It is important to note, however, that they carry a dual representation, adding an idealized 'stick-and-node' representation of the structure. They are also capable of representing structural loads and load combinations and the abstract behavior of connections, as connection releases, as are needed for analyses used to gain building code approval. These capabilities provide engineers with direct interfaces for running structural analysis applications. Figure 9.8-2 in the Penn National case study (Chapter 9) shows a model of a shear wall in a BIM tool and the results of an in-plane lateral load analysis of that wall.

Energy analysis has its own special requirements: one dataset set for representing the external shell for solar radiation; a second set for representing the internal zones and heat generation usages; and a third set for representing the HVAC mechanical plant. Additional data preparation by the user, usually an energy specialist, is required. By default, only the first of these sets are represented in a typical BIM design tool.

Lighting simulation, acoustic analysis, and air flow simulations based on computational fluid dynamics (CFD) each have their own particular data needs. While issues related to generating input datasets for structural analysis are well understood and most designers are experienced with lighting simulations (through the use of rendering packages), the input needs for conducting other kinds of analyses are less understood and require significant setup and expertise.

Providing the interfaces for preparing such specialized datasets is an essential contribution of the special-purpose environmental analysis building models reviewed in Section 5.3.1. It is likely that a suite of preparation tools for performing detailed analyses will emerge embedded within future versions of primary BIM design tools. These embedded interfaces will facilitate checking and data preparation for each individual application, as has been done for preliminary design. A properly implemented analysis filter will: (1) check that the minimum data is available geometrically from the BIM model; (2) abstract the requisite geometry from the model; (3) assign the necessary material or object attributes; and (4) request changes to the parameters needed for the analysis from the user.

The commonly used analysis/simulation applications for detailed design are shown in Table 5-3. Both public data exchange formats and direct, proprietary links with specific BIM design tools are listed. The direct links are built using middleware public software interface standards, such as ODBC or COM,

Table 5-3 Some of the common analysis/simulation applications and their exchange capabilities.

		Import Formats					Export Formats					
	Application	CIS/2	IFC	DXF/	SDNF	SAT	GBXML	CIS/2	IFC	DXF	SDNF	**Direct Links**
Structural Analysis	SAP200,ETABS	•	•					•	•			Revit® Structures
	STAAD-Pro	•						•				Tekla Structures, Bentley
	RISA			•						•	•	Revit® Structures
	GT-STRUDL	•			•			•				
	RAM							•		•	•	Revit® Structures
	Robobat	•	•									Revit® Structures
Energy Analysis	DOE-2											
	EnergyPlus		•				•		•	•		Ecotect
	Apache			•								IES
	ESP-r			•						•		Ecotect
Mechanical Equipment Simulation	TRNSYS											
	Carrier E20-II											
Lighting Analysis/ Simulation	Radiance			•			•					ArchiCAD®
Acoustic Analysis	Ease			•								
	Odeon			•								
Air Flow/CFD	Flovent			•		•						
	Fluent											
	MicroFlo											IES
Building Function Analysis	EDM Model Checker		•									
	Solibri		•									

or proprietary interfaces, such as ArchiCAD®'s GDL or Bentley's MDL. These exchanges make portions of the building model accessible for application development.

CIS/2, which is the most commonly used public data exchange format, is the result of intense development effort by the structural steel industry (see Chapter 3 for details). It provides extensive exchange coverage for structural analysis applications but only for steel structures. Efforts have been undertaken to make the IFC schema supportive of structural analysis, and to a lesser degree, energy analyses. Initial work has been done to enable IFC models to carry annual solar radiation gains, but not lighting, acoustic, or airflow simulations. Such tailoring of the IFC model can be expected as BIM technology becomes more widely adopted.

A uniform direct exchange format to support all analysis types is not likely to be developed, because different analyses require different abstractions from the physical model, with properties that are specific to each analysis type. Most analyses require careful structuring of the input data by the designer or the engineer who prepares the model.

The above review focuses on quantitative analysis dealing with the physical behavior of buildings. Less complex but still complicated criteria must also be assessed, such as fire safety and access for the disabled. Recently, the availability of neutral format (IFC) building models has facilitated two products supporting rule-based model checking. Solibri considers itself to be a spell and grammar checking tool for building models. EDM ModelChecker™ provides a platform for undertaking large-scale building code checking and other forms of complex configuration assessments. EDM is the platform used in CORENET, the Singapore automated building code checking effort (CORENET 2007). A similar building code effort is underway in Australia (Ding et al. 2006).

Solibri (Solibri 2007) has implemented the Space Program Validation application for GSA (GSA 2006) and is in the process of developing additional testing for circulation layout. One aspect of Space Program Validation for the area derivation of one space is shown in Figure 5-6. The application compares the program areas against the ones in the layout, based on the ANSI-BOMA area calculation method, to determine compliance with the space program. Such assessment applications dealing with both qualitative and quantitative assessments will become more widely used as standard representations become more available.

Some BIM design tools also provide space programming assessment capabilities. Revit® has a space planning assessment capability, and ArchiCAD® has a plug-in of Trelligence Affinity™ that offers similar capabilities.

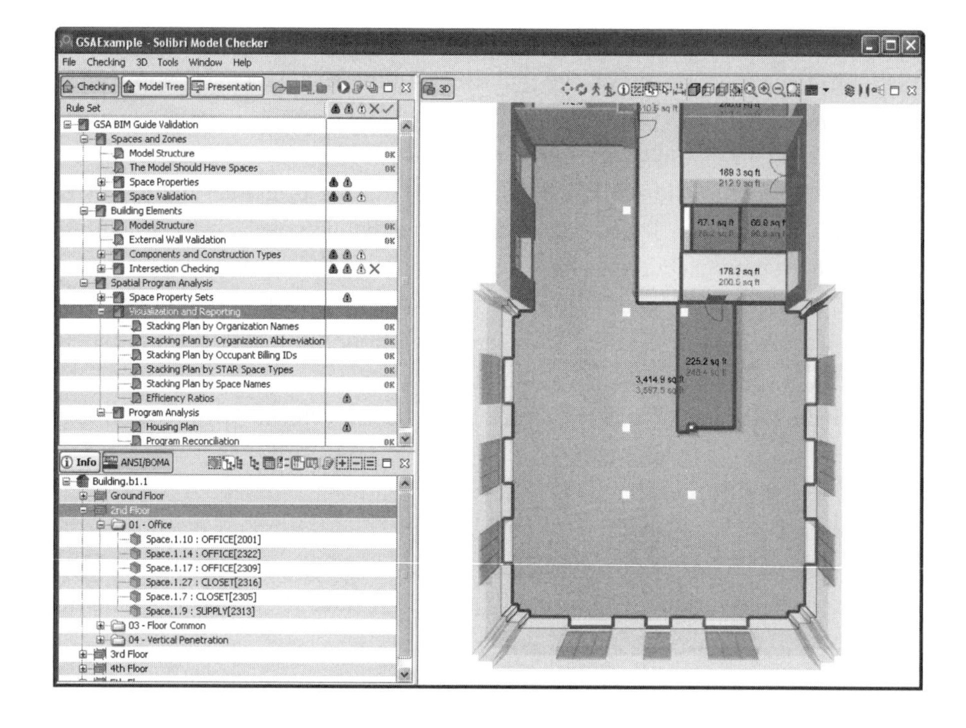

FIGURE 5-6 Example derivation of the ANSI-BOMA space area, for comparison with the specified program area. *Image provided courtesy of the Office of the Chief Architect, Public Buildings Service, U.S. General Services Administration.*

Improving Organizational Performances within Facilities

While the performance of a building shell is obvious and directly tied to design and construction, buildings are also built to house various healthcare, business, transportation education or other functions. The constructed space can contribute to the efficient functioning of the operations carried out within the building. These are obvious in manufacturing facilities, where the layout of operations is well understood to have an effect on efficient production. The same logic has been applied to hospitals, based on the recognition that doctors and nurses spend a significant time each day walking. More recently, issues of developing space layouts that can support varied emergency procedures in trauma units and intensive care facilities have also been studied.

The processing time in airports is something we all face and can be affected by airport planning. As the workforce becomes more oriented toward creative production, the open, friendly work environments found in Silicon Valley will become more commonplace everywhere. The increasing percentage of GDP devoted to healthcare indicates that improvements that can be generated through improved design—associated with new procedures— are an area worthy of intense analysis and study. Whether architects take up such analytical capabilities, clearly, the integration of building designs with models of organizational processes, human circulation behavior, and other related phenomena will become an important aspect of design analysis.

These issues are generally driven by owner recognition of need, and are discussed in Section 4.5.7. Motivation for such studies being undertaken as specialized design services is addressed in Section 5.4.1.

Cost Estimation

While analysis and simulation programs attempt to predict various types of building behavior, cost estimation involves a different kind of analysis and prediction. Like the previous analyses, it needs to be applicable at different levels of design development, taking advantage of the information available and making normative assumptions regarding what is missing. Because cost estimation addresses issues relevant to the owner, contractor, and fabricator, it is also discussed from these varying perspectives in Chapters 4, 6, and 7 respectively.

Until recently, the product or material units for a project were measured and estimated through manual counting and area calculations. Like all human activities, these involved errors and took time. However, building information models now have distinct objects that can be easily counted, and along with volumes and areas of materials, can be automatically computed, almost instantaneously. The specified data extracted from a BIM design tool can thus provide an accurate count of the building product and material units needed for cost estimation. The Jackson Mississippi Federal Courthouse case study (Chapter 9) provides tables indicating the variations generated by different firms producing manual bills of materials for concrete foundations, as a component of the overall cost estimate. A fuller review of cost estimating systems is provided in Section 6.6.

Cost estimation integrated with a BIM design tool allows designers to carry out value engineering while they are designing, considering alternatives as they design that make best use of the client's resources. Following traditional practices and removing cost items at the end of a project, when ease of making changes is the criterion rather than the most effective changes, usually leads to poor results. Incremental value engineering while the project is being developed allows practical assessment throughout design.

Collaboration

Throughout design, collaborative work is undertaken between the design team and engineering and technical specialist consultants. This consultative work involves providing the appropriate project information regarding the design, its use and context to the specialists to review, and gaining feedback/advice/changes. The collaboration often involves team problem-solving, where each participant only understands part of the overall problem. Traditionally, these collaborations have relied on drawings, faxes, telephone calls and physical meetings. The move to electronic documents and drawings offers new options for electronic transfer, email exchanges and Web conferencing with online model and drawing reviews.

Most major BIM systems include support for model and drawing review and online markups. New tools display 3D building models or 2D drawings for review, without the complexity of full model generation capabilities. These view-only applications rely on formats similar to external reference files used in drafting systems, but are quickly becoming more powerful. A sharable building model in a neutral format, such as VRML, IFC, DWF or Adobe®3D, is easy to generate, compact for easy transmission, allows mark-ups and revisions, and enables collaboration via Web conferences. Some of these model viewers include controls for managing which objects are visible and for examining object properties. In the near future, clients will demand take away copies of such models for extended personal review and assessment.

It is important to recognize, if only in retrospect, the difficulty inherent in reading and understanding 2D sections and details. Comparatively, most everyone is able to read and understand a 3D model, which allows for a more inclusive and intuitive planning and review process. This is particularly important for clients that lack the experience needed for interpretatiing 2D drawings. Regular reviews with all of the parties involved in a design or construction project can be undertaken using 3D BIM models along with tools like Webex®, GoToMeeting® or Microsoft's Live Meeting®. Conference participants may be distributed worldwide and are limited only by work/sleep patterns and time-zone differences. With voice and desktop image sharing tools—in addition to the ability to share building models—many issues of coordination and collaboration can be resolved.

Collaboration takes place minimally at two levels: first among the parties involved, using web meeting and desktop displays like those described above. The other level involves project information sharing. In addition, the opportunity of close collaboration between consultants requires definition of data exchange structures that are able to support that collaboration and maintain consistency between models as design revisions are made. To explain the idea, we consider the case of structural analysis, based on the description of the tools and exchange formats presented earlier. This example can be extended to other types of shared design information.

Once the building model's system is ready for analysis and a data exchange format is available, a data exchange can be initiated. As reviewed in Chapter 3, workflows can be articulated to a fine-grained set of information exchanges and dialogues. At the basic level, the exchange formats can be reduced to two types:

1) As a one-way flow from the BIM design tool to the analysis application. This flow is diagrammed in Figure 5-7. The designer passes the existing structural network and its loads (if known) and any constraints regarding sections to the structural designer. The structural designer adds assumptions about connections, load conditions and other structural behavior. Based on initial analysis results, the structural engineer may try different alternatives, and then propose changes back to the designer. The return is made externally through drawings and sketches, and all updates are entered manually by the designer in the BIM tool. These changes may involve location changes, member size changes, detailed bracing layouts, and other design aspects that were not initially addressed. Because such changes are often detailed and tedious, they may be easily misinterpreted. Later updates of the structure are reanalyzed using the same process, with the engineer inserting the analysis assumptions each time in a new analysis run.

2) As a two-way flow, where the design application supports flow both to the analysis tool and also accepts results back. Here, the flow to the structural engineer is the same as before. After the analysis and needed iterations, the analyst passes proposed changes back to the BIM design application digitally. The system automatically presents these as model changes which the designer can accept or reject. This requires that the design application can detect the updated members and their properties, match them to the initial dataset and

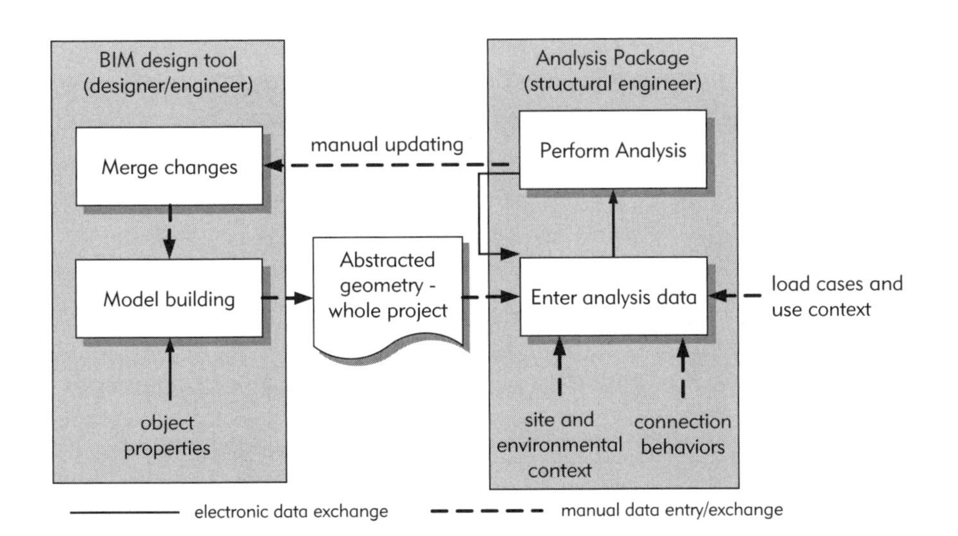

FIGURE 5-7
Analysis information flows based on a one-way flow into the analysis package.

make the proper updates. This flow is diagrammed in Figure 5-8. Later analysis iterations send just modifications back to the structural engineer, requiring only incremental changes in the analysis dataset. Assumptions from the previous analyses are maintained and the connection and loading assumptions only have to be updated for those members that were modified.

The one-way workflow is closer to the form of current practice where the structural engineer sketches notes on the drawings regarding changes, but all updates are physically implemented by the design team. The second workflow is parallel to practices where the structural engineer can make changes directly to the information on a drawing set's structural layers. The second workflow is significantly more efficient, allowing iterations to be made in minutes instead of days and can support step-by-step or even automated improvement and optimization, allowing full application of a structural engineer's expertise.

Two-way flows require special preparation of both the BIM design tool and the receiving analysis application. In order to match up the changed objects received back from the analysis application, a discriminating object-ID must be generated by the design application to be carried within the analysis application and returned with the updated dataset. The BIM design tool must generate the IDs for the objects that are to be passed back and forth. These are then matched in the design application against existing objects' IDs to determine which objects have been modified, created or deleted. Similar matching is required in the analysis application. These two-way capabilities have been realized in the interfaces with some structural analyses. All building data models, IFC and CIS/2, support the definition of a globally unique ID (GUID).

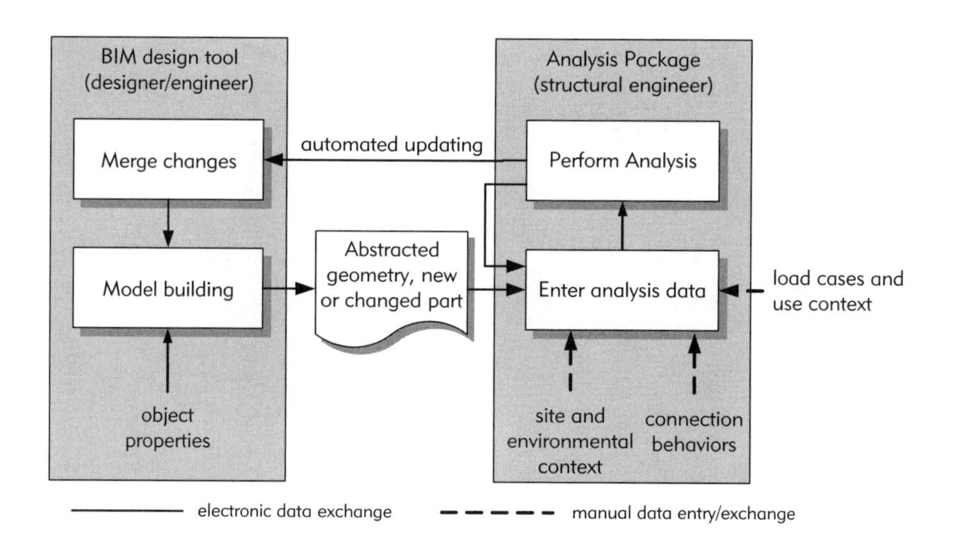

FIGURE 5-8
The information flows supporting two-way exchange of analysis data.

In summary, effective collaboration using two-way workflows can generally be achieved between BIM design applications and structural analyses. Effort is still required to create effective two-way exchanges in most other analysis areas.

The rationale for quicker iterations between designers and consultants is part of the lean design philosophy. Long iterations result in both sides multi-tasking, often on multiple projects. Multi-tasking results in lost time remembering issues and context of the designs on each return to a project, and makes human errors more likely. Longer iterations lead to higher levels of multi-tasking, whereas shorter cycles allow continuous work on projects. The result is less wasted time and better progress on each design task.

Experimental Design Using a Design 'Workbench'

While the previous sections outline standard practices regarding design development in large building projects, there is also another important use of analysis applications for innovative buildings that require a high level of design information development (see section 5.2.1). This is in the more experimental mode, where architects and/or engineers explore new structural systems, new energy distribution systems, new building environmental control systems, new materials or new construction methods, or other aspects of a building that may or may not work and be effective. These are examples of in-the-field design research.

An example of such explorations is the Al Hamra Firdous Tower to be built in Kuwait City, shown in Figure 5-9. With the aim of maximizing Gulf views and minimizing solar heat gain on the office floors, a quarter of each floor plate is chiseled out of the south face, shifting from west to east over the height of the tower. The result of this operation reveals a rich, monolithic stone at the south wall framed by the graceful, twisting ribbons of torqued walls and defines the iconic form of the tower. Extensive studies were undertaken to minimize solar gain. The panelization of the interior irregular surface required special study, shown in Figure 5-10. The layouts were managed using Visual Basic scripts; Digital Project was the design platform.

The case studies in Chapter 9 provide two other outstanding examples: the Beijing Aquatic Center, whose structural system is absolutely unique, and was developed through a large number of iterations optimizing different parts of the structural elements; and the San Francisco Federal Building, with its commitment to using natural ventilation in a highly innovative and engineered manner. Each of these designs relied on hundreds of analysis runs, in one case looped within an optimization tool. For undertaking such work, there is no straightforward approach; each project and experimental design problem poses its own questions.

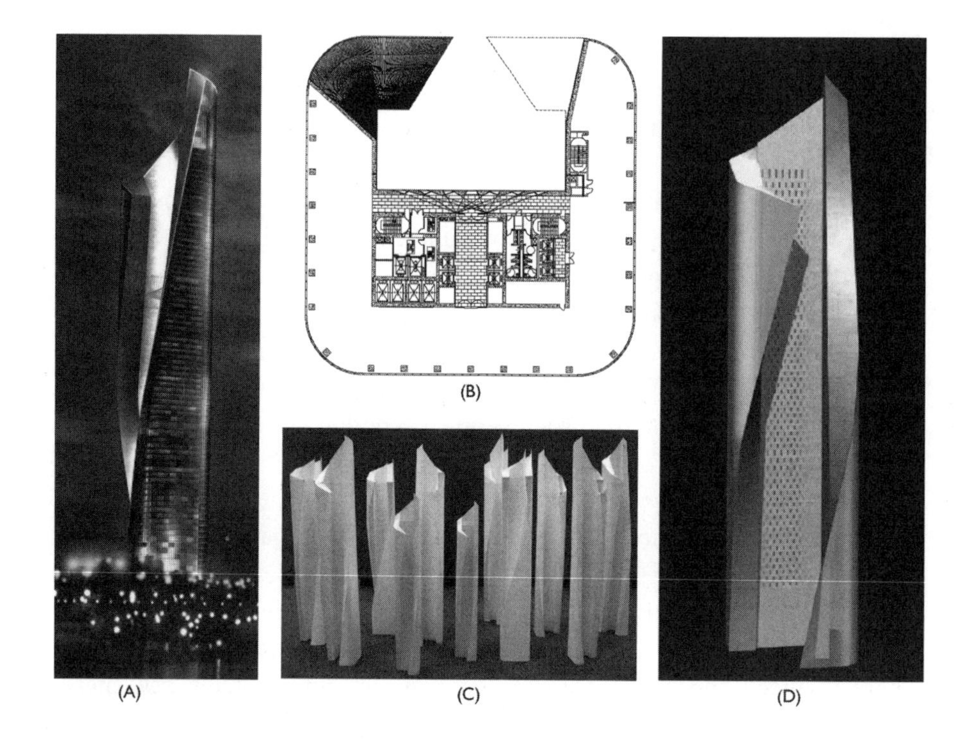

FIGURE 5-9
A) A rendering of the 412 meter Al Hamra Firdous Tower in Kuwait City by SOM. B) A typical floor plan showing the "chiseled out", south-facing section of the floor plate that varies by floor. C) Case study models showing the" slice" from different views. (D) Model of internal core of tower.
Images provided courtesy of SOM.

For structures, experimental design can rarely be achieved through the use of stick models. Rather, it requires 3D finite element models. For energy studies, standard building analysis tools cannot address certain aspects of materials, mechanical systems, or the control systems of interest. For example, double wall systems with adaptive automatic controls cannot be analyzed in a straightforward manner within DOE-2 or Energy+. More creative methods of analysis must be undertaken. In each case, multiple tools are needed, and often, they must be organized in such a way that the output of one analysis supplies the input for another. General mathematical tools, such as MatLab® or Mathematica®, are often essential.

One area of particular interest is the method of fabrication to support curved surfaces. Frank Gehry and Sir Norman Foster have shown that the development of freeform shapes in architecture can be accomplished by defining the external shell digitally and planning its fabrication in terms of the material shapes making up the form. Often these firms outline the fabrication of pieces used to construct the pieces, using computer numerical control (CNC) machinery. CNC equipment is required for economic reasons because every shape is unique. This has come to be known as 'design-for-fabrication', popularized by Kieran and

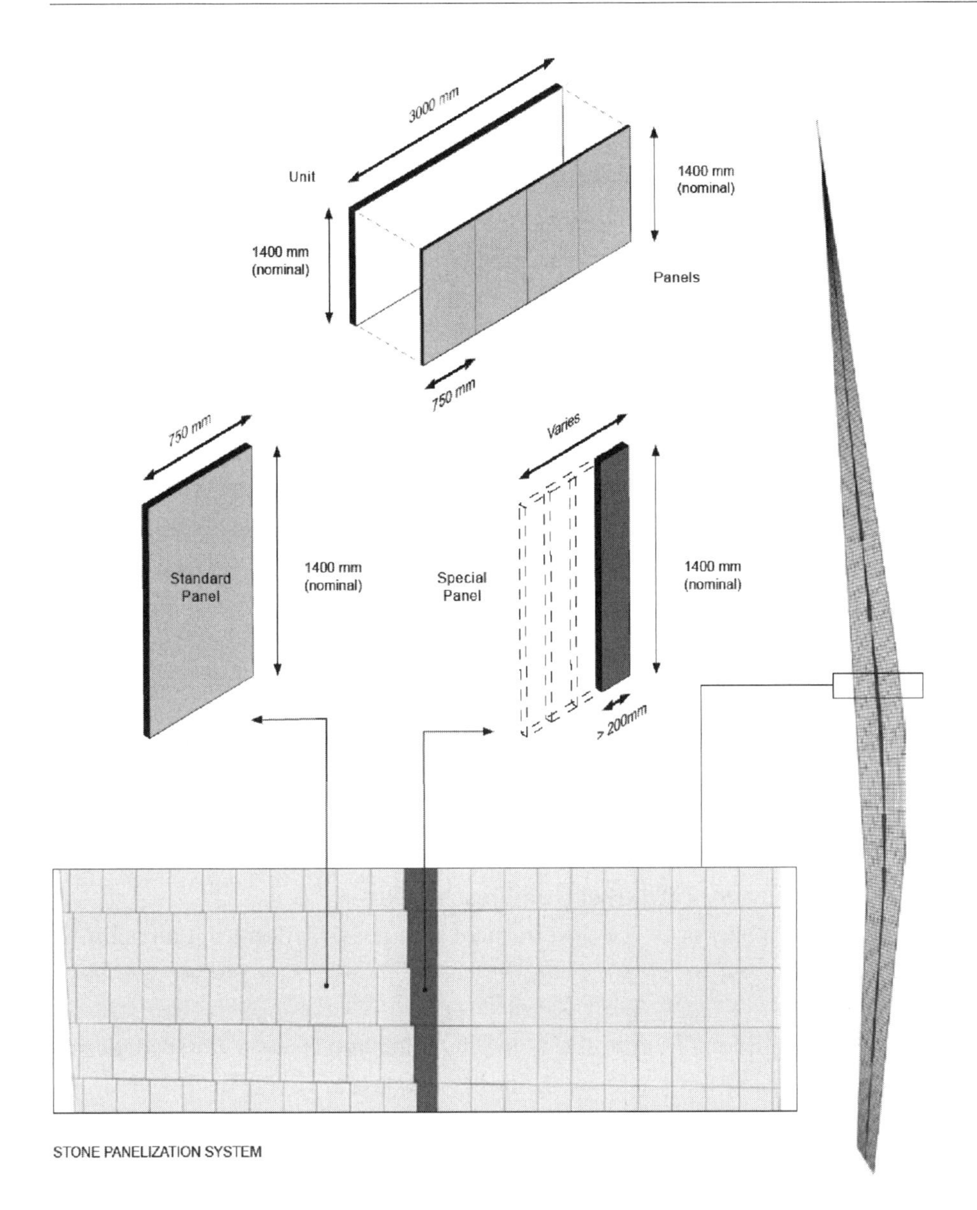

FIGURE 5-10
The panelization layout for the Al Hamra Firdous Tower. The panels were controlled using Visual Basic scripts in Digital Project.
Image provided courtesy of SOM.

Timberlake's book, *Refabricating Architecture* (Kieran and Timberlake 2003). How to design pieces of a building to take advantage of CNC production methods is a high-demand area of exploration, in both architecture schools and innovative offices.

In addition to the Firdous Tower, another example of such work is presented in Chapter 2, Figure 2-7, of the Dublin Ireland Football (soccer) stadium. The structural pieces are controlled by the parametric structure, allowing their forms to be automatically generated and later fabricated.

Each method of fabrication has its own associated set of processes with specialized demands for the digital data that supports it. Fabrication rules and best practices address relevant issues associated with various steps in the design process, such as the conversion geometry to CNC machine instructions, methods of connection, finish qualities, tolerances, and material sizes. These and other production processes are described in the book *Digital Design and Manufacturing* (Schodek et al. 2005).

An exploration of new forms often benefits from an automated search over a large and abstract set of alternatives, e.g., optimization, as opposed to selecting from a few manually generated alternatives. The challenge is to develop a well-defined range of designs that can be explored automatically and to identify good solutions by establishing well-defined objective functions. This is greatly facilitated by defining a parametric closed model (meaning there is a closed, possibly huge, but finite space of possible configurations). For example, simulated annealing software is available from Taygeta Scientific® and is incorporated into Mathematica®. This method was used in the structural optimization of the Beijing Aquatics Center (see Chapter 9). Another such product is modeFRONTIER (Esteco 2007).

Until now, design optimization has been the exclusive capability of only a few firms, most notably Ove Arup and Partners. The opportunity now exists for a much wider application.

5.3.3 Construction-Level Building Models

Designers can approach the development of a construction-level model in at least two different ways:

(1) As traditionally conceived, the building model is a detailed design expressing the intent of the designer and the client. In this view, the contractors are expected to develop their own independent construction model and documents.

(2) As a partially detailed model to be further detailed for use in all aspects of construction, planning, and fabrication. In this view, the design model is the starting point for elaboration by the construction team.

The main reason why the first approach has traditionally been adopted by architects is to eliminate liability for construction issues by taking the approach that they are not providing construction information but only design intent. This is apparent in the text disclaimers that commonly appear on architectural drawings, which transfer responsibility for dimensional accuracy and correctness to the contractors. Of course, technically this means that the contractor or fabricators should develop their models from scratch, reflecting the intent of the designer, and requiring repeated rounds of submittals, design reviews, and corrections.

The authors consider such practices—based strictly on design intent—to be inherently inefficient and irresponsible to clients. We encourage designers to take the second view, providing their model information to fabricators and detailers and allowing them to elaborate the design information as needed to both maintain the design intent and refine the design for fabrication. Appropriate contracting methods to allow this are reviewed in Chapter 4.

The structural engineer's model of the USC School of Cinematic Arts provides an excellent example of this approach. As can be seen in Figure 5-11, the structural engineer has provided all of the structure geometry with cast-in-place concrete rebar and steel connection details. The different fabricators can all refine their details using the same model; coordination between the different systems is ensured.

Almost all existing tools for generating building information models support a mixture of full 3D component representation, 2D representative sections,

FIGURE 5-11
A view of a design engineer's Tekla Structures model of the USC School of Cinematic Arts.
The model contains details for three subcontractors – structural steel, rebar fabricator and cast-in-place concrete – and enables the engineer to ensure design coordination among these systems. (See color insert for full color figure.)
Image provided courtesy of Gregory P. Luth & Associates, Inc.

plus symbolic 2D or 3D schematic representations, such as centerline layouts. Pipe layouts may be defined in terms of their physical layout or as a centerline logical diagram with pipe diameters annotated alongside them. Similarly, electrical conduit can be placed in 3D or defined logically with dotted lines. As reviewed in Chapter 2, the building models resulting from this mixed strategy are only partially machine-readable. The level of detail within the model determines how machine-readable it is and the functionality that it can achieve. Automated clash checking can only be applied to 3D solids. Decisions regarding the level of detail required of the model and its 3D geometry of elements must be made as construction level modeling proceeds.

Today, recommended construction details supplied by product vendors cannot yet be defined in a generic form allowing insertion into a *parametric* 3D model. This is because of the variety of underlying rule systems built into the different parametric modelers (as described in Section 2.2) Construction details are still most easily supplied in their conventional form, as drawn sections. The potential benefits for supplying parametric 3D details, to strengthen vendor control of how their products are installed and detailed, has large implications regarding liability and warrantees. This issue is developed in Chapter 8. On the designers' side, however, the current reliance on 2D sections is both a rationale to not undertake 3D modeling at the detail level, and a quality control handicap to be overcome.

Building System Layouts

One of the major productivity benefits associated with BIM is in the area of mechanical, electrical, and plumbing design and contracting. In traditional construction methods, architects and MEP consultants logically define these systems in terms of components and general layout (using centerline approximations). Detailed layout and fabrication is left for subcontractors to complete for each respective system. Each trade must execute a detailed design, fabricate the parts, measure the actual spaces created following the onsite installation of preceding systems, make corrections (either to shop drawings or to already fabricated components), and finally install their systems. The resulting work process is lengthy, error prone, and wasteful.

The alternative is to simultaneously 'pull' 3D detailed designs for all systems by using BIM for collaboration and relying on virtual construction of the building in the computer before it is built onsite. This enables a much higher degree of prefabrication and pre-assembly of integrated system units. These units may be delivered just-in-time for onsite installation. This process is described in detail in Chapter 7. The Camino Medical Group Building and the GM Production Plant case studies in Chapter 9 are excellent examples of simultaneous pulled detailed design and fabrication processes. The benefits associated with this method involve a reduced need for onsite crews and space for laydown areas; and prefabricated units can be larger, which greatly reduces onsite fabrication time and cost. The implication for designing in this method is that all system layouts must be prepared in 3D before any fabrication begins. Modularization, planning for componentization, and simplifying onsite installation are additional design considerations.

Different construction types and building systems involve different kinds of expertise for detailing and layout. Curtain walls, especially for custom designed systems, involve specialized layout and engineering. Precast concrete and structural steel are other areas that involve specialized design, engineering, and fabrication expertise. Mechanical, electrical, and plumbing systems require sizing and layout, usually within confined spaces. In these cases, specialists involved in the design require specific design objects and parametric modeling rules to lay out their systems, size them, and specify them. Later in the production process, the layouts are further refined to support automatic analyses at one level and automatic fabrication and quantity takeoff at the other.

Specialization, however, requires a careful approach for integration in order to realize efficient construction. The designers and the fabricators/constructors for each system are typically separate and distinct organizations. While 3D layout during the design phase carries many benefits, if it is undertaken too early it may result in multiple iterations which will reduce or eliminate these benefits. On the other hand, if it is carried out too late, the project may be delayed. Prior to selecting a fabricator, the architects and MEP engineers should only generate "suggested layouts." After the fabricator is selected, the production objects may be detailed and laid out; and this layout may differ from the original due to production preferences or advantages that are unique to the fabricator. Therefore the level of detail pursued in each system depends on contractual arrangements.

Fortunately, the boundary between system design and fabrication is always in flux. BIM tools will be most effective when used in parallel—and as seamlessly as possible—by all system designers and fabricator subcontractors. We believe that BIM tools provide strong advantages for design-build contractual arrangements for building systems and that the use of construction detail level models—where design models are used directly for fabrication detailing—will become more prevalent due to cost and time savings.

In practice, each building system can be laid out separately, sharing only 3D reference geometry with the other systems to serve as a guide. This works well, allowing layout coordination to be managed without requiring full editing interoperability. Both the host BIM design tools and the specialized building system design applications require effective reference geometry for importing as well as exporting the system layouts that are to be used by system fabricators to guide their layouts.

Numerous applications are available to facilitate operations within or in concert with the primary BIM design tools used by an A/E firm or consultant. A representative sample is shown in Table 5-4, which contains a list of mechanical and HVAC, electrical, piping, elevators and trip analyses and site planning applications. These support areas are undergoing rapid development by specialized building system software developers. The software under development is also being integrated with major BIM design tools and acquired by BIM vendors. As a result, BIM vendors will be able to offer increasingly complete building system design packages.

Readers interested in more detailed discussion of the role of BIM in fabrication for construction are referred to Chapter 7, which focuses exclusively on these aspects.

Drawing and Document Production

Drawing generation is an important BIM production capability, and will remain so for another decade or so. At some point, drawings will stop being

Table 5-4 Building system layout applications.

Building System	Application
Mechanical & HVAC	Carrier E20-II HVAC System Design Bentley Building Mechanical Systems Vectorworks Architect ADT Building Systems Autodesk Revit® Systems
Electrical	Bentley Building Electrical Vectorworks Architect Autodesk Revit® Systems
Piping	Vectorworks Architect ProCAD 3D Smart Quickpen Pipedesigner 3D Autodesk Revit® Systems
Elevators/Escalators	Elevate 6.0
Site Planning	Autodesk Civil 3D Bentley PowerCivil Eagle Point's Landscape & Irrigation Design
Structural	Tekla Structures Autodesk Revit® Structures Bentley Structural

the design information of record and instead the model will become the primary legal and contractual source of building information. However, today every design firm of record still needs to produce schematic design, design development, and construction drawings, to fulfill contract requirements, to satisfy building code requirements, for contractor/fabricator estimation, and to serve as the contract documents between designer and contractors. These documents have important uses, beyond their contractual ones. Drawings are used during construction to guide layout and work. In the Penn National case study in Chapter 9, existing contractual requirements enforced the use of drawings, even where they were no longer needed for some of these functions. General drawing production requirements from BIM tools are presented in Chapter 2, Section 2.2.3.

A single model representation both guarantees consistency and automates most aspects of drawing production. This is realized by the single model being used to generate all plans, sections, elevations, structural, mechanical, electrical and other systems drawings. Supported by appropriate libraries, construction documentation production time can be significantly reduced.

With the development of BIM and its report generating capabilities, once the legal restrictions on the format of drawings is eliminated, options arise that can further improve the productivity of design and construction. Already, fabricators that have adopted BIM tools are developing new drawing and report generation layouts that better serve specific purposes. These apply not only to rebar bending and bills of material, but also layout drawings that take advantage of the 3D modeling of BIM tools. An aspect of BIM research is the development of specialized drawings for different fabricators and installers. An excellent example is provided in Figure 5-12. New representations facilitating easy interpretation of research results during design is another area where research is enhancing BIM capabilities.

The long-term goal is to completely automate the production of drawings from a model. However, a close look at special conditions makes evident that various special cases arise in most projects that are themselves so rare that planning for them and preparing template rules is not worth the effort. Thus review for completeness and layout of all drawing reports prior to release is likely to remain a needed task for the foreseeable future.

Specifications

A fully detailed 3D model or building model does not yet provide sufficiently definitive information for constructing a building. The model (or historically, the corresponding drawing set) omits technical specifications of materials, finishes, quality grades, construction procedures, and other information required

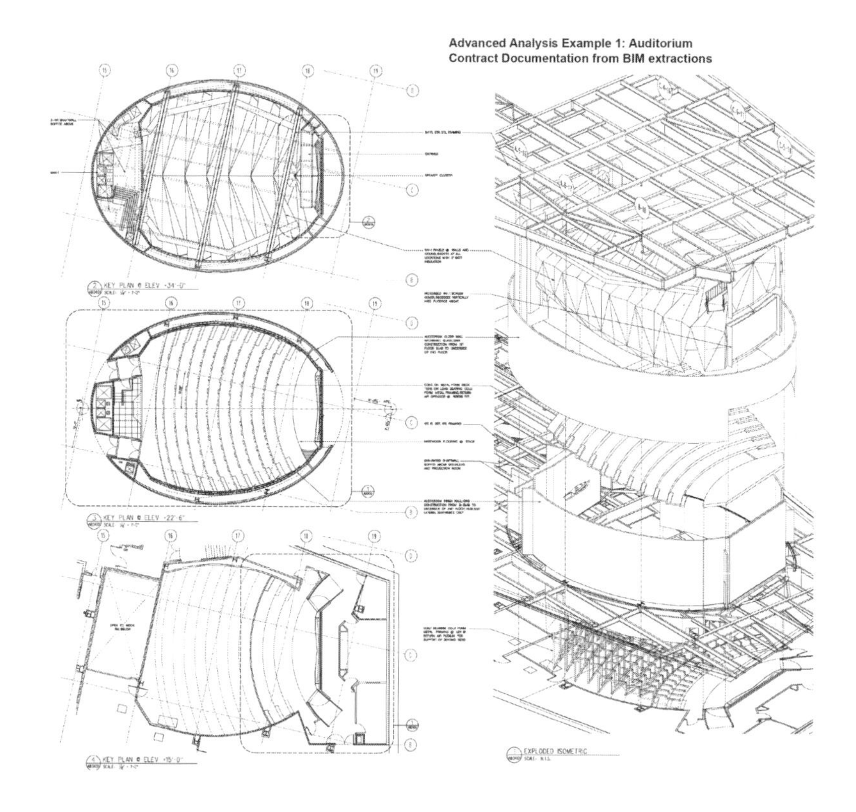

FIGURE 5-12 Detailed layout of the auditorium at the Merck Research Laboratories Boston. Associated drawings included panel fabrication layout. The design was especially complicated because of the skewed structural grid.
Image provided courtesy of KlingStubbins.

for managing the realization of a desired building outcome. This additional information is packaged as the project specifications. Specifications are organized according to types of materials within a project and/or classes of work. Standard specification classifications are Uniformat® (of which there are two slightly different versions) or Masterformat®. For each material, type of product, or type of work, the specification defines the quality of the products or materials and identifies any special work processes that need to be followed.

Various IT applications are available for selecting and editing the specifications relevant to a given project, and in some cases, to cross-link them with relevant components in the model. One of the earliest specification systems to cross-reference with a BIM design model was e-Specs®, which cross-links with objects in Revit®. e-Specs maintains consistency between the reference object and the specification. If the reference-object is changed, the user is notified that the relevant specification must be updated. Specifications can also be associated with library objects, so that a spec is automatically applied when the library object is incorporated into the design.

Uniformat defines a document structure that was conceived as a companion to a construction drawing set. One limitation of this tool is the specification structure covers broad areas with multiple possible applications within a given building project. Logically, this limits links to one-way functions, because a single specification clause applies to multiple but somewhat diverse objects in the design. One cannot directly access the objects that a spec paragraph applies to. This limitation restricts the management of specification quality. The Construction Specification Institute (the owner of Uniformat®) is decomposing the structure of Uniformat to support a bi-directional relationship between building objects and specifications. The new classifications, called Omniclass®, will lead to a more easily managed structure for specification information of model objects (OmniClass 2007).

5.3.4 Design-Construction Integration

The historical separation of design from construction did not exist in medieval times and only appeared during the Renaissance. Throughout long periods of history, the separation was minimized through the development of close working relationships between construction craftsmen, who in their later years would work 'white collar jobs' as draftsmen in the offices of architects (Johnston 2006). But in recent years, that link has weakened. Draftsmen are chiefly junior architects and the communication channel between field craftsmen and the design office has atrophied. In its place, an adversarial relationship has arisen, largely due to the risks associated with liabilities when serious problems arise.

To make matters worse, the complexity of modern buildings has made the task of maintaining consistency between increasingly large sets of drawings extremely challenging, even with the use of computerized drafting and document control systems. The probability of errors, either in intent or from inconsistency, rises sharply as more detailed information is provided. Quality control procedures are rarely capable of catching all errors, but ultimately, all errors are revealed during construction.

A building project requires design not only of the built *product* but also design of the *process* of construction. This recognition lies at the heart of design-construction integration. It implies a design process that is conscious of the technical and organizational implications inherent in how a building and its systems are put together as well as the aesthetic and functional qualities of the finished product. In practical terms, a building project relies on close collaboration between experts situated across the spectrum of building construction knowledge, as well as particularly close collaboration between the design team and the contractors and fabricators. The intended result is a designed product and process that are coherent and integrates all the relevant knowledge.

Different forms of procurement and contracting are reviewed in Chapters 1 and 4. While the contractor perspective is given in Chapter 6, here we consider teaming from the designer's perspective. Below, we list a few of the benefits of integration:

- Early identification of long lead-time items and shortening of the procurement schedule (see the GM Production Plant case study in Chapter 9).
- Value engineering as design proceeds, with continuous cost estimates and schedules, so that tradeoffs are integrated fully into the design rather than after-the-fact in the form of 'amputations.'
- Early exploration and setting of design constraints related to construction issues. Insights can be gained from contractors and fabricators so that the design facilitates constructability and reflects best practice, rather than making changes later with added cost or accepting inferior detailing. By designing initially with fabrication best-practices in mind, the overall construction cycle is reduced.
- Facilitating identification of the interaction between erection sequences and design details and reducing erection issues early on.
- Reducing the differences between the construction models developed by designers and the manufacturing models needed by fabricators, thus eliminating unnecessary steps and shortening the overall design/production process.
- Significantly shortened cycle times for fabrication detailing, reducing the effort required for design intent review and consistency errors.

Part of the design-construction collaboration involves (and requires) deciding when the construction staff is to be brought on. Their involvement can begin at the project's outset, allowing construction considerations to influence the project from the beginning. Later involvement is justified when the project follows well-tried construction practices or when programmatic issues are important and do not require contractor or fabricator expertise. Increasingly, the general trend is to involve contractors and fabricators earlier in the process, which often results in the gaining of efficiencies that would not be captured in a traditional design-bid-build plan.

5.3.5 Design Review

In the new methods of practice outlined above, the separation between the design model and fabrication model is greatly reduced and often carried out in an

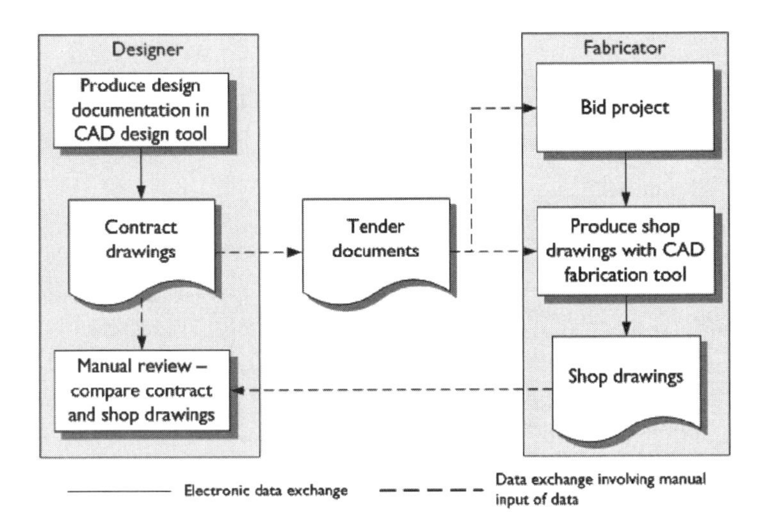

FIGURE 5-13 Traditional design review process, including alternative equipment selection.

overlapping timeframe. However realized, the final arrangement results in two building system models, one capturing the design intent and used to coordinate all systems, and fabrication (or shop) models of specific systems that capture the production design and detailing needed for fabrication and erection.

Evolution of the design model to the fabrication model inevitably involves additions and changes. These changes must be reviewed by the design team to verify that the design intent has not been lost. Two kinds of reviews are required: (1) replacement of one piece of equipment or manufactured piece with another, which may have different shape and connections (we assume that the specification review and acceptance is handled separately); (2) that the geometry or placement of manufactured and made-to-order pieces are consistent with the placement and geometry of all other components.

In traditional paper-based methods, these two checks require the comparison of the design (contract) documents with the fabricator's shop drawings, often by overlaying the two sets of drawings on a light table. The process is diagramed in Figure 5-13. Because of different layouts, formats and conventions, these comparisons are arduous and traditionally can take a week or more.

In the new process (Figure 5-14), upfront collaboration between designer and fabricator allows system products to be selected early, so that layout can be done once in a consistent manner. With fully 3D models, shop model review is reduced to loading the two 3D models into a reviewing system to check placement and that the important surfaces overlap. Examples of this

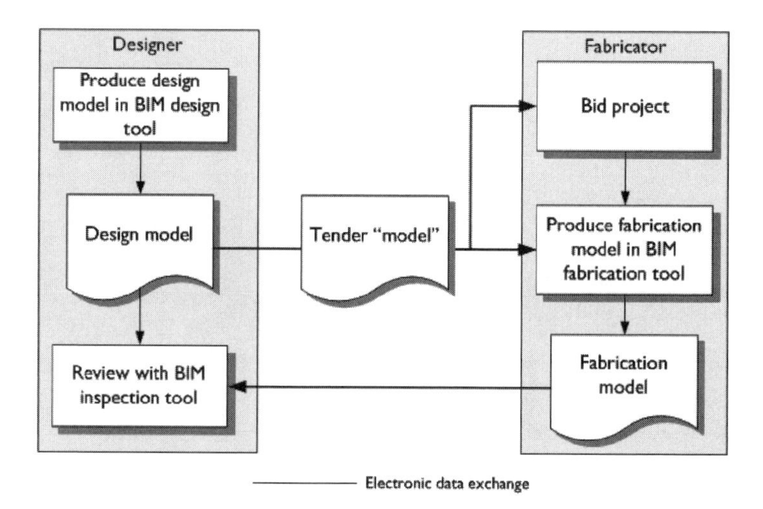

FIGURE 5-14
A model-based design review of fabrication models.

process are presented in the GM Flint Michigan and also the San Francisco Federal Building case study by Morphosis, both in Chapter 9. Coincident surfaces are automatically highlighted for the reviewers' attention. This reduces the time and effort required to identify errors. In the future, we expect that the overlap reporting will be further augmented by rule-based checking to identify those objects that have moved or that have had their finished surfaces moved beyond some preset tolerance, thus further automating the review process.

5.4 BUILDING ELEMENT MODELS AND LIBRARIES*

BIM involves the definition of a building as a composed set of objects. BIM design tools provide different pre-defined libraries of fixed geometry and parametric objects. These are typically generic objects based on standard onsite construction practices that are appropriate for early-stage design (see Table 2-1). As a design is developed, object definitions become more specific, elaborated with expected or targeted performances, such as for energy, sound, and cost, etc. Visual features are also added to support rendering. Technical and performance requirements can be outlined so that object definitions specify what the final constructed or purchased product should achieve. This product specification then becomes a guide for selecting or

*This section was adapted from information provided by James A. (Andrew) Arnold, courtesy of Tectonic Partners.

constructing the final object. Previously, different models or datasets were hand-built for these different purposes and not integrated. Now it is possible to define an object once and use it for multiple purposes. The challenge is to develop an easy-to-use and consistent means for defining object instances appropriate for the current stage of design and supporting the various uses identified for the stage. Later, the specification is superseded by the selected product. Thus, multiple levels of object definition and specification are needed. These range from early stage design using generic objects to fabrication-level detailing of the final object as it will be implemented, whether as a constructed component or commercial product. Throughout this process, objects undergo a sequence of refinements of shape material properties used to support analyses, simulation, cost estimation, and other uses. Some issues of managing object properties are reviewed in Section 2.2.2. Over time, we expect these sequences to be better defined as phases, which are different from SD, DD and CD, to become more structured and part of regular practice. At the end of construction, the building model will consist of hundreds or thousands of building element models that can be transferred to a facility management organization to support operations and management (see Chapter 4).

For reference, there are over 10,000 building product manufacturers in North America. Each manufacturer produces a few to tens of thousands of products resulting in potentially hundreds of thousands of products and product applications for fulfilling a broad range of architectural expression.

Building Element Models (BEMs) are 2D and 3D geometric representations of physical products such as doors, windows, equipment, furniture, fixtures, and high-level assemblies of walls, roofs, ceilings, and floors at the various levels of detail needed, including specific products. For design firms involved in particular building types, parametric models of space types may also be carried in libraries, such as for hospital operating suites or radiation treatment rooms, to enable their re-use across projects. We consider these spatial and construction assemblies to also be BEMs. Over time, the knowledge encoded in these model libraries will become a strategic asset. They will represent 'best-practices,' as firms incrementally improve and annotate them with information based on project use and experiences. The risk for errors and omissions will decrease as firms realize greater success in developing and using high quality models from previous use.

5.4.1 Object Libraries

It is anticipated that BEM libraries will reference useful information for a range of contexts and applications throughout the project delivery and facility maintenance life-cycle. Developing and managing BEMs introduces new challenges for AEC firms, because of the large number of objects, assemblies, and object families that need to be structured for access, possibly involving multiple office locations.

Organization and Access

A review of current BIM design tools shows that they have each defined and implemented a heterogeneous set of object types, using unique object families (see Table 2-1). These objects will need to be accessed and integrated into projects using standard nomenclature for cross-product interpretation. This would enable them to support interoperability, interfacing with cost estimation, analysis, and eventually building code and building program assessment applications, among others. This includes naming, attribute structure, and possibly the designation of topological interfaces with other objects reflected in the rules used to define them. This may involve translation of objects to a common structure or defining a dynamic mapping capability that allows them to maintain their 'native' terms but also allows them to be interpretable with synonym and hyponym relations. A hyponym is a more restricted or more general term then the target one. Roof structure, for example, is a hyponym for space frame.

The complexity and company investment needed to develop BEM content suggests the need for tools for management and distribution that allow users to find, visualize, and use BEM content. Classifications, such as CSI Masterformat or Uniformat, are orthogonal to BEM. A door is a door, whether indexed by one classification or another. The new Omniclass™ classifications, now in draft form by the CSI, are expected to provide more detailed object-specific classifications and accessing mechanisms (Omniclass 2007). Given these new tools for indexing BEM classifications, it should be possible to organize BEMs for access by any number of classification schemas. A well designed library management system should support this flexibility for navigating a classification tree of BEM models.

5.4.2 Portals

Public and private portals are emerging in the marketplace. Public portals provide content and promote community through forums and indexes to resources, blogs, etc. The content tools primarily support hierarchical navigation, search, download, and in some cases upload for BEM files. Private portals permit object sharing between firms and their peers that subscribe to joint sharing agreements under control of server access and management. Firms or groups of firms that understand the value in BEM content and the value/cost relation in different application areas may share BEMs or jointly support their development. Private portals enable firms to share common content and also protect content that encodes specific, proprietary design knowledge.

Table 5-5 lists several portal sites. Most content is generic. One portal, BIMWorld, specializes in building product manufacturer specific content. Though its coverage is thin, it demonstrates that building product manufacturers are recognizing the importance of distributing product information in BEM format. It provides fully parametric objects with topological connectivity for Revit®, for ADT, and to a lesser degree for Bentley and ArchiCAD®. The Form Fonts EdgeServer™ product is an example of server technology that supports controlled sharing between peers.

The Google 3D Warehouse is a public repository for SketchUp content. This service could force a re-thinking of the library metaphor given the tools, technologies, and business opportunities it provides. 3D Warehouse incorporates:

- The ability for anyone to create a segmented area of the warehouse;
- The ability for anyone to create a schema and classification hierarchy to support library search;
- Free storage and other back-end services (reliability, redundancy etc.);
- The ability for a developer to link from a Web page to a model in 3D Warehouse, thereby putting up a *storefront* that uses 3D Warehouse as a back-end.

These capabilities are enabled by technologies that make use of semantic modeling for search, broadband access, and maturing tools and standards for application interaction, as well as to create new business opportunities. For example, McGraw Hill Sweets has begun experimenting with 3D Warehouse by creating a McGraw Hill Sweets' Group and placing Sweets-certified manufacturer BEM models in SketchUp format in the Warehouse. This Google distributed service technology for storage and search, combined with Sweets domain information and know-how for developing an AEC-specific domain model, is an example of new business opportunities enabled by technology. Google Warehouse objects are described as being for early design and have the editing and content capabilities and limitations of SketchUp. The formats of the objects in these different portals are those formats supported by the portal's content platform.

5.4.3 Desktop/LAN Libraries

Private libraries are desktop software packages designed to manage BEM content and closely integrate with the user's desktop file management system. They automate the loading of BEMs into a standalone library from a BIM tool,

Table 5-5 Comparison features of public BEM product portals.

Portals	Firm Developed BEM	Public, Private or Peer to Peer	Version Management	Navigation	Selection	Product Configura-tion	BEM Modeling Guidelines	Ranking/ Annotation	Content Platform	Website:
Autodesk Revit® 9.0 Library	No	Public	No	By Revit® Categories	Yes	No	No	No	Revit®	http://revit.autodesk.com/library/html/index.html
Revit City	No	Public	By Revit® software release	By CSI Masteerformat 04, Revit City organization, Keyword	Yes	No	No	Rating	Revit®	http://www.revitcity.com/downloads.php
Autodesk Revit® User Group	No	Public	by Revit® software release	By Revit® category, Revit® Release, Unit of Measure, Manufacturer, Author	Yes	No	No	No	Revit®	http://www.augi.com/revit.exchange/rpcviewer.asp
Objects Online	No	Public	No	By ObjectsOnline Category	Yes	No	No	No	ArchiCAD®, AutoCAD®, and SketchUp	http://www.objectsonline.com/customer/home.php

Google Warehouse	Yes	Public	No	Ability to create classification schema and add search tags	Yes	No	No	Rating	SketchUp	http://sketchup.google.com/3dwarehouse/
BIMWorld	No	Public	No	By CSI Masterformat 04, by Uniformat, by Manufacturer	Yes	Link to mfr-provided configurator in a few cases	No	No	Revit®, AutoCAD®, Microstation, SketchUp	http://www.bimworld.com/
Form Fonts	Yes	Public, peer-to-peer	No	By CSI Masterformat, by Keyword, platform, manufacturer	Yes	No	No	No	SketchUp, also Auto-CAD®, Revit®, ArchiCAD®, 3DMax, Lightwave	http://www.formfonts.com
BIM Content Manager	Yes	Private	No	By Revit® Category, folder, project, CSI Masterformat, Uniformat, Keyword, manufacturer	Yes	No	No	Yes	AutoCAD®, Revit®	http://www.digital-buildingsolutions.com/
BIM Library Manager	Yes	Private	No	By Revit® category	Yes	No	Yes	No	Revit®	http://www.tectonicbim.com

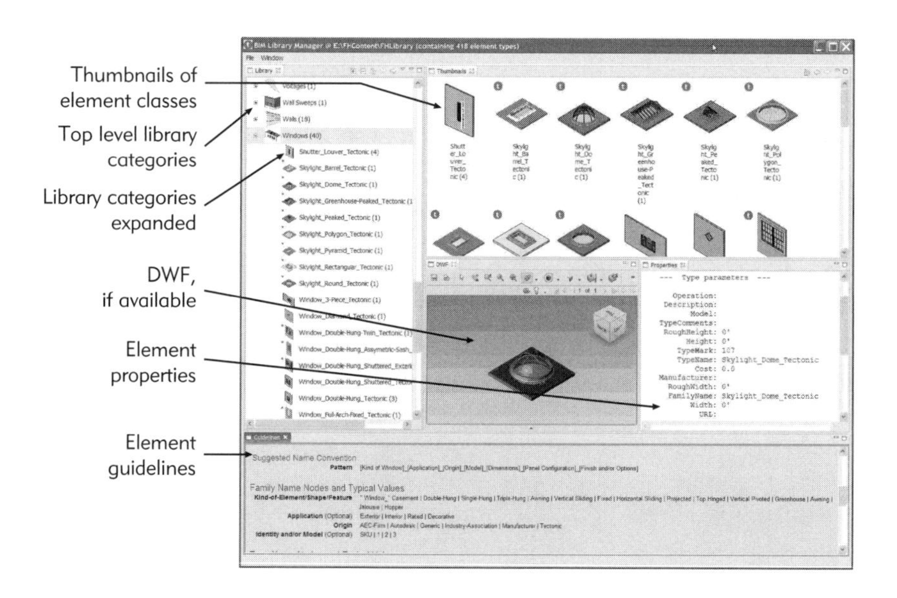

FIGURE 5-15 Multi-level structure of Tectonic Network's Element Library. *Image provided courtesy of Tectonic Partners, Inc.*

such as Revit®, or the user's own file system or a corporate network. They provide taxonomy for classification and definition of property sets upon entry and later for searches and inspection for retrieval. They assist searching, for example based on 3D visualization, inspection of categories, types, and property sets. The companies providing such tools also plan public portals for sharing BEMs across firms (file upload and download, community tools, etc.) and distributing manufacturer specific BEMs for building products.

One example of these products is Tectonic Partner's BIM Library Manager (Arnold 2007). It includes best practice guidelines for BEM modeling on the Revit® platform and supports incremental and dynamic elaboration of the underlying BEM as designers obtain information and attribute data from different sources. Figure 5-15 shows the Tectonic BIM Library Manager. The tree view of Revit® Families shows all families in the file system that are automatically loaded into the Library. BIMworld has introduced a similar product, called BIMContentManager. It works in conjunction with their portal. Additional products are being developed by both companies around these library capabilities.

5.5 CONSIDERATIONS IN ADOPTION FOR DESIGN PRACTICE

Moving the base representation of building design from a set of drawings, even if produced digitally, to a building model has many potential direct benefits: automatically consistent drawings, identification and removal of 3D spatial conflicts, automatic and accurate preparation of bills-of-material, improved support for analysis, cost and scheduling applications, and others. Three-dimensional modeling throughout the entire design process facilitates coordination and design review; and these capabilities lead to more accurate design drawings, faster and more productive drawing production, and improved design quality.

5.5.1 BIM Justification

While BIM offers the potential to realize new benefits, these benefits are not free. The development of a 3D model, especially one that includes information that supports analyses and facilitates fabrication, involves more decisions and incorporates more effort than the current set of construction documents. Considering the inevitable additional costs of implementing new systems, retraining staff, and developing new procedures, it is easy to assume that the benefits may not seem worthwhile. Most firms that have taken these steps, however, have found that the significant initial costs associated with the transition result in productivity benefits at the construction document level. Even the initial transition to producing consistent drawings from a model is worthwhile.

In the existing business structure of the construction industry, designers are usually paid a fee calculated as a percentage of construction cost. Success in a project is largely intangible, involving smoother execution and fewer problems, improved realization of design intent – and realizing a profit. With the growing awareness of the capabilities offered by BIM technology and practices, building clients and contractors are exploring new business opportunities (see Chapters 4 and 6). Designers can begin to offer new services that can be added to the fee structure. These services can be grouped into two broad areas:

1) Concept design development:

Performance-based design using analysis applications and simulation tools to address:

- sustainability and energy efficiency,
- cost and value analysis during design,
- programmatic assessment using simulation of operations, such as in healthcare facilities.

2) Integrating design with construction:

- Improved collaboration with the project team: structural, mechanical, electrical engineers, steel, MEP, precast and curtainwall fabricators. BIM use among a project team improves design review feedback, reduces errors, lowers contingency issues, and leads to faster construction.

- Expedited construction, facilitating offsite fabrication of assemblies, and reducing field work.
- Automation in procurement, fabrication and assembly and early procurement of long lead-time items.

Additional services during the concept design phase can have major owner benefits regarding reduced construction costs, lower operating costs, and improving organizational productivity and effectiveness. Comparing initial costs with operating costs is notoriously difficult, with varying discount rates, varied maintenance schedules and poorly tracked costs. However, studies from Veterans Administration hospitals have found that less than eighteen months of functional operations of a VA hospital are equal to its construction costs, (see Figure 5-16) meaning that savings in hospital operations, even with higher first cost, can be hugely beneficial. The VA has also found that the lifetime fully amortized costs of energy are equal to one eighth of construction costs (See figure 5-16) and this percentage is likely to increase. In addition, the VA has found that fully discounted plant operating costs (including energy and building security) are roughly equal to construction costs. There are many other cost items available from (Department of Veteran Affairs 2007). These examples provide an indication of the reduction in operating costs and increases in performance that building owners/operators will be seeking.

The benefits of integrating BIM design with construction are already well articulated in Section 5.3.4.

BIM Design Productivity Benefits

One way to indirectly assess the production benefits of a technology such as BIM is according to the reduction of errors. These are easily tracked by the number of Requests for Information (RFI s) and Change Orders (COs) on a project. These will always include a component based on the client's change of mind or changes in external conditions. However, changes based on internal consistency and correctness can be distinguished and their numbers on different

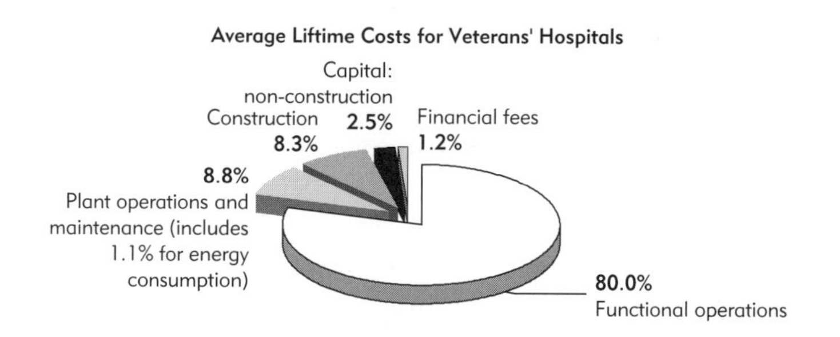

FIGURE 5-16 The various components of the lifetime capital and operating costs of a veteran's hospital. *Image provided courtesy of Veteran's Administration (Smoot 2007).*

projects collected. These indicate an important benefit of BIM and have been reported in several of the case studies in Chapter 9.

Another way to assess technology benefits is in terms of productivity. Labor productivity is the total cost in terms of labor hours and salaries to realize some task. It is an often under-applied consideration for design firms. More often, decisions are made according to the qualitative benefits outlined above, and the anticipated future benefits – as a change in the culture of construction. If a technology can reduce the number of hours required to produce a given result, for example construction documentation, then the return on investment can be assessed by calculating the value of the reduction in labor costs and comparing it with the various investments required.

Design firms are often not familiar with methods of assessing productivity. An initial step in making such an assessment is to establish a baseline for comparison. Few firms keep track of the unit costs associated with design development and construction drawing detailing, for example, based on building floor area, facade area, or project type. These can provide a baseline metric to evaluate the costs or benefits of a transition to new design technologies (such a method is described by (Thomas et al. 1999).

The second step is to estimate the productivity gain of the new technology, in this case BIM. Apart from the productivity enhancement figures provided by various BIM vendors (Autodesk 2004), there is little data available within design firms that have already adopted BIM or even in available research literature. Research into the productivity gain for producing structural engineering drawings with rebar detailing has yielded gains between 21% and 59%, depending on the size, complexity, and repetitiveness of the structures (Sacks and Barak 2006). A few figures are also provided in the case studies in Chapter 9. Of course, benefits for a particular design firm are necessarily speculative until real projects are undertaken. An assessment should distinguish time saved weighted according to the average wage of those doing the work and its percentage of the firm's annual labor cost. This will provide a weighted productivity gain. The resulting percentage can be multiplied by the annual direct labor costs for design activities to compute the annual benefit.

The last step is to calculate the investment costs of adoption. The largest cost will be the labor cost of training time, which should include both direct costs for time spent and also the 'learning curve cost' of initially reduced productivity as people learn to use the new tools. Hardware and software costs can be estimated in consultation with a BIM vendor. Productivity benefits will grow to their full extent over time. Finally, the total annual benefit divided by the total cost should provide a quick measure of the annual return on investment and the time needed to recoup the cost.

Section 2.3.1 provides guidelines for the selection of BIM tools. Modeling tools are not only for internal usage. Another consideration is the needs of companies that are frequent design partners. Ideally, if there are certain dominant working relations, decisions should be made with some level of coordination.

It also should be recognized that a single BIM tool is not necessarily ideal. Some firms decide not to limit themselves to a single model generation tool, but rather to support multiple BIM products, recognizing that some tools have non-overlapping benefits.

5.5.2 Training and Deployment

BIM is of course a new IT environment, requiring training, system configuration, library and document template setup, and adaptation of design review and approval procedures, often combined with new business practices. These need to be developed incrementally, side-by-side with existing production methods, so that learning problems do not jeopardize the completion of current projects.

We encourage preparation of a detailed deployment plan for any firm considering making a change to BIM; adoption should not be treated as an ad hoc activity. The more grounded the plan is in relation to a company's strategic goals, the more successful adoption is likely to be. The following sections address a range of issues to be considered in the deployment plan.

Training usually starts with one or a small number of IT specialists that both plan for system configurations and introduce a training program for the rest of the firm. System configuration includes hardware selection (BIM tools demand powerful workstation hardware), server set up, plotting and printing configurations, network access, integration with reporting and project accounting, setup of libraries (described in Sec. 5.4.1), and other company-specific system issues.

Early projects should focus on the basic skills needed for modeling buildings and producing drawings, including incrementally compiling object libraries and getting the basics down before undertaking more advanced integration efforts. After the basics of project management have been realized, the door is open to a variety of extensions for taking advantage of the multiple integration and interoperability benefits that BIM offers.

An important note of caution during the early phase of BIM adoption is to avoid providing too much model detail too soon. Because methods of project definition and detailing are partially automated in BIM, it is possible, if details are defined too quickly, for a design concept to be misinterpreted. Detailed models are easy to realize while still in the conceptual design phase but may lead to errors and client misunderstanding by inadvertently making overreaching decisions that become hard to reverse. It is important for BIM users to understand this issue and to manage the level of detailing more explicitly than would be done by hand. A reconsideration of the level of detail provided to consultants and collaborators has also been found to be worthwhile. These parties can be brought into discussions earlier or later, depending on their roles. Detailed MEP 3D layout should not be done until later in the process to avoid multiple revisions. On the other hand, curtain wall consultants and fabricators may be brought in earlier to help plan structural connections and detailing.

Architects represent only one component of an overall design team. Collaboration requires a number of engineering, mechanical, or other specialty consultants. The default initial integration arrangement is to rely on drawings in the conventional manner. Very quickly, however, the extra steps required for producing drawings leads to the desire for model-based exchanges (the issues involved are reviewed in detail in Chapter 3). Data exchange methods must be worked out on a company-by-company basis. Model-based coordination using web conferencing is a straightforward and very effective means of managing projects (see Section 5.3.2).

5.5.3 Phased Utilization

A wide range of integration capabilities are reviewed at the start of this chapter and in the expanded set of design services discussed in Section 5.4.1. These should be kept in mind as new projects offer opportunities to move BIM integration to new levels. In addition to external services discussed earlier, other services can be undertaken in almost any context. Among these are:

- Integration with cost estimation to allow continuous tracking throughout project development.
- Integration with specifications for better information management.
- Development of proprietary company libraries of detailing, room configurations, and other design information to facilitate the transfer of specialized staff knowledge to corporate knowledge.

Each type of integration involves its own planning and development of workflows and methods. For example, data exchange methods require definition, testing, and coordination among participants. Taking a step-by-step approach will allow for incremental training and adoption of advanced services without undue risks, which will lead to radically new capabilities within the overall design firm.

5.6 NEW AND CHANGED STAFFING WITHIN DESIGN FIRMS

The greatest challenge in implementing new design technologies is the intellectual transition in getting senior design team leaders to adopt new practices. These senior staff, often partners, have decades of experience with clients, design development procedures, design and construction planning and scheduling, and project management that represent part of the core intellectual property within any successful firm. The challenge is to engage them in the transition in a way that enables them to realize both their own expertise and also the new capabilities that BIM offers them.

Among the several potentially effective ways to address this challenge:

- Team partners with young BIM-savvy design staff who can integrate the partner's knowledge with the new technology.
- Provide one-on-one training one day a week or on a similar schedule.
- Host a charette for design teams that includes training for partners in a relaxed offsite location.

Similar transition issues exist with other senior staff, such as project managers and similar methods may be used to facilitate their transition. No method is sure-fire. The transition of a design organization is largely cultural. Through their actions, support, and expression of values, senior associates communicate their attitudes toward new technology to the junior members within the organization.

A second major challenge in any design firm will be the changed composition of staff with respect to skills. Because BIM most directly enhances productivity for design documentation, the proportion of hours spent on any project shifts away from construction documentation. Within a typical practice, a designer skilled in BIM can realize the intention and detailing of a project with much less outside drawing or modeling support than was previously required. Details, material selections, and layouts only need to be defined once and can be propagated to all drawings where they will eventually be visible. As a result, the number of junior staff members working on construction documentation will be reduced. A good example of the way in which the workload for a project is shifting in an architectural practice that has already adopted BIM can be seen in Table 5-6. This data was reported by a principal architect at a large design firm (Birx 2005). While the total labor hours are reduced, the total cost did not change substantially due to a shift toward a more experienced labor staff.

Table 5-6 Shifting demand for design skills on a typical project.

	Project Hours		
Professional Grade	**Pre-BIM**	**Post-BIM**	**Change**
Principal	32	32	0%
Project manager	128	192	33%
Project architect	192	320	40%
Architect 1	320	192	−67%
Intern architect	320	96	−233%
Total	992	832	−19%

Although the need for entry-level architects is reduced, drawing cleanup, model detailing, and integration and coordination of multiple building subsystems will continue as important and valuable tasks.

BIM technology has new associated overhead costs beyond that of software investment. As firms already know, system management, often under the management of the Chief Information Officer (CIO), has become a crucial support function for most firms. IT dependency expands as it supports greater productivity in the same way that electricity has become a necessity for most kinds of work. BIM inevitably adds to that dependency.

As design firms adopt BIM, they will need to assign responsibility for the two much-expanded roles that will be crucial to their success:

1) Systems Integrator—this function will be responsible for setting up exchange methods for BIM data with consultants inside and outside the firm. It also involves setup of libraries (as described in Section 2.2.4) and templates for company use. The applications may be limited to a single set that are used in every project or a variable set that is selected according to the type of project and the consultants involved.

2) Model Manager—while the protocols for version control and managing releases are well developed and understood within the drawing document-based world (whether paper or virtual), options are different and more open-ended with BIM. There may be a single master model or a set of federated ones. Since models are accessible 24/7, releases can potentially be made multiple times a day. As a result, the potential for model corruption exists. Because a project model is a high-value corporate product, maintaining its data integrity justifies explicit management. The model manager determines the policies to be followed for establishing read-and-update privileges, for merging consultants' work and other data into a master model, and for managing model consistency across versions.

Dealing with model review and releases and managing the consistency of models will require special attention until a set of conventions becomes standard. The model manager role must be assigned for each project.

5.7 NEW CONTRACTUAL OPPORTUNITIES IN DESIGN

Design services have traditionally been offered at a fee for services rendered, with the fee proportional to the costs of construction. The final set of construction documents defines the design to the level needed for competitive bidding, cost estimation, and conveying "design intent." Current AIA contract forms (AIA 1994) seek to limit the responsibility of the designer for actual construction issues, placing that responsibility instead on the contractor. On the other hand, there have been some explorations by architects to provide facility management and other downstream services in order to capture more of the value latent in architectural drawings (Demkin 2001).

While there has been longstanding criticism of the architect's standard fee structure (Stephens 2007), the advent of BIM provides what may possibly be the ultimate challenge to current business models. Current business models stipulate that the designers are responsible for design intent but not for providing reliable construction information. This has led some contractors to develop entirely new building information models that are rebuilt in a separate BIM environment, essentially using the design model as an underlay. The new model serves as the reference model for construction, into which various fabrication-level models of separate building systems are integrated. This model, which must be checked and approved by the architect, is used for final coordination, fabrication, and construction.

A well-known example of this type of arrangement is the Denver Art Museum designed by Daniel Libeskind and built by M.A. Mortenson. Interestingly, the contractor was the primary awardee of an American Institute of Architects prize for the Stellar Architecture BIM Award from the Technology in Architectural Practice Committee of the AIA in 2006. Libeskind's design was represented as a form·Z model of the building geometry, which only represented design intent. Its geometry served as the basis for a structural steel model produced using Tekla Structures. A variety of additional models of different building subsystems were compiled in various other software packages and coordinated with the structural model using Naviswork's Jetstream software.

In contrast to this clean partition between design and fabrication and its obvious replication of modeling, there are examples of integrated models such as the Flint GM plant design by Ghafari and Associates, presented in the Chapter 9. In this case, a single model was used as the design coordination model, although fabrication models were developed in different systems to provide the specialist functionality needed for different types of fabrication.

In the Denver Art Museum, the value of the architectural model lay entirely with its expression of design intent, while the fabrication models dealt entirely with constructability, details of functional performance, construction costs and scheduling aspects. On the other hand, the GM engine plant design model included both the issues of design intent and also the issues of constructability, functional performance, costs and schedule. Thus the services provided in each case were very different, and justified different fee structures.

In addition to the extra services described in Section 5.4.1, design firms can offer other services in a BIM-driven construction environment, including:

- Calculate estimated facility operating costs (energy units) for final design, or on a release-by-release basis.
- Prepare models appropriate for code reviews by authorities offering BIM enabled code review.
- Provide model management services for the contractor so that the model can be used for procurement, scheduling and other services of the contractor's choosing.
- Coordinate fabrication level systems for spatial configuration.
- Provide geometric detail information for CNC fabrication of specific subsystems.
- Prepare performance metrics for building commissioning.
- Provide detailed as-built information models for facility management and operations.

We encourage design firms to explicitly define the business and service model that they plan on carrying out with respect to each project by articulating the services to be offered in such a way that the client gains a good understanding of the scope of design services, including at what point the project ends and when responsibility is transferred to the contractors. Some of the different forms of contracting are reviewed in Section 4.4.

Each new service option has its own requirements that must be identified carefully in terms of the scope and limits of that level of service. In addition, issues associated with legal liabilities must be considered. In summary, these types of services provide new options for leveraging BIM technology to enhance design value in construction and for generating additional revenue for designers.

Chapter 5 Discussion Questions

1. Thinking about the level of information needed for cost estimation, scheduling and purchasing, outline your recommendation regarding the level of detail that should be defined in a design model. Consider and recommend what the role of designers should be in supporting these activities.
2. Other buildings built for the Beijing Olympics, beyond the Aquatic Center, used BIM. Much information is available on the Beijing Olympics and associated web sites. For any one of the other buildings, review and report how the design was carried out, and how information was carried for fabrication and construction.
3. Consider any specific type of building system, such as hung ceiling systems, or an off-the–shelf curtainwall system. For that system, identify how it could be supported by automation tools for its custom adaptation to a particular project. How could its fabrication be facilitated? Identify which levels of automation are practical today and which not.
4. Obtain the recommended set of details for installing a manufactured door, window or skylight. Examine and identify, using paper and pencil, the variations that might apply the detail. List these variations as a specification for what an automated parametric detailer needs to do.
5. Propose a new service for a design firm, based on the capabilities of BIM. Outline how the service would be of value to the owner. Also outline a fee structure and the logic behind that structure.
6. Conceptual design is often undertaken in such non-traditional BIM tools as form·Z or Maya. Lay out the alternative design development process utilizing one of these tools, in comparison to one of the new BIM tools. Assess the costs and benefits of both development paths.

CHAPTER 6

BIM for the Construction Industry

6.0 EXECUTIVE SUMMARY

Utilizing BIM technology has major advantages for construction that save time and money. An accurate building model benefits all members of the project team. It allows for a smoother and better planned construction process that saves time and money and reduces the potential for errors and conflicts. This chapter explains how a contractor can obtain these benefits and what changes to construction processes are desirable.

Perhaps the most important point is that contractors must push for early involvement in construction projects, or seek out owners that require early participation. Contractors and owners should also include subcontractors and fabricators in their BIM efforts. The traditional design-bid-build approach limits the contractor's ability to contribute their knowledge to the project during the design phase, when they can add significant value.

While some of the potential value of a contractor's knowledge is lost after the design phase is complete, significant benefits to the contractor and the project team can still be realized by using a building model to support a variety of construction work processes. These benefits can ideally be achieved by developing a model in-house with the collaboration of subcontractors and fabricators; having a consultant develop a model is also possible.

The level of detail of the information in a building model depends on what functions it will be used for. For example, for accurate cost estimating, the model must be sufficiently detailed to provide the material quantities needed for cost evaluation. For 4D CAD schedule analysis, a less detailed model is adequate, but it must contain temporary works (scaffolding, excavation) and show how the construction will be phased (how deck pours will be made, etc.).

One of the most important benefits is derived from close contractor coordination that can be achieved when all of the major subcontractors participate in using the building model for detailing their portions of the work. This permits accurate clash detection and correction of clashes before they become problems in the field. It enables increased offsite prefabrication which reduces field cost and time and improves accuracy. Each of these uses of a building model is discussed in detail and examples are illustrated in the case studies in Chapter 9.

Any contractor that is contemplating the use of BIM technology should be aware that there is a significant learning curve. The transition from drawings to a building information model is not an easy one because almost every process and business relationship is subject to some change in order to exploit the opportunities offered by BIM. Clearly, it is important to plan these changes carefully and to obtain the assistance of consultants who can help guide the effort. At the end of the chapter we provide suggestions for making the transition and identify what problems can be anticipated.

In the absence of owner or designer-driven BIM efforts, it is vital that contractors establish leadership in the BIM process if they are to gain the advantages for their own organization and better position themselves to benefit from industry-wide BIM adoption.

6.1 INTRODUCTION

This chapter begins with a discussion of the various types of contractors and how BIM can provide benefits for their specific needs. It then goes into depth on important application areas that apply to most contractors. These include:

- Clash detection
- Quantity takeoff and cost estimating
- Construction analysis and planning
- Integration with cost and schedule control and other management functions
- Offsite fabrication
- Verification, guidance, and tracking of construction activities

It follows with a discussion of the contractual and organizational changes that are needed to fully exploit the benefits that BIM offers. It concludes with some thoughts on how BIM can be implemented in a construction company.

6.2 TYPES OF CONSTRUCTION FIRMS

There is a tremendous range of construction companies, from large companies that operate in many countries and offer a wide range of services to small companies that have individual owners who work on one project at a time and provide a highly specialized service. There are far more of the latter (small-scale companies) than the former, and they perform a surprisingly large percentage of the total construction volume. Data for 2004 is shown in Figure 6-1. It shows that a large percentage of firms were composed of just 1 to 19 people (91.6%), but a majority of construction employees worked in firms larger than 19 people (61.6%). A very small percentage of firms (0.12%) had over 500 workers, and they employed 13.6% of the workforce. Average firm size was 9 employees.

When we look at the building industry, the range of contractors is also very large in terms of the services they offer. The bulk of the industry consists of contractors who start with a successful bid, self-perform some of the work, and hire subcontractors for specialized services. Some contractors limit their service to managing the construction process. They hire subcontractors for all construction work. At the other end of the spectrum are design-build firms that take responsibility for both the design and construction processes but subcontract the bulk of the construction work. Almost all contractors end their responsibilities when construction is complete, but there are some that offer services in the turnover and management phases of the finished building

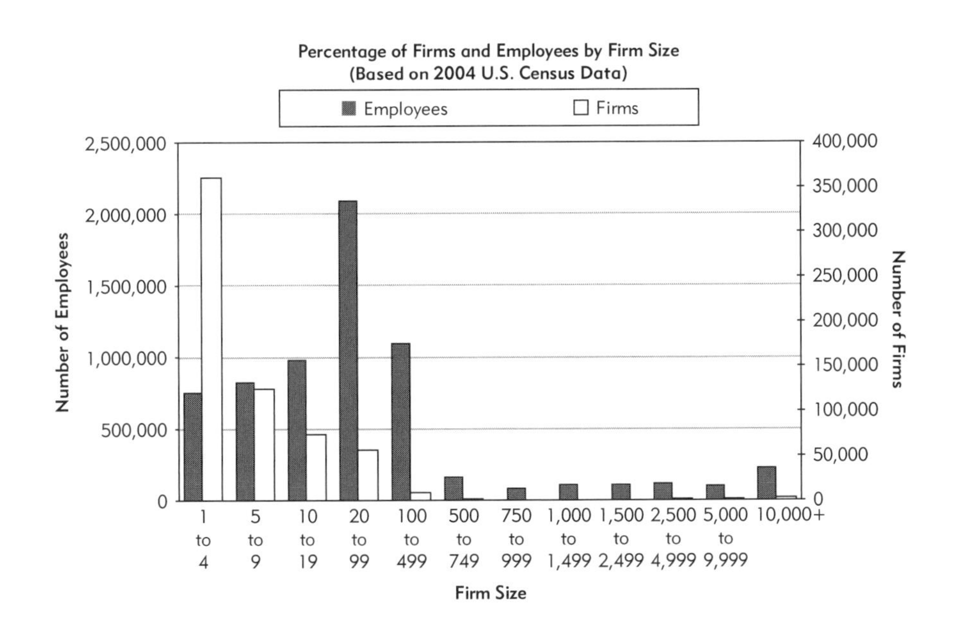

FIGURE 6-1 Distribution of 751,098 construction firms and total employees by size of firm for 2004.
Source: US Census Bureau, NAICS 23 – Construction.

(build-operate-maintain). Figure 6-2 shows the percentage of firms in each major sector of the construction industry in 2004. It shows that a majority of all firms fall in the specialty trade category (mainly small subcontractors).

Home builders differ from most other construction companies in that they act as developers: buying the land and applying for zoning changes, planning and constructing the infrastructure, and designing and building the homes that are sold. Home builders range in size from large public firms that build thousands of homes each year to individuals that build just one home at a time.

Fabricators of components produced offsite function as a hybrid between manufacturers and contractors. Some fabricators, such as precast concrete manufacturers, produce a range of standard products as well as custom items designed for a given project. Steel fabricators fall into the same category. A third group includes specialty fabricators that manufacture structural or decorative items from special steel, glass, wood, or other materials.

Finally, there are many types of subcontractors that specialize in one area or type of work, such as electrical, plumbing, or mechanical detailing. The general contractor selects these subcontractors based on competitive bids or they are pre-selected based on previous business relationships that have demonstrated effective collaboration. The specialized construction knowledge of these subcontractors can be very valuable during design, and many of them perform design as well as construction services. The percentage of work done by subcontractors varies widely depending on the type of work and contract relationship.

A typical project team organization is illustrated in Figure 6-3. There are many options for the organization of the project team. One is for the owner to hire a construction manager (CM), who then advises the owner or architect on

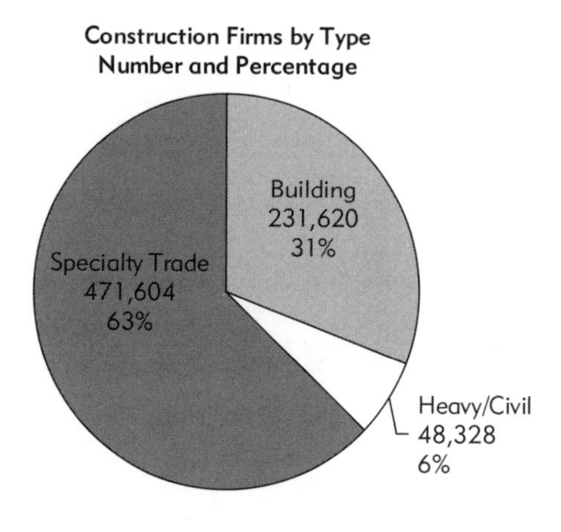

FIGURE 6-2 Percent of firms in each major construction sector, 2004. US Census Bureau, NAICS 23 – Construction.

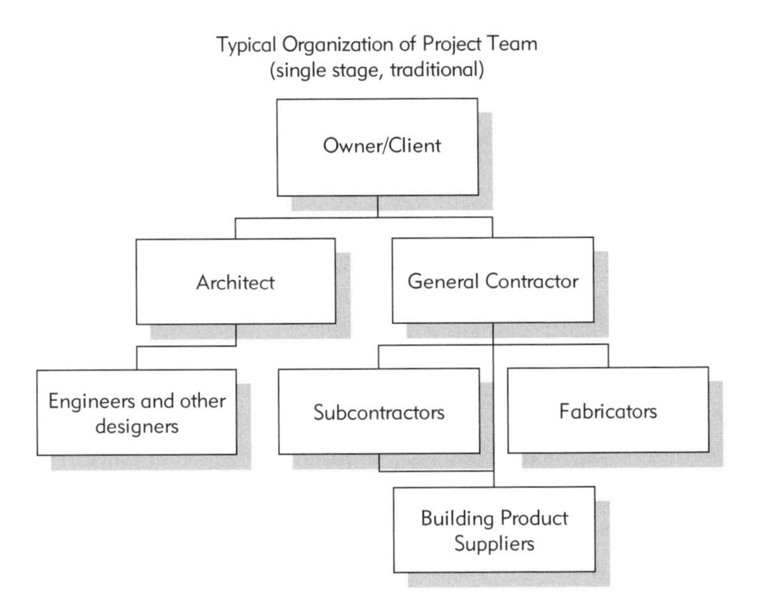

FIGURE 6-3 Typical organization of a project team for a building project.

the construction of the project but rarely assumes the risks associated with cost overruns.

The design-build (DB) firm is an important variation of the "typical" organization shown above (see Chapter 1 Section 1.1.2 for additional discussion of DB). The DB organization assumes responsibility for both design and construction. It serves as the single point of responsibility for nearly all problems associated with the project after an agreement has been reached on project scope and the total budget and schedule are established. The DB model reduces risk for the client because it eliminates disputes associated with determining which firm is responsible for design errors or construction problems. The use of BIM in a DB firm can be very advantageous because early integration of the project team is possible and expertise is available for building the model and sharing it with all team members. This important advantage, however, cannot be achieved if the DB firm is organized along traditional disciplines and the designers work with 2D or 3D CAD tools that produce drawings or other documents that are merely handed-off to the construction group when the design is complete. In this case, much of the value that BIM brings to the project is lost, because the building model must be created after the design is complete. While this can still provide some value (see discussion following), it overlooks one of the major benefits of BIM for a construction organization—the ability to overcome the lack of true integration between design and construction. This lack of integration is the Achilles' heel of many projects.

6.3 INFORMATION CONTRACTORS WANT FROM BIM

Given the diversity of contractor types described above, it is not surprising that there is a wide range of processes and tools currently in use across the industry. Larger firms typically use computer-based systems for almost all of their key work processes, including: estimating, construction planning and scheduling, cost control, accounting, procurement, supplier and vendor management, marketing, etc. For tasks related to the design, such as estimating, coordination and scheduling, paper plans and specifications are the typical starting point, even if the architect used 2D or 3D CAD systems for the design. These require contractors to manually perform quantity takeoffs to produce an accurate estimate and schedule, which is a time-consuming, tedious, error-prone and expensive process. For this reason, cost estimates, coordinated drawings, and detailed schedules are often not performed until late in the design process.

Fortunately, this methodology is beginning to change, as contractors are recognizing the value of BIM for construction management. By using BIM tools, architects are potentially able to provide models that contractors can use for estimating, coordination, construction planning, fabrication, procurement, and other functions. At a minimum, the contractor can use this model to quickly add detailed information. To permit these capabilities, a building model would optimally provide contractors with the following types of information:

- ***Detailed building information*** contained in an accurate 3D model that provides graphic views of a building's components comparable to that shown in typical construction drawings and with the ability to extract quantity and component property information.
- ***Temporary components*** to represent equipment, formwork and other temporary components that are critical to the sequencing and planning of the project.
- ***Specification information associated with each building component*** with links to textual specifications for every component that the contractor must purchase or construct.
- ***Analysis data related to performance levels and project requirements*** such as structural loads, connection reactions and maximum expected moments and shear, heating and cooling loads for tonnage of HVAC systems, targeted luminance levels, etc. This data is for fabrication and MEP detailing.
- ***Design and construction status*** of each component to track and validate the progress of components relative to design, procurement,

installation, and testing (if relevant). This data is added to the model by the contractor.

No BIM tool today comes close to satisfying this list of requirements, but this list serves to identify the information needs for future BIM implementations. Today, most BIM tools support the creation of information in the first and second items in the list.

An accurate, computable, and relatively complete building model that includes the above information is needed to support critical contractor work processes for estimating, coordinating trades and building systems, fabricating components offsite, and construction planning. It is important to note that each new work process often requires that the contractor add information to the model, since the architect or engineer would not traditionally include means and methods information such as equipment or production rates, which are critical for estimating, scheduling and procurement. Contractors use the building model to provide a base structure to extract information and will add construction-specific information as-needed to support various construction work processes.

Additionally, if the scope of work for the contractor includes turnover or operations of the facility, links between BIM components and owner control systems, such as maintenance or facility management, will facilitate the handover process to the owner at the end of the project. The building model needs to support representation of information related to all of these processes.

6.4 PROCESSES TO DEVELOP A CONTRACTOR BUILDING INFORMATION MODEL

While use of BIM technology is increasing rapidly, it is in the early stages of implementation and contractors are now utilizing many different approaches to leverage this new technology. Most design teams are not creating models for every project, which has led to contractors taking ownership of the modeling process. Even when architectural use of BIM becomes commonplace, contractors will need to model additional components and add construction-specific information to make building models useful to them. Consequently, many leading-edge contractors are creating their own building models from scratch to support estimating, 4D CAD, procurement, etc. Figure 6-4 shows a common workflow of a contractor creating a building information model from 2D paper drawings.

Note that, in some cases, the contractor is building a 3D model that is only a visual representation of the project. It does not contain parametric components or relations between them. In these cases, use of the model is limited to

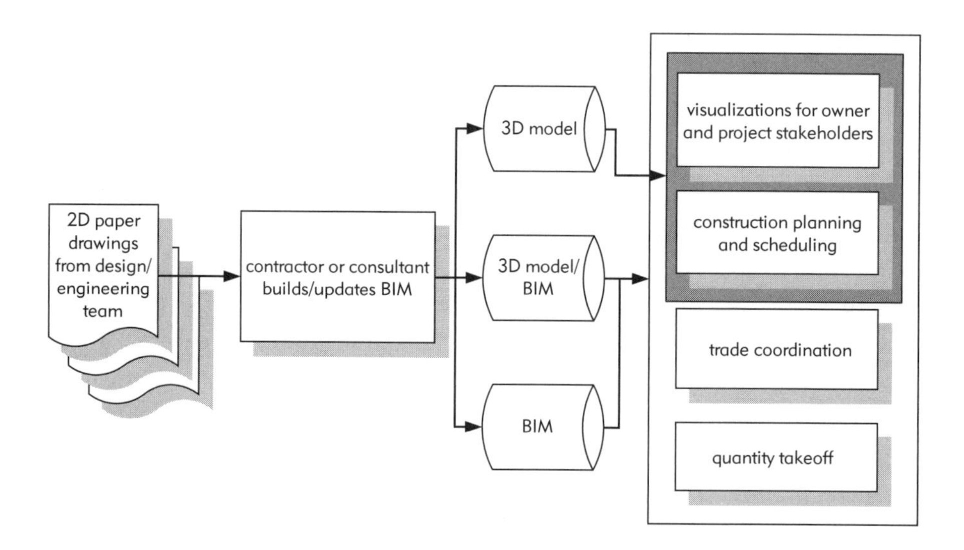

FIGURE 6-4 BIM process flow for a project where the contractor builds the construction model from 2D drawings and then uses it for quantity takeoff, construction planning, and clash detection.

clash detection, visualization, and visual planning, such as 4D, because the 3D model does not define discrete quantifiable components to support quantity takeoff or trade coordination. In other cases, contractors may build a hybrid 3D/parametric model that includes some BIM components, which enable some coordination and quantity takeoff. When contractors do produce a full building model, they can leverage it for multiple purposes.

Another approach for implementing BIM is illustrated in Figure 6-5. In this case, the project team collaborates on a model – 3D, BIM, or hybrid – in an environment that is suited to their practice. Alternatively, if a specific organization works in 2D, the contractor or consultant can convert the 2D to 3D/BIM so that their work can be entered into the shared model. Typically, the contractor or the consultant manages the integration of these various models, which are developed independently by different members of the project team but then merged into a collaborative model. The shared model can be used by the project team for coordination, planning, quantity takeoff, and other functions. While this approach does not take advantage of all the tools that a full-featured building information model supports, it does reduce costs and time compared to traditional practices. The shared 3D model becomes the basis for all construction activity and allows for much greater accuracy than 2D drawings.

As the practice and use of BIM increases, new processes will evolve. The case studies in Chapter 9 highlight a variety of ways in which contractors are adapting their work process to leverage BIM. In the following sections, we discuss specific modeling processes.

Home builders provide a good example of how a design-build effort can benefit from the use of BIM technology. When developing designs for model homes,

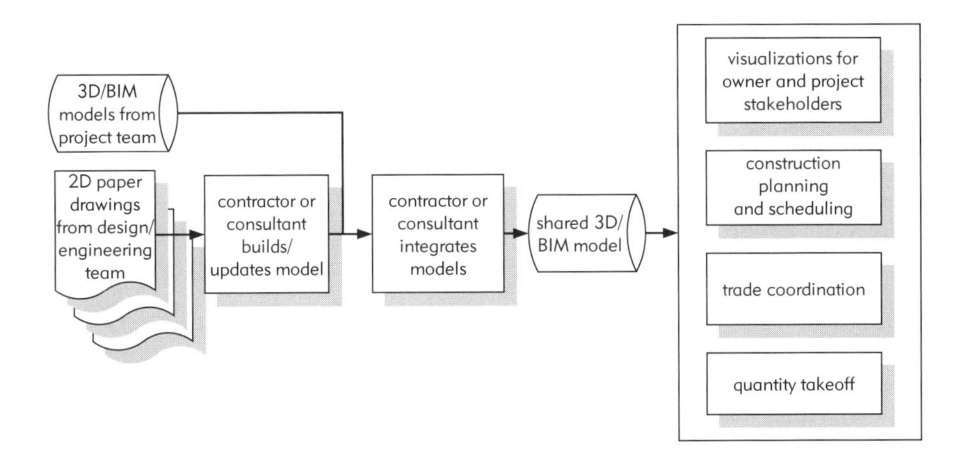

FIGURE 6-5 Process flow for a project, where the architect and other designers and sub contractors use 3D modeling tools (or have a consultant develop a 3D model from 2D drawings) and contribute to a shared 3D model.

a building information model can provide rapid feedback on the quantity and cost implications of a design change. When a buyer requests design modifications to a model home, this capability can provide fast visual and cost feedback and allow the prospective buyer to quickly reach an agreement with the builder. This kind of rapid response to clients' needs is of great value, especially for construction companies that provide customized building options based on systematic methods of construction†.

4D CAD tools allow the contractor to simulate and evaluate the planned construction sequence and share it with others in the project team. Objects in the building model should be grouped according to the phases of construction and linked to appropriate activities in a project schedule. For example, if a concrete deck will be placed in three pours, then the deck must be detailed into three sections so that this sequence can be planned and illustrated. This applies to all objects needed for these three pours: concrete, steel, embeds, etc. In addition, the excavation areas and temporary structures such as scaffolding and lay-down areas should be included in the model. This is a key reason why contractor knowledge is beneficial when defining a building model. If the model is built by the architect or the contractor while the building is still being designed, the contractor can provide rapid feedback regarding constructability, sequencing, and estimated construction cost. Early integration of this information is of great benefit to the architect and owner.

† Three examples are the high-tech office buildings provided by the Beck Group, small scale steel buildings provided by <<http://www.butlermfg.com/steel_bld_ctr/>> or <<http://www.steelbuildings.com>> and precast parking structures designed, manufactured and built by Finfrock. Each of these companies has developed sophisticated BIM applications integrated with cost-estimating systems. The trend they represent, exploiting BIM to provide a competitive advantage by providing customized but yet 'off-the-shelf' buildings, is detailed in Chapter 7.

6.5 REDUCTION OF DESIGN ERRORS USING CLASH DETECTION

A critical work process for any contractor is trade and system coordination. Today, most clash detection is performed manually by overlaying individual system drawings on a light table to identify potential conflicts. Similarly, contractors use traditional 2D-CAD tools to overlay CAD layers to visually and manually identify potential conflicts. These manual approaches are slow, costly, prone to error, and depend on the use of up-to-date drawings. To overcome these problems, some organizations use custom-written applications for automatically detecting clashes between drawing entities on different layers. Automatic detection of conflicts is an excellent method for identifying design errors, where objects either occupy the same space (a hard clash) or are so close (a soft clash) that there is insufficient space for access, insulation, safety, etc. In some publications, the term 'clearance clash' is used instead of 'soft clash'. The terms are synonymous.

BIM-based clash detection provides many advantages over traditional 2D coordination methods like overlays on a light table or automated 3D checks. Use of a light table is time consuming, error prone and requires that all drawings be current. 3D clash detection relies on 3D geometry models for identifying geometric entities often return a large number of meaningless clashes. Second, if the 3D geometries are not solids. the clash detection tool cannot detect clashes between objects within other objects. It can only detect clashes between surfaces. Furthermore, qualification of clashes into meaningful categories for the contractor is greatly inhibited due to lack of semantic information embedded in the 3D geometry models. A clash between surfaces could be a wall abutting a wall or a pipe running through a wall. The contractor has to verify and review each of these potential clashes.

In contrast, BIM-based clash detection tools allow automatic geometry-based clash detection to be combined with semantic and rule-based clash analysis for identifying qualified and structured clashes. BIM-based clash detection tools allow contractors to selectively check clashes between specified systems, such as checking for clashes between mechanical and structural systems, because each component in the model is associated with a specific type of system. Consequently, the clash detection process can be performed at any level of detail and across any number of building systems and trades. A BIM-based clash detection system can also utilize these component classifications to more readily perform soft clash analyses. For example, the contractor can search for conditions in which the clearance or space between mechanical components and the sub-floor is less than two feet. These types of clash detection analyses are only possible with well-defined and structured building models.

Regardless of the model's accuracy, the contractor must ensure that the building is modeled with an appropriate level of detail. It must have sufficient details for piping, ducts, structural steel and attachments, and other components, so that clashes can be accurately detected. If the detailing is inaccurate, a significant number of problems will not be found until the building is constructed, at which time they could be costly and time consuming to resolve. Proper detailing of the model by subcontractors or other project team members responsible for the design of these systems is required. These subcontractors need to participate in the model development process as early as possible. Ideally, resolution would take place in a common project site office, where a large monitor can be used to display each problem area and each discipline can contribute their expertise to the solution. Agreed upon changes can then be entered into the appropriate design model prior to the next clash detection cycle. Figure 6-6 shows a snapshot of two employees from the contractor and subcontractor using a building information model to support MEP coordination. This was done in a trailer at the job site. The case study of the Camino Group Medical Building in Chapter 9 is a good example of early subcontractor participation in detailing a 3D model used for clash detection and other functions.

There are two predominant types of clash detection technologies available in the marketplace: 1) clash detection with BIM design tools and 2) BIM integration tools that perform clash detection. All major BIM design tools include some

FIGURE 6-6 Snapshot of contractors and subcontractor using a building information model to support MEP coordination.
Courtesy of Swinerton, Inc.

clash detection features that allow the designer to check for clashes during the design phase. But the contractor often needs to integrate these models and may or may not be able to do so successfully within the BIM authoring tool due to poor interoperability or the number and complexity of objects.

The second class of clash detection technologies can be found in BIM integration tools. These tools allow users to import 3D models from a wide variety of modeling applications and visualize the integrated model. Examples of this are Navisworks' JetStream package (Navisworks 2007) and Solibri Model Checker (Solibri 2007). The clash detection analyses that these tools provide tend to be more sophisticated, and they are capable of identifying more types of soft and hard clashes. The drawback is that identified clashes cannot be fixed immediately because the integrated model is not directly associated with the original model. In other words, the information flow is one-way and not bi-directional. An exception to this statement is the Solibri Model Checker and Issue Locator which do have the ability to provide feedback in the originating building model for Architectural Desktop (from Autodesk) and ArchiCAD (from Graphisoft). These changes must also be introduced into the originating systems or upstream modeling tools. Then new files are generated when the integrated model is updated and a new clash detection analysis is performed. The updating process can cause errors and delays that must be further coordinated through careful file management among the project team.

6.6 QUANTITY TAKEOFF AND COST ESTIMATING

There are many types of estimates that can be developed during the design process. These range from approximate values early in the design to more precise values after the design is complete. Clearly, it is undesirable to wait until the end of the design phase to develop a cost estimate. If the project is over budget after the design is complete, there are only two options: cancel the project or apply value engineering to cut costs and possibly quality. As the design progresses, interim estimates help to identify problems early so that alternatives can be considered. This process allows the designer and owner to make more informed decisions, resulting in higher quality construction that meets cost constraints. Just as significant, using a BIM approach can reduce the time needed to achieve a high quality building by improving design and construction collaboration and accuracy.

During the early design phase, the only quantities available for estimating are those associated with areas and volumes, such as types of space, perimeter lengths, etc. These quantities might be adequate for what is called a *parametric cost estimate,* which is calculated based on major building parameters. The parameters used depend on the building type, e.g., number of parking spaces and floors for a parking garage, number and area of each type of commercial space, number of floors, quality level of materials for a commercial building, location of building, etc. Unfortunately these quantities are not generally available in early design packages (such as SketchUp), because they do not define object types, such as those created by a BIM package. Therefore, it is important to move the early design model into BIM software to allow for quantity extractions and approximate cost estimates. An example of this type of system is the DProfiler modeling and estimating system from Beck Technology (see additional description of this system in Chapter 4 and the Hillwood Commercial Project case study in Chapter 9).

As the design matures, it is possible to rapidly extract more detailed spatial and material quantities directly from the building model. All BIM tools provide capabilities for extracting counts of components, area and volume of spaces, material quantities, and to report these in various schedules. These quantities are more than adequate for producing approximate cost estimates. For more accurate cost estimates prepared by contractors, problems may arise when the definitions of components (typically assemblies of parts) are not properly defined and are not capable of extracting the quantities needed for cost estimating. For example, BIM software might provide the linear feet of concrete footings but not the quantity of reinforcing steel embedded in the concrete; or the area of interior partition walls but not the quantity of studs in the walls. These are problems that can be addressed, but the approach depends on the specific BIM tool and associated estimating system.

It should be noted that while building models provide adequate measurements for quantity takeoffs, they are not a replacement for estimating. Estimators perform a critical role in the building process far beyond that of extracting counts and measurements. The process of estimating involves assessing conditions in the project that impact cost, such as unusual wall conditions, unique assemblies, and difficult access conditions. Automatic identification of these conditions by any BIM tool is not yet feasible. Estimators should consider using BIM technology to facilitate the laborious task of quantity takeoff and to quickly visualize, identify, and assess conditions, and provide more time to optimize prices from subcontractors and suppliers. A detailed building model is a risk-mitigation tool for estimators that can significantly reduce bid costs, because it reduces the uncertainty associated with material quantities. The One Island East Office Tower case study in Chapter 9 is an excellent example of this.

Estimators utilize a variety of options to leverage BIM for quantity takeoff and to support the estimating process. No BIM tool provides the full capabilities of a spreadsheet or estimating package, so estimators must identify a method that works best for their specific estimating process. Three primary options are:

1. Export building object quantities to estimating software
2. Link the BIM tool directly to the estimating software
3. Use a BIM quantity takeoff tool

Each of these options is discussed in detail below.

6.6.1 Export Quantities to Estimating Software

As previously noted, most BIM tools offered by software vendors include features for extracting and quantifying BIM component properties. These features also include tools to export quantity data to a spreadsheet or an external database. In the U.S. alone, there are over 100 commercial estimating packages and many are specific to the type of work estimated. At the present time, however, surveys show that MS Excel is the most commonly used estimating tool (Sawyer and Grogan 2002). For many estimators, the capability to extract and associate quantity takeoff data using custom Excel spreadsheets is often sufficient. This approach, however, may require significant setup and adoption of a standardized modeling process.

6.6.2 Directly Link BIM Components to Estimating Software

The second alternative is to use a BIM tool that is capable of linking directly to an estimating package via a plug-in or third-party tool. Many of the larger estimating software packages now offer plug-ins to various BIM tools. These include: Sage Timberline via Innovaya (Innovaya 2007), U.S. Cost (Rundell 2006; U.S.Cost 2007); and Graphisoft Estimator (VICO 2007). These tools allow the estimator to associate components in the building model directly with assemblies, recipes, or items in the estimating package. These assemblies or recipes define what steps and resources are needed for construction of the components on site or for the erection or installation of prefabricated components. Assemblies or recipes often include references to the activities needed for the construction, e.g., form, place rebar, place concrete, cure, and strip forms. The estimator is able to use rules to calculate quantities for these items based on the component properties or manually enter data not extracted from the building information model. The assemblies may also include items representing necessary resources such as labor, equipment, materials, etc. and associated time and cost expenditures. As a result, all information required to develop a complete cost estimate and detailed list of basic activities can be used for construction planning. If this information is related to the BIM components, it can be used to generate a 4D model. The graphic model can also be linked to the estimate to illustrate the model objects associated with each line item within that estimate. This is very helpful for spotting objects that have no cost estimate associated with them.

This approach works well for contractors who have standardized on a specific estimating package and BIM tool. Integrating BIM component information from subcontractors and various trades, however, may be difficult to manage if different BIM tools are used. There are clear benefits to this highly integrated approach, but one potential shortcoming is the need for the contractor to develop a separate model. Of course, if the architect is not using BIM, then a contractor model is a necessity. When this is not the case, it is more efficient for the designer's model to provide the starting point for the contractor once the team has agreed on component definitions. If the project team is standardized on a single software vendor platform, this method may be suitable. This requires either a design-build approach or a contract that integrates the main project participants from the beginning of the project. Once again, early integration and collaboration are the keys to effective use of BIM technology. The AGC "BIM Guidelines for Contractors" emphasizes this point (see discussion in 6.8 below).

6.6.3 Quantity Takeoff Tool

A third alternative, shown generically in Figure 6-7, is to use a specialized quantity takeoff tool that imports data from various BIM tools. This allows estimators to use a takeoff tool specifically designed for their needs without having to learn all of the features contained within a given BIM tool. Examples of these are: Exactal (Exactal 2007), Innovaya (Innovaya 2007), and OnCenter (OnCenter 2007). These tools typically include specific features that link directly to items and assemblies, annotate the model for 'conditions,' and create visual takeoff diagrams. These tools offer varying levels of support for automated extraction and manual takeoff features. Estimators will need to use a combination of both manual tools and automatic features to support the wide range of takeoff and condition checking they need to perform.

Additional changes to the building model require that any new objects be linked to proper estimating tasks so that accurate cost estimates can be obtained from the building model, depending on the accuracy and level of detail already modeled.

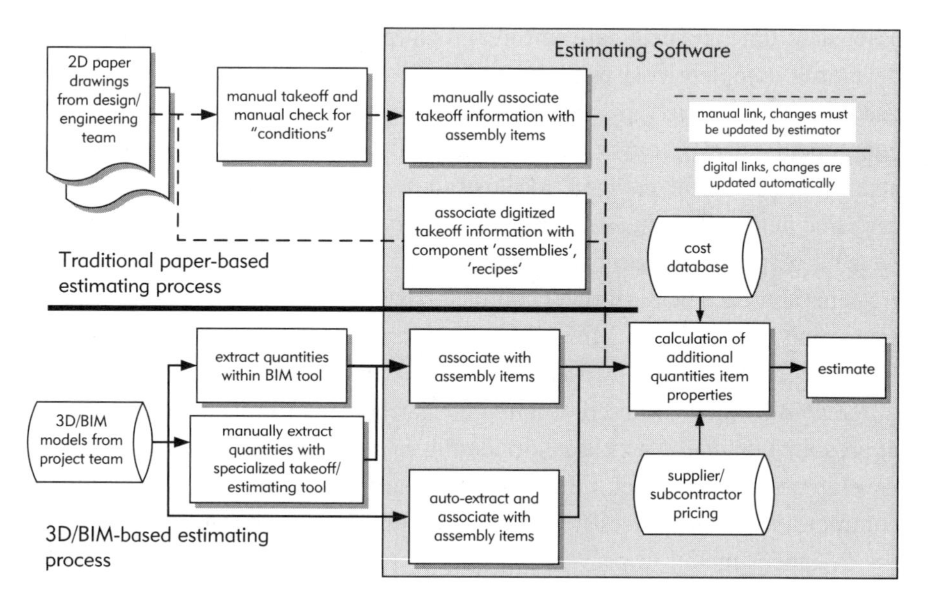

FIGURE 6-7 Conceptual diagram of a BIM quantity takeoff and estimating process.

6.6.4 Guidelines and BIM Implementation Issues to Support Quantity Takeoff and Estimating

Estimators and contractors should understand how BIM can support specific estimating tasks by reducing errors and improving accuracy and reliability within the estimate. More importantly, they can benefit from the ability to respond rapidly to changes during critical phases of the project, a challenge many estimators face on a daily basis. Here are some guidelines to consider:

- ***BIM is only a starting point*** for estimating. No tool can deliver a full estimate automatically from a building model. If a vendor advertises this, they don't understand the estimating process. Figure 6-8 illustrates that a building model can provide only a small part of the information needed for a cost estimate (material quantities and assembly names). The remaining data comes either from rules or manual entries provided by a cost estimator.
- ***Start simple***. If you are estimating with traditional and manual processes, first move to digitizers or on-screen takeoff to adjust to digital takeoff methods. As estimators gain confidence and comfort with digital takeoff, consider moving to a BIM-based takeoff.
- ***Start by counting***. The easiest place to start is estimating the tasks that involve counting, such as doors, windows, and plumbing fixtures. Many

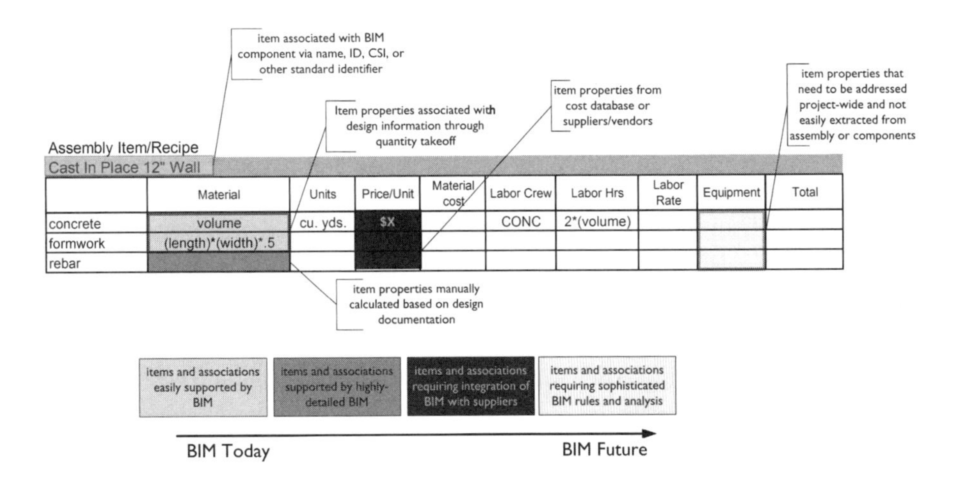

FIGURE 6-8 Example of how BIM component definitions relate to estimating assembly items and recipes.
It's important to note that BIM provides only a subset of the information estimators need to compute cost, and BIM components provide takeoff information but often lack the detailed capability of automatically computing labor, job (non-permanent) material and equipment costs.

BIM tools provide scheduling functionality and simple functions to query and count specific types of components, blocks, or other entities. These can also be verified and validated.

- ***Start in one tool, and then move to an integrative process***. It's easiest to start by doing takeoff in the BIM software or a specialized takeoff application. This limits potential errors or issues with respect to translating data and moving model data from one application to another. Once the estimator is confident that the data provided by a single software package is accurate and valid, then the model's data can be transferred to a secondary takeoff tool for validation.
- ***Set expectations***. The level of detail in the BIM takeoff is a reflection of the level of detail in the overall building model. If rebar isn't included in the building model, these values won't be auto-calculated. The estimator needs to understand the scope of the model information and what is represented.
- ***Start with a single trade or component type*** and work out the kinks.
- ***Automation begins with standardization***. To fully leverage BIM, designers and estimators will need to coordinate methods to standardize building components and the attributes associated with those components for quantity takeoff. In addition, in order to generate accurate quantities of sub components and assemblies, such as the studs inside a wall, it is necessary to develop standards for these assemblies.

6.7 CONSTRUCTION ANALYSIS AND PLANNING

Construction planning and scheduling involves sequencing activities in space and time, considering procurement, resources, spatial constraints, and other concerns in the process. Traditionally, bar charts were used to plan projects but were unable to show how or why certain activities were linked in a given sequence; nor could they calculate the longest (critical) path to complete a project. Today, schedulers typically use Critical Path Method (CPM) scheduling software such as Microsoft Project, Primavera SureTrak, or P3 to create, update, and communicate the schedule using a wide variety of reports and displays. These systems show how activities are linked and allow for the calculation of critical path(s) and float values that improve scheduling during a project. Specialized software packages that are better suited to building construction, such as Vico Control, enable schedulers to do line-of-balance scheduling. Sophisticated planning methods for resource-based analysis, including resource-leveling and scheduling with consideration of uncertainty, such as Monte Carlo simulation, are also available in some of the off-the-shelf packages. Other software tools are available for detailed schedules for a short time period of one or two weeks that need to consider individual subs, the availability of materials, etc.

Traditional methods, however, do not adequately capture the spatial components related to these activities, nor do they link directly to the design or building model. Scheduling is therefore a manually intensive task, and it often remains out-of-sync with the design and creates difficulties for project stakeholders to easily understand the schedule and its impact on site logistics. Figure 6-9 shows a traditional Gantt chart which illustrates how difficult it is to evaluate the construction implications of this type of schedule display. Only people thoroughly familiar with the project and how it will be constructed can determine whether this schedule is feasible. Two types of technologies have evolved to address these shortcomings.

The first is 4D CAD, which refers to 3D models that also contain time associations. 4D CAD tools allow schedulers to visually plan and communicate activities in the context of space and time. 4D animations are movies or virtual simulations of the schedule.

The second approach is to use analysis tools that incorporate BIM components and construction method information to optimize activity sequencing. These tools incorporate spatial, resource utilization, and productivity information. These two approaches are discussed in the following sections.

6.7.1 4D models to support construction planning

4D models and tools were initially developed in the late 1980s by large organizations involved in constructing complex infrastructure, power, and process projects in which schedule delays or errors impacted cost. As the AEC

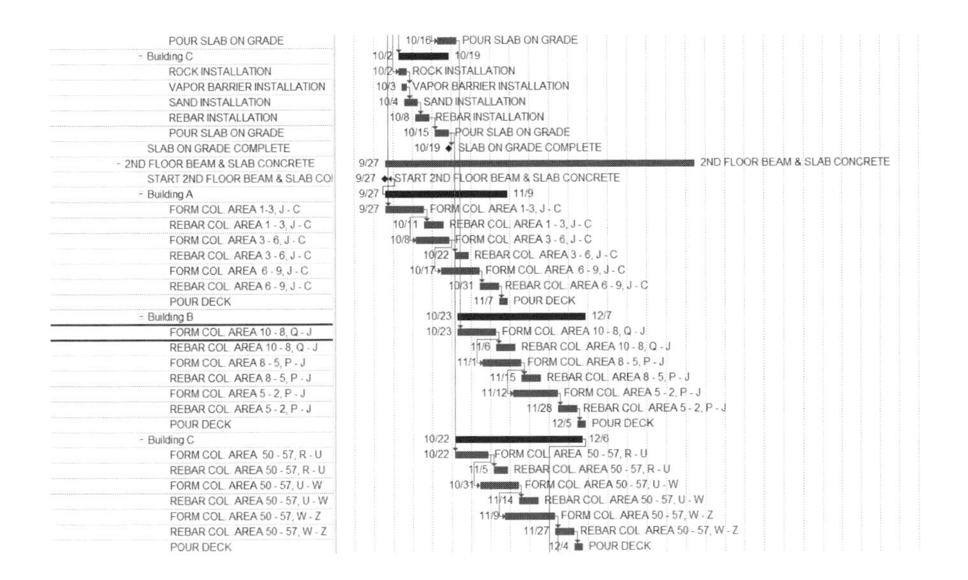

FIGURE 6-9 Sample Gantt chart of a construction schedule for a project involving three buildings and multiple floors and areas.
Assessing the feasibility or quality of a schedule based on a Gantt chart is often difficult for many project participants and requires manually associating each activity with areas or components in the project since there are no visual associations with the referenced areas, such as "Area 10" to a drawing or diagram.

industry adopted 3D tools, construction organizations built manual 4D models and combined snapshots of each phase or period of time in the project. Custom and commercial tools evolved in the mid to late 1990s, facilitating the process by manually creating 4D models with automatic links to 3D geometry, entities, or groups of entities for construction activities (see Figures 6-10, 6-11 and 6-12). BIM allows schedulers to create, review, and edit 4D models more frequently, which has lead to the implementation of better and more reliable schedules. The following sections discuss the benefits of 4D models and the various options schedulers have when producing them.

6.7.2 Benefits of 4D Models

4D simulations function primarily as communication tools for revealing potential bottlenecks and as a method for improving collaboration. Contractors can review 4D simulations to ensure that the plan is feasible and efficient as possible. The benefits of 4D models are:

- ***Communication:*** Planners can visually communicate the planned construction process to all project stakeholders. The 4D model captures both the temporal and spatial aspects of a schedule and communicates this schedule more effectively than a traditional Gantt chart.

FIGURE 6-10
4D view of construction of Vancouver Convention Center showing foundation and structural steel erection.
A tower crane was included in the model to review crane reach, clearances and conflicts. (See color insert for full color figure.) *Courtesy Pacific Project Systems Inc., MTC Design/3D, (4D modeling); Musson Cattell Mackey Partnership, Downs/Archambault & Partners, LMN Architects (architects); Glotman Simpson Consulting Engineers (structural engineers); PCL Constructors Westcoast Inc (CM)*

- ***Multiple stakeholder input:*** 4D models are often used in community forums to present to laypersons how a project might impact traffic, access to a hospital, or other critical community concerns.
- ***Site logistics:*** Planners can manage laydown areas, access to and within the site, location of large equipment, trailers, etc.
- ***Trade coordination:*** Planners can coordinate the expected time and space flow of trades on the site as well as the coordination of work in small spaces.
- ***Compare schedules and track construction progress:*** Project managers can compare different schedules easily, and they can quickly identify whether the project is on track or behind schedule.

Above all, 4D CAD requires that an appropriate 3D model of the building be linked to a project schedule that, in turn, provides start and end dates and floats for each object. There are a number of systems that provide these linkage capabilities.

The above considerations make the use of 4D CAD a relatively expensive process to setup and manage during a project. Prior experience and knowledge of

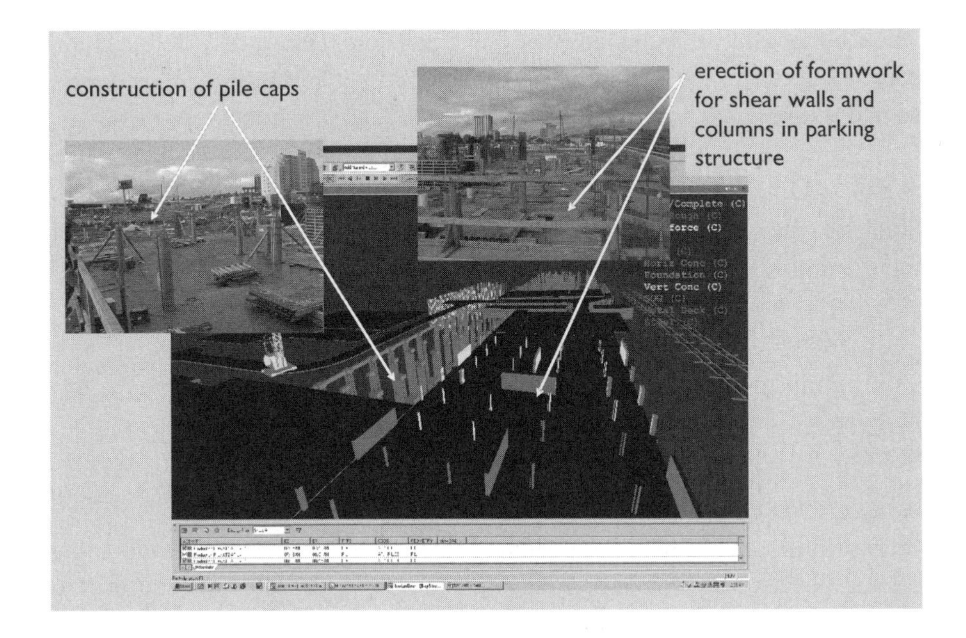

FIGURE 6-11 A snapshot of a 4D model and photos from the project site.
The project team used the model to support zone management and plan concurrent activities of foundation and concrete work. While a 4D model supports communication of sequencing of such work, the model did not include formwork and other temporary components which do impact the ability to perform work in the field. *Courtesy of DPR Construction.*

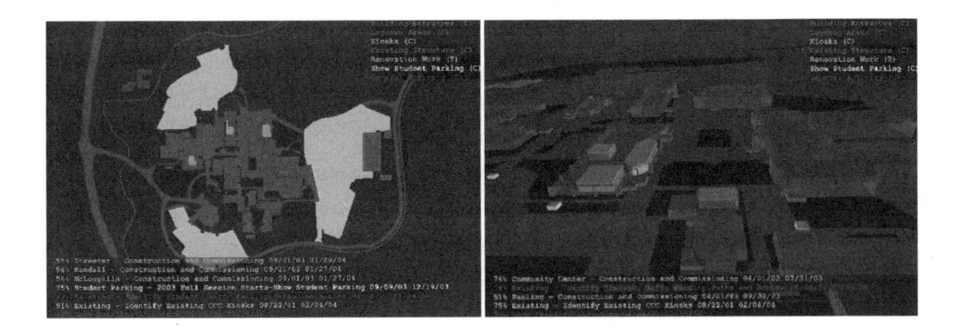

FIGURE 6-12 4D snapshots of a campus wide project showing various construction activities occurring throughout the campus to both landscape, road, and facilities.
These images help a contractor to communicate with the owner and the campus community about impacts to parking, roads, and access to specific buildings. *Courtesy of DPR Construction.*

the level of detail needed to produce an accurate linked schedule are necessary to achieve the full benefits associated with this tool. When used properly however, the associated cost and time benefits have been found to far exceed the initial implementation cost. For a good example, see the Camino Group Medical Building case study in Chapter 9. On this project 4D CAD analysis showed that it would be necessary to build the office building before the garage in order to have adequate access to the building site. This had not been realized when the project started and required a restructuring of the design and construction process.

6.7.3 4D Modeling Processes

Similar to the options estimators have, schedulers can choose from a variety of tools and processes to build 4D models:

1. Manual method using 3D or 2D tools
2. Built-in 4D features in a 3D or BIM tool
3. Export 3D/BIM to 4D tool and import schedule

Manual, CAD-Based Methods

Construction planners have been building 4D models manually for decades using colored pencils and drawings, with different colors for different sequences to show the progression of work over time. With the advent of CAD, planners transferred this process to CAD drawings that utilize colored-fills, shading, and the ability to turn CAD entities on and off. In some cases, where the model included naming conventions or component attributes related to the construction schedule, the process could be automated. In most cases, planners worked with a third party to create high-end movies or rendered animations to visually demonstrate the schedule. These animations are visually appealing and a great marketing tool, but they are not adequate planning or scheduling tools. Because they are produced manually, it remains difficult to change, update, or do rapid real-time scenario planning. When the schedule's details change, the planner must re-synchronize the 4D image manually with the schedule and create a new set of snapshots or animations. Because of these manual update requirements, the use of these tools is normally limited to the initial stages of design when visualization of the construction process is desired for the client or some outside agency.

BIM Tools with 4D capability

One way to generate 4D snapshots is through features that automate filtering of objects in a view based on an object property or parameter. For example, in Revit each object can be assigned to a 'phase' that is entered as text, such as "June 07" or "existing" and order these phases as desired. Users can then apply filters to show all objects in a specified phase or previous phases. This type of 4D functionality is relevant for basic phasing and generation of 4D snapshots but does not provide direct integration with schedule data. Additionally, features to interactively playback a 4D model common in specialized 4D tools are not provided.

Most BIM tools don't have built-in 'date' or 'time' capabilities, and require specific 4D modules or add-on tools to directly link to schedule data. Table 6-1 provides a brief overview of both built-in 4D features and add-on 4D functionality available for the popular BIM tools.

Due to the shortcomings inherent in manual and CAD/BIM-based 4D modeling tools, several software vendors began offering specialized 4D tools for producing 4D models from 3D models and schedules. These tools facilitate the production and editing of 4D models and provide the scheduler with numerous features for customizing and automating production of the 4D model. Typically, these tools require that data from a 3D model be imported from a CAD or BIM application. In most cases, the extracted data is limited to geometry and a minimal set of entity or component properties, such as 'name', 'color', and a group or hierarchy level. The scheduler imports relevant data into the 4D tool, then 'links' these components to construction activities, and associates them with types or visual behaviors. Figure 6-13 illustrates two approaches to creating the 4D model. The top part shows how a series of snapshots of the construction process can be created from 2D drawings. The lower portion illustrates how a true 4D model can be created from a 3D model linked to a construction schedule using specialized 4D software. Figure 6-14 shows the types of data sets that are used by 4D software to generate the 4D model.

Table 6-1 Selected BIM tools with 4D capability

Company	Product	Remarks
Bentley	ProjectWise Navigator	See Bentley in table 6-2 below
Autodesk	Revit Architecture	Revit is the only BIM tool with some built-in 4D capability for basic 4D phasing. Each Revit object includes parameters for "phasing" that allow users to assign a "phase", to an object and then use Revit's view properties to view different phases and create 4D snapshots. It is not possible to 'playback' a model, however. Via the API, users can link to scheduling applications and exchange data with tools like MS Project to automate some 4D entry.
Gehry Technologies	Digital Project	An add-on product, Construction Planning and Coordination, allows users to link 3D components to Primavera or MS Project activities with their associated data and generate 4D simulation analysis. Construction related objects need to be added (and removed when appropriate) to DP model. Changes to Primavera or MS Project schedule are propagated to linked DP model.
Graphisoft	VICO Constructor Simulation Module for ArchiCAD	See VICO in table 6-2 below

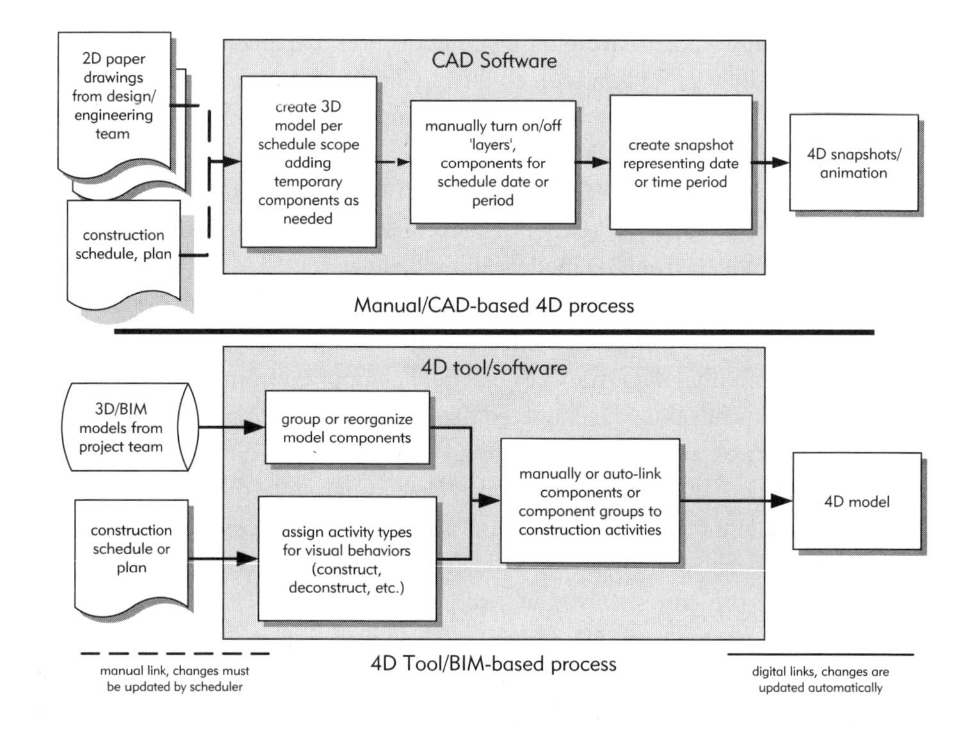

FIGURE 6-13 Diagram showing two different 4D modeling processes. The manual process is typically done within an available CAD, BIM, or visualization software. Specialized 4D software eliminates some steps, and provides direct links to the schedule and building model thus making the process faster and more reliable.

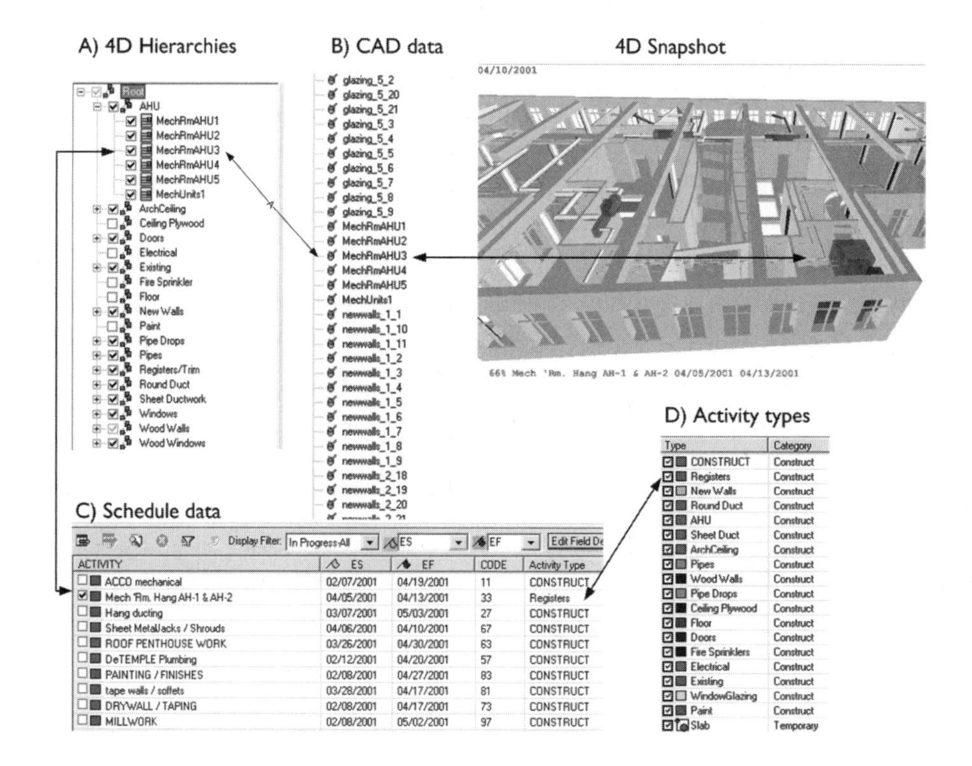

FIGURE 6-14 Diagram showing the key data interfaces of a 4D model.
A) 4D hierarchy or grouping of components related to activities in the schedule.
B) Organization of CAD data provided by design and engineering organizations.
C) Schedule data that can be illustrated hierarchically but is typically a set of activities with properties, such as start and finish dates.
D) Activity types that define the visual behavior of the 4D model.

Table 6-2 4D Tools

Company	Product	Remarks
Bentley	ProjectWise Navigator	This is a stand alone application that provides a series of services for: Importing multiple 2D and 3D design files from many sources (DWG, DGN, DWF, etc.) Reviewing 2D drawings and 3D models concurrently Following links between data files and components Reviewing interferences (clashes), and Viewing and analyzing schedule simulations
CommonPoint	Project 4D ConstructSim	Common Point 4D includes some specialized 4D features such as conflict analyses, adding laydown objects, animation, and custom features to create 4D movies. The 4D linking process includes drag-drop manual linking and automatic linking. Users can distribute a 4D viewer to team members. Common Point 4D only imports "vrml" geometry via exporters to popular BIM tools with limited object meta-data.; and users can import from all the major schedule software products. The higher-end ConstructSim supports more data-rich import of more formats with high-end analyses, organization, and status visualization features.
Innovaya	Visual Simulation	Links any 3D design data in DWG with either MS Project or Primavera scheduling tasks and shows projects in 4D. Generates simulation of construction process. Synchronizes changes made to either the schedule or to 3D objects. Uses color codes to detect potential schedule problems such as objects assigned to two concurrent activities or not assigned to any activity.
Navisworks	JetStream Timeliner	The Timeliner module includes all the features of JetStream's visualization environment and supports the largest number of BIM formats and best overall visualization capabilities. The Timeliner supports automatic and manual linking to imported schedule data from a variety of schedule applications. Manual linking is tedious and not user-friendly and there are few custom 4D features.
Synchro ltd.	Synchro 4D	This is a powerful new 4D tool with the most sophisticated scheduling capabilities of any of the 4D software. The tool requires deeper knowledge of scheduling and project management than the other tools to take advantage of its risk and resource analysis features. The tool includes built-in tools to visualize risk, buffering, and resource utilization in addition to basic 4D visualization.
VICO Software	Virtual Construction	Virtual Construction 5D construction planning system consisting of Constructor, Estimating, Control and 5D Presenter. The building model is developed in Constructor and objects are assigned recipes that define the tasks and resources needed to build or fabricate them, quantities and costs are calculated in Estimator, schedule activities are defined and planned using lie of balance techniques in Control and then the 4D construction simulation is visualized in Presenter. As an alternative to using Control, schedule dates can be imported from Primavera or MS Project.

Here are some things to consider when evaluating specialized 4D tools listed in Table 6-2:

- ***BIM Import capabilities:*** What geometry or BIM formats can users import and what types of object data can the tool import, e.g., geometry, names, unique identifiers, etc.? In some cases the tools only import geometry, geometry names, and hierarchy. This may be sufficient for basic 4D modeling, but other data may be needed so users can view object properties or filter or query based on this data.

- ***Schedule import capabilities:*** What schedule formats does the tool import and are the formats native files, text files? Some scheduling applications like Primavera work with a database. If so, the tool will need to support connections to the database and extraction of the schedule data.
- ***Merge/Update for 3D/BIM building model:*** Can users merge multiple files into a single model and update portions or all of the model? If a project involves models created in multiple BIM tools, the 4D modeling process will require import and merging these models into one tool. Thus, the 4D tool must provide this capability.
- ***Reorganization:*** Can you reorganize the data after it has been imported? (see discussion in following section). Tools that support easy reorganization of model components will greatly expedite the modeling process.
- ***Temporary components:*** Can users add (and later remove) temporary components such as scaffolds, excavation areas, storage areas, cranes, etc. to the 4D model? In many cases, users have to create these components and import them with the model geometry. Ideally, the 4D tool would have a library to allow users to quickly add these components.
- ***Animation:*** Can you simulate detailed crane simulations, or other installation sequences? Some 4D tools allow users to "move" objects over a specified time period to allow visualization of equipment movement.
- ***Analysis:*** Does the tool support specific analyses such as time-space conflict analysis, a feature found in Common Point, to identify activities happening in the same space.
- ***Output:*** Can users easily output multiple snapshots for specified periods of time or create movies with pre-defined views and time periods? The custom output features will facilitate sharing the model with project team.
- ***Automatic linking:*** Can users automatically link building components to schedule items based on fields or rules. This is useful for projects with standard naming conventions.

6.7.4 BIM-Supported Planning and Scheduling Issues and Guidelines

While the mechanics of the planning and scheduling process may vary depending on the planner's tools, there are several issues that any planner or 4D modeling team should consider when preparing and developing a 4D model.

Model Scope

If the model is for marketing or a design competition, its life will be relatively short. The appropriate level of detail depends on what the client has requested. If the team intends to use the model for the duration of the project, then a plan should outline when to migrate from a 90-day or higher-level schedule containing perhaps 100–300 activities to a detailed, 1-week to 3-week look-ahead schedule containing thousands of activities. Teams may start with constructing 'shells' of buildings and then replace these buildings with detailed interiors.

Level of Detail

The level of detail is affected by the size of the model, the time allotted for building it, and what critical items need to be communicated. An architect may build a highly-detailed wall system to support a rendering for comparing materials. The contractor may also elect to represent this system using a single component, because the critical issues are sequencing of the floors or wall sections, not the wall system's sequence of installation. In other cases, the sequencing of detailed components, such as a sophisticated structural earthquake system, may require a more detailed model for each installation step. The construction tasks required to build a given object may also require multiple activities, e.g. a foundation footing object requires excavation, forming, placing rebar, placing concrete, curing concrete, and stripping forms.

Planners can use a single component to represent multiple activities. A single wall section can be used to show formwork, rebar concrete pour, concrete finishing, and wall finishes. The team can apply multiple activities and activity types to a single component.

Re-organization

4D tools often allow the scheduler to re-organize or create custom groupings of components or geometric entities. This is an important feature because the way that the designer or engineer organizes a model is not usually sufficient for relating components to activities. For example, the designer may group systems of components for ease of duplicating when creating the model, such as a column and a footing. The planner, however, will organize these components into zones of slabs or footings. Figure 6-14 shows a design hierarchy and a 4D hierarchy for two different organizations of a model. This ability tore-organize is critical for developing and supporting a flexible and accurate 4D model.

Temporary components

The building model should reflect the construction process so that even temporary structures, excavation details, and other features that exist during construction can be shown in the 4D simulation. Figure 6-15 shows a 4D model that contains scaffolding to help construction planners evaluate safety and constructability issues. The scaffolding is necessary because it will influence spatial constraints for people and equipment.

Decomposition and Aggregation
Objects shown as a single entity, such as a slab, may need to be broken into portions to show how they will be constructed. Another issue that planners face is how to break up specific components, such as walls or roofs, that a designer or engineer would model as a single component but the planner would divide or break-up into zones. Most specialized tools do not provide this capability, and the planner must perform these 'break-ups' within the 3D/BIM tool.

Schedule Properties
Early start and completion dates are often used for 4D simulation. It may be desirable however, to explore other dates, such as a late start or finish or a leveled start or finish, to view the impact of alternative schedules on the visual simulation of the construction process. Additionally, other schedule properties are valuable in the 4D modeling process that are often project-specific. For example, in one study a team associated specific activities with the number of hospital beds that were either taken out-of-service or made operational so that the team could visualize, at any time, the number of hospital beds available and ensure that a minimum number could remain in use. It is also possible to code each activity

FIGURE 6-15
A 4D model snapshot showing scaffolding. Adding temporary equipment is often critical for determining the feasibility of the schedule; the details allow subcontractors and planners to visually assess safety and constructability issues. (See color insert for full color figure.) *Image provided courtesy of M.A. Mortenson, Inc.*

with a property titled 'Area,' or 'Responsibility' so that the model can show who is responsible for certain activities and quickly identify trades working near each other to improve coordination.

6.8 INTEGRATION WITH COST AND SCHEDULE CONTROL AND OTHER MANAGEMENT FUNCTIONS

During the construction process, organizations use a variety of tools and processes to manage and report on the project's status. These range from schedule and cost control systems to systems for accounting, procurement, payroll, safety, etc. Many of these systems report or rely on design and building-component information, yet they are not typically linked or associated with design drawings or BIM. This leads to redundant efforts of manually entering design information and identifying problems associated with the synchronization of various systems and processes. BIM software can provide vital support for these tasks, because it has detailed quantity and other component information that can be linked to other applications. Furthermore, contractors and project stakeholders can gain new insights by leveraging a graphic model to visually analyze project progress and highlight potential or existing problems. Some examples of how organizations are using 3D/BIM to support these tasks are:

- ***Project Status:*** Each component can have a field 'status,' and depending on the project, values may be 'in design,' 'approved for construction review,' 'in fabrication,' etc. These fields can then be associated with colors so that the team can quickly determine the status of the facility and identify bottlenecks or areas that are behind schedule.
- ***Procurement Purchasing:*** Since BIM objects define what needs to be purchased, it is possible to make purchases directly using the BIM tool. At the time this book was written (early 2007), this capability was in a very early stage of development. This capability will certainly improve, as product manufacturers develop models of their products that can be stored on Internet servers and found using search systems. A good example of a BIM procurement application has been developed by 1st Pricing (1stPricing 2007). Using downloadable plug-ins, it allows procurement within AutoCAD, ArchiCAD, Architectural Desktop, TurboCAD, and soon Revit. This product provides real-time quotes on doors and windows delivered to the job site based on zip code. Other types of components are being added to the system.

- ***Procurement tracking:*** Another important issue is the procurement status of services and material. Often, schedules consist of large numbers of construction activities, which makes it difficult to relate parallel design and procurement activities. By tracking the status of these activities, planners can perform queries to easily identify gaps in the procurement process as they relate to design and construction. By linking the schedule to a building information model, it is also possible to visualize where procurement delays are likely to impact the building. For example, if a long-lead item is scheduled to be installed in two months and the procurement process is not yet complete, the team can address the issue quickly to prevent further downstream delays. A visual link to a building model helps to better predict the impact that procurement delays will have on construction.
- ***Safety management:*** Safety is a critical issue for all construction organizations. Any tool that supports safety training, education, and reveals unsafe conditions is valuable to the construction team. A visual model allows teams to assess conditions and identify unsafe areas that might otherwise go unrealized until the team is in the field. For example, on a theme park project, a team modeled envelopes for testing rides to ensure that no activities were taking place during the testing period within the test envelope. Using 4D simulation, they identified a conflict and resolved it ahead of time.

6.9 USE FOR OFFSITE FABRICATION

Offsite fabrication requires considerable planning and accurate design information. It is becoming more common for contractors to fabricate components offsite to reduce labor costs and risks associated with onsite installation. Today, many types of building components are produced and/or assembled offsite in factories and delivered to the site for installation. BIM provides the capability for contractors to input BIM component details directly, including 3D geometry, material specifications, finishing requirements, delivery sequence, and timing, etc. before and during the fabrication process. In this section, the benefits from the perspective of the contractor are discussed. The benefits from the perspective of the fabricator are explained in more detail in Chapter 7.

Coordination of subcontractors' activities and designs constitutes a large part of a contractor's added-value to a project. Contractors able to exchange accurate BIM information with fabricators can save time by verifying and validating the model. This reduces errors and allows fabricators to participate earlier in the pre-planning and construction process.

There are excellent examples of close coordination and exchange of models between contractors and fabricators in the steel and sheet metal industries. As discussed in Chapter 7, many steel fabricators leverage 3D technologies to manage and automate the steel fabrication process. The adoption of product model exchange formats, such as the CIS/2 format (explained in detail in Chapter 3) (CIS/2 2007), greatly facilitates the exchange of information between design and engineering, contractors, and fabricators. These conditions allow project teams to coordinate and optimize the sequence of steel or sheet metal. In Chapter 9, the benefits of a close digital relationship between a contractor and a fabricator are captured in several case studies.

The structural steel industry is well-positioned to leverage BIM due to the efforts of the AISC (AISC 2007) and the development of the CIS/2 (CIS/2 2007) format. Other standards are being developed for precast concrete but are not yet in commercial use. The National BIM Standards effort (NIBS 2007) considers how to use building information models to provide information for fabrication. The NBIMS is reviewed in Chapter 3. Further details, including BIM technology requirements and available software products, are discussed in Chapter 7.

6.10 USE OF BIM ONSITE: VERIFICATION, GUIDANCE, AND TRACKING OF CONSTRUCTION ACTIVITIES

Contractors must field-verify the installation of building components to ensure that dimensional and performance specifications are met. When errors are found, the contractor must spend further time rectifying them. The building model can be used to verify that actual construction circumstances match those shown in the model. Note that even when a project team creates an accurate model, human error during installation remains a possibility, and catching these errors as they occur or as soon as possible has great value. An example of this occurred on the Letterman Digital Arts Center (LDAC) in San Francisco, where the project team built a complete model after the project had been designed and subsequently documented a field error in a report (Boryslawski 2006) as described in the following excerpt:

> "During one of the daily rounds of onsite photography, we recognized a critical error shown in the positioning of concrete formwork, which was quickly confirmed by referencing the BIM. This error occurred when the formwork

layout person measured from a column that was off the standard grid to the edge of the concrete slab. Pouring more concrete in this complex post-tension slab construction would have had serious consequences not only for the contractor but also for the entire project, as there were three more floors to be built above this floor. The problem was solved just as the concrete was being poured, saving what would have most definitely been a major expense."

In this situation, the intimate knowledge gained by virtually building the project allowed the team to discover these field errors. The team combined traditional field-verification processes of daily site walks with model reviews to detect potential field errors.

More sophisticated techniques are evolving to support field verification, guide layout, and track installation. Some examples of these are:

- ***Laser scanning technologies:*** Contractors can use laser technologies, such as laser measurement devices that report data directly to a BIM tool, to verify that concrete pours are situated in exactly the correct location or that columns are properly located. Laser scanning can also be used effectively for rehabilitation work and capturing as-built construction details. Demonstrations are being undertaken now (2007) by the GSA to assess this use (GSA 2007).
- ***Machine-guidance technologies:*** Earthwork contractors can use machine-guided equipment to guide and verify grading and excavation activities driven by dimensions extracted from a 3D/BIM model.
- ***GPS technologies:*** Rapid advances in GPS and the availability of mobile GPS devices offer contractors the ability to link the building model to global-positioning-systems to verify locations. Systems developed at Carnegie-Mellon University and used by transportation departments to facilitate delivery of information to field workers on road or bridge construction are managed through the coordination of GPS and 2D/3D/BIM, enabling field crews to quickly find related information based on their location.
- ***RFID tags:*** Radio Frequency Identification (RFID) tags can support the tracking of component delivery and installation onsite. BIM components that include references to RFID tags can automatically update with links to field scanning devices and provide contractors with rapid feedback on field progress and installation.

The use of BIM in the field will increase dramatically as mobile devices and methods to deliver BIM information to field workers becomes commonplace.

6.11 IMPLICATIONS FOR CONTRACT AND ORGANIZATIONAL CHANGES

The above descriptions of BIM-supported work processes for contractors emphasize the advantages of early and continual collaboration of the project team so that key project participants are involved in the development of the virtual model. Contractors of all types that integrate their practice around BIM, as opposed to traditional 2D CAD, will reap the greatest advantages. Projects that involve designers as well as general and major subcontractors, by incorporating constructability, cost, and construction planning knowledge earlier in the process, will experience project-wide benefits for all team members. The advantages that an integrated, collaborative, BIM-supported approach can bring will make it a favored and widely used method in the future.

This organizational approach will, of course, require new contracts that encourage close collaboration and sharing of information, as well as a sharing of the technology's associated benefits. A new approach to sharing risks and setting fees may also be required, because the increased emphasis on early collaboration means that efforts by team members and the benefits they produce may change. Advanced owners are already experimenting with new ideas for exploring how to better incorporate contractor involvement through a BIM-driven process. Some of these are discussed in the case studies in Chapter 9.

The Associated General Contractors (AGC) is closely following the implications of BIM for their members. In October, 2006, the AGC published a 48-page document titled, "The Contractor's Guide to BIM," (AGC 2006) which is available at their web site (free for their members, at a cost for others). The report is based on first-hand experience provided by contractors that have already used BIM. The guide discusses the implementation of BIM using 2D drawings produced by the design team and contrasts this with the faster and more accurate process of starting with a 3D/building model generated by the design team. The guide suggests that an experienced digital modeler can create a building model from 2D drawings in one-to-two weeks at a cost of 0.1% to 0.5% of the total construction costs. Contractors must balance these costs with the many potential benefits of BIM, as discussed earlier in this chapter.

With respect to changes in management responsibilities, "The Contractor's Guide to BIM" (AGC 2006) says:

> "Whether the design is issued in the form of 2D printed documents or a 3D electronic media or in a combination of both, the responsibilities of the members of the project team remain unchanged. The important issue is to ensure that project team members thoroughly understand the nature and exactitude of the information that is being conveyed."

It adds:

> "Contractors and Construction Managers need to recognize that coordination whether with BIM technology or a light table is a core service not an added service. BIM tools that can facilitate a great deal of coordination are now available and when applied appropriately they can reduce the cost and time of construction. The question is not whether BIM will be used on a project, but to what extent it will be used. It is known that BIM coordination improves communication, which decreases construction cost and time, thus reducing risk. Contractors and Construction Managers have a responsibility to evaluate the costs of various implementation processes and provide the results of this evaluation to Owners and design teams in quantifiable terms.
>
> As the leaders of construction coordination, Contractors and Construction Managers have a responsibility to encourage and facilitate the sharing and distribution of BIM technology on a project. They must also understand and convey the nature of the information that is being shared. Appropriate contract language that will foster the open sharing of BIM information must be developed. The contract language can not alter the relationships of the project team members or change their responsibilities beyond their ability to perform. As an example, if a designer approves an electronic file prepared by a detailer, and this file contains a dimensional inaccuracy, the designer must be protected to the same extent that they would had the approval document been a printed drawing."

Finally, while the guide does not recommend specific contract changes for accommodating BIM, it suggests that all parties agree to rely on the model (as opposed to 2D drawings in cases where the two representations do not agree); it suggests that all members of the team be given access to and take responsibility for their part of the model; and it recommends that an audit trail be maintained that tracks all changes made to the model. Clearly, this is an area that is rapidly evolving with the use of BIM tools.

6.12 BIM IMPLEMENTATION

Contractors working in close collaboration with project teams during the design phase will encounter fewer barriers to BIM adoption compared to contractors working in a design-bid-build environment. In the latter case, the collaboration process does not start until the job has been awarded to the low bid contractor; in the former, the contractor is involved with design decisions and can contribute construction knowledge to the design. The same applies to the subcontractors that participate in the project.

In both cases, the contractor needs to understand how the use of 3D/BIM, rather than 2D drawings, can be used to support coordination, estimating, scheduling, and project management. A good implementation plan involves making sure that management and other key staff members acquire a thorough understanding of how BIM supports specific work processes. This should be done at a company-wide level, although any particular project could be used as a starting point. If the architects and other designers on the company's projects are not all using BIM technology, it will be necessary for the contractor to build models that are appropriate for the above functions. This will expose them to a deeper understanding of model building and the required standards—for colors, objects, construction knowledge, etc.—that need to be incorporated into the model. Training can be obtained from BIM software firms or from specialized consultants. The cost of building the model, estimated at about 0.1% of construction cost (see the Camino Group Medical Building case study in Chapter 9 for discussion of model building cost), will be more than offset by the eventual savings in errors, shortened project durations, better use of prefabrication options, fewer workers in the field, and improved collaboration among the team. This topic is discussed in greater detail for subcontractors and fabricators in Chapter 7.

Chapter 6 Discussion Questions

1. There is tremendous variation in the size and type of construction companies. In 2004, what percent of firms were composed of 1 to 9 people? In what sector were a majority of these firms?
2. What are the main advantages of design-build over design-bid-build contracts? Why does the use of BIM favor the design-build contract? For public projects, why are design-bid-build contracts often preferred (see also Chapter 1, Section 1.1.2)?
3. From the contractor's point-of-view, what kinds of information should a building model contain? If the architect uses BIM to design a building, what information needed by the contractor is NOT likely to be present?

(Continued)

4. What approaches are available to develop a building model that can be used by the contractor? What are the limitations and benefits of each approach?
5. What level-of-detail is needed in a building model for useful clash detection? What are the reasons for detecting soft as opposed to hard clashes? What role do subcontractors play in the clash detection process?
6. What are the main advantages and limitations of using BIM for preparing a cost estimate? How can an estimator link the building model to an estimating system?
7. What are the basic requirements for performing a 4D analysis of a construction schedule? What are the contractor's options for obtaining the information needed to carry out this analysis? What major benefits can be obtained from this analysis?
8. How can BIM be linked to cost and schedule control systems? What advantages does this provide?
9. What are the main advantages of using BIM for procurement? Why is it still difficult to do this?
10. What are the requirements for using a building model for off-site fabrication? What types of exchange standards are needed for fabrication of steel members?
11. What types of organizational and contractual changes are needed for effective BIM use?

CHAPTER 7

BIM for Subcontractors and Fabricators

7.0 EXECUTIVE SUMMARY

Buildings have become increasingly complex. They are one-of-a-kind products requiring multi-disciplinary design and fabrication skills. Specialization of the construction trades and economies of prefabrication contribute to increasingly larger proportions of buildings' components and systems being pre-assembled or fabricated off-site. Unlike the mass production of off-the-shelf parts, however, complex buildings require customized design and fabrication of 'engineered-to-order' (ETO) components, including: structural steel, precast concrete structures and architectural facades, curtain walls of various types, mechanical, electrical and plumbing (MEP) systems, timber roof trusses, and reinforced concrete tilt-up panels.

By their nature, ETO components demand sophisticated engineering and careful collaboration between designers to ensure that pieces fit within the building properly without interfering with other building systems. Design and coordination with 2D CAD systems is error-prone, labor intensive and relies on long cycle-times. BIM addresses these problems in that it allows for the 'virtual construction' of components and coordination among all building systems prior to producing each piece. The benefits of BIM for subcontractors and fabricators include: enhanced marketing and rendering through visual images and automated estimating; reduced cycle-times for detailed design and production; elimination of almost all design coordination errors; lower engineering and detailing costs; data to drive automated manufacturing technologies; and improved pre-assembly and prefabrication.

Accurate, reliable, and ubiquitous information is critical to the flow of products in any supply chain. For this reason, BIM systems can enable leaner construction methods if harnessed across an organization's many departments or through the entire supply chain. The extent and depth of these process changes depends on the extent to which building information model databases are integrated within and between organizations.

To be useful for fabrication detailing, BIM tools need at least to support parametric and customizable parts and relationships, provide interfaces to management information systems, and be able to import building model information from building designers' BIM tools. Ideally, they should also offer good information model visualizations and export data in forms suitable for automation of fabrication tasks using computer-controlled machinery.

Within the chapter, the major classes of fabricators and their specific needs are discussed. For each fabricator type, appropriate BIM software tools are listed and the leading tools are surveyed. Finally, the chapter provides guidance for companies planning adoption of BIM. To successfully introduce BIM into a fabrication plant with its own in-house engineering staff, or into an engineering detailing service provider, adoption must begin with setting clear, achievable goals with measurable milestones. Human resource considerations are the leading concern; not only because the costs of training and setup of software to suit local practices far exceed the costs of hardware and software, but also because the success of any BIM adoption will depend on the skill and good will of the people tasked with using the technology.

7.1 INTRODUCTION

The professional gap between designers and builders that opened during the European Renaissance has continued to widen over the centuries, while building systems have grown increasingly complex and technologically advanced. Technical drawings and specifications on paper were the essential medium for communicating designer intent to builders, thereby bridging the gap.

Over time, builders became more and more specialized and began to produce building parts offsite, first in craft shops and later in industrial facilities, for subsequent assembly onsite. As a result, designers no longer wielded control over the entire design; expert knowledge for any given system lay within the realm of specialized fabricators. Builders and fabricators are now required to prepare their own drawing sets, called shop drawings, for two purposes: to develop and detail the designs for production and, no less importantly, to communicate their construction intent back to the designers for approval.

In fact, the two-way cycle of communication is not simply a review but an integral part of designing a building. Even more so, this has become the case where multiple systems are fabricated and their design must be integrated consistently. Drawings are used to coordinate the location and function of various building system parts. This is the case today for all but the simplest buildings.

In traditional practice, paper drawings and specifications prepared by fabricators for designers fulfill additional vital purposes. They are a key part of commercial contracts for the procurement of fabricators' products. They are used directly for installation and construction, and they are also the primary means for storing information generated through the design and construction process.

For subcontractors and fabricators, BIM supports the whole collaborative process of design development, detailing, and integration. In many recorded cases, BIM has been leveraged to enable greater degrees of prefabrication than were possible without it, by shortening lead times and deepening design integration. As noted in Chapter 2, object-based parametric design tools had already been developed and used to support many construction activities, such as structural steel fabrication, before the earliest comprehensive BIM tools became available.

Beyond these short term impacts on productivity and quality, BIM enables fundamental process changes, because it provides the power to manage the intense amount of information required of 'mass customization,' which is a key precept of lean production (Womack and Jones 2003).

As the use of lean construction methods (Howell 1999) becomes widespread, subcontractors and fabricators will increasingly find that market forces will compel them to provide customized prefabricated building components at price levels previously appropriate for mass-produced components.

After defining the context for our discussion (Section 7.2), this chapter describes the potential benefits of BIM for improving various facets of the fabrication process, from the perspective of the subcontractor or fabricator responsible for making and installing building parts (Section 7.3) to the fundamental process changes to be expected (Section 7.4). BIM system requirements for effective use by fabricators are listed and explained for modeling and detailing in general (Section 7.5). Detailed information is provided for a number of specific trades (Section 7.6). Significant software packages for fabricators are listed, and pertinent issues concerning the adoption and use of BIM are discussed (Section 7.7).

7.2 TYPES OF SUBCONTRACTORS AND FABRICATORS

Subcontractors and fabricators perform a very wide range of specialized tasks in construction. Most are identified by the type of work they do, or the type of components they fabricate. For a discussion of the ways in which they can exploit BIM, the degree of engineering design required in their work is a useful way of classifying them. Looking beyond bulk raw materials, building components can be classified as belonging to one of three types:

1. **Made-to-stock components,** such as standard plumbing fixtures, drywall panels and studs, pipe sections, etc.
2. **Made-to-order components,** such as pre-stressed hollow-core planks*, and windows and doors selected from catalogs.
3. **Engineered-to-order components,** such as the members of structural steel frames, structural pre-cast concrete pieces, facade panels of various types, custom kitchens and other cabinet-ware, and any other component customized to fit a specific location and fulfill certain building functions.

The first two classifications are designed for general use and not customized for specific applications†. These components are specified from catalogs. Most BIM systems enable suppliers to provide electronic catalogs of their products, allowing designers to embed representative objects and direct links to them in building information models. The suppliers of these components are rarely involved in their installation or assembly onsite. As a result, they are rarely involved directly in the design and construction process. For this reason, this chapter focuses on the needs of designers, coordinators, fabricators, and installers of building components of the third type: engineered-to-order (ETO) components.

7.2.1 Engineered-To-Order (ETO) Component Producers

ETO producers typically operate production facilities that manufacture components that need to be designed and engineered prior to actual production. In most cases, they are subcontracted to a building's general contractor or, in the case of a project being executed by a construction management service company, they are subcontracted to the owner. The subcontract typically encompasses detailed design, engineering, fabrication, and erection of their products.

---

*Hollow-core planks are pre-engineered but can be custom-cut to arbitrary lengths.

†They are distinguished in that the second type is only produced as needed, usually for commercial or technological reasons, such as high inventory costs or short shelf-life.

Table 7-1 Engineered-to-order building components and their annual market volume in the U.S.

Engineered-to-order component fabricator/ designer/coordinator		Value of shipments or services in 2002 ($1M)
Structural steel buildings		$ 3,079
	Fabricators	$2,897
	Detailers	$182
Precast concrete		$ 4,131
	Structural products	$3,180
	Architectural facades	$951
Curtain walls		$1,707*
Timber trusses (floor and roof trusses)		$4,487
Reinforcing bars for concrete		$1,934

*Estimate based on new construction for office and commercial buildings.

Source: 2002 Economic Census, U.S. Census Bureau, U.S. Department of Commerce. (U.S. 2004).

Although some companies maintain large in-house engineering departments, their core business is fabrication. Others outsource part or all of their engineering work to independent consultants (dedicated design service providers – see below). They may also subcontract erection or installation of their product onsite to independent companies.

Some examples of ETO producers are provided in Table 7-1 along with estimations of their respective market volume in the U.S. In addition, there are building construction trades that do not function exclusively as ETO producers but offer significant ETO component content as part of their systems. Examples are: plumbing, heating, ventilation and air-conditioning (HVAC), elevators and escalators, and finish carpentry.

7.2.2 Design Service Providers

Design service providers offer engineering services to producers of engineered-to-order components. They perform work on a fee basis and generally do not participate in actual fabrication and onsite installation of the components they design. Service firms include: structural steel detailers, precast concrete design and detailing engineers, and specialized facade and curtain-wall consultants, among others.

Designers of tilt-up concrete construction panels are a good example of such providers. Their expertise in engineering, designing, and preparing shop-drawings enables general contractors or specialized production crews to make large reinforced concrete wall panels in horizontal beds onsite and then lift (or tilt) them into place. This onsite fabrication method can be implemented by relatively small contracting companies, by virtue of the availability of these design service providers.

7.2.3 Specialist Coordinators

Specialist coordinators provide a comprehensive ETO product provision service by bringing together designers, material suppliers, and fabricators under a 'virtual' subcontracting company. The rationale behind their work is that they offer flexibility in the kinds of technical solutions they provide, because they do not have their own fixed production lines. This type of service is common in the provision of curtain walls and other architectural facades.

The 100 11th Avenue, New York, case study (see Chapter 9) is a good example of this kind of arrangement. The designers of the facade system assembled an *ad hoc* 'virtual' subcontractor composed of a material supplier, a fabricator, an installer and a construction management firm.

7.3 THE BENEFITS OF A BIM PROCESS FOR SUBCONTRACTOR FABRICATORS

Figure 7-1 shows the typical information and product flow for ETO components in building construction. The process has three major parts: project acquisition (preliminary design and tendering), detailed design (engineering and coordination), and fabrication (including delivery and installation). The process includes cycles that allow the design proposal to be formulated and revised, repeatedly if necessary. This typically occurs at the detailed design stage, where the fabricator is required to obtain feedback and approval from the building's designers, subject not only to their own requirements but also to the coordination of the fabricator's design with other building systems also in development.

There are a number of problems with the existing process. It is labor intensive, with much of the effort spent producing and updating documents. Sets of drawings and other documents have high rates of inaccuracies and inconsistencies, which are often not discovered until erection of the products onsite. The same information is entered into computer programs multiple times, each time for a distinct and separate use. The workflow has so many intermediate points for review that rework is common and cycle-times are long.

Leveraging BIM can improve the process in several ways. First, BIM can improve the efficiency of most existing steps in the 2D CAD process by increasing productivity and eliminating the need to manually maintain consistency across multiple drawing files. With deeper implementation, however, BIM changes the process itself by enabling degrees of prefabrication that remain prohibitive in

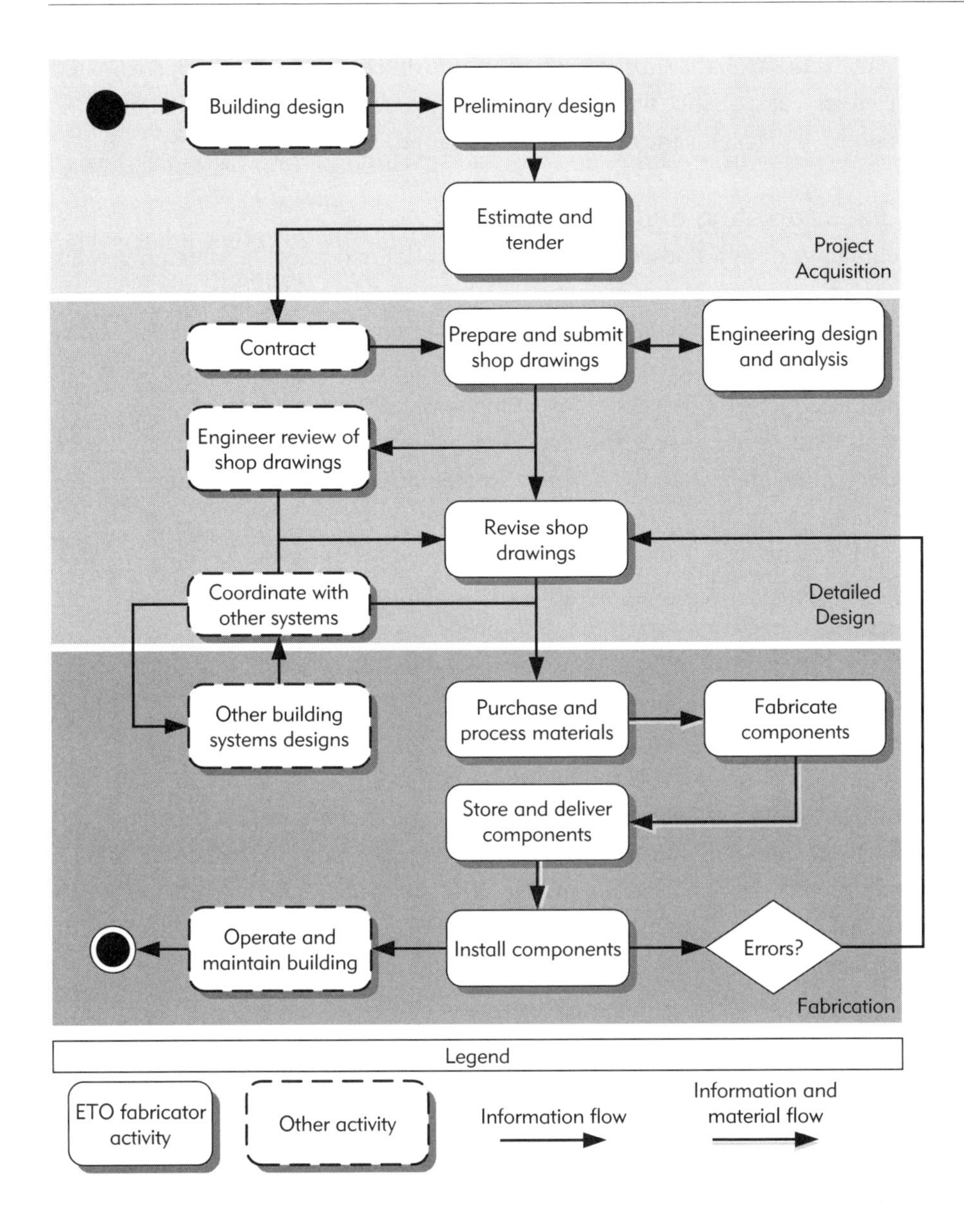

FIGURE 7-1 Typical information and product flow for a fabricator of ETO components.

coordination costs with existing information systems. When implemented in the context of lean construction techniques, such as with *pull flow*‡ control of detailing, production, and installation, BIM can substantially reduce lead times and make the construction process more flexible and less wasteful.

‡Pull flow is a method for regulating the flow of work in a production system whereby production at any station is signaled to begin only when an 'order' for a part is received from the next station downstream. This is in contrast to traditional methods where production is 'pushed' by command from a central authority. In this context, pull flow implies that detailing and fabrication of components for any particular building section would begin only a short preset time before installation became possible for that section.

In Section 7.3, the short-term benefits are first explained in an approximated chronological sequence with reference to the process map shown in Figure 7-1. Section 7.4 discusses the more fundamental process change.

7.3.1 Marketing and Tendering

Preliminary design and estimating are essential activities for obtaining work for most subcontractor fabricators. To win a project with a profitable price requires precision in measuring quantities, attention to detail, and the ability to develop a competitive technical solution – all of which demand significant time investments by the company's most knowledgeable engineers. Generally, not all tenders are successful, and companies are required to estimate more projects than are eventually performed, making the cost of tendering a sizable part of the company's overhead.

BIM technology aids engineers in all three of these areas: developing multiple alternatives, detailing solutions to a reasonable degree, and measuring quantities.

For marketing purposes, the persuasive power of a building information model for a potential client is not limited to its ability to provide a 3D or photorealistic image of a proposed building design, as is the case for software that is limited to 3D geometric modeling. Its power lies in its ability to adapt and change designs parametrically and better exploit the embedded engineering knowledge, allowing for more rapid design development for satisfying clients' needs to the greatest extent possible. The following excerpt from the Penn National case study describes the story of a precast concrete estimator's experience using a BIM tool to develop and sell a design for a parking garage:

> "To give you some background on this project, we started it as a design-build project for one of the salesmen. Bill modeled the entire garage (240′ wide × 585′ long × 5 supported levels), without connections or reinforcing, in 8½ hours. It is composed of 1,250 pieces. We sent PDF images to the owner, architect, and engineer.
>
> The next morning we had a conference call with the client and received a number of modifications. Bill modified the model by 1:30 PM. I printed out the plan, elevations, and generated a Web viewer model. I sent these to the client at 1:50 PM via email. We then had another conference call at 2:00 pm. Two days later, we had the project. The owner was ecstatic about seeing a model of his garage. Oddly enough, it's supposed to be 30 miles from our competitor's plant. In fact, their construction arm is who we will be contracted to.
>
> We figured it would have taken 2 weeks in 2D to get to where we were in 3D. When we had the turnover meeting (a meeting we have to turnover scope from estimating to engineering, drafting and production) we projected the

model on a screen to go over the scope of work. It went just as we envisioned. It was exciting to see it actually happen that way."

This example underscores how shortened response times – obtained through the use of BIM – enabled the company to better address the client's decision making process.

The project referenced in this excerpt – the Penn National Parking Structure – is documented in further detail in the case studies in Chapter 9. Alternative structural layout configurations were considered. For each, the producer automatically extracted a quantity takeoff that listed the precast pieces required. These quantities enabled the provision of cost estimates for each, allowing the owner and general contractor to reach an informed decision concerning which configuration to adopt.

7.3.2 Reduced Production Cycle-Times

The use of BIM significantly reduces the time required to generate shop drawings and material takeoffs for procurement. This can be leveraged in three ways:

- To offer a superior level of service to building owners, for whom late changes are often essential, by accommodating changes later in the process than is possible in standard 2D CAD practice. Making changes to building designs that impact fabricated pieces close to the time of fabrication is very difficult in standard practice. Each change must propagate through all of the assembly and shop drawings that may be affected and must also coordinate with drawings that reflect adjacent or connected components to the piece that changed. Where the change affects multiple building systems provided by different fabricators or subcontractors, coordination becomes far more complex and time-consuming. With BIM tools, the changes are entered into the model and updated erection and shop drawings are produced automatically. The benefit is enormous in terms of time and effort required to properly implement the change.
- To enable a "pull production system" where the preparation of shop drawings is driven by the production sequence. Short lead-times reduce the system's 'inventory' of design information, making it less vulnerable to changes in the first place. Shop drawings are produced once a majority of changes have already been made. This minimizes the likelihood that additional changes will be needed. In this 'lean' system, shop drawings are produced at the last responsible moment.
- To make prefabricated solutions viable in projects with restricted lead-times between the contract date and the date demanded for the

commencement of onsite construction, which would ordinarily prohibit their use. Often, general contractors find themselves committing to construction start dates with lead-times that are shorter than the time required to convert conventional building systems to prefabricated ones, due to the long lead-times needed for production design using 2D CAD. For example, a building designed with a cast-in-place concrete structure requires, on average, two to three months to complete the conversion to precast concrete and produce the first pieces needed. In contrast, BIM systems shorten the duration of design to a point where more components with longer lead-times can be prefabricated.

These benefits derive from the high degree of automation that BIM systems are capable of achieving, when attempting to generate and communicate detailed fabrication and erection information. Parametric relationships between building model objects (that implement basic design knowledge) and their data attributes (that enable systems to compute and report meaningful information for production processes) are the two features of BIM systems that make these improvements possible. This technology is reviewed in further detail in Chapter 2.

A reduction in cycle time can be achieved by exploiting automation for the production of shop drawings. The extent of this benefit has been explored in numerous research projects. In the structural steel fabrication industry, fabricators reported almost a 50% savings in time for the engineering detailing stage (Crowley 2003). The General Motors Production Plant case study (Chapter 9) documents a project with a 50% reduction of overall design-construction time compared to traditional design-bid-build projects (although some of this reduction can be attributed to other technologies that were used in addition to 3D models of the structural steel).

An early but detailed evaluation of lead-time reduction in the case of architectural precast concrete facade panels was performed within the framework of a research project initiated by a consortium of precast concrete companies (Sacks 2004). The first Gantt chart in Figure 7-2 shows a baseline process for engineering the design of an office building's facade panels. The benchmark represents the shortest theoretical duration of the project using 2D CAD, if work had been performed continuously and without interruption. The benchmark was obtained by reducing the durations measured for each activity in the actual project to the net number of hours that the project team worked on them. The second Gantt chart shows an estimated timeline for the same project, if performed using an available 3D parametric modeling system. In this case, the reduction in lead-time decreased from the baseline minimum of 80 working days to 34 working days.

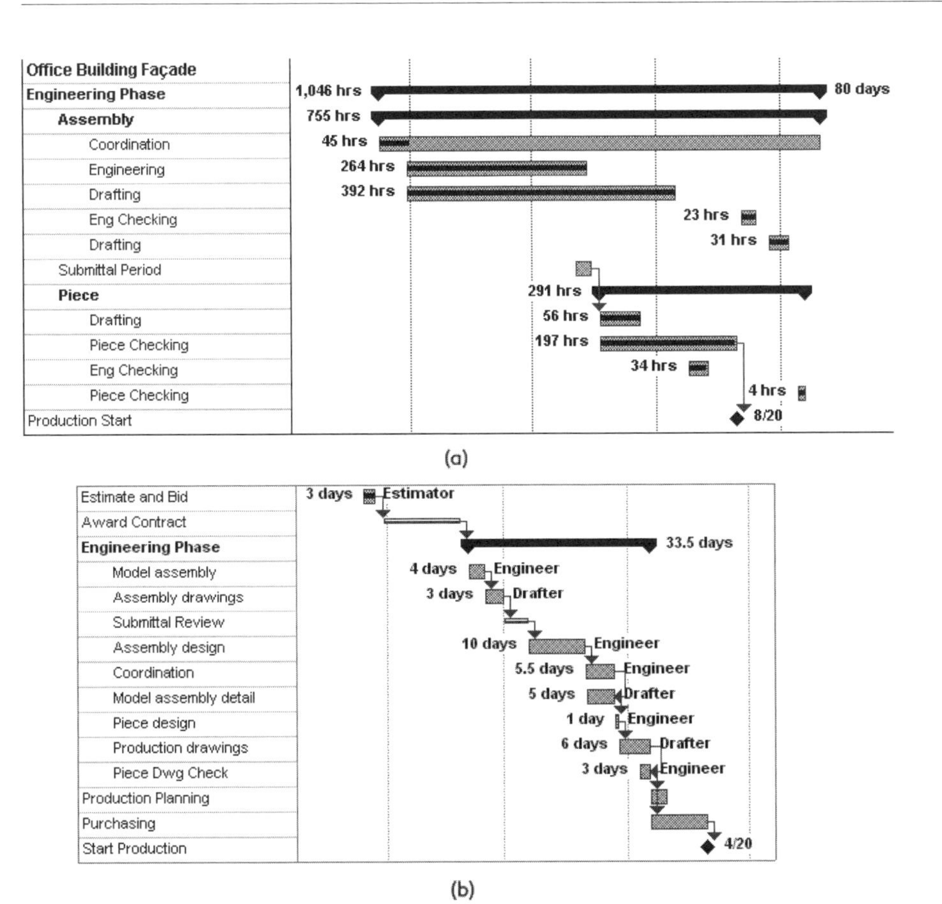

FIGURE 7-2 a) A benchmark of production lead-time for engineering design and detailing of architectural precast facade panels using 2D CAD; and b) an evaluation of a comparable lead-time using 3D parametric modeling (Sacks 2004). Reproduced from the Journal of Computing in Civil Engineering 18(4), by permission of the American Society of Civil Engineers.

7.3.3 Reduced Design Coordination Errors

In the introduction to this chapter, we mentioned the need for fabricators to communicate construction intent to designers. One of the reasons for this is that the information obtained through the submittal and approval process is essential to the design team as a whole. It allows the team to identify potential conflicts inherent in the design. A physical clash between two components, where they are destined to occupy the same physical space, is the most obvious problem. It is termed a *hard clash*. *Soft clashes* occur when components are placed too close to one another, albeit not in physical contact, such as rebars that are too close to allow for the proper placing of concrete or pipes that require adequate space for insulation. Constructability problems are a third type, where certain components obstruct the construction or erection of other components by restricting or preventing access[§].

[§]The latter two – soft clashes and constructability clashes – are sometimes referred to as clearance clashes.

When design coordination is incomplete – in any given situation – the conflicts are discovered during installation of the second component. Regardless of who carries the legal and fiscal liability for the resulting rework and delays, the fabricators inevitably suffer. Construction is leaner when work is predictable and uninterrupted.

BIM offers numerous technical benefits that improve design coordination at all stages. Of particular interest to fabricators is the ability to create integrated models of potentially conflicting systems at production-detail levels. A common tool for conflict detection is Navisworks Jetstream software (Navisworks 2007), which imports models from various formats into a single environment for identifying physical clashes. The clashes are identified automatically and reported to the users (this application is discussed in Chapter 6, Section 5, and is apparent in the Camino Medical Group Building in Case Study 9.3).

Current technology limitations prevent the resolution of clashes directly using this system. Technically, it is not possible to make corrections in the integrated environment and then port them back to the originating modeling environments. Once the team has decided upon a solution for a conflict identified in the review software, each trade must then make the necessary changes within their individual BIM software. Repeating the cycle of importing the models to the review software enables close to real-time coordination, especially if the detailers for the trades are co-located, as they were in the Camino Medical Group Building case study discussed below. In future systems it should be possible to report the clash back to each trade's native BIM tool by using the component IDs (see Chapter 3 for a detailed explanation of these interoperability issues.)

To avoid design coordination conflicts, the best practice is for detailed design to be performed in parallel and within collaborative work environments involving all of the fabricating trades. This avoids the almost inevitable need for rework in the detailed design, even when conflicts in the completed designs have already been identified and resolved. Essentially, this was the process adopted in the Camino Medical Group Building project by DPR Construction and its trade subcontractor partners (see Chapter 9). Detailers for plumbing, HVAC, sprinkler systems, electrical conduits, and other systems were collocated in a site office and detailed each of their systems in close proximity with one another and in direct response to the progress of fabrication and installation of the systems onsite. Almost no coordination errors reached the job site itself.

Another significant waste occurs when inconsistencies appear within the fabricator's *own* drawing sets. Traditional sets, whether drawn by hand or using CAD, contain multiple representations of each individual artifact. Designers and drafters are required to maintain consistency between the various drawings as the design development progresses and further changes are made. Despite quality control systems of various kinds, entirely error free drawing sets are rare. A detailed study of drawing errors in the precast concrete industry, covering some 37,500 pieces from various projects and producers, showed that the costs of design coordination errors amount to approximately 0.46% of total project costs (Sacks 2004).

Two views of drawings of a precast concrete beam are shown in Figure 7-3. They serve as a good example of how discrepancies can occur. Figure 7-3 (a) shows a concrete beam in an elevation view of the outside of the building; and Figure 7-3 (b) shows the same beam in a piece fabrication shop drawing. The external face of the beam had brick facing, which is fabricated by placing the bricks face down in the mold. The shop drawing should have shown the back of the beam up, i.e. with the bare concrete (internal to the building) face up in plan view. Due to a drafting oversight, the inversion was not made and the beam was shown with the external face up, which resulted in all eight beams in this project being fabricated as 'mirror images' of the actual beams needed. They could not be erected as planned – see Figure 7-3 (c) – which resulted in expensive rework, reduced quality, and construction delays.

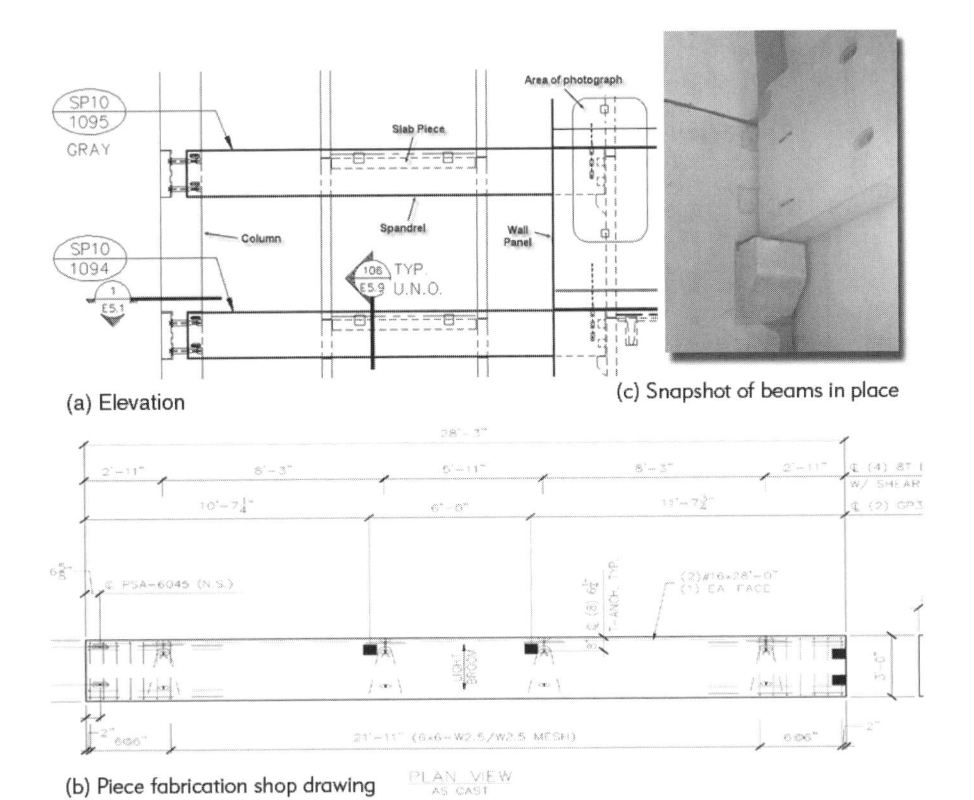

FIGURE 7-3 Drawing inconsistency for a precast concrete spandrel beam: a) elevation, b) piece fabrication shop drawing drawn in mirror image in error, and c) the beams in place with mismatched end connection details. (Sacks et al. 2003) Reproduced from the Journal of the Precast/Prestressed Concrete Institute 48(3), with permission of the Precast/Prestressed Concrete Institute.

7.3.4 Lower Engineering and Detailing Costs

BIM reduces direct engineering costs in three ways:

- Through the increased use of automated design and analysis software
- Almost fully automated production of drawings and material takeoffs
- Reduced rework due to enhanced quality control and design coordination

One major difference between BIM and CAD is that building information objects can be programmed to display seemingly 'intelligent' behaviors. This means that the preprocessing of data for analysis software of various kinds, from thermal and ventilation analyses to dynamic structural analyses, can be performed directly from BIM data or within the BIM tool itself. For example, most BIM tools used for structural systems enable the definition of loads, load cases, support conditions, material properties, and all other data needed for structural analyses, such as finite element analysis.

It also means that BIM systems can allow designers to adopt a top-down design development approach, where the software propagates the geometric implications of high-level design decisions to its constituent parts. For example, the fine details of shaping pieces to fit to one another at connections can be carried out by automated routines based on pre-made custom components. The work of detailing the designs for production can, to a large extent, be automated. Apart from its other benefits, automated detailing directly reduces the number of hours that must be consumed to detail ETO components and to produce shop drawings.

Most BIM systems produce reports, including drawings and material takeoffs, in a highly automated fashion. Some also maintain consistency between the model and the drawing set without explicit action on the part of the operator. This introduces savings in the number of drafting hours needed, which is particularly important to fabricators that previously spent the lion's share of their engineering hours on the tedious task of preparing shop drawings.

Various estimates of the extent of this direct productivity gain for engineering and drafting with the use of BIM tools have been published (Autodesk 2004; Sacks 2004), although few recorded measurements are available. One set of large-scale experiments was undertaken for the case of preparing construction drawings and detailing rebar for cast-in-place reinforced concrete structures using a BIM tool with parametric modeling, customizable automated detailing routines, and automated drawing preparation (Sacks and Barak 2007). The buildings had previously been detailed using 2D CAD, and the hours worked were recorded. As can be seen in Table 7-2, the reduction in engineering and drafting hours for the three case study projects fell in the range

Table 7-2 Experimental data for three reinforced concrete building projects.

Hours worked	Project A	Project B	Project C
Modeling	131	191	140
Reinforcement detailing	444	440	333
Drawing production	89	181	126
Total 3D	**664**	**875**	**599**
Comparative 2D hours	1,704	1,950	760
Reduction	61%	55%	21%

of 21% to 61%. (Figure 7-4 shows axonometric views of the three cast-in-place reinforced concrete structures modeled in the study.)

7.3.5 Increased Use of Automated Manufacturing Technologies

Computer-controlled machinery for various ETO component fabrication tasks has been available for many years. Examples include: laser cutting and drilling machines for structural steel fabrication; bending and cutting machines for fabricating reinforcing steel for concrete; saws, drills and laser projectors for timber truss manufacture; water jet and laser cutting of sheet metal for ductwork; pipe cutting and threading for plumbing; among others. However, the need for human labor to code the computer instructions that guide these machines proved to be a significant economic barrier to their use.

2D CAD technology provided a platform for overcoming data input barriers by allowing third-party software providers to develop graphic interfaces, where users could draw the products rather than coding them alphanumerically. In almost every case, the developers found it necessary to add meaningful

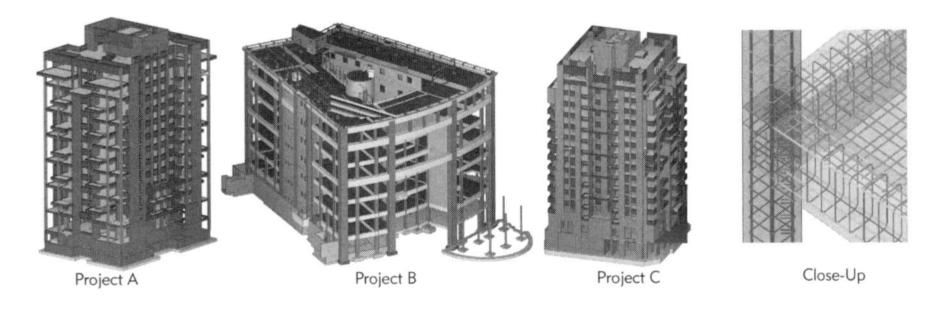

FIGURE 7-4 Axonometric views of projects A, B, and C.
These models, prepared as part of an experiment to evaluate 3D modeling productivity, contain complete rebar details. The close-up image shows detailed rebars in a balcony slab and supporting beams.

information to the graphics that represented the pieces to be fabricated by creating computable data objects that represented building parts. They could then automatically generate parts and material takeoffs, resulting in what might be called 'building part information modeling' applications.

The parts, however, continued to be modeled separately for each fabrication stage. When changes were made to building systems, operators had to manually revise or reproduce the part model objects to maintain consistency. Apart from the additional time required, manual revision suffers the drawback that inconsistencies may be introduced. In some cases, such as for the structural steel fabrication industry, software companies addressed this problem by developing top-down modeling systems for updating within assemblies and parts, so that a change would propagate almost entirely automatically to the affected pieces. These developments were constrained to certain sectors, such as the structural steel industry, where market size, the scale of economic benefit from use of the systems, and technological advances made investment in software development economically viable. These applications evolved into fully object-oriented 3D parametric modeling systems.

BIM tools model every part of a building using meaningful and computable objects, and so provide information from which the data forms required for controlling automated machinery can be extracted with relative ease. Unlike their 2D-CAD based predecessors, however, they also provide the logistical information needed for managing the fabrication processes, including links to construction and production schedules, product tracking systems, etc.

7.3.6 Increased Pre-Assembly and Prefabrication

By removing or drastically reducing the overhead effort required to produce shop drawings, BIM tools make it economically feasible for companies to prefabricate a greater variety of pieces for any building project. Automatic maintenance of geometric integrity means that making a change to a standard piece and producing a specialized shop drawing or set of CNC instructions demands relatively little effort. More structurally diverse buildings, such as the Walt Disney Concert Hall in Los Angeles (Post 2002) or the Beijing Aquatics Center (Chapter 9), become possible and increasingly more of the standard parts of buildings can be prefabricated economically.

The trend toward prefabrication is encouraged by the relative reduction in risk associated with parts not fitting properly when installed. Each trade's perception of that risk, or of the reliability of the design as a whole, is strongly influenced by the knowledge that all other systems are similarly and fully defined in 3D and reviewed together.

With few exceptions, 2D CAD did not give rise to new fabrication methods[‖], and it did little to aid the logistics of prefabrication offsite. BIM tools, on the other hand, are already enabling not only greater degrees of prefabrication than could be considered without them but also prefabrication of building parts that were previously assembled on site. For example, in the Camino Medical Group Building (described in detail in Chapter 9), large sections of pipes and plumbing fixtures were pre-assembled on stud frames and then rolled into place.

Because the tools support close coordination between building systems and trades, integrated prefabrication of building modules that incorporate parts of multiple systems is now feasible. Construction of the Staffordshire Hospital in the UK provided an excellent example of the use of integrated building modules with parts from multiple systems (Court et al. 2006, Pasquire et al. 2006). Figure 7-5 shows how components of HVAC, plumbing, sprinkler, electrical, and communication systems can be assembled together in a module for simple installation in the ceiling of a corridor onsite. Coordinating the physical and logistical aspects of integration to this degree is only possible given the richness and reliability of the information provided by BIM tools.

7.3.7 Quality Control, Supply Chain Management, and Lifecycle Maintenance

Numerous avenues for applying sophisticated tracking and monitoring technologies in construction have been proposed and explored in various research projects.

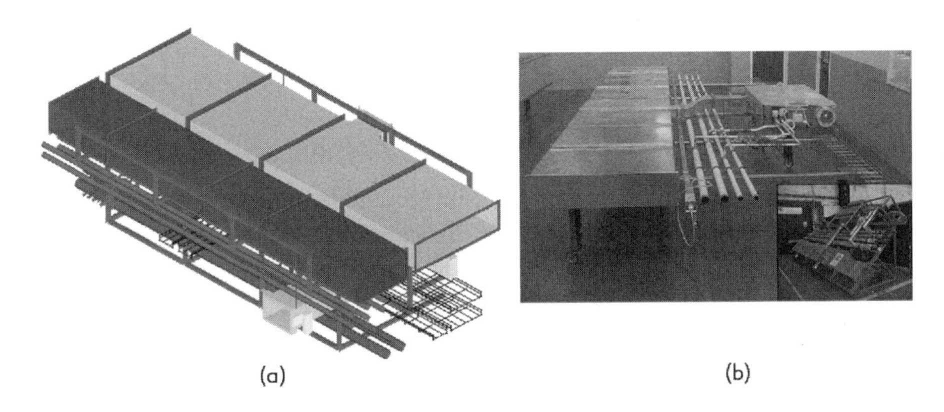

(a) (b)

FIGURE 7-5 Prefabricated ceiling services modules with parts of HVAC, electric, and plumbing systems all installed together.
a) Shows a 3D model view (Court et al. 2006) and b) shows factory prototypes (Pasquire et al. 2006). *Images courtesy of Crown House Technologies, Laing O'Rourke, UK.*

[‖]One notable exception is the BAMTEC system in which entire carpets of rebar, with customized bar diameters and lengths, are welded together and brought to site in rolls.

They include: the use of radio-frequency ID (RFID) tags for logistics; comparing as-built structures to design models with laser scanning (LADAR); monitoring quality using image processing; and reading equipment 'black box' monitored information to assess material consumption. Many more are described in the 'Capital Projects Technology Roadmap' devised by FIATECH.

For manufacturers of ETO products for construction, three main areas of application will be:

- Monitoring of the production, storage, and delivery of components to the site using GPS and RFID systems;
- Supporting the installation or erection of components and quality control using LADAR and other surveying technologies;
- Providing life-cycle information about components and their performance using RFID tags and sensors.

A common thread that runs through all of these proposed systems is the need for a building model to carry the information against which monitored data can be compared. The quantity of data that is typically collected by automated monitoring technologies is such that sophisticated software is required to interpret them. For this interpretation to be meaningful, the designed state of the building product, both geometry and other product and process information, must be available in a computer-readable format.

7.4 BIM-ENABLED PROCESS CHANGE

As we have described in earlier chapters, BIM's primary contribution for general contractors, subcontractors, and fabricators is that it enables *virtual construction.* From the perspective of those directly responsible for producing buildings, whether onsite or in offsite fabrication facilities, this is not just an improvement but a new way of working. For the first time, construction managers and supervisors can practice putting the pieces together before they actually commit to the labor and materials. They can explore product and process alternatives, make changes to parts, and adapt the construction procedures in advance. And they can perform all of these activities in close collaboration with one another across different trades continuously and as construction progresses, allowing them to cope with unforeseen situations as they develop or deal with new changes introduced by owners and designers.

Despite the fact that BIM software tools, as a whole, are not mature enough to make virtual construction simple and commonplace, best practices by leading construction teams throughout the world are already resulting in the process changes described below. Teams, such as those engaged in the Camino Medical Group Building and the GM Production Plant projects (case studies reported in Chapter 9), are setting the direction for progress. They are succeeding, not because they were expert at operating any one or other software, but as a result of the integrated way they exploited BIM technology to build virtually and in a collaborative fashion early in the project.

7.4.1 Leaner Construction

In the manufacturing world, lean production methods evolved to meet individual clients' demands for highly customized products, without the waste inherent in traditional methods of mass production (Womack and Jones 2003). In general, the principles developed apply to any production system, but given the differences between production of consumer products and building construction, adaptation of the manufacturing implementations was needed.

Lean construction is concerned with process improvement, so that buildings and facilities may be built to meet the clients' needs while consuming minimal resources. This requires thinking about how work *flows,* with an emphasis on identifying and removing obstacles and bottlenecks. Lean construction places special focus on workflow stability. A common cause of long construction durations are the long buffer times introduced by subcontractors to shield their own productivity where quantities of work made available are unstable and unpredictable. This occurs because subcontractors are reluctant to risk wasting their crews' time (or reducing their productivity) in the event that other subcontractors fail to meet their commitments to complete preceding work on time, or in case materials are not delivered when needed, or design information and decisions are delayed, etc.

One of the primary ways to expose waste and improve flow is to adopt *pull flow control,* in which work is only performed when the demand for it is made apparent downstream in the process, with the ultimate pull signal provided at the end of the process by the client. Work flow can be measured in terms of the overall cycle-time for each product or building section, the ratio of activities that are completed as planned, or the inventory of work in progress (known as 'WIP'). Waste is not only material waste but process waste: time spent waiting for inputs, rework, etc[1].

---

[1]Readers interested in a brief introduction to the concepts of lean thinking are referred to the work of Womack and Jones (2003); references and links to the extensive literature on the subject of lean construction specifically can be found at the Web site of the International Group for Lean Construction (www.iglc.net).

BIM facilitates leaner construction processes that directly impact the way subcontractors and fabricators work in four ways:

1. Greater degrees of prefabrication and pre-assembly driven by the availability of error-free design information resulting from virtual construction (the ways in which BIM supports these benefits are described in Section 7.3.5) translates to reduced duration of onsite construction and a **shortened product cycle-time** from the client's perspective.
2. Sharing models is not only useful for identifying physical or other design conflicts; shared models that are linked to planned installation timing data using 4D CAD techniques enable exploration of construction sequences and interdependencies between trades. Careful planning of production activities at the weekly level is a key tenet of lean construction. It is commonly implemented using the 'Last Planner™' system (Ballard 2000), which filters activities to avoid assigning those which may not be able to be carried out correctly and completely. Thus, a priori identification of spatial, logical, or organizational conflicts through step-by-step virtual construction using BIM **improves workflow stability.**
3. **Enhanced teamwork:** the ability to coordinate erection activities at a finer grain among different trades means that traditional interface problems – involving the handover of work and spaces from team to team – are also reduced. When construction is performed by better integrated teams, rather than by unrelated groups, fewer and shorter time buffers are needed.
4. When the gross time required for actual fabrication and delivery is reduced – due to the ability to produce shop drawings faster – fabricators are able to reduce their lead times. If lead times can be reduced far enough, then fabricators will be able to reconfigure their supply to sites more easily to take advantage of the improved pull flow. This extends beyond just-in-time delivery to just-in-time production, a practice that substantially **reduces inventories of ETO components** and their associated waste: costs of storage, multiple-handling, damaged or lost parts, shipping coordination, etc. Also, because BIM systems can generate reliable and accurate shop drawings at the last responsible moment – even when late changes are made – fabricators of all kinds can be more responsive to clients' needs, because pieces are not produced too early in the process.

7.4.2 Less Paper in Construction

When CAD was adopted initially, electronic transfers became a partial alternative to communicating paper drawings. The more fundamental change that BIM introduces is that drawings are relegated from the status of information archive to that of communicating medium, whether paper or electronic. In cases where BIM serves as the sole reliable archive for building information, paper printouts of drawings, specifications, quantity takeoffs, and other reports primarily serve to provide more easily legible access to the information.

For fabricators exploiting automated production equipment, as described in Section 7.3.5, the need for paper drawings largely disappears. For example, parts of timber trusses that are cut and drilled using CNC machines are efficiently assembled and joined on beds, where the geometry is projected from above using laser technology. Productivity for the assembly of complex rebar cages for precast concrete fabrication improves when the crew consults a color-coded 3D model, which they can manipulate at will on a large screen, instead of interpreting traditional orthogonal views on paper drawings. The delivery of geometric and other information to structural steel erectors onsite using PDAs that graphically display 3D VRML steel models (translated from CIS/2 models by NIST's software) is a similar example (Lipman 2004).

The need for paper reports is greatly reduced as information from BIM fabrication models begins to drive logistics, accounting, and other management information systems and is aided by automated data collection technologies. It is, perhaps, only the slow pace of legal and commercial change that prevents this chapter from being titled 'Paperless Construction.'

7.4.3 Increased Distribution of Work

The use of electronic building information models means that communication over long distances is no longer a barrier to the distribution of work. In this sense, BIM facilitates increased outsourcing and even globalization of two aspects of construction work that were previously the domain of local subcontractors and fabricators.

First, it is possible for design, analysis, and engineering to be carried out more easily by geographically and organizationally disperse groups. In the structural steel industry, it is becoming commonplace for individuals, armed with powerful 3D parametric detailing software, to become freelancers providing services to fabricators that have greatly reduced their in-house engineering departments. Outsourcing of 3D modeling and plant engineering work to India in sectors like aerospace, automotive and industrial machinery is already common.

Second, better design coordination and communication means that fabrication itself can be outsourced more reliably, including shipping parts over long distances. In the case study describing the building at 100 11th Ave in New York City (Chapter 9), accurate BIM information enabled the production of facade components in China for installation in New York.

7.5 GENERIC BIM SYSTEM REQUIREMENTS FOR FABRICATORS

In this and the following section, we define the system requirements that ETO component fabricators, design service providers, and consultants should require from any software they are considering. This section defines generic requirements common to all and places special emphasis on the need for fabricators to participate actively in compiling comprehensive building information models as part of collaborative project teams. The following section (7.6) expands the list of requirements to include specialized needs of specific types of fabricators.

Note that the most basic required properties of BIM systems, such as support for solid modeling, are not listed, because they are essential for all users and almost universally available. For example, while fabricators require solid modeling tools for clash detection and volumetric quantity takeoffs, they are provided in all BIM software because section views cannot be produced automatically without them.

7.5.1 Parametric and Customizable Parts and Relationships

The ability to automate design and detailing tasks to a high degree – and for building information models to remain coherent, semantically correct, and accurate even as they are manipulated – are cornerstones for reaping the benefits of BIM for fabricators. Creating models would be excessively time-consuming and impractical if operators were required to generate each and every detailed object individually. It would not only be time-consuming but also highly error-prone if operators were required to actively propagate all changes, from building assemblies to all of their detailed constituent components.

For these reasons, fabricators must have software systems that support parametric objects and relationships between objects at all levels (parametric objects and relationships are defined in Chapter 2) in order to use BIM. The structural steel connection shown in Figure 7-6 illustrates this requirement. The software selects and applies an appropriate connection according to its pre-defined rules. (Setup and selection of rule sets for a project may be done by the engineer of record or by the fabricator, depending on the accepted practice.) If the profile shape or parameters of either of the connected members are subsequently changed, the geometry and logic of the connection updates automatically.

An important aspect to evaluate is the degree to which customized parts, details, and connections can be added to a system. A powerful system will

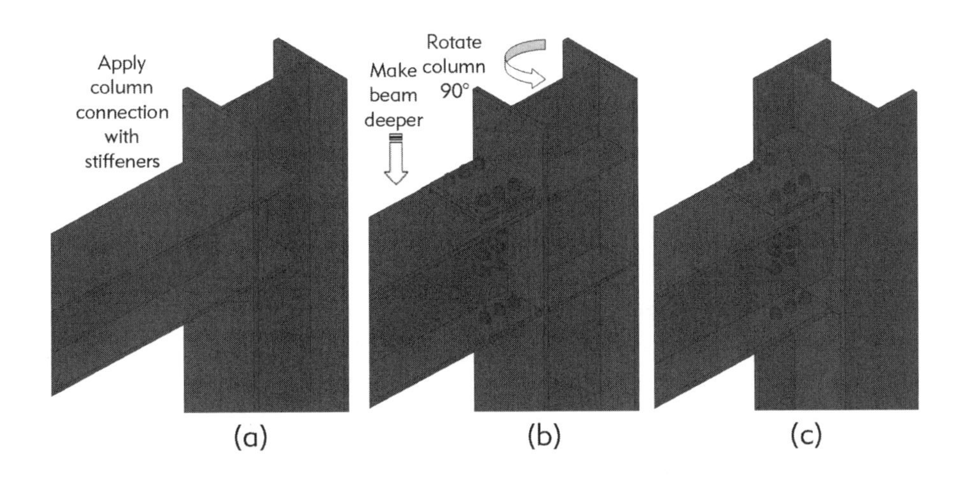

FIGURE 7-6 Structural steel connection in Tekla Structures. The software applies the connection selected by the operator (A to B) and automatically updates the customized connection when the beam is made deeper and the column is rotated (B to C).

support: nesting of parametric components within one another; modeling of geometric constraints, such as 'parallel to' or 'at a constant distance from;' and application of generative rules that determine whether a component will be created in any given context.

7.5.2 Reporting Components for Fabrication

The ability to automatically generate production reports for each individual ETO component in a building is essential for fabricators of all kinds. Reporting may include: preparation of shop drawings; compiling CNC machinery instructions; listing constituent parts and materials for procurement; specifying surface finish treatments and materials; and listing hardware required for installation on site, etc.

In prefabrication of any type of ETO component, it is important to be able to group the components in different ways to manage their production (i.e. procurement of parts, preparation of forms and tools, storage, shipping and erection). Precast concrete parts and fabricated formwork pieces for cast-in-place concrete are commonly grouped according to their molds, so that single molds can be used for multiple parts with minor modifications between each use. Reinforcing bars must be produced and bundled in groups according to their association with building elements.

To support these needs, BIM tools should be able to group components according to criteria specified by operators on the basis of both their geometric information and meta-data. In the case of geometric shapes, the software should be able to distinguish between parts on the basis of the degree to which the pieces are similar or dissimilar. For example, timber trusses might be given a primary identifier for grouping those trusses with the same overall shape and configuration, while a secondary identifier could be used to distinguish sub-groups of one or more trusses with minor differences within the primary group. If a generic truss family were given the type identifier '101,' then a sub-group of a few trusses within the generic '101 family' might include a particular member with a larger profile size that is otherwise the same as a '101' and might be named sub-family '101-A'.

In some applications, prefabricated ETO components will require that some of the constituent parts be delivered loose to the job site, such as weld plates for site joints. These too must be grouped and labeled to ensure delivery to the right place at the right time. Where parts must be cast into or bolted onto the building's structure, they may need to be delivered in advance to other subcontractors or even to other fabricators. All of this information must be generated and applied to the objects, preferably automatically, within the BIM tool.

7.5.3 Interface to Management Information Systems

A two-way interface to communicate with procurement, production control, shipping, and accounting information systems is essential in order to fully leverage the potential benefits detailed earlier in the chapter. These may be stand-alone applications or parts of a comprehensive enterprise resource planning (ERP) suite. To avoid inconsistencies, the building information model should be the sole source for part lists and part production details for the full operation. Fabrication is performed over time, during which changes may continue to be made to the building's design. Up-to-date information regarding changes made to pieces in the model must be available to all of a company's departments at all times, if errors are to be avoided. Ideally, this should not be a simple file export/import exchange but an online database link. Minimally, the software should provide an application programming interface, so that companies with access to programming capability can adapt data exchanges to the requirements of their existing enterprise systems.

Where building information models are integrated with other management systems, automated tracking systems for ETO components, from production through storage, delivery, erection, and operation become feasible. Systems exploiting bar-code tracking are common, while the more powerful radio-frequency identification (RFID) technology has been shown to be feasible for only some ETO component types (Ergen et al. 2007).

7.5.4 Interoperability

By definition, subcontractors and fabricators provide only part of a building's systems. The ability to communicate information between their BIM systems and those of the designers, general contractors, and other fabricators is essential. Indeed, one may conceive of a comprehensive building information model as consisting of the full set of system models maintained in the distinct BIM tools of the numerous design and construction trades, even if there is no one unified database.

The technical aspects of interoperability are discussed thoroughly in Chapter 3, including both its benefits and limitations. Suffice it to say that for the purposes of BIM tool selection by subcontractors and fabricators, the capability to import and export models using an appropriate industry exchange standard should be considered mandatory. Which standard is most important depends on the industry sector: for structural steel the CIS/2 format is essential; for most other sectors the IFC format will likely be most useful.

7.5.5 Information Visualization

A 3D building model view is a very effective platform for entering and visualizing management information, particularly for erectors and general contractor staff outside the fabricator's organization. Customizable functions for generating model displays that are colored according to a variety of production status data are essential.

Two good examples are the use of 4D CAD techniques for micro-planning of a construction operation and the use of a model interface to pull the delivery of prefabricated parts to the job-site in a just-in-time configuration. In the first, a building model that included the structural members and the resources (cranes) and activities was used for step-by-step planning and simulation of the erection sequence for steel and precast concrete elements for an underground subway station roof in London, UK (Koerckel and Ballard 2005). Careful planning was essential so that the project team could meet a strict 48-hour time limit for erection, during which train traffic was suspended. For a detailed description of 4D CAD techniques and benefits, please refer to Section 7 in Chapter 6.

The second example is illustrated by Figure 7-7 (a), which shows a spread of drawings laid out by a site supervisor to select precast pieces for delivery. The effort of coordinating between multiple sets of drawings and lists and the resultant human errors would be eliminated if the site supervisor could simply point and click on a color-coded model to compile delivery lists, as shown in Figure 7-7 (b).

7.5.6 Automation of Fabrication Tasks

The selection of a BIM software tool should reflect the opportunities and plans for automation of the fabrication tasks. These vary with each building system. Some companies will already have CNC machines of different kinds, such as rebar bending and cutting machines, laser cutters for steel profiles or plates, or sophisticated conveyor and casting systems for precast concrete. For some fabricators, these technologies may be drivers for adopting BIM; for others, they will be new, and BIM will enable their introduction. In either case, it is important to consider

(a) (b)

FIGURE 7-7 a) Drawings spread by an erection supervisor to select precast pieces for delivery to the site. b) A model of a precast structure that has been color-coded according to availability of pieces for delivery to the site. Different colors (shown here in grayscale) are used to represent pieces that have been erected, pieces available, pieces currently in production, and pieces ready for production.

the information requirements and the interfaces that are supported by the BIM software.

7.6 MAJOR CLASSES OF FABRICATORS AND THEIR SPECIFIC NEEDS

This section describes specific requirements for fabricators of various kinds. It also provides a short list of software packages (available at the time of publication) for each class of fabricator. The software packages are listed in Table 7-3 along with explanations of their functionality for each domain and sources for additional information.

7.6.1 Structural Steel

With steel construction, the overall structure is divided into distinct parts that can be easily fabricated, transported to the site, erected, and joined, using minimal material quantities and labor, all under the necessary load constraints defined by the structural engineers.

Simply modeling the structure in 3D with all detailing of nuts, bolts, welds, plates, etc. is not sufficient. The following are additional requirements that should be met by steel detailing software:

- **Automated and customizable detailing of steel connections.** This feature must incorporate the ability to define rule-sets that govern the ways in which connection types are selected and parametrically adapted to suit specific situations in structures.

Table 7-3 BIM Software for Subcontractors and Fabricators.

BIM Software	Building System Compatibility	Functionality	Source for information
Tekla Structures	Structural steel, Precast Concrete, CIP reinforced concrete. Mechanical, Electrical, Plumbing, Curtain walls	Modeling, analysis pre-processing, fabrication detailing. Coordination	www.tekla.com
SDS/2 Design Data	Structural steel	Fabrication detailing	www.dsndata.com
StruCAD	Structural steel	Fabrication detailing	www.acecad.co.uk
Revit Structures	Structural steel, CIP reinforced concrete	Modeling, analysis pre-processing	www.autodesk.com/revit
Revit Systems	Mechanical, Electrical, Plumbing and piping	Modeling	www.autodesk.com/revit
3d+	Structural steel		3dplus.cscworld.com/
Structureworks	Precast concrete	Modeling, fabrication detailing.	www.structureworks.org
Revit Building	Curtain walls	Modeling	www.autodesk.com/revit
aSa Rebar Software	CIP reinforced concrete	Estimating, detailing, production, material Tracking, accounting	www.asarebar.com
Allplan Engineering	Structural steel, CIP reinforced concrete, Precast concrete	Modeling, detailing rebar	www.nemetschek.com
Allplan Building Services	Heating, Ventilation, Air-conditioning (HVAC), Mechanical, Electrical, Plumbing (MEP)	Modeling	www.nemetschek.com
Catia (Digital Project)	Curtain walls	Modeling, FEM analysis, parsing production data for CNC	www.3ds.com
Graphisoft ArchiGlazing,	Curtain walls	Modeling	www.graphisoft.com
SoftTech V6 Manufacturer	Curtain walls	Modeling and fabrication detailing	www.softtechnz.com
CADPIPE Commercial Pipe	Piping and plumbing	Modeling and fabrication detailing	www.cadpipe.com
CADPIPE HVAC and Hanger	HVAC ducts	Modeling and fabrication detailing	www.cadpipe.com
CADPIPE Electrical and Hanger	Electrical conduits, cable trays	Modeling, detailing	www.cadpipe.com
Quickpen PipeDesigner	Piping and plumbing	Modeling, fabrication detailing,	www.quickpen.com
Quickpen DuctDedesigner	HVAC	Modeling and fabrication detailing	www.quickpen.com
Bentley Building Mechanical Systems	HVAC ducts and piping	Modeling	www.bentley.com
Graphisoft Ductwork	HVAC ducts	Modeling	www.graphisoft.com
CAD-Duct	HVAC ducts, Piping, Electrical containment	Modeling and fabrication detailing	www.cadduct.com
SprinkCAD	Fire sprinkler systems	Modeling and detailing	www.sprinkcad.com

- **Built-in structural analysis capabilities, including finite element analysis.** Alternatively, as a minimum, the software should be able to export a structural model, including the definition of loads in a format that is readable by an external structural analysis package. In this case, it should also be capable of importing loads back to the 3D model.
- **Output of cutting, welding, and drilling instructions directly to computer numerically controlled (CNC) machinery.**

Available software (see Table 7-3): Tekla Structures, SDS/2 Design Data, StruCAD, 3d+.

7.6.2 Precast Concrete

Information modeling of precast concrete is more complex than modeling structural steel, because precast concrete pieces have internal parts (rebar, prestress strands, steel embeds, etc.), a much greater freedom in shapes, and a rich variety of surface finishes. These were among the reasons why BIM software tailored to the needs of precast concrete became available commercially much later than for structural steel.

The specific needs of precast concrete fabrication were researched and documented by the Precast Concrete Software Consortium (PCSC) (Eastman et al. 2001). The first two needs specified above (section 7.6.1) for structural steel – automated and customizable detailing of connections and built-in structural analysis capabilities – apply equally to precast concrete. In addition, the following requirements are specific to precast concrete:

- The ability to model pieces in a building model with geometric shapes different from the geometry reported in shop drawings. All precast pieces are subject to shortening and creep, which means their final shape is different than that which is produced. Precast pieces that are eccentrically pre-stressed become cambered when pre-stress cables are released after curing. The most complex change occurs when long precast pieces are deliberately twisted or warped. This is commonly done with long double tee pieces in parking garages and other structures to provide slopes for drainage, by setting the supports of one end at an angle to those of the other end. The pieces must be represented with warped geometry in the computer model, but they must be produced in straight pre-stressing beds. Therefore, they must be rendered straight in shop drawings. This requires a relatively complex geometric transformation between the assembly and the shop drawing representations of any intentionally deformed piece.
- Surface finishes and treatments cannot simply be applied to faces of parts but often have their own distinctive geometry, which may require

subtraction of volume from the concrete itself. Stone cladding, brick patterns, thermal insulation layers, etc. are all common examples. Special concrete mixes are used to provide custom colors and surface effects but are usually too expensive to fill the whole piece. The pieces may be composed of more than one concrete type, and the software must support the documentation of volumes required for each type.

- Specialized structural analyses of individual pieces – to check their resistance to forces applied during stripping, lifting, storage, transportation, and erection, which are different to those applied during their service-life in a building – are required. This places special emphasis on the need for integration with external analysis software packages and an open application programming interface.
- The grouping of a precast piece's constituent parts must be done according to the timing of their insertion: cast into the unit at time of fabrication, cast into or welded onto the building foundation or structure, or supplied loose (bundled with the piece) to the site for erection.
- Output of rebar shapes in formats compatible with fabrication control software and automated bending and cutting machines.

Available software (see Table 7-3): Tekla Structures, Structureworks.

7.6.3 Cast-In-Place (CIP) Reinforced Concrete

Like precast concrete, cast-in-place reinforced concrete has internal components that must be modeled in detail. All of the requirements for structural analysis, generating and reporting rebar shapes for production and placing, and for measuring concrete volumes, are equally valid for cast-in-place concrete.

Cast-in-place (CIP) concrete, however, is quite different from both structural steel and precast concrete, because cast-in-place structures are monolithic. They do not have clearly defined physical boundaries between components, such as with columns, beams, and slabs. Indeed, whether the concrete volume at the components' intersection is considered part of one or part of the other component's joint framing is determined based on the reporting needs. Likewise, the same rebars may fulfill a specific function within one member and a different function within a joint, such as with top steel in a continuous beam that serves for shear and crack resistance within the span but also as moment reinforcement over the support.

Another difference is that cast-in-place concrete can conveniently be cast with complex curved geometries, with curvature in one or two axis directions and variable thicknesses. Although non-uniform multi-curved surfaces are rare, domes are not uncommon. Any company that encounters curved concrete surfaces in its construction projects should ensure that the descriptive geometry engine of any modeling software can model such surfaces and the solid volumes they enclose.

A third difference is, unlike steel and precast components, that CIP concrete structures are partitioned differently for analysis and design than for fabrication. The locations of pour stops are often determined in the field and do not always conform to product divisions, as envisioned by the designers. Nevertheless, if the members are to be used for construction management as well as for design, they must be modeled both ways.

Each of these scenarios requires a different multi-view approach to modeling objects than is available in most BIM software packages that currently offer some functionality for CIP concrete modeling. The ability to switch between distinct but internally consistent representations of 3D concrete geometry and idealized members for structural analysis, as provided in Revit Structures, represents a first step in the right direction.

Lastly, CIP concrete requires layout and detailing of formwork, whether modular or custom designed. Some modular formwork manufacturing companies do provide layout and detailing software, which allows users to graphically apply standard formwork sections to CIP elements in 3D. The software then produces the detailed bills of material required and the drawings to aid laborers in erecting the modular forms. 'ELPOS' and 'PERI CAD,' provided by PERI of Germany, are two examples (http://www.peri.de/ww/en/pub/company/software/elpos.cfm). Unfortunately, the existing applications are based on CAD software representations, mostly using 2D views. The suppliers are more likely to provide BIM-integrated solutions as demand grows.

Available software (see Table 7-3): Tekla Structures, Revit Structures, aSa Rebar Software integrated with Microstation, Nemetschek Allplan Engineering.

7.6.4 Curtain Walls and Fenestration

Curtain walls include any wall closing system that does not have a structural function, i.e. it does not carry gravity loads to the foundations of a building. Among custom-designed and fabricated curtain walls – essentially involving ETO components – aluminum and glass curtain walls are typical. They can be classified as *stick systems, unit systems,* or *composite systems*. In this chapter, *fenestration* includes all window units that are custom-designed for fabrication and installation in a specific building, with profiles of steel, aluminum, timber, plastic (PVC), or other materials.

Stick systems are built in-situ from metal profiles (usually aluminum), which are attached to the building frame. They are similar to structural steel frames in that they are composed of longitudinal extruded sections (vertical mullions and horizontal transoms) with joints between them. Like precast facade panels, their connections to the structural frame must be detailed explicitly for every context. They place a unique requirement on modeling software, because they are highly susceptible to changes in temperature, which cause expansion and contraction; as such, their joints must be detailed to allow for free movement without compromising their insulating or aesthetic functions. Joints with appropriate degrees of freedom and sleeves to accommodate and hide longitudinal movement are common. Stick systems require only assembly modeling, with no need for piece fabrication detail drawings. The ability to plan erection sequences in order to accommodate tolerances is critical.

Unit systems are composed of separate prefabricated pieces installed directly onto the building's frame. A key feature for modeling is the need for high accuracy in construction, which means that dimensional tolerances for the building's structural frame should be modeled explicitly.

Composite systems include unit and mullion systems, column cover and spandrel systems, and panel (strong back) systems. These require not only detailed assembly and piece fabrication details but must also be closely coordinated with a building's other systems.

Curtain walls are an important part of any building information model, because they are central to all analyses of building performance other than overall structural analysis (i.e. thermal, acoustic, and lighting). Any computer simulation that can be performed on a model will need the relevant physical properties of the curtain wall system and its components – not only its geometry. Models should also support local wind and dead load structural analyses for the system components.

Most curtain wall modeling routines that are commonly available in architectural BIM systems allow for preliminary design only and have no functionality for detailing and fabrication. The 100 11th Ave., New York, case study in Chapter 9, which presents a complex curtain wall designed for a residential building, serves as a good example of this type of use. On the other hand, widespread manufacturing detailing software systems for curtain walls and fenestration, such as the DeMichele Group packages, are intended for modeling individual windows or curtain wall sections, without compiling them into whole building models. Due to the nature of the steel and aluminum profiles used in most curtain walls, some companies have found mechanical parametric modeling tools, such as Solidworks and Autodesk Inventor, to be more useful.

Available software (see Table 7-3): Digital Project (Catia), Tekla Structures, Revit Building, Nemetschek Allplan, Graphisoft ArchiGlazing, SoftTech V6 Manufacturer.

7.6.5 Mechanical, Electrical and Plumbing (MEP)

Three distinct types of ETO component systems are included in this category: ducts for HVAC systems; piping runs for water and gas supply and disposal; and routing trays and control boxes for electrical and communication systems. These three systems are similar both in nature and in the space they occupy within a building, but they also depend on specific requirements for detailing and fabrication software.

Ducts for HVAC systems must be cut from sheet metal sections, fabricated in units that can be conveniently transported and maneuvered into position, and then assembled and installed in place at a building site. Duct units are three dimensional objects and often have complex geometries.

Piping for supply and disposal of various liquids and gases is composed of extruded profiles that also incorporate valves, bends, and other equipment. While not all piping is engineered-to-order, any sections that require cutting, threading, or other treatments that must be done in a workshop prior to delivery are considered ETO components. In addition, spools of piping components that are pre-assembled as complete units prior to delivery and/or installation are also considered pre-engineered, even if most or all of their constituent parts are off-the-shelf components.

Although electrical and communication cables are largely flexible, the conduits and trays that carry them may not be, which means their layout must be coordinated with other systems.

The first and most generic requirement for these systems to be supported by BIM is that their routing in space must be carefully coordinated. Routing requires easy-to-follow or color-coded visualization and system tools for identifying clashes between systems. Figure 7-8, which was prepared by a general contractor (the Mortenson Company) for coordination purposes, is an excellent example of how a building's MEP systems can be modeled, checked, and prepared for fabrication, production, and installation.

Although physical clash detection is available in most piping and duct software, in many cases soft clash detection is also needed. Soft clash detection refers to certain requirements, where minimum clear space must be maintained between different systems, such as the minimum distance between a hot water pipe and electrical cables. The software must allow users to set up rules that define verifiable spatial constraints between different pairs of systems when clash checks are performed.

A second generic requirement is the grouping of objects for production and installation logistics. Numbering or labeling components must be performed on three levels: a unique part ID for each piece; a group ID for installation spools; and a production group ID that the system assigns based on the collection of identical or largely similar parts for fabrication or procurement. Grouping

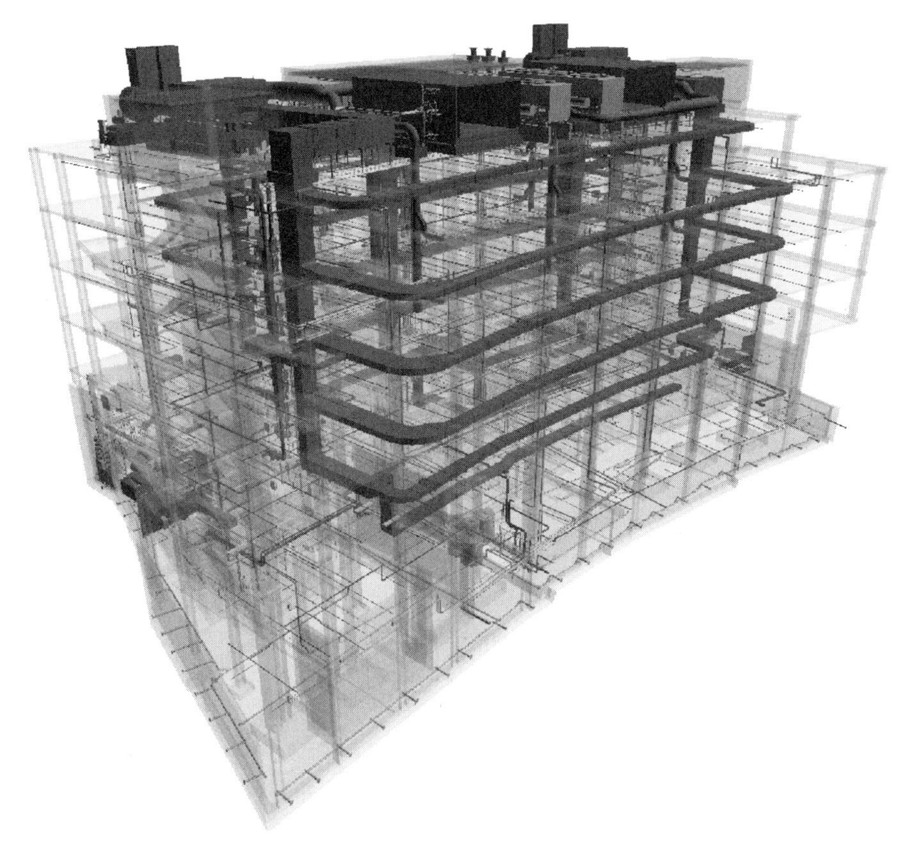

FIGURE 7-8
A model view showing a building's MEP systems with transparent building structure components, prepared by a general contractor (Mortenson) for construction coordination.
(See color insert for full color figure.) *Image provided courtesy of Mortenson.*

of parts for site delivery, with collections of separate components belonging to duct runs and pipe spools, is particularly important. If any part is missing or cannot fit into place due to dimensional changes or fabrication errors, productivity degrades and the workflow is disrupted. To avoid this, BIM systems must provide material takeoff lists and seamless integration with logistics software for labeling schemes to allow complete and correct collections of parts to be pulled to the work-face at the right time. One technology for aiding this is the use of bar-codes to track pipe spools and duct sections. A less mature method is the use of radio frequency identification (RFID) tags.

Unique BIM requirements for each of the systems are as follows:

- Most duct sections are fabricated from flat sheet metal. Software should generate cutting patterns – unfolded from 3D geometric shapes – and translate the data into a format appropriate for plasma cutting tables or other machinery. The software should also offer optimization of the nesting pattern to minimize off-cut waste.

- Piping spools are commonly represented in symbolic isometric drawings. Software should enable display in multiple formats, including full 3D representation, line representation, and symbolic form, as well as 2D plans, sections, and isometric views. In addition, it should automatically generate spool assembly drawings with bill of material data.

Software applications capable of generating detailed models and fabrication information for MEP systems were made available earlier than for other building systems. This was mainly because ducts, pipes, etc, are generally composed of distinct parts, which have standard geometries that are independent of local conditions at the interfaces between parts. Solid modeling and Boolean operations were not needed, and self-contained parametric parts could be added by programming purpose-built routines. It was therefore possible to provide fabrication level modeling on the basis of generic CAD software, which lacked more sophisticated parametric and constraint modeling capabilities.

The drawback of CAD-based applications, as opposed to BIM based applications, is that CAD platforms do not maintain logical integrity when changes are entered. Neighboring duct sections should adjust when changes are made to individual sections or to a duct run as a whole. When a duct or pipe that penetrates a slab or wall moves, the hole in the slab or wall should either also be moved or healed if it is no longer needed. Many MEP applications lack the import and export interfaces needed for industry-wide interoperability, such as support for IFC models.

Subcontractors and fabricators are likely to continue using CAD-based tools, such as those listed below, because the BIM software packages that offer MEP capabilities (at the time of this book's publication) – such as Revit Systems and Bentley Building Mechanical Systems – do not extend to the production of detailed fabrication drawings. For this reason, it is important to ensure that any CAD-based tool is capable of supporting file formats that can be uploaded into design coordination programs like Navisworks Jetstream (Navisworks 2007).

Available software (see Table 7-3): Quickpen (PipeDesigner and DuctDesigner), CADPIPE (HVAC, Commercial Pipe, Electrical, Hanger), CAD-Duct, SprinkCAD, Revit Systems, Bentley Building Mechanical Systems, Graphisoft Ductwork.

7.7 ADOPTING BIM IN A FABRICATION OPERATION

A robust management strategy for the adoption of BIM must concern aspects beyond software, hardware, and the training of engineering staff, because of its range of impact on workflows and people.

BIM systems are a sophisticated technology that impacts every aspect of a fabrication subcontractor's operations, from marketing and estimating through engineering, procurement of raw materials, fabrication, shipping to installation onsite, and maintenance. BIM does not simply automate existing operations that were previously performed manually or using less sophisticated software, it enables different workflow patterns and production processes.

BIM systems directly improve engineering and drafting productivity. Unless a company experiences sustained growth in sales volume through the adoption period, the number of people needed for these activities will be reduced. Downsizing may be threatening to employees whose energy and enthusiasm is critical for changing work procedures. A thorough plan should account for this impact by considering and making provisions for all staff, both those selected for training and those for whom other tasks may be found. It should aim to secure involvement and commitment at an early stage.

7.7.1 Setting Appropriate Goals

The following guideline questions may help in setting goals for an effective adoption plan and for identifying the actors inside and outside the company who should be party to the plan. They apply equally to fabrication companies with in-house detailing capabilities and to companies that specialize in providing engineering detailing services.

- How can clients (building owners, architects, engineering consultants, and general contractors) benefit from fabricators' enhanced proficiency using BIM tools? What new services can be offered that presently are not? What services can be made more productive, and how can lead-times be shortened?
- To what degree can building model data be imported from upstream sources, such as from architects' or other designers' BIM models?
- How early in the process will models be compiled? At the tendering stage or once a contract has been won? What is the appropriate level of detail for a tendering stage model?
- If a tendering model has been prepared, how much of the information compiled is useful for the engineering and detailing phase that follows if the project is won?
- How and by whom will the company's standard engineering details and drawing templates be embedded in custom library components in the software? Will libraries be compiled at the time of adoption or incrementally as-needed for the first projects modeled?

- Can BIM offer alternative modes of communicating information within the company? This requires open discussion with different departments to ascertain real needs. Asking a production department head, "How do you want your shop drawings to look?" may miss the point in a BIM adoption, where alternative forms of presenting the information may be possible. Viewing, manipulating and querying models on screen is a viable addition to traditional drawings. People need to be informed of the new possibilities.
- How will information be communicated to consultants in the submittal process? BIM-capable architects and engineering consultants are likely to prefer to receive the model rather than drawings. How will review comments be communicated back to the company?
- How will the information generated be communicated within the company and between the production and logistics departments? To what degree will building information models be used to generate or display management information? What is needed (software, hardware, programming) to integrate BIM systems with existing management information systems, or will new management systems be adopted in parallel? Most BIM software vendors provide not only fully functional authoring versions but also limited functionality viewing or reporting versions at lower prices than the full package. Such versions are likely to be adequate for production or logistics departments and personnel.
- What is the appropriate pace of change? This will depend on freeing-up the time of those individuals committed to the company's BIM adoption activities.
- How and to what degree will the existing CAD software be phased out? How much buffer capacity should be maintained during the adoption process? Are there any clients or suppliers who will not move to BIM and may therefore require that a limited CAD capacity be maintained?
- What are the needs and capabilities of any suppliers to whom engineering work is outsourced? Will they be expected to adapt? Will the company provide them some support in making the transition to BIM, or will they be replaced with BIM-savvy engineering service providers?

7.7.2 Adoption Activities

Once software and hardware configurations have been selected, the first step will be to prepare a thorough adoption plan, starting with definitions of the goals to be achieved and selection of the right staff to lead the adoption, both as managers and as first learners. Ideally, the adoption plan will be developed together with or by the selected leaders in close consultation with key people from the production and logistics departments company-wide. The plan should detail timing and personnel commitments for all of the following activities:

- **Training engineering staff to use the software.** A word of caution: 3D object modeling is sufficiently dissimilar in concept from CAD drawing that some experienced CAD operators find the need to 'unlearn' CAD behavior a serious barrier to effective use of BIM software. As with most sophisticated software, proficiency is built with practice over time; staff should not be trained until the organization can ensure that they can devote time to continued use of the software in the period immediately following the training.
- **Preparation of custom component libraries, standard connections, design rules, etc.** For most systems and companies, this is a major task, but on the other hand it is a key determinant of the level of productivity that can be achieved. Different strategies can be considered. Custom components can be defined and stored incrementally as-needed on the first projects performed; a large proportion of the libraries can be built ahead of time; or a mixed approach is possible. Larger companies may elect to dedicate a specially trained staff member to compile and maintain part libraries, because parametric modeling libraries are considerably more complex and sophisticated than those used with 2D CAD.
- **Customization of the software to provide drawing and report templates suitable for the company's needs.**
- Immediately after training, the 'first learners' can be tasked with **'ghosting' a project.** This involves attempting to model a project that is being produced in parallel using the standard CAD software. Ghosting provides an opportunity to explore the breadth of a real project, while not bearing responsibility to produce results according to production schedules. It also reveals the limitations of training and the degree of customization that will have been achieved.
- **Seminars and/or workshops for those impacted but who are not direct users** – other departments within the company, raw material and processed product suppliers, providers of outsourced services, and clients – to inform them of the capabilities, enlist their support, and solicit ideas for improved information flows that may become possible. In one such seminar at a precast concrete company, the manager of the rebar cage assembly shop was asked to comment on various options for shop drawing dimensioning formats. Instead, he responded by asking if he could have a computer for

3D viewing of rebar cages color-coded by bar diameters, which he felt would enable his team to understand the cages they were to tie in a fraction of the time they currently needed to interpret 2D drawing sets.

7.7.3 Planning the Pace of Change

The introduction of new BIM workstations should be phased. The personnel undergoing training are likely to remain unproductive during their training and less productive than with CAD tools during the early period, as they progress along a learning curve. The first people trained are also likely to be unproductive for a longer period than most others, because they will have to customize the software to suit company-specific products and production practices. In other words, there is likely to be a need for *additional* personnel at the early stages of adoption, followed by a fairly sharp drop. This can be seen in the total number of personnel needed, as shown in the last row of each adoption plan in Table 7-4.

Table 7-4 shows a feasible plan for a phased replacement of a company's existing 18 CAD workstations with 13 BIM workstations. It lists the numbers of CAD and BIM workstations planned for operation in each of the first four periods following the introduction of BIM software. It is based on estimates for two unknowns: the degree of expected productivity gain and anticipated rate

Table 7-4 Staged adoption of BIM workstations for a fabricator's engineering department.

Adoption periods	Start	P1	P2	P3	P4
Plan ignoring growth in work volume					
Equivalent CAD workstations required	18	18	18	18	18
CAD workstations operating	18	18	13	3	
CAD workstations saved			5	15	18
BIM workstations added		3	6	2	
BIM workstations operating		3	9	11	11
Total workstations	18	21	22	14	11
Plan considering growth in work volume					
Equivalent CAD workstations required	18	18	19	20	21
CAD workstations operating	18	18	14	5	
CAD workstations saved			5	15	21
BIM workstations added		3	6	3	1
BIM workstations operating		3	9	12	13
Total workstations	18	21	23	17	13

of growth in business volume, if any. The rate of growth in volume can be expressed conveniently in terms of an equivalent number of CAD workstations needed to cope with the volume (the table shows two options, ignoring and considering growth in work volume). The rate of productivity gain used to prepare this table is 40% and is based on the number of hours required to produce the same output using BIM as would be produced using CAD. In terms of drawing production, that translates to 60% of the hours currently spent using CAD. This is a conservative estimate based on available measures from research, as detailed in Section 7.3.4 above.

Table 7-4 also demonstrates how downtime for training and reduced productivity at the start of the learning curve can be accounted for. A simplifying assumption in this regard is that the BIM workstations introduced in each period will only become fully productive in the period that follows. Thus, there is no reduction in CAD workstations in the first adoption period, despite the addition of three BIM workstations. In the second period, the reduction in CAD workstations is five and is equal to the number of BIM workstations that become productive (three, the number added in the preceding period) divided by the productivity ratio (3/60%=5).

The increase in personnel needed during the first adoption period may be ameliorated by outsourcing or by overtime, but it is likely to be the main cost item in a BIM adoption cash flow plan and usually significantly more costly than the software investment, hardware, or direct training costs. Companies may decide to stagger the adoption gradually to reduce its impact; indeed, planning period durations may be reduced over time (integrating new operators is likely to be smoother once more colleagues have made the conversion and as the BIM software becomes more deeply integrated in day-to-day procedures). In any event, from a management perspective, it is important to ensure that the resources needed for the period of change will be recognized and made available.

7.7.4 Human Resource Considerations

In the longer term, the adoption of BIM in a fabricator's organization is likely to have far-reaching effects in terms of business processes and personnel. Achieving the full benefits of BIM requires that estimators, who are commonly among the most experienced engineers in a fabrication organization, be the first to compile a model for any new project, because it involves making decisions about conceptual design and production methods. This is not a task that can be delegated to a draftsperson. When projects move to the detailed design and production stages, it will again be the engineers who are capable of applying the correct analyses to models and, at least, the engineering technicians who will determine the details. For trades such as electrical, HVAC and piping, communications, etc., detailing should be done in close collaboration with a general contractor and other trades to ensure constructability and correct sequencing of work, which again requires extensive knowledge and understanding of the domain.

As observed in Chapter 5 (BIM for the design professions), here too the skill set required of BIM operators is likely to result in a decline of the traditional role of drafting. Companies should be sensitive to this in their adoption plan, not only for sake of the people involved but because BIM adoption may be stifled if the wrong people are expected to pursue it.

7.8 CONCLUSIONS

In purely economic terms, subcontractors that fabricate engineered-to-order components for buildings may have more to gain from BIM than any other participant in the building construction process. BIM directly supports their core business, enabling them to achieve efficiencies that fabricators in other sectors, such as the automotive industry, have achieved through the application of computer-aided modeling for manufacturing.

There are numerous potential benefits for fabricators. These include: enhanced marketing and tendering; leveraging the ability to rapidly produce both visualizations and accurate cost estimates; reduced production cycle-times, allowing fabrication to begin at the last responsible moment and accommodate late changes; reduced design coordination errors; lower engineering and detailing costs; increased use of automated manufacturing technologies; increased pre-assembly and prefabrication; various improvements to quality control and supply chain management resulting from the integration of BIM with ERP systems; and much improved availability of design and production information for life-cycle maintenance.

While almost all fabricators and subcontractors can benefit from better coordination between their work packages and those of their peers, each trade can benefit in more specific ways, depending on the nature of their work. In this chapter, BIM practices were described in detail for a small number of trades: structural steel, precast concrete, cast-in-place concrete, curtain wall fabrication, and MEP trades. This is not meant to imply that BIM cannot be used effectively for other trades; we encourage every trade to consider and develop its opportunities, whether through organized group action or persistent trial-and-error by individual companies.

Chapter 7 Discussion Questions

1. List three examples of engineered-to-order (ETO) components of buildings. Why do fabricators of ETO components prepare shop drawings?
2. What is the difference between made-to-stock and made-to-order components? Provide examples of each in the construction context.
3. How can BIM reduce the cycle-time for marketing, detailed design, fabrication, and erection of ETO components in construction? Select one type of component and use its process to illustrate your answers.
4. Why are pre-assembled integrated system modules, such as those described in Section 7.3.6, very difficult to provide using traditional CAD systems? How does BIM resolve the problems?
5. What are the ways in which BIM can facilitate lean construction?
6. What are the features of BIM systems that enable 'push of a button' changes to details of the kind shown in Figure 7.6?
7. Imagine that you are assigned responsibility for the adoption of BIM in a company that fabricates and installs HVAC ducts in commercial and public buildings. The company employs six detailers who use 2D CAD. Discuss your key considerations for adoption and outline a coherent adoption plan, citing major goals and milestones.
8. What are the features of building information models, and what are the process benefits they bring, as opposed to 2D drawing practices, that make global procurement of ETO components possible and more economical?

CHAPTER 8

The Future: Building with BIM

8.0 EXECUTIVE SUMMARY

BIM is not a thing or a type of software but a human activity that ultimately involves broad process changes in construction.

Already, a wide variety of owners are demanding BIM and changing contract terms to enable it. New skills and roles are developing. Successful pilot implementations in construction are leading to corporate-wide uptake by pioneering contractors; and construction contractors are implementing sophisticated ERP systems. A survey conducted in early 2007 found that 74% of US architectural firms are already using 3D modeling and BIM tools, although only 34% of those use it for *intelligent modeling*. BIM-standard efforts—such as the National BIM Standards in the US—are gathering steam; and the public is increasingly demanding greener buildings. BIM and 4D CAD tools are becoming common in construction site offices. The lack of appropriately trained professional staff, rather than the technology itself, is the current bottleneck to widespread implementation.

The technology trends include the development of automated checking for code conformance and constructability using building information models. Some vendors have expanded the scope of their BIM tools, while others offer more discipline-specific functionality, such as construction management functions. It is becoming more common for building product manufacturers to provide 3D catalogs; and BIM is helping to make globalization of fabrication for increasingly complex building sub-assemblies economically viable.

But BIM is a work in progress. As it develops and its use becomes more widespread, the extent of its impact on the way in which buildings are built will become more apparent. In this chapter, we first extrapolate from these trends to the short-term future. The next five years are likely to see much broader adoption of basic BIM tools. BIM will contribute to a higher degree of prefabrication, greater flexibility and variety in building methods and types, fewer documents, far fewer errors, less waste, and higher productivity. Building projects will perform better, thanks to better analyses and exploration of more alternatives, fewer claims, and fewer budget and schedule overruns. These are all improvements on existing construction processes.

Numerous societal, technical, and economic drivers will determine the development of BIM in the mid-term future (10–12 years). The latter part of this chapter identifies the drivers and obstacles in the timeframe leading up to 2020. We reflect on the likely impacts of the drivers on BIM technology, on the design professions, on the nature of construction contracts and the synergy between BIM and lean construction, on education and employment, and on statutory and regulatory processes.

The big picture is that BIM will facilitate early integration of project design and construction teams, making closer collaboration possible. This will help make the overall construction delivery process faster, less costly, more reliable, and less prone to errors and risk. This is an exciting time to be an architect, an engineer, or any other AEC industry professional.

8.1 INTRODUCTION

BIM is beginning to change the way buildings look, the way they function, and the ways in which they are built. Throughout this book, we have intentionally and consistently used the term BIM to describe an activity (*building information* ***modeling***), rather than an object (as in *building information* ***model***). This reflects our belief that BIM is not a thing or a type of software but a human activity that ultimately involves broad process changes in construction. In this chapter, we aim to provide two perspectives on the future of building using BIM: *where BIM is taking the AEC industry*, and *where the AEC industry is taking BIM*.

We begin with a short introduction describing the conception and maturation of BIM until the present (2007). We then provide our perspectives on what the future holds. The forecast is divided into two timeframes: a fairly confident forecast of the near future that looks ahead to the next five years (until 2012) and a more speculative long-term forecast looking ahead to the year 2020. The near-term forecast reflects current market trends—many discussed in earlier chapters of this book—and then reviews current research. The long-term forecast relies on analyses of likely drivers and a fair amount of intuition. Beyond 2020, potential advances in hardware and software technologies as well as business practices, make it impossible to predict anything reliably, and so we refrain from speculation.

After 2020, construction industry analysts will reflect, with the benefit of hindsight, on the process changes that will have occurred by 2020. They will likely find it difficult to distinguish definitively between such influences as BIM, lean construction, and performance-driven design. In the absence of each other, these techniques could, theoretically, flourish on their own. Their impacts, however, are complementary in important ways, and they are being adopted simultaneously. Practical examples of their synergies are apparent in some of the case studies in the following chapter (such as the GM Production Plant, the Camino Group Medical Building, and the Federal Office Building). We address some of these synergies in sections 8.2 and 8.3.

8.2 THE DEVELOPMENT OF BIM UP TO 2007

BIM technology has crossed the boundary between research concept and viable commercial tool, and it is set to become as indispensable to building design and construction as the proverbial tee square or hammer and nail. The transition to BIM, however, is not a natural progression from computer-aided drafting (CAD). It involves a paradigm shift from drawing to modeling, and it facilitates—and is facilitated by—a concurrent shift from traditional competitive project delivery models to more collaborative practices in design and construction.

The concept of computer modeling for buildings was first proposed when the earliest software products for building design were being developed (Bijl and Shawcross 1975; Eastman 1975; Yaski 1981). Progress toward BIM was restricted first by the cost of computing power and later by the successful widespread adoption of CAD. But idealists in academia and the construction software industry persisted, and the research needed to make BIM practical continued to move forward. The foundations for object-oriented building product modeling were laid throughout the 1990s (Gielingh 1988; Kalay 1989; Eastman 1992). Parametric 3D modeling was developed both in research and by software companies for specific market sectors, such as structural steel. Current BIM tools are the fulfillment of a vision that has been predicted, by many, for at least two decades.

BIM technology will continue to develop rapidly. Just as the concepts of how BIM tools should work drove their technological development, a renewed vision of the future of building with BIM—emphasizing workflows and construction practices—is now needed. Readers who are considering the adoption of BIM tools for their practices and educators teaching future architects, civil engineers, contractors, building owners, and professionals, should all understand not only the current capabilities but also the future trends and their potential impacts on the building industry.

8.3 CURRENT TRENDS

Market and technology trends are good predictors of the near-term future in any field, and BIM is no exception. The trends observed reveal the potential direction and influence BIM will have in the construction industry. The following paragraphs outline the trends that influence our forecast. They are summarized in the box on the following page.

Sophisticated owners are beginning to demand BIM and to change contract terms to enable it. The General Services Administration (GSA) of the US federal government, representing a sophisticated owner, demands the use of BIM models that are capable of supporting automated checking to determine whether the design meets program requirements. Civilian construction for the DOD is requiring that all of their projects be based on BIM. Sutter Health, a California medical services provider with a multi-billion dollar construction program, is actively encouraging the use of BIM by its providers as an integral part of its lean construction practices. The Swire Properties One Island East case study (Chapter 9) is an example of a project in which an enlightened owner of a major skyscraper demanded the use of BIM. Statsbygg, an owner/developer company responsible for procuring all public construction for the Norwegian government, has begun to demand BIM use. All of these owners are motivated by the economic benefits they perceive to be inherent in building with BIM. Indeed, the heavy engineering and process plant industries have relied on 3D modeling for engineering, procurement, and construction (EPC) for over a decade.

New skills and roles are developing. The productivity gain for the documentation stage of precast and cast-in-place concrete structures has been measured in case studies and researched in numerous contexts, and has been found to be in the range of 30%–40%. Although reliable numerical data is not yet available for architectural design, the trend observed is similar, and the implication is downsizing of drafting staff in building design practices of all kinds.

BIM Process and Technology Trends

Process trends

- Owners are demanding BIM and changing contract terms to enable its use.
- New skills and roles are developing.
- In a recent survey, 25% of US architectural firms reported using BIM tools for 'intelligent modeling.'
- Successful pilot implementations in construction are leading to corporate-wide uptake by pioneering contractors.
- The benefits of running an integrated practice are broadly recognized.
- Construction contractors are implementing sophisticated ERP systems.
- Standards efforts are gathering steam.
- Green building is increasingly demanded by clients.
- BIM and 4D CAD tools are becoming common tools in construction site offices.

Technology Trends

- Automated checking for code conformance and constructability using building information models is becoming available.
- Major BIM tools are adding functionality and integrating capabilities of other products, providing even richer platforms for use.
- Vendors are increasingly expanding their scope and providing discipline-specific BIM tools.
- Building product manufacturers are beginning to provide 3D catalogs.
- BIM tools with construction management functions are becoming increasingly available.
- BIM is encouraging the globalization of prefabrication for increasingly complex building sub-assemblies

This may be compensated, to a degree, by the addition of modeling roles by architects and engineers.

25% of US architectural firms surveyed in 2007 reported already using BIM tools for 'intelligent modeling.' More specifically, 74% of the firms surveyed reported using 3D/BIM tools, but only 34% of those claimed they use it for 'intelligent modeling' (i.e. not simply for the generation of 2D drawings and visualizations) (Gonchar 2007). Five years ago, the use of intelligent modeling was rare.

Successful pilot implementations in construction are leading pioneering contractors to re-engineer their processes, by taking corporate-wide advantage of these early benefits. Pilot projects that made early intensive use of what were still imperfect BIM tools—and showed dramatic success—have indicated the nature of the technology's impact on construction. Among the cases studies in Chapter 9, the GM Production Plant showed how BIM can facilitate design-build projects; the Camino Group Medical Building showed how BIM is essential in enabling lean pull flow control for detailing of MEP systems, resulting in a high degree of pre-assembly; and the Penn National parking structure showed how prefabrication can be almost entirely error-free. A recent survey sponsored by the Construction User's Round Table indicates, "A growing proportion of early adopters report plans to transform their organizational strategy" (CIFE 2007).

The benefits of integrated practice are broadly recognized. Leading architectural firms are beginning to recognize that future building processes will require integrated practice of the whole construction team and will be facilitated by BIM. All members of the building team, not only the engineering consultants but contractors and fabricators, are recognized to have valuable input for design. This is leading to new forms of partnerships, with more design-build projects, more construction firms incorporating their own design offices, and more innovative and intensive teaming. The GM Flint Plant and the 100 11th Ave. New York City case studies exemplify this trend.

Construction contractors are implementing sophisticated ERP systems. A parallel—but as yet unrelated—development is the increasing adoption of enterprise resource planning (ERP) software systems by general contractors. These systems, which are common in other industries but are now being adopted in construction, connect purchasing, accounting, inventory, and project planning at the company level for multiple projects (Grilo and Jardim-Goncalves 2005). Once the back-office systems are automated and in place, construction organizations will begin to integrate these systems with their CAD, 3D, and BIM systems (Eastman et al. 2002). Several European companies are beginning to integrate their 3D/BIM tools with Oracle ERP and model server technologies (Dunwell 2007, Neuberg 2007).

Standards efforts are gathering steam. In the US, the National Institute for Building Sciences (NIBS) has begun facilitating industry definition of a set of National BIM Standards, which aims to smooth out data exchange within specific construction workflows. Organizations like the Association of General Contractors (AGC) have already published the *BIM Guidelines,* which provides advice on how to best leverage the technology. In 2006, the American Institute of Steel Construction amended its code of standard practice to require that a 3D model, where it exists, be the representation of record for design information. All major BIM tool vendors now support, to a lesser or greater degree, some form of IFC standard exchange.

Green building is increasingly demanded by a public conscious of the threats of climate change. BIM helps building designers achieve environmentally sustainable construction, by providing tools for the analysis of energy needs and for accessing and specifying building products and materials with low environmental impact. The use of natural ventilation in the San Francisco Federal Building, as described in the case study in Chapter 9, provides a good example of this synergy.

BIM and 4D CAD tools are becoming commonplace in construction site offices. Over the past decade, 4D tools have gradually moved from the research lab (McKinney et al. 1996; McKinney and Fischer 1998) to the construction office and site (Haymaker and Fischer 2001; Schwegler et al. 2000; Koo and Fischer 2000). BIM use is evident onsite in both the One Island East and the Camino Medical Group Building case studies in Chapter 9. Today, all major BIM tool vendors provide 4D functionality, and several smaller companies also sell 4D tools.

Automated checking for code conformance and constructability using building information models has become available. In Singapore, part of the design checks of building code compliance required for building licenses are already automated. Innovative companies, such as Solibri and EPM, have developed model-checking software (Jotne 2007; Solibri 2007) using IFC files and are intent on extending their capabilities. Coordination between complex building systems using superimposed 3D models is becoming common.

Vendors are increasingly expanding their scope and providing discipline-specific BIM tools. Major BIM vendors are adding discipline-specific interfaces, objects, design rules, and behaviors to the same base parametric modeling engine (witness 'XXX Building/Architect', 'XXX Structure', 'XXX MEP', etc.). These vendors have also extended the scope of their software capabilities by acquiring structural analysis applications. One such vendor has purchased a building systems coordination application; another is developing a sophisticated contractor site management application.

Building product manufacturers are beginning to provide 3D catalogs. Products as diverse as JVI mechanical rebar splices, Andersen windows, and many others can be downloaded as 3D objects and inserted parametrically into models from several online sites.

Construction management functions are being integrated into BIM tools. The extension of 4D CAD to include cost—what is called *5D CAD*—and further extension to incorporate additional management parameters to *nD CAD* are already being undertaken by various solution providers. These promise to offer better insight into how projects can be built feasibly and reliably. The concept of *virtual construction* is no longer familiar only to the research community. It is increasingly being used and appreciated in practice, as indicated by the recent *Virtual Design and Construction Survey* (VDC) (CIFE 2007). Constructor 2007 (VicoSoftware 2007) is an example of this trend; Innovaya (Innovaya 2007) is another.

BIM is helping to make the fabrication of increasingly complex building sub-assemblies economically and globally viable. Large curtain wall system modules are already being fabricated in China, at costs and quality that cannot be matched in the developed world (see the 100 11th Ave. New York City case study in Chapter 9 for an example). The need for transport time allowances means that lead-times for design are short, and the modules must be fabricated right the first time. BIM produces reliable and error-free information and shortens lead-times.

The process and technology trends outlined above were formative in our attempt to look ahead at the future of building with BIM, in this chapter's following sections. BIM, however, is not developing in a vacuum. It is a computer-enabled paradigm change, and so its future will also be influenced by developments in Internet culture and by other similar and less predictable drivers.

8.4 VISION 2012

Recent years have witnessed the realization of many of the ideas of BIM visionaries, and the next five years will see increasing numbers of successful implementations, changes in the building industry, and new trial uses and extensions of what can be achieved with BIM, beyond its use today. This period will see the transition of BIM from an early adoption technology to accepted mainstream practice; and the transition will impact all building professionals and participants. But the greatest impact will be on the individual practitioner, who will need to learn to work, design, engineer, or build with BIM.

8.4.1 Impact on the Design Professions: Shifting Services and Roles

Designers will experience productivity gains and deliver higher quality services. In the next five years, building designers will continue to adopt BIM, and by the end of the period, 60–70% of firms will have worked on a project making full use of BIM, compared to the 25% that use it today (Gonchar 2007). The two main drivers for broad adoption will be: 1) client demand for enhanced quality of service; and 2) productivity gain in preparing documentation. The competitive advantage BIM provides will motivate individual firms to adopt BIM, not only for the sake of internal improvement but to gain a competitive advantage in the marketplace.

The most significant shift for design firms will be in the quality and nature of their services. Currently, designers mostly rely on experience and rule-of-thumb judgments regarding cost, functional performance, and energy and environmental impacts of their designs. Most of their time and effort is spent producing project documents and on meeting explicit owner requirements. Some of the case studies and sections in this book note how early adopters of BIM are beginning to move towards performance-based design, by utilizing tools to better inform their design decisions. Design firms (with a push from clients) will begin to broaden their scope of services to include detailed energy and environmental analyses, operations analyses within facilities (such as for healthcare), and value engineering throughout the design process, based on BIM-driven cost estimates; and these are just a few of the possibilities. Initially, these services will be market differentiators. Later, some will become common practices for all. As these firms develop their new technical environments and expertise with BIM, the late adopters or non-BIM design firms will find it increasingly difficult to compete and will be forced to either adapt or die.

Architecture and engineering firms will face a workplace with changing roles and activities. Junior architects will be expected to demonstrate proficiency with BIM as a condition of employment, in the same way that CADD proficiency has been required since the 1990s. Some downsizing will occur among staff members dedicated to document producing activities. New roles will emerge with titles such as the *building modeler* or *model manager,* requiring design and technical know-how. The model manager will work with the project team to update the building model, guarantee origin, orientation, naming and format consistency, and to coordinate the exchange of model components with internal design groups and external designers and engineers.

As detailing and documentation production phases become increasingly automated in various areas of engineering, cycle times for processing will be significantly reduced. These trends were already witnessed in the Penn National and the 100 11th Ave. New York case studies (see Chapter 9). Little's Law (Hopp and Spearman 1996), which relates cycle times and levels of work-in-progress to throughput, explains that for any given workload, reducing cycle times means that the level of work in progress is reduced. The implication is that firms should be able to reduce the number of projects they have in active design at any given time in their practices. Thus, some of the waste inherent in moving employees' attention from one project to another at frequent intervals may be reduced.

BIM tools and processes will facilitate the trend first enabled by the Internet to divest and outsource services, leading to further empowerment of small firms with a highly technical and BIM-skilled staff. Increased opportunities will exist to provide freelance technical or very specialized design services in response to the ever-growing complexity of building systems and materials. Four of the ten case studies in Chapter 9 benefited from outside specialist advisors contributing to the use of BIM (the San Francisco Federal Office Building, 100 11th Avenue New York City, the Beijing National Aquatics Center, and the One Island East Hong Kong project). Consortia of specialist design firms are able to collaborate around a common building model, often achieving outstanding team results in shorter times than was ever possible with drawings. This makes it both efficient and practical for such firms to provide new design and performance analyses and/or production advice under the leadership of the primary design firm, which may be a large innovative firm or a small firm with high design and coordination skills. In some ways, we may see an acceleration of the trend described in Section 5.3 and a similar evolution of design services that we saw over the last forty years in contracting services. The contracting design firm will do a reduced amount of work but will coordinate and integrate the work of multiple specialist advisors. These trends are evident today and will grow incrementally to respond to the increasing complexity of design services.

Although much will change, many aspects of building design will stay rooted in current practice. In the short term, the majority of clients, local regulatory authorities, and contractors will continue to demand drawings and paper documentation for projects. Many non-leading design practices will only use BIM to generate consistent drawings, for team communication, and for hand-off to contractors. Only a minority of firms will have shown the way to integrate building performance capabilities with standard general design functions.

8.4.2 Impact On Owners: Better Options, Better Reliability

Owners will experience changes in the quality and nature of services available and an overall increased reliability of the project budget, program compliance, and delivery schedule. Many owners are already experiencing this. Chapter 4 and several of the case studies describe owners who were introduced to new processes and deliverables. Within the next five years, owners can expect the changes in the design professions—discussed in the previous section—to translate into more offerings by service providers to deliver a building information model and to perform services related to analyzing, viewing, and managing the model.

In the early project phases, owners can expect to encounter more 3D visualizations and conceptual building information models with programmatic analysis (see Chapter 5 for a discussion of these tools). Three-dimensional models and 4D simulations produced from building information models are far more communicative and informative to lay people than technical drawings. With the increasing availability of 3D-based internet technologies, like earth viewers and virtual communities, owners will have more options to view project models and use them for marketing, sales, and evaluation of designs in the site context. Building information models are also far more flexible, immediate, and informative than computer-renderings of buildings produced using CAD technologies; and they enable owners and designers to generate and compare more design options early in the project, when decisions have the most impact on the project and life-cycle costs.

These technical developments will have different impacts for different owners, depending on their business incentives. Owners who build to sell will find that they can demand and achieve much shorter design durations for conceptual design and construction documentation. On the other hand, for owners who have an economic interest in the life-cycle costs and energy efficiency of their buildings, the conceptual design stage will provide the opportunity for an in-depth study of the behavior of each alternative building design. Savvy owners—with the perception that conceptual-level models can be developed and evaluated rapidly—are likely to express greater demand for design quality. In an effort to optimize building design, they will demand thorough exploration of more alternatives, in terms of construction cost, sustainability, energy consumption, lighting, acoustics, maintenance, and operations.

During this time period, more advanced analysis and simulation tools will emerge as options for specific types of facilities, such as healthcare, public access areas, stadiums, transit facilities, civic centers, and educational centers. Figure 8-1 shows an example of a tool that allows healthcare owners and their

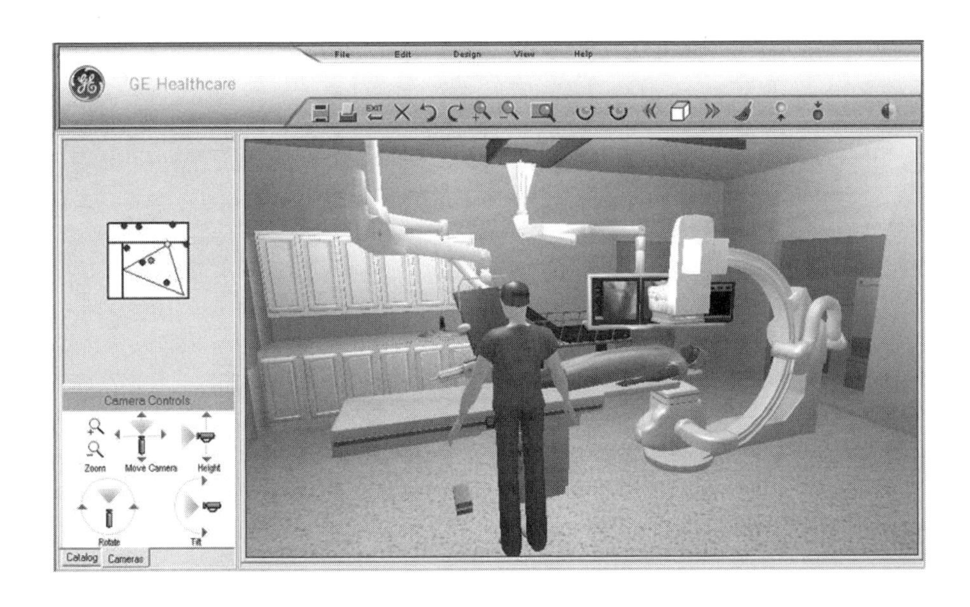

FIGURE 8-1 Example of a component-based simulation of an operating room, allowing the owners and designers to compare different equipment. The equipment components include parameters and behaviors, ensuring that proper clearances and distances are maintained. *Image provided courtesy of View22 and GE Healthcare.*

designers to compare different configurations of hospital rooms with different equipment. Since the actual occupants and users are central to assessing and evaluating any design, tools that work integrally with a BIM system to provide intelligent configuration capabilities will become more widespread.

Similarly, sophisticated construction clients will drive the development of automated design review software for different building types. These will assess a given building design at different stages of development and according to different preset guidelines. For example, the GSA is already extending its program area checking tool to other aspects of design and other building types. One program beginning to be implemented allows for circulation assessments of various layout options during conceptual design. It focuses on courthouses, which have major circulation and security requirements. An early example of this type of testing is shown in Figure 8-2. Other public or private organizations can be expected to develop similar protocols for other building types, such as hospitals and schools.

For first-time (and often one-time) construction clients, different and less desirable possibilities may occur. They may not be familiar with BIM and its potential uses and, as a result, inadequately engage the design team in assessing the project's more subtle goals regarding function, cost, and time-to-delivery. If

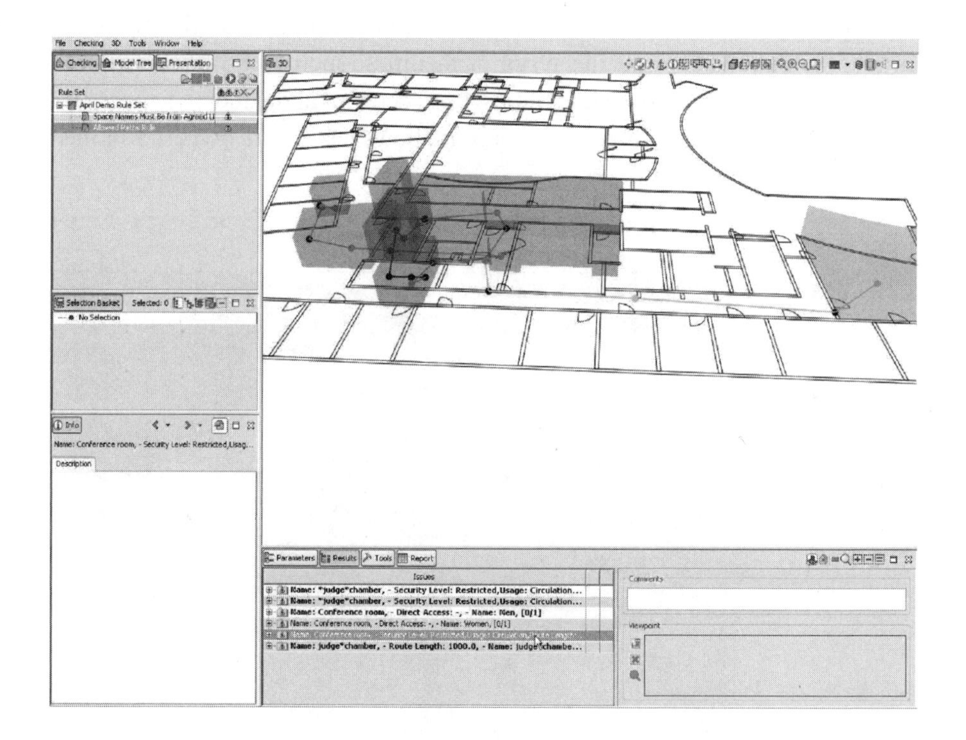

FIGURE 8-2
A courthouse circulation path that fails to provide secure access between a trial jury room and a restroom.
The system checks the security zones of all possible paths, which may span multiple floors (GT-COA 2007; Solibri 2007).

designers are not disciplined, they can develop fairly detailed designs rapidly and create building models that appear convincing and appealing. If the vital stage of conceptual design is short-circuited, premature production level modeling can lead to a lot of rework later in the process. In the worst cases, inadequately designed buildings that do not meet the clients' needs may be built. Like any powerful technology, BIM too is open to abuse. Building clients unfamiliar with the capabilities that BIM technology offers are advised to educate themselves and select knowledgeable design consultants in order to obtain professional design service that exploits the technological capabilities of BIM to achieve the desired objectives of the project.

Smart owners, on the other hand, will demand a faster and more reliable building process from their design and construction teams. The use of design-bid-build for private construction will continue to decline, as owners realize that an integrated team is the best way to obtain value from BIM technology. They will start measuring BIM teams by the number of change orders and delays on prior projects (get it right the first time). The number of claims should go down as the design and construction process becomes more efficient. Lawyers and expert witnesses will become a smaller part of the owner's life (and fears). Time and cost contingencies will shrink. Clients that build frequently will look for teams of design and construction professionals that have BIM experience and know how to leverage these tools with lean processes.

As the use of 4D and BIM-coordination by contractors becomes more commonplace, owners will increasingly appreciate the power of these tools to improve budget and schedule reliability as well as overall project quality. They will begin to require status reports, schedules, and as-builts in BIM formats. More owners will seek out model managers or require that their construction manager perform this task and facilitate the model management network, either as a central model repository or one that is distributed across the project teams. Thus, owners will increasingly face the challenge of managing and storing the building information models. They may also need to address intellectual property issues, if individual members of collaborative teams contribute proprietary information (Thomson and Miner 2006).

Post-construction, owners will consider whether or not to utilize the model for facility management, as discussed in Chapter 4. If they choose to do so, they will need to learn how to update and maintain it. During this time period, we can expect increased use and maturation of BIM-based facility management products. We will see the first cases of building information models integrated with building monitoring systems for comparing and analyzing predicted and observable building performance data, which will provide owners and operators with better tools for managing their building operations.

8.4.3 Impact on Construction Companies: BIM at the Construction Site

Construction companies, for competitive advantage, will seek to develop BIM capabilities both in the field and in the office. They will use BIM for 4D CAD and for collaboration, clash detection, client reviews, production management, and procurement. In many ways, they will be in a better position than most other participants in the construction supply chain for leveraging the short-term economic benefits of ubiquitous and accurate information.

Chapters 6 and 7 explained how BIM can contribute to reducing construction budgets and schedules, as a result of better quality designs (i.e., fewer errors) and by enabling greater degrees of prefabrication. A positive effect of the ability to develop design details fairly early in the process is that rework, which commonly results from unresolved details and inconsistent documentation, is mostly eliminated. These effects have already been reported in case studies—such as the Camino Group Medical Building and the 100 11th Ave. New York project in Chapter 9 and in pioneering projects like the Denver Art Museum Extension—but they are likely to become widespread only toward the end of the time-window considered in this section (2012).

Some mechanical parametric modeling software companies may develop products for different types of construction fabrication that are designed for and integrated with NC fabrication equipment. This will allow new custom-fabricated products to become part of construction, including molded plastic panels, novel kinds of ductwork, and others.

This period should see increasingly smoother transitions from design models to construction models. Software *wizards*—using parametric templates of work packages with embedded construction methods—will be applied, to rapidly compile a construction model from a design model. Ideas like the *recipes* in Constructor 2007 software (Vico Software 2007) are an early indication of what can be expected. For example, a parametric template for a post-tensioned flat slab will lay out the formwork design and determine labor and equipment inputs, material quantities, and delivery schedules based on a generic slab object in a design model. A resulting construction model can be analyzed for cost, equipment, and logistic constraints and for schedule requirements; and the alternatives can be compared. Thus, construction planning will be greatly enhanced. The parametric templates will also serve as a repository for corporate knowledge; in as far as they will embed an individual company's way of working into these software applications.

The role of the *building modeler* will be an issue among contractors and fabricators, due to the mixed roles of senior staff and the complexity of some

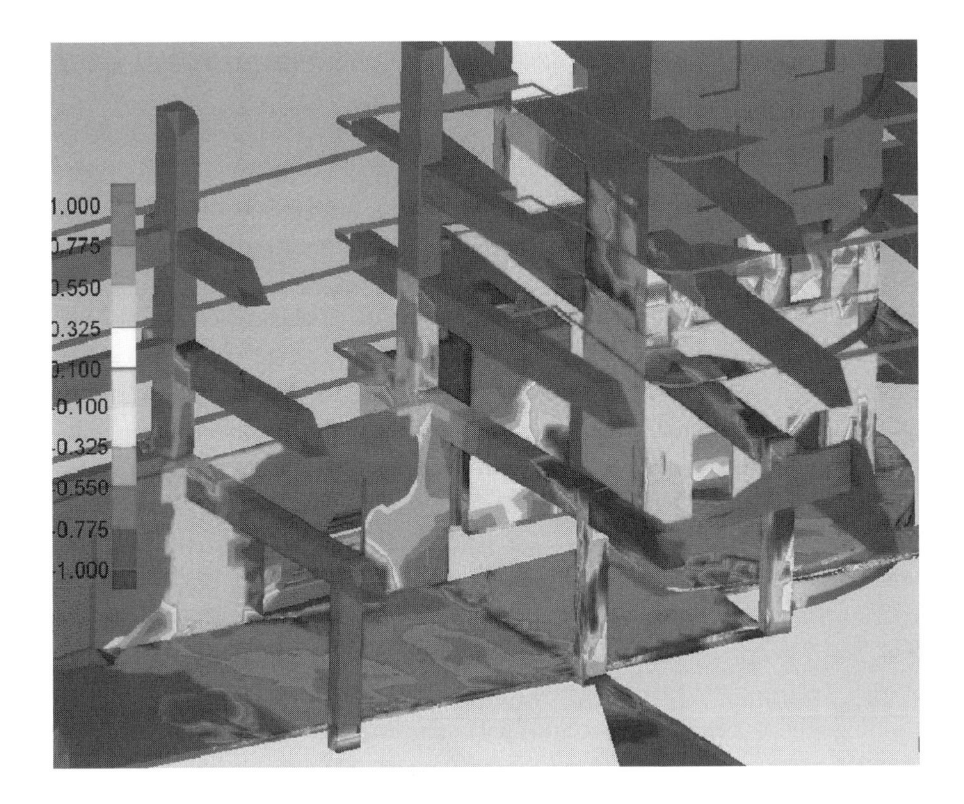

FIGURE 8-3 Laser scanning point cloud data can be mapped onto BIM objects to show deviations of the as-built geometry from the designed geometry. The colors represent the degree of deviation from the planned (gray) surfaces, according to the scale at the left of the figure (see color insert for full color figure). *Image courtesy of Elsevier (Akinci et al. 2006).*

detailing systems. As third party engineering service detailers for precast, reinforced concrete, and other systems gain proficiency in BIM, they will become the de facto building modelers in the same way as in the steel fabrication business. Also, BIM sets the scene for the use of automated surveying and other data collection technologies. For example, laser scanning can produce *point cloud surveys* of existing physical geometry. Interpretation of the point clouds for generating meaningful building objects that can be used in a building information model is possible, but it is a time-consuming endeavor that limits the technology's use for most situations. When matched to a 3D model, as shown in Figure 8-3, point cloud data can be used directly to highlight deviations of the as-built geometry from the designed geometry (Akinci et al. 2006).

8.4.4 Impact on Construction Contracting: Closer Collaborations Among Designers and Contractors

As we have noted in Chapter 6, BIM provides considerably more advantages in the context of design-build type procurement arrangements. As design and construction companies gain experience with BIM, recognition of the added-value that can be achieved will push them to move building procurement from design-bid-build to negotiated contracts, cost-plus, design-build, and augmented design arrangements.

Some construction companies will expand their services in the areas of model development and management. Others may provide package services for full building delivery through a leveraged use of BIM technology.

Both the Internet and BIM tools will facilitate increasing degrees of globalization in construction, not only in the design and parts supply but in the fabrication of engineered-to-order components of increasing complexity. The fabrication of the steel and glass panels of the curtain wall system for the 100 11th Avenue, New York City case study project is an early example. The accuracy and reliability of production data prepared using BIM allowed building products and assemblies that would traditionally be procured locally to be made anywhere in the world. Curtain wall panels are one example. Large modular pre-fabricated utility systems or complete bathroom units may be others. Competition in the construction fabrication area will spread globally.

8.4.5 Impact on Construction Education: Integrated Education

Leading schools of architecture and civil engineering have already begun teaching BIM to undergraduates in their first year, and that trend is likely to spread in parallel with the adoption of BIM in the design professions. The third authors' experience to date in teaching BIM is that students are able to grasp the concepts and become productive using BIM tools more quickly than they were with CAD tools.

The current lack of trained personnel remains a barrier to BIM adoption, forcing many companies to retrain experienced CAD operators in the new tools. Because BIM requires different ways of thinking about how designs are developed and building construction is managed, retraining requires not only learning but the unlearning of old habits, which is difficult. New graduates, whose entire undergraduate experience was influenced by their familiarity with BIM and its use for the full range of student projects, are likely to have a profound influence on the way companies of all kinds deploy BIM. Inevitably, a good deal of innovation in work practices is to be expected.

8.4.6 Impact on Statutory Authorities: Planting the Seeds of Online Access and Review

One possible impact of the Internet is its ability to empower the public at-large to participate in statutory processes, such as the approval or rejection of building plans. Posting building designs for public review, however, is rare. One of the reasons may be that the accepted format of drawings is not accessible to the average citizen. If navigable 3D models of proposed buildings were placed within a realistic depiction of their context and posted online, a more democratic public review process would be feasible.

Visual inspection is already technically possible within the Google Earth® environment, but the idea can be extrapolated to envisage the merging of multiple information sources to create a virtual environment in which design and approval takes place using BIM. Geographical Information Systems (GIS) are commonplace in many municipal jurisdictions and utility services. The data includes topographical conditions, infrastructure facilities, existing structures, environmental and climate conditions, and statutory requirements. The challenges of interoperability for this kind of GIS information are not quite as complex as they are for exchanging intelligent building models; it may become possible and economically viable for jurisdictions to provide models in packages for individual project sites, which could be delivered to building designers for use directly within their BIM authoring tools.

Green or sustainable construction practices are likely to get a boost from BIM, because building information models can be analyzed for compliance with energy consumption standards, for their use of green construction materials, and for other factors included in certification schemes like LEED. The ability to objectively and automatically assess building models will make the enforcement of new regulations more practical. Some building codes already require that energy analyses be performed on all buildings, to comply with standards for energy consumption. The use of performance-based standards, as opposed to prescriptive standards, is likely to increase. The first energy calculation tools integrated within BIM tools are already available, which means that BIM will facilitate the push for sustainable buildings.

8.4.7 Impact on Project Documentation: On-Demand Drawings

The importance of drawings is expected to decline as BIM becomes ubiquitous on construction sites, but drawings are unlikely to disappear until digital display technologies are flexible and hardy enough for everyday use onsite (this is discussed in the medium-term forecast below). One function of drawings in today's construction industry is for documentation of business transactions in the form of appendices to construction contracts. Already, however, there are indications that BIM models can better serve this purpose, partly because of their improved accessibility to non-professionals.

Because drawings can be produced on-demand from the model using customized formats, the development of better onsite documentation for crews and installers will lead to new capabilities. Isometric views with sequential assembly views and bills of material will facilitate crew operations. An early example is presented in Figure 5-12.

A technical and legal hurdle that must be resolved is the notion of *signing* a digital model, or even its individual components. Another is the issue of whether access to models in the future, as applications develop and old versions are no longer supported, will remain reliable. Both of these issues have been resolved in other business fields, and the economic drivers are strong enough to ensure that they will be resolved for building information models too. Solutions may take advantage of advanced encryption technologies, third-party archiving of original model files, neutral view-only formats, and other techniques. In practice, a growing number of project participants already choose to build according to models, rather than drawings. Legal practice will have to keep pace with commercial practice.

8.4.8 Impact on BIM Tools: More Integration, More Specialization, More Information

Building information model generation tools still contain broad room for improvement and enhancement, in terms of the breadth of their coverage of the building construction domain. They can also expand on the types of parametric relationships and constraints they support. These tools will incorporate increasingly comprehensive families of building parts and products.

The ready availability of BIM platforms will encourage a new wave of plug-ins that will emerge over the next five years. Several areas of new product development are likely. One may be the emergence of better tools for architectural conceptual design, integrating aspects of DProfiler (Beck Technology 2007), Trelligence Affinity (Trelligence 2007), and Ecotect (Ecotect 2007), as discussed in Chapter 5. Another likely area will be layout and fabrication tools for new materials and building surfaces. Yet others might include new support software for store layout, fixturing, interior office layout, and detailing by the many design-related trades that serve building owners or lessees, etc.

Increased **integration of analysis interfaces** within design modeling software is technically feasible and desirable. Competition between vendors of leading BIM authoring tools interested in providing comprehensive suites of software products is already evident, because the issue of interoperability remains insufficiently resolved. Vendors can build BIM software suites either by buying up analysis software providers or by forming alliances that enable an analysis pre-processor to run directly from their interfaces. The trend began with embedded structural analysis software and is likely to continue with energy and acoustic analyses, estimating, code checking, and planning compliance.

Because of the large and growing size of BIM project files and the difficulties inherent in managing model exchanges, there will be a growing **demand**

for BIM servers with the potential for managing projects at the *object* level, rather than at the file level. These may be offered by a variety of companies, including BIM authoring software companies, existing project collaboration Web service-providers, and new startups. The technology for such exchanges already exists within BIM systems that enable multiple users to access the model simultaneously, by locking individual objects; it is necessary to port that capability to a larger and more functionally complete database environment. Given that transactions are primarily incremental updates of objects and their parameters (as opposed to complete model exchanges), the actual amount of data that needs to be transferred is fairly small, certainly much smaller than equivalent sets of CAD files.

Model viewer software such as DWF viewers, Tekla's and Bentley's Web viewers, 3D PDF, and others are becoming important tools, due to their simplicity. They are likely to begin offering more extensive information than just graphics and basic object IDs and properties. A wide variety of applications—including quantity takeoffs, basic clash checking, and even procurement planning—can be used as information consumers only; they do not need to update information to BIM models. As such, they may be able to run directly from DWF-style files. These simplified file formats may be exploited by a variety of *output only* third party plug-ins for use with Web interfaces.

New tools for locating and inserting building product and assembly models, called *building element models* (BEMs) (Arnold 2007), are under rapid development. Two development issues are **semantic searching and compatibility of BEMs to multiple BIM platforms.** Today, we are already able to search the Web and find building products based on user defined criteria, if one knows product names and/or standard material names. Semantic searching will enable searches that accept a broad range of synonyms, with methods that understand class and inheritance relations and can deal with combinations of attributes. The underlying problems of semantic representation can be found in all industries. Since these challenges cross industries, an unprecedented amount of money is being invested in semantic search technology (Bader et al. 2003). AEC practitioners should look forward to tools that leverage BIM semantics to organize content in several ways and provide users with the ability to develop customized semantic searches. For example:

- Find an automatically controlled louver window shading system that can span between six-foot–on-center mullions;
- Find all products that are applied in a particular context across multiple projects.

These capabilities will gradually become available. More powerful search and selection capabilities are likely to become the market distinguishers for different commercial e-business Web sites. The introduction of these capabilities will begin by 2012, but full capabilities will not yet be realized.

As noted in Chapter 2, building model authoring tools currently incorporate a variety of parametric modeling capabilities. As a result, an object with parametric rules developed for one system cannot be translated and imported into another without losing parametric behavior. This restricts the development of effective BEMs for use in different BIM tools. These restrictions will whittle away as more complex translation capabilities are developed incrementally. BEMs that rely on fixed shape geometries, such as bathroom fixture and door hardware, are already available, as described in Chapter 5. Future extensions will support parametrically varying alternative shapes, such as:

- Assemblies with varying layouts and shape-based on context, such as structural waffle slabs or acoustic ceiling systems;
- Topologically varying shapes, such as stairways and railings.

Eventually, automatic embedding of 3D details, such as for exterior walls, roofing systems, etc. will be provided.

With increasing amounts of information available electronically and as building information models incorporate more process annotations, information visualization will become central to the overall work process. Multi-display environments or **interaction information workspaces** (Liston et al. 2000; Liston et al. 2001) will become common in the office and onsite. Many of the case studies in Chapter 9 show examples of the changing office environment with SmartBoards and large displays. New environments, such as the iRoom shown in Figure 8-4, enable project teams to interact with the building information model and the entire information space. Team members can simultaneously view the model, the schedule, specifications, tasks, and relationships between these views.

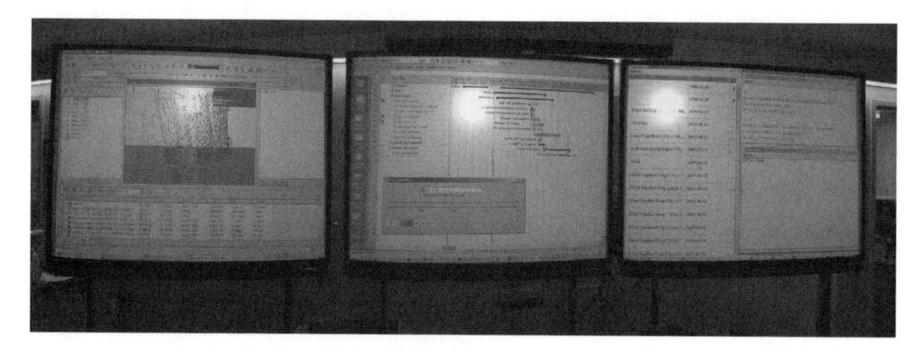

FIGURE 8-4 Sample of a multi-display workspace with related views of the project model. (left screen) A 4D view of the project; (middle screen) The schedule; and (right screen) Component property list and specification information. *Image provided courtesy of CIFE, Stanford University.*

8.4.9 Impact on Research: Model Analysis, Simulation, and Work Processes

The trends described in section 8.1 were loosely grouped in the areas of process and technology. The need for research relates to both design and construction processes and to the interdependent technologies that BIM depends on. New technology leads to process changes; and process change gives rise to new tools.

BIM and the Internet level the playing field in terms of access to building information at both the project and industry-wide levels. Information flow becomes near instantaneous, and collaboration among all concerned within a project can become synchronous, which is a paradigm change from traditional asynchronous workflows. Traditional workflows with sequential generation, submittal, and reviews of drawings—which can be iterative and wasteful due to rework—are no longer appropriate. The professional and legal constructs that have evolved in relation to these workflows are equally unsuitable for collaborative design and construction processes, with shortened cycle times and closely integrated information flows.

While academic research has a role to play in defining new concepts and measures of information flow that promote integrity and value, it is likely that trial-and-error efforts by industry pioneers—driven by practical imperatives—will be the primary source of new BIM workflows. New contractual forms, job descriptions, commercial alignments, and procurement arrangements will need to be synthesized, tested, and refined. Such efforts will support and stimulate the development of new tools in both academia and industry. Some of the directions for the latter are outlined below.

Maintaining integrity across different design models (e.g., architectural vs. structural vs. construction) will be imperative, as changes are made to the different models by their respective disciplines. Unfortunately, in the short term, interoperability tools like the IFCs will not support coordination beyond visual inspection and the identification of physical clashes in geometry. Managing changes across different systems—involving loads (structural or thermal) or other performance relations—will be an important and limiting problem. One approach to resolve this relies on artificially intelligent agents capable of performing comparisons between different models, including across disciplines. These expert system agents would need to track changes within models and then identify logical changes made to one model that ought to be propagated to other models, with or without negotiation between designers. Research will need to determine the nature of the relationships between building objects that are implemented in different discipline-specific systems.

The need to develop production building code checkers and other types of customizable design review tools will lead to the recognition that hard-coding such rules is not the best way to define and implement them. As with other software applications, hard-coding generates tools that are too expensive to write and debug and are inflexible for making changes. Instead, high-level and special purpose rule definition languages will emerge, facilitating the general development of rule-checking in buildings. At first, they may deal with simpler application areas, such as circulation assessment. Subsequently, enhancements will allow for interior room layout and assessment and finally for construction assembly and detailing. These languages will allow non-programmers to write and edit checking-rules in a more direct manner. Two types of back-end tools may be implemented to interpret and run the same languages: (1) for implementation as a stand-alone checker, possibly on a Web server; and (2) embedded directly into a BIM design tool, allowing checking while designing.

The development of these languages would facilitate the implementation of design assessment tools in a wide variety of areas, for different building types, a range of clients, and for building code agencies.

Research is needed to address the various types of model geometry needed for different types of analyses. While most people are familiar with the need for stick models for structural analysis, few are aware of the need for tessellation structures of single bounded surfaces to represent separately managed energy zones within a building. Automatic methods for tessellation are needed for pre-processing models for energy analysis. Another type of geometric abstraction will be necessary for enclosing spatial volumes for computational fluid dynamics. Such models use heuristics to determine which geometric features are required for capturing essential air flows. Further development of automated geometry abstractions is needed if these analyses are to move into everyday use.

Research on the integration of multiple types of analysis will need significant improvement. For example, to show the interaction of heat flows with natural convection, the output from a simulation of energy radiation to internal materials within a space will be used as input for a computational fluid dynamics CFD model. Multi-criteria optimization methods are available, such as genetic algorithms of various kinds, but utility functions that can express the integrated performance of buildings with respect to different functions will be needed. Developing these relationships would allow parametric models to automatically vary to search for performance objectives dealing with weight, solar gain, energy use, and other objectives. This would enable new levels of comprehensive performance-based design that are not possible today.

In the same way that semiconductor fabrication plants undertake for-hire chip fabrication, prefabrication plants for construction may support custom numerical control (CNC) fabrication with little or no manual input for precast concrete, steel welded systems, and a few types of exterior carbon fiber reinforced plastics.

Fabrication plants will rely on model data provided by the designers to generate CNC instructions, needing only minimal checking by the component producer. This will reduce the costs associated with custom fabrication, bringing them closer to that of standard construction and spreading their capital investments over many projects. A variety of research will become the basis for this capability:

- CNC fabrication languages and new automated production methods
- Checking and verification languages to allow for the validation of candidate design instruction
- Increased levels of production automation in fabrication technology. CNC machines will need to be adapted from current machine tools and integrated with material handlers and automated assemblers.

Another area of research that is already of interest addresses the question of how to determine the best methods for delivering model data directly to the *work face* at construction sites. This research concerns hardware, software, database architectures, and human-machine interfaces. Although PDAs, tablet PCs, and mobile phones are all widely available and will become increasingly useful for presenting BIM information onsite, paper documents are still the most common technology in use today.

Lastly, throughout the period during which basic BIM research and development was pursued, the research community generated many conceptual applications for building models that could not be implemented in practice, because the BIM tools were not mature enough or in widespread use. Examples include: automated control of construction equipment, such as cranes, robotic pavers, and concrete surface finishers; automated data collection for performance monitoring; construction safety planning; electronic procurement and logistics; and many others. While there are still hurdles to overcome—for example, how standardized building products and services can be modeled for use in multiple BIM environments and comprehensive modeling capabilities for cast-in-place reinforced concrete—implementation of some of these applications may become commercially viable once building information models for construction management are more common.

8.4.10 Vision 2012: The Limitations

Given the relative inertia of the construction industry and its fragmented structure (Chapter 1) BIM adoption will not be complete by the end of this timeframe. Paper drawings—or at least 2D drawing formats which can be communicated electronically—will remain common forms of construction documentation. Only minor shifts can be expected in the costs of construction during this period. Indeed, full adoption of BIM in any firm requires two to three years to become effective. Thus, it is unlikely that significant industry-wide productivity gains will be observed by 2012. Local effects may be dramatic; building forms once considered impractical—due to either technical or budget constraints—will become common. Successful early adapters in both design and construction will profit from their foresight until the rest of the industry catches up.

8.5 DRIVERS OF CHANGE AND BIM IMPACTS UP TO 2020

In looking beyond the five year horizon, we start by identifying both the drivers for change that are likely to motivate people and organizations involved in building construction and the obstacles they are likely to face. With these in mind, we have tried to assess developments in the areas of BIM technology, the ways in which building information is delivered, design services, building product specifications, code-checking, construction management practice, employment, professional roles, and the integration of building information into business systems.

8.5.1 Economic, Technological, and Societal Drivers

There are a number of economic, technological, and societal factors that are likely to drive the future development of BIM tools and workflows. These will include: globalization, specialization, and the commoditization of engineering and architectural services; the move to lean construction methods, the increasing use of design-build and fast track projects; and the demand for facility management information.

Globalization resulted from the elimination of barriers to international trade. In construction, the possibility of moving the production of building parts to more cost-effective locations will increase demand for highly accurate and reliable design information, so that pieces can be shipped great distances with a high degree of confidence that whey will fit correctly when installed.

Specialization and commoditization of design services is another economic driver that will favor BIM. As niche skills, such as producing renderings or performing sets of structural analyses, are better developed and defined—and long-distance collaboration more accepted—BIM will enable the delivery of special services.

Design-build and fast-track construction projects demand close collaboration between the design and construction functions. Such collaborations will drive the adoption and development of BIM. Finally, the commercial interests of software vendors and competition between them are fundamental drivers that will compel the enhancement and development of BIM systems.

Perhaps the most important economic driver for BIM systems and their adoption will be the intrinsic value that their quality of information will provide to building clients. Improved information quality, building products, visualization tools, cost estimates, and analyses will lead to better decision-making during design. The value of building models for maintenance and operations is all likely to initiate a snowball effect, where clients demand the use of BIM on their projects. (This is already happening at the GSA).

Technical progress in computing power, remote sensing technologies, computer-controlled production machinery, distributed computing, information exchange technologies, and other technologies will open new possibilities that software vendors will exploit to their own competitive advantage. Another technical area that may introduce further developments that influence BIM systems is euphemistically referred to as *artificial intelligence*. BIM tools are convenient platforms for a renaissance of expert system developments for a range of purposes, such as code checking, quality reviews, intelligent tools for comparing versions, etc. design guides, and design wizards. Many of these efforts are already underway but will take another decade to become standard practice. Information standardization is another driver for progress. Consistent definitions of building types, space types, building elements, and other terminology will facilitate e-commerce and increasingly complex and automated workflows. It can also drive content creation and aid in the management and use of parametric building component libraries, both private and public. Ubiquitous access to information, including component libraries, makes the use of computable models more attractive for a wide variety of purposes.

Among societal and cultural drivers, the demand for sustainable construction may be the most significant factor. Architects and engineers will be tasked with providing much more energy-efficient buildings that utilize recyclable materials, which means that more accurate and extensive analyses will be needed. BIM systems will need to support these capabilities.

The increasing power of computer and global-positioning system (GPS) technology will allow for greater use of building information models in the field, which will enable faster and more accurate construction. GPS guidance is already an important component of automated earthworks equipment control systems; similar developments can be expected in construction.

8.5.2 Obstacles to Change

As a counterpoint to the drivers mentioned above, there are numerous obstacles to progress that BIM faces. These include: technical barriers, legal and liability issues, regulation, inappropriate business models, resistance to changes in employment patterns, and the need to educate large numbers of professionals.

Construction is a collaborative endeavor, and BIM enables closer collaboration than CAD; however, this will require that workflows and commercial relationships support an increase in the sharing of both liabilities and rewards. BIM tools and IFC file formats do not yet adequately address support for the management and tracking of changes to models; nor are contract terms sufficiently developed to handle these collective responsibilities.

The distinct economic interests of designers and contractors are another possible barrier. In construction business models, only a small portion of the economic benefits of BIM can accrue to designers. The major payoffs will go to contractors and owners. A mechanism does not yet exist for rewarding designers that provide rich information models. Nonetheless, BIM developers cater specifically to design professions. Although the development of BIM software is capital intensive, software vendors will have to assume the commercial risk of developing sophisticated tools for construction contractors.

The major technical barrier is the need for mature interoperability tools. While Moore's Law in practice suggests that hardware will not be a barrier, the development of standards has been slower than expected; and the lack of effective interoperability continues to be a serious impediment to collaborative design.

8.5.3 Development of BIM Tools

What effect will these trends have on the future development of BIM tools? Apart from improvements to the human-machine interfaces that can be expected of any software, BIM tools will experience significant enhancements in the following areas:

- Improved import and export capabilities using protocols like IFCs. The market will demand this, and software vendors will comply. But given their commercial interests, they will also pursue a second option:
- Each BIM authoring tool will expand its repertoire of applications, enabling increasingly complex buildings to be designed and built using a family of related tools built on the same platform without the need for data translation and exchange.
- 'Lite' BIM tools for specific building types, such as single-family residential housing, have been available for some time. If their data can be imported to professional BIM tools, they may reach a point where owners are able to virtually 'build' their dream buildings or apartments and then transfer them to professionals for actual design and construction.

- There will be movement away from desktop applications to internet-based interactions that employ BIM and integrate Web-based content, from services to building element models and analysis tools. The Tectonic BIM Library Manager and QTO plug-in are early examples of AEC services based on an advanced software architecture that supports this paradigm (Arnold 2007).
- BIM tools that support products involving complex layout and detailing are also expected to come to market, in much the same way that HVAC equipment companies developed software in the 1980s for selecting system components (e.g., Carrier, Trane). These specialized tools will experience widespread use, because they can potentially carry special product warrantees that are only honored if the product is detailed using these tools. The degree of success or impact that these applications will have is not yet evident.

8.5.4 Role of Drawings

Drawings are fundamentally paper-based in format. Drawing symbols and formatting conventions have evolved primarily because paper is a two-dimensional medium; orthographic projections were essential for measuring distances on paper. If and when digital displays become sufficiently cheap and flexible to suit the conditions of work onsite, paper printouts of drawings will likely disappear. Once drawings are no longer printed, there will be no clear reason to maintain their formatting conventions. In the face of the superior medium of 3D building information models, they may finally disappear altogether.

In the design domain, visualization formats will replace drawing types, with different formats developed for each of the parties involved: owners, consultants, bankers and investors, and potential occupants. These formats may include standard walkthrough views with audio and tactile feedback added to the visual content. User controlled walkthroughs will support further interrogation of the model. For example, a client may want spatial data or a developer may want to query rental rates. Integration of these services into the fee structure will add value to architectural services.

8.5.5 Design Professions: Providing New Services

Increasingly, architects will provide the integration environment needed for modern practice. This includes: multi-screen conference rooms, supporting parallel projection of physical design, schedules, procurement tracking, and other aspects of project planning. Contractors will also provide these services, and competition will determine which firms will provide these services in the future. In most projects, it is offered by both, first with the architect in the lead for design coordination and then moving to the contractors for coordinating construction.

Most projects will be managed in a federated manner on BIM servers, with separate models dealing with the scope of expertise for each profession. Better coordination tools will be available for maintaining consistency across federated model sets, but the role of *model manager* will be as essential as any other professional service. Models will support a growing number of analyses run on derived views for energy, structures, acoustics, lighting, environmental impacts, and fabrication. They will also support a variety of automated checks, by responding to building codes, material design handbooks, product warrantees, functional analyses of organizational operations within the structure, and operations and maintenance procedures.

8.5.6 Integrated Design/Build Services and Agreements

Over the next decade, we expect many innovations in project delivery mechanisms. New forms of contracts will be explored, possibly based on Limited Liability Corporations (LCCs) or the Australian form of relationship contracting. The balancing of risk and rewards will become part of the equity relationship with clients, and plans will explicitly state the distribution of benefits as well as penalties. An early example of such an effort is the Sutter Health Incentive pooling plan presented in the Camino Medical Group Building case study in Chapter 9.

Another possible area of innovation is a more explicit definition of workflows for supporting project development and completion. An option provided by the workflow exchanges defined in the National BIM Standard is that they will be referred to in contracts—describing which information flows will be used, at what stages of the project, and who they will be exchanged between—based on a working process agreed to during contract negotiations. This outline of the work plan will make the collaboration model explicit and determine when engineering consultants, fabricators, and others will become involved. Such an agreement will, in turn, affect staffing requirements for each party for the project's duration and provide a workflow of the project that can be tracked and eventually supported by additional automation.

8.5.7 Building Product Manufacturers: Intelligent Product Specs

As BIM becomes ubiquitous, designers will prefer to specify building products that offer information to be inserted directly into a model in electronic form, including hyperlinked references to the suppliers' catalogs, price lists, etc. Rudimentary electronic building product catalogs available today will evolve into sophisticated and intelligent product specs, including information that enables structural, thermal, lighting, LEED compliance, and other analyses, in addition to the data now used for specifying and procuring products.

The basic challenges for realizing high levels of semantic search will have been addressed, and new capabilities that allow for searches based on color, textures, and shape will become available. The import and exchange of parametric objects will have become an old issue, with only fine-grained enhancements still being explored.

8.5.8 Construction Regulation: Automated Code-Checking

Checking building design models for compliance with code requirements and planning restrictions is an area that will be developed further over the next ten years. This functionality could be provided in one of two ways:

- Application service providers sell/lease code-checking software plug-ins embedded in BIM software tools. The plug-in extracts local requirement data from online databases maintained by the service provider, as a service to local jurisdictions. Designers check their designs continuously as they evolve.
- External software directly checks a neutral model file, such as an IFC file, for code compliance. The designer exports the model and the check is run on the IFC model.

Both developments are possible, although the former has an advantage for users; providing feedback directly to the model will make fixing problems easier than receiving an external report that needs interpretation before edits can be made. Because design is an iterative process—and designers will want to obtain feedback, make changes, and check again—it may be preferred.

The latter case will be required to guarantee final compliance with the code. With the proper XML link, external software can also provide input for the source BIM tool. In both cases, the files containing encoded planning and code requirement rules should be small and generalized format files that are easily maintained.

8.5.9 Lean Construction and BIM

Lean construction (Koskela 1992; LCI 2004) and BIM are likely to progress hand-in-hand, because they are complementary in several important ways. When applied to building design, lean thinking implies: reduced waste through the elimination of unnecessary process stages that provide no direct value to the client, such as with producing drawings; concurrent design to eliminate errors and rework, as far as possible; and shortened cycle durations. BIM enables all of these goals.

The need to efficiently produce highly customized products for discerning consumers is a key driver of lean production (Womack and Jones 2003). An essential component is the reduction of cycle-times for individual products, because it helps designers and producers better respond to clients' (often changing) needs. BIM technology can play a crucial role in reducing the duration of both design and construction, but its main impact is felt when the design phase duration is efficiently collapsed. Rapid development of conceptual designs, strong communication with clients through visualization and cost estimates, concurrent design development and coordination with engineering consultants, error-reduction and automation in producing documentation, and facilitated prefabrication all contribute to this effect. Thus, BIM will become an indispensable tool for construction, not only because of its direct benefits but because it enables lean design and construction.

8.5.10 Construction Companies: Information Integration

The next step for construction will be the integration of specialized enterprise resource planning (ERP) software with construction building information models. Models will become the core information source for quantities of work and materials, construction methods, and resource utilization. They will play a pivotal role in enabling the collection of automated data for construction control. Early versions of these integrated systems will appear before 2012 in the form of plug-in software added to architectural oriented BIM authoring tools. Applications for construction management that are added in this way are liable to be limited in functionality, due to the fundamental differences between object classes, relationships, and aggregations needed for design and construction.

Only later will fully purpose-built applications mature. They may be developed in a combination of three ways:

- Vendors of production detailing systems will add objects to model work packages and resources, with built-in parametric functions for rapid detailing according to company practices. Built into these systems will be applications for construction planning (scheduling, estimating, budgeting and procurement) and construction management (purchasing, production planning and control, quality control). The result will be highly detailed models for construction management. Multiple projects will be managed in company-wide setups with multiple models.
- As extensions to standard ERP systems, with specific 'live' links added to BIM models. These applications will have transparent interfaces to BIM tools but remain external to them.

- As entirely new ERP applications built for construction, with tightly integrated construction-specific building model functionality as well as business and production management functions, such as accounting, billing, and order tracking.

Regardless of the route taken, far more sophisticated tools for construction management—capable of integrating functions across a company's separate projects—will result. The ability to balance labor and equipment assignments across multiple projects and coordinate small-batch deliveries are examples of the kinds of benefits that may be achieved.

Once building information models integrated with ERP systems are commonplace, the use of automated data collection technologies, such as LADAR (laser scanning), GPS positioning, and RFID tags, will also become common, both for construction and work monitoring and logistics. These tools will replace existing measuring methods for layout of large-scale buildings; building to the model will become standard practice.

Globalization trends along with BIM-enabled integration of highly-developed design and commercial information—facilitating prefabrication and preassembly—will cause the building construction industry to be closer aligned with other manufacturing industries, with a minimum of activity performed onsite. This does not imply mass production but lean production of highly customized products. Each building will continue to have unique design features, but BIM will enable their prefabrication in ways that ensure compatibility when all parts are delivered. As a result, foundations may become the major component of work still performed onsite.

8.5.11 BIM Skills and Employment: New Roles

Because BIM is a revolutionary shift away from drawing production, the set of skills needed is quite different. Whereas drafting demands familiarity with the language and symbols of architectural and construction drawings, BIM demands a very good understanding of the way buildings come together. Drafting is the laborious act of expressing ideas on two-dimensional media, whether paper or screen; modeling is akin to actually building the building. Therefore, it makes sense for skilled architects and engineers to model directly, rather than instruct others to do it for them only as a matter of record. When BIM is managed as if it is simply a more sophisticated version of CAD, its power to enable rapid exploration and evaluation of design alternatives is overlooked.

Early experience in teaching BIM to undergraduate civil engineering students has demonstrated that it is much easier to learn BIM compared to learning the combined skills of preparing orthographic engineering drawings and operating CAD tools. BIM appears fairly intuitive to students, and it more closely resembles their perception of the world. If undergraduate engineering and architectural schools begin teaching BIM skills within the first years of professional training, it will only be a matter of time before design professionals are able to create and manage their own BIM models. If such education were to begin today—as it has in some schools—it would take five to ten years for BIM savvy professionals to become commonplace in design offices and construction companies.

Until that happens, the ranks of BIM operators will likely be drawn from drafters, detailers, and designers able to make the conceptual transition. Junior staff members are more likely to transition successfully, so most design firms will maintain the split between designers and documenters. Only when designers manipulate models directly, however, will the traditional split be blurred. As with most sophisticated new technologies, those skilled in their use during early adoption will benefit from the imbalance of supply and demand, and they will command premium salaries. This effect will mitigate over time, although in the long term, greater productivity through BIM will cause the average wage of building design personnel to rise.

Naturally, as the years pass, interfaces will become more intuitive, and professional users who have grown up with SimCity® and other gaming environments will be better equipped to operate BIM systems. At that point, designers modeling directly will become prevalent.

While these roles are directly centered on current BIM tools, the workbench environment enabled by integrating sustainability, cost estimation, fabrication, and other technologies with BIM tools will lead to new kinds of specialized roles. Already, energy-related design issues are commonly dealt with by specialists within a design team. Value engineering with new materials is another example. In these cases, we will see many new roles emerge in both design and fabrication. They will address the growing diversity of specialized issues that a generalist designer or contractor cannot address, and this will lead to further subcontracting of both design and fabrication services.

Chapter 8 Discussion Questions

1. What kinds of new architectural and construction services will the use of BIM enable?
2. How might the roles of architects, engineers, and contractors change as a result of BIM? What new educational requirements are needed for employees of architectural, engineering, and construction contracting firms to allow better utilization of BIM and support for the new roles?
3. How will building owners benefit from the use of BIM before and after the construction of a building?
4. What effect might BIM have on the potential for globalization of design and construction?
5. How can BIM facilitate lean design and construction practices? In your answer, cite possible effects of BIM on rework in design, cycle time, and levels of work-in-progress fabrication and construction.
6. How can building product manufacturers take advantage of BIM in the marketing and procurement of their products? What technological developments are needed in BIM and other support systems?
7. In your view, what are the three most important areas that require research and development to further advance the use of BIM technology? These may be technological, organizational, or practice oriented.

CHAPTER 9

BIM Case Studies

9.0 INTRODUCTION TO BIM CASE STUDIES

In this chapter we present ten case studies of projects in which BIM played a significant role. They represent the experiences of owners, architects, engineers, contractors, fabricators, and even construction crews—all pioneers in the application of BIM. The case studies listed in Table 9-1 represent a broad range of buildings for different functions including manufacturing, government facilities, justice, medical, sports, residential, offices, parking, retail, and commerce. The largest building is over 1.5 million square feet; the smallest is a 32,000 square foot curtain wall facade.

Taken as a whole, the case studies cover the use of BIM across all phases of the facility delivery process (as shown in Figure 9-1) by a wide range of participants (see Table 9-2). Each demonstrates a diverse set of benefits to various organizations, resulting from the implementation of BIM tools and processes. Table 9-3 indexes the case studies according to the benefits listed in Section 1.5 of Chapter 1. The wide variety of software used is shown in Table 9-4.

No single project has yet realized all or even a majority of BIM's potential benefits, and it is doubtful that all of the benefits that the technology enables have been discovered or even identified. Each case study presents the salient aspects of the BIM process and focuses on the ways each team used the available tools to maximum benefit. We also highlight the many lessons that these teams learned as they encountered challenges in implementing the new technologies and processes.

Many of the projects were in progress at the time of writing, preventing a full review or complete assessment of the benefits. Naturally, research was

Table 9-1 The location of case study projects.

No.	Project	Location
1	General Motors Production Plant	Flint, Michigan, US
2	Coast Guard Facility Planning	Multiple locations
3	Camino Group Medical Building	Mountain View, California, US
4	National Aquatics Center	Beijing, China
5	Federal Office Building	San Francisco, California, US
6	100 11th Avenue Apartments	New York City, NY, US
7	One Island East Office Tower	Hong Kong, China
8	Penn National Parking Structure	Grantville, PA, US
9	Hillwood Commercial Project	Dallas, TX, US
10	Federal Courthouse	Jackson Mississippi, US

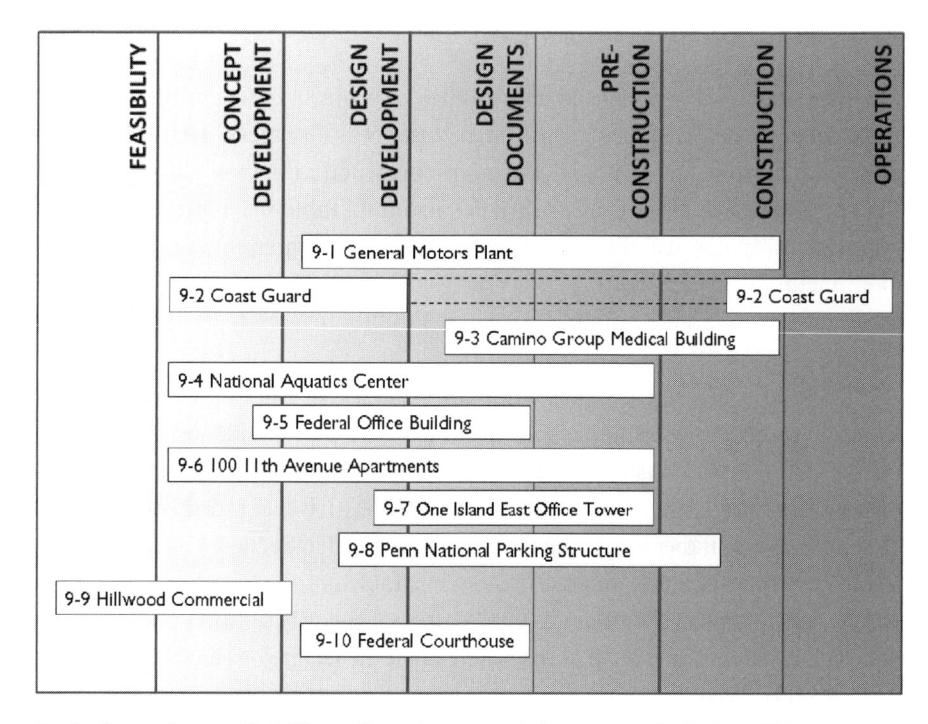

FIGURE 9-1 **Case study guide showing various project phases.**

limited to the availability of various participants and their willingness to disclose information. Architecture, engineering, construction, fabrication, and real estate development are competitive fields, and organizations are often reluctant to disclose their enterprise expertise. Nevertheless, most organizations and individuals were extremely helpful and made significant efforts to share their stories and provide images, information, and important insights.

Table 9-2 Participant index for the case studies.

Participants	9.1 General Motors Production Plant	9.2 Coast Guard Facility Planning	9.3 Camino Group Medical Building	9.4 Beijing National Aquatics Center	9.5 San Francisco Federal Building	9.6 100 11th Avenue NYC	9.7 One Island East Office Tower	9.8 Penn National Parking Structure	9.9 Hillwood Commercial Project	9.10 Jackson Federal Courthouse
Owner/developer	•	•	•		•		•		•	•
Architect	•	•		•	•	•	•			•
Engineer	•		•	•	•		•	•		•
Contractor	•		•	•	•		•		•	
Subcontractor/ Fabricator	•		•		•	•	•	•		
Facility Operations/End users	•	•								•

Table 9-3 Benefit checklist for the case studies. The benefits correspond to the list in Section 1.5, Chapter 1.

Project Phase	Benefits	9.1 General Motors Production Plant	9.2 Coast Guide Facility	9.3 Camino Group Medical Building	9.4 Beijing National Aquatics Center	9.5 SF Federal Office Building	9.6 100 11th Avenue NYC	9.7 One Island East Office Tower	9.8 Penn National Parking Structure	9.9 Hillwood Commercial Project	9.10 Jackson Federal
Feasibility study (Chapter 4)	Support for project scoping, cost estimation		•							•	
Concept design (Chapters 4 and 5)	Scenario planning		•							•	
	Early and accurate visualizations				•		•		•		
	Optimize energy efficiency and sustainability					•					

(Continued)

Table 9-3 Continued

Project Phase	Benefits	9.1 General Motors Production Plant	9.2 Coast Guide Facility	9.3 Camino Group Medical Building	9.4 Beijing National Aquatics Center	9.5 SF Federal Office Building	9.6 100 11th Avenue NYC	9.7 One Island East Office Tower	9.8 Penn National Parking Structure	9.9 Hillwood Commercial Project	9.10 Jackson Federal
Integrated Design/ construction (Chapter 5)	Automatic maintenance of consistency in design	•		•	•	•	•	•	•		•
	Enhanced building performance and quality				•	•		•		•	•
	Checks against design intent					•			•		•
	Accurate and consistent drawing sets	•		•	•	•	•	•	•		•
Construction execution/ coordination (Chapters 6 and 7)	Earlier collaboration of multiple design disciplines	•		•	•	•	•				•
	Synchronize design and construction planning	•		•			•	•	•		
	Discover errors before construction (clash detection)	•		•		•		•	•		
	Drive fabrication and greater use of prefabricated components	•		•	•		•		•		
	Support lean construction techniques	•		•							
	Coordinate/ synchronize procurement	•		•	•		•				
Facility operation (Chapter 4)	Lifecycle benefits regarding operating costs					•					
	Lifecycle benefits regarding maintenance/		•								

Table 9-4 Software checklist for the case studies.

Application Area	Software Tool	9.1 General Motors Production Plant	9.2 Coast Guard Facility Planning	9.3 Camino Group Medical Building	9.4 Beijing National Aquatics Center	9.5 SF Federal Office Building	9.6 100 11th Avenue NYC	9.7 One Island East Office Tower	9.8 Penn National Parking Structure	9.9 Hillwood Commercial Project	9.10 Jackson Federal Courthouse
Model Generation Tools											
Architecture	ArchiCAD		•								
	Bentley Architecture*	•				•					
	Digital Project						•	•			
	ONUMA Planning System		•								
	Revit Building										•
Structural	Bentley Structural*				•						
	Revit Structures			•							•
	SDS/2	•									
	Tekla Structures					•			•		
MEP	CADDuct			•							
	Design Series	•									
	IntelliCAD	•									
	Piping Designer 3D			•							
	Revit Systems										•
BIM Related Tools											
2D	AutoCAD	•		•		•	•	•	•		•
	SprinkCAD			•							
3D	Architectural Desktop			•							
	3D Studio Max				•						
	FormZ					•					
	Rhino				•		•				
Manufacturing	CATIA						•				
	Solidworks						•				
Database	MySQL		•								

Table 9-4 Continued

Application Area	Software Tool	9.1 General Motors Production Plant	9.2 Coast Guard Facility Planning	9.3 Camino Group Medical Building	9.4 Beijing National Aquatics Center	9.5 SF Federal Office Building	9.6 100 11th Avenue NYC	9.7 One Island East Office Tower	9.8 Penn National Parking Structure	9.9 Hillwood Commercial Project	9.10 Jackson Federal Courthouse
Analysis Tools											
Structural	RAM	•									
	Robot						•				
	STAAD Pro								•		
	Strand				•		•				
Estimating	Dprofiler									•	
	US Cost										•
Clash detection	Navisworks	•		•							•
Energy	EnergyPlus					•					

*Bentley Architecture and Bentley Structures are based on the Microstation Triforma platform

Figure 9-1 and Tables 9-2, 9-3, and 9-4 are guides for readers to both compare the case studies and to quickly find a case study that matches a reader's special interest. They provide an index according to the following four classifications:

1. Project phase – Figure 9-1.
2. Participants – Table 9-2.
3. Benefits experienced – Table 9-3.
4. BIM and other software tools used – Table 9-4.

9.1 FLINT GLOBAL V6 ENGINE PLANT EXPANSION

BIM to shorten the design-construction cycle using lean construction methods

General Motors (GM) recently developed and tested innovative design and construction procedures at two of their automotive plants in Michigan. Driven by the need to better respond to the ever-changing car market in flexible ways,

GM aimed to minimize design and construction time and postpone commitment to construction projects until the last minute. Principles from lean construction and a thorough integration of BIM technology were the means to achieve these goals.

Traditionally, the automobile industry has been open to new BIM approaches, especially since similar techniques have already proven themselves in manufacturing. Parametric modeling for mechanical parts has already reached a high degree of maturity in comparison to parts modeled for buildings. GM has aggressively implemented a new delivery system, requiring all participants to engage in BIM technology.

The client set a very tight schedule for the completion of the new Flint project. Consequently, they sought a company with sufficient experience in the design of automotive manufacturing plants and the use of BIM technology. To stay within the limited available time frame, the project team adopted the Design-Build (DB) approach, as it has proven effective at reducing project durations. The project team was formed around the goal of employing BIM throughout the project's life-cycle. Based on that commitment, modeling capabilities became the primary criteria for the procurement of services. Ghafari Associates LLC was selected to take the Architecture-Engineer (AE) role in the DB team.

Ghafari Associates and GM had previously collaborated on several projects. The most prominent project was a new plant in Lansing Delta Township (LDT). GM's LDT assembly complex was built with the shortest construction time to date, which was 20% faster than the five-year-old Lansing Grand River facility. Innovation doesn't always come easy, but the steep learning curve of the LDT project and the determination of the team members culminated in a picture-book second project in Flint. Although Flint is receiving the credit in this case-study, it was due to the procedures developed at LDT and the experience gathered in the process that enabled the Flint Plant's success. For the LDT project, GM insisted on 2D paper-based delivery, which resulted in considerably more time and effort being spent interfacing between the 2D plans and the 3D model.

Important lessons were learned from the LDT project. In order to gain further benefits from BIM, GM and Ghafari established the following objectives:

- All project team members must be committed to using digital models to optimize the benefits of the BIM approach. The efficiencies associated with improved interfaces between partners or processes occur only if all project participants use BIM tools.

- Subcontractors for all trades need to continuously provide input to the 3D model and refine their 3D models, from fabrication to field installation.
- Where necessary, Ghafari would need to act as a consultant to the less experienced firms in order to facilitate production of 3D models by all specialty trades.

Many of the aspects that enabled the engine plant extension were not based on new technologies. Ghafari's advanced technology manager, Samir Emdanat, observed that the primary technological aspects were just as achievable two years earlier. The greater difficulty was in distributing the building model and conveying the know-how for manipulating it to all members of the project team. At the start of the project, GM waived the requirement to use a single platform and allowed the team to choose best-in-class software for each of the design, detailing, and fabrication tasks.

9.1.1 The Project

The existing 760,000 sf Flint Engine South Plant was delivered in 2000. The 442,000 sq. ft. plant extension, reported here, was constructed in 2005 and began production in September 2006.

The design of an automotive plant is focused on meeting functional requirements centered on the manufacturing process, versus aesthetic ones. Automotive facilities have intricate infrastructures comprising not only HVAC ducts, electrical installations, and piping but also assembly lines for the manufacturing process.

General Motors was especially serious about restricting construction time while streamlining the budget and maintaining quality. The additional manufacturing facility had to be designed and delivered under a fast-track schedule of just 40 weeks. A conventional DB project would have taken 60 weeks or more, and a Design-Bid-Build (DBB) approach would have taken considerably more than 60 weeks.

It was also clear to the entire team that these requirements could not be met with a conventional serial 2D paper-based workflow. It would be necessary to compress the timeline of the critical path. This could only be done by running design, engineering, and construction activities in parallel. This resulted in modifications to the serial exchange of building models, with frequent and concurrent exchanges of the building model to keep all teams up-to-date. The success of this approach becomes clear, considering that the project was delivered within 35 weeks, a 12.5% time saving on an already stringent schedule. Various timelines are shown in Figure 9-1-1.

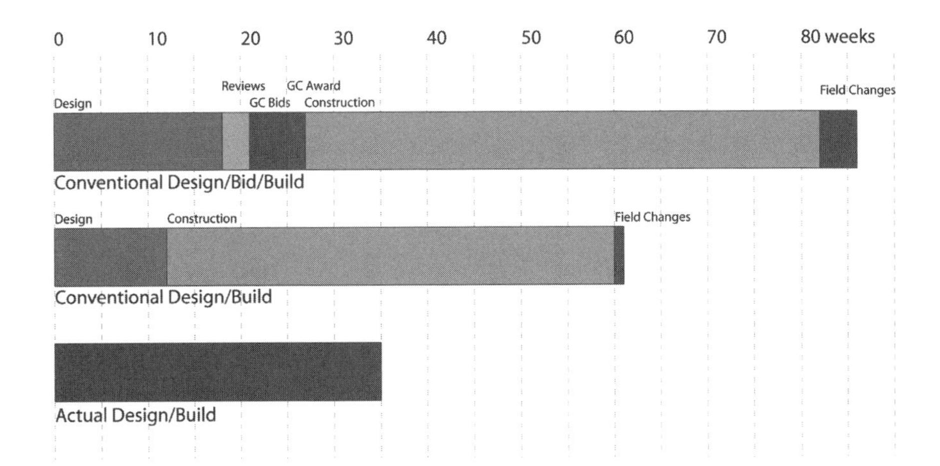

FIGURE 9-1-1 Conventional Design-Bid-Build (80+ weeks; 60 weeks if fast-tracked); Conventional Design-Build (60 weeks).

9.1.2 The Project Team

Ghafari

Since Yousif B. Ghafari founded the company in 1982, their success has been closely linked to effectively adopting CAD technology in an engineering and multi-disciplinary practice. Continuing this tradition, the firm has taken a leading role as practitioners of BIM technology. A high proportion of Ghafari's customers are from the automotive sector, where the benefits of 3D are well known. Additionally, Ghafari has developed into a firm in which technological support for cross-disciplinary communication is sought after and integrated into all business processes.

Ghafari created the Advanced Technology Group as an independent unit within the firm that is responsible for researching and defining best practice solutions for BIM workflows. One of their main guidelines is not the use of BIM simply for technology's sake but to assess which solutions are practical for each individual project. The case is especially convincing when BIM can be carried through to the construction phase. Emphasis is given to integrating partner firms, since many of the advantages are not achievable if only a subset of the involved parties is using BIM.

Over the years, Ghafari developed special know-how in 3D and 4D modeling, digital document exchange, and interference control. Prominent strategies include the frequent exchange of documents within the company and among subcontractors and the desire to reduce requests for information (RFI) and field changes. Streamlining these processes has resulted in improved time and cost efficiency and ultimately lower bids. Since multiple platforms and software packages are used throughout the company, interoperability is an important feature. Thus, Ghafari is a proactive supporter of open standards.

Contractors

The lead contractor for the Flint V6 Plant Extension, the Ideal Group, consists of six member companies. The group had previously worked with General Motors on several construction projects. While they have not always been active in applying BIM, the companies were willing to extend their know-how in this direction. No BIM project management tools were used by the Ideal Group. They were primarily responsible for holding trade contracts and processing payment requests.

Ghafari and the lead contractor, the Ideal Group, started the design process immediately, in order to establish the required documents needed to prepare a GMP (Guaranteed Maximum Price) contract within a few weeks. The DB team then initiated the procurement process for the subcontractor services while taking benefit of the input provided by GM. The ability and willingness to exploit the BIM approach to the fullest extent was an important element in selecting the subcontractors. The project team, including Ghafari, the Ideal Group, and the major subcontractors was set at the outset of the project. Most

Table 9-1-1 Factual data for the Flint Global V6 Engine Plant Expansion Project.

Hard Facts	
Project	Flint Global V6 Engine Plant Expansion
Location	Flint MI, USA
Client	General Motors Corp.
AE	Ghafari Associates, LLC
Contract Type	Design-Build
General Contractor	The Ideal Group, Inc. (Construction)
Contractors	Dee Cramer, Inc. (HVAC) John E. Green Company (Mechanical) Douglas Steel Fabrication Corp. (Steel)
Software	RAM, RAM Advance, SDS/2, Bentley (TriForma suite and Project Wise), AutoCAD, Design Series, Quick Pen, IntelliCAD, Navisworks (see Table 9-1-2 for uses)
Neutral Formats	SDNF, CIS/2
BIM Areas	3D BIM collaboration, interference detection, as-built documents,lean construction, just-in-time delivery, component fabrication
Area	442,000 sf
Duration	35 weeks
Cost	Not disclosed

of the subcontractors (shown in Table 9-1-1) came from specialized fields in which BIM-related technologies had already been very successful. In particular, the steel manufacturers had already integrated BIM into their workflows. Similar developments had taken place for HVAC, piping, and industrial installations. This is due, in part, to the fact that these industries deal with a restricted set of well-defined elements.

In addition, for the purpose of integrating the partner firms into the BIM process, Ghafari co-located all of the detailing teams to a single site and offered training and consulting to those project team members who were not familiar with BIM technology and its associated processes. This was in addition to software vendors' specific training. Because of the single project location, modeling issues and practices were addressed and resolved faster.

9.1.3 The Design Process

3D Direct Digital Workflow

Apart from the time constraints, an additional challenge identified early-on was the need to issue the steel mill order within 3 weeks of the construction start date. If this deadline was missed, the mill-rolling cycle would result in a six-week delay for the steel delivery or an increase in cost for buying off-the-shelf products. The solution was to completely eliminate paper-based delivery by handing the BIM model directly to the fabricator.

Eliminating the traditional 2D CAD paper-based workflow enabled a higher degree of parallelism by removing sequential steps. Additionally, overhead time for drawing layout, printing, and delivering was also eliminated. Of course, collaboratively working on a BIM model also opened-up other time-saving opportunities, such as collision detection and prefabrication, which given an appropriate model, could be implemented with relative ease.

Weekly meetings were held where the combined models where checked for interference and connection alignments with on-hand representatives from all firms. The working out of necessary alterations at the time they were identified was vital to the project's success.

As-Built Before Construction

Once the initial structural design was complete, an iterative process was initiated between the AE and the steel fabricator (Table 9-1-2). This process was based on a combined release of the 3D models prepared by individual parties. The fabricator received the model, extracted the steel quantities, and modeled the construction details. The AE reintegrated the steel fabrication model into the design model and checked for interferences. Using this approach, the

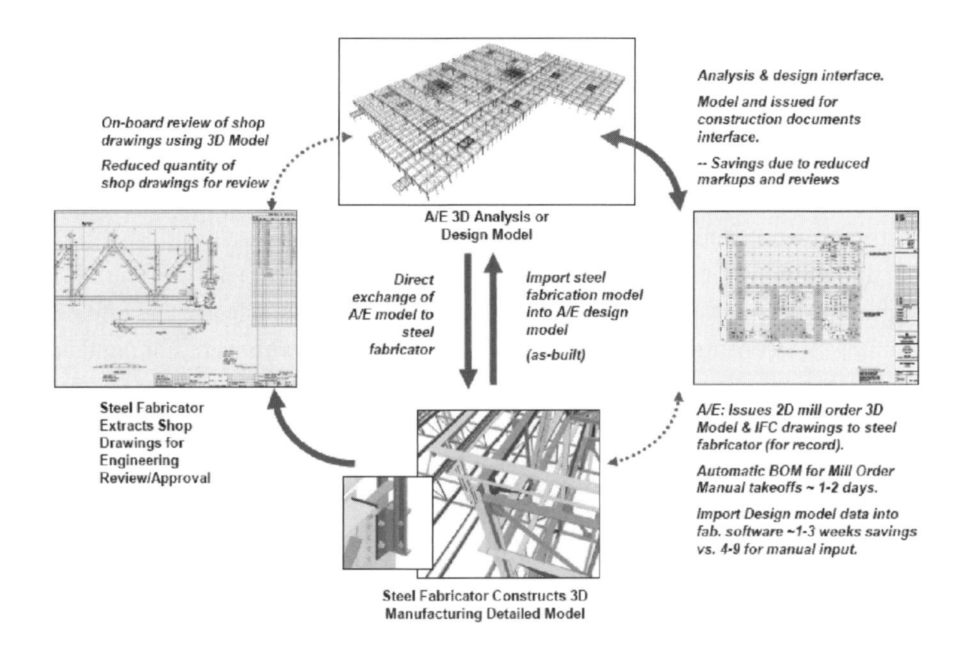

FIGURE 9-1-2 Iterative process for the structural steel refinement. The steel order was issued within three weeks of the project's start date. The model was declared 'as-built' earlier in the process.

structural design was finalized within a three-week timeframe, Mechanical-Electrical-Plumbing (MEP) design began in parallel, and subsequent activities were based on the as-built structural model before construction even started. In this way, the design and shop modeling phases were merged.

Most internal reviews were performed using digital data, which introduced savings in overhead that are typically associated with paper-based workflows, although extraction shop drawings were still needed for approval purposes.

Not only did the project team have access to the as-built model of the steel structure in advance of construction, but the laborious production of as-built documentation at the end of the project was eliminated. Tight integration of the fabricator and several iterations of design reviews effectively reduced costly change orders to almost zero.

Lean design approach

The application of 3D digital workflows was a key factor in the implementation of supply chain management strategies and lean design principles and construction processes.

Conceptually, during the lifecycle of a construction project, a project team is responsible for transforming labor and material into a building. In other words, design and construction can be viewed as a series of activities, where some add value and others do not, as shown in Figure 9-1-3. There are numerous time-consuming, non-value-adding activities in the design process, such as correction

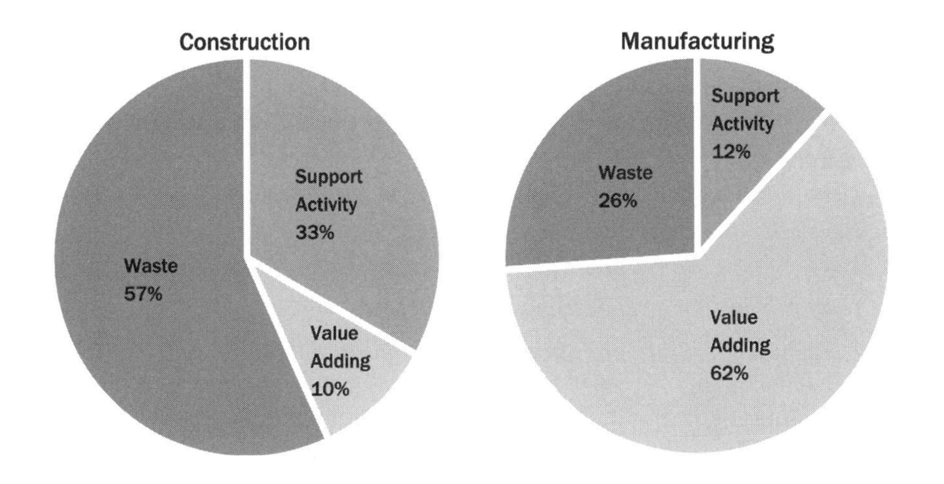

FIGURE 9-1-3 These two charts illustrate the differences between the construction and manufacturing industries' current business practices. (Source: Construction Industry Institute).

of errors and rework, the physical handling and organization of documents, and transportation, inspection, and movement during the construction process.

As a leader in lean manufacturing, GM encourages the adoption and implementation of lean principles throughout the design and construction process of all of its projects.

One problem inherent in a 2D paper-based workflow is the design team must follow sequential steps and make decisions based on known priorities. Often, the effect of a decision regarding a specific system or component on other systems is not immediately known. The natural outcome of this approach is extended duration for the design process, which is identified by frequent rework and corrections, a hefty volume of untraceable paper documents, and undetected

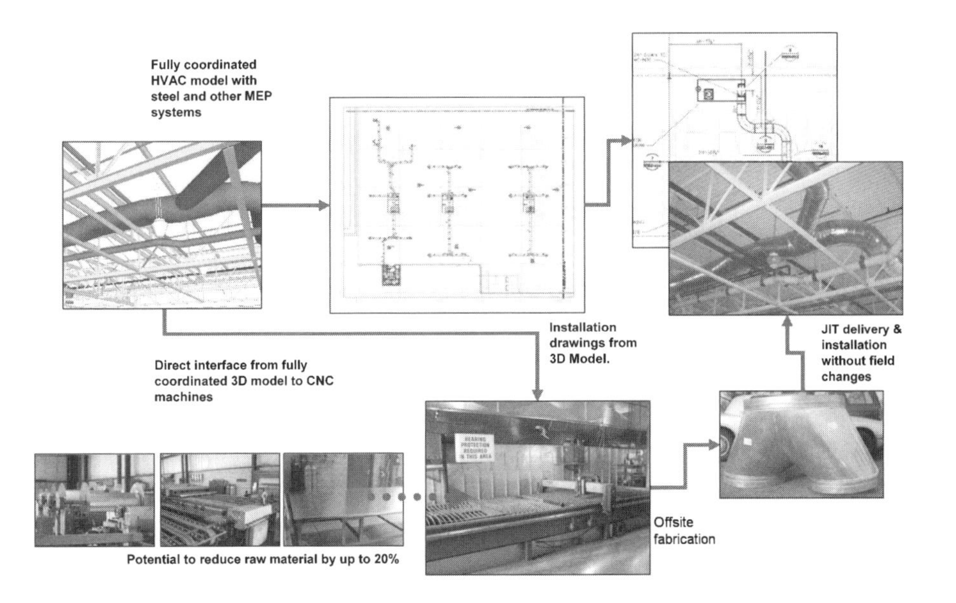

FIGURE 9-1-4 HVAC value stream. 3D design models were shared with the sheet metal contractor, who refined them to include installation-level detailing. They then used the models to support CNC cutting equipment. HVAC was installed with zero changes in the field.

interferences and pending decisions that ultimately must be addressed on the jobsite. Solving interferences after they are detected in the field costs owners large amounts of money, extends the project duration, and reduces the quality of the end product.

Using lean construction concepts, the objective was to optimize the process by:

1. Improving the value-adding processes, and
2. Reducing the number of non-value adding activities.

These objectives were made possible by the continuous study of processes aiming to:

- Smooth interfaces among the parties involved in the process, and
- Improve the flow of resources and information.

Motivated by these goals, the team was able to completely rebuild the traditional processes and generate new ones for constructing a building extension of high quality and acceptable cost in a safe environment.

In this project, BIM tools helped the team apply lean design and construction concepts in order to realize significant benefits. Using a 3D direct-digital workflow allowed the team to make instant and concurrent decisions through collaboration with other entities and eliminate interferences using the Navisworks automated collision detection tool. The design team was responsible for managing the timely flow of information among various parties, in order that they might interact more efficiently. Furthermore, the design team was responsible for reviewing installation-level models constructed by subcontractors. This allowed for the resolution of all potential issues and interferences prior to the start of construction. It also facilitated a change-order-free construction process.

Model exchange

Though most processes were made leaner by adopting digital document exchanges, some overhead was introduced by the necessity of managing multiple application tool formats and data exchange needs, as listed in Table 9-1-2. The smooth exchange of data capability is an essential feature of BIM that enables numerous applications to communicate seamlessly. The full realization of smooth data exchange, however, is an ideal that has not yet been realized. While some neutral format approaches are starting to demonstrate success, it often remains necessary to devise workarounds, particularly when a direct translation does not exist or cannot deliver the needed results.

Table 9-1-2 Model exchange workflows – Ghafari's know-how provided the basis for ensuring seamless data exchanges.

Model exchange workflows	
Steel Structure	RAM Advance → SDNF → custom code → TriForma → SDS/2
	SDS/2 → CIS/2 → TriForma
	TriForma → AutoCAD
	AutoCAD models available in TriForma – no translations
Piping	Design Series from early schematic to fabrication isometrics
HVAC	TriForma then traced over in IntelliCAD for CNC interfaces
	IntelliCAD models back to TriForma
Fire Protection	Design Series for coordination
Architecture	Bentley Architecture for TriForma through CDs
Collision Detection	Navisworks for all coordination and interference checking, but same models were also shared in CAD formats in 3D for all team participants

Ghafari had already acquired substantial expertise in managing data exchanges, and the project team benefited from their knowledge. Though they generally rely on out-of-the-box solutions to save on the development and maintenance of software solutions, Ghafari found that some adaptations for addressing specific project requirements were occasionally needed, and these were made using the respective Application Programming Interfaces (APIs). Additionally, once a workflow was laid out and tested, automation was deployed to eliminate repetitive and error-prone tasks. It was due to these experiences that Gharfari was held responsible for ensuring that all translations were seamless.

An important element for communicating design changes and for implementing the newest models was a provision for a file exchange and collaborative server specialized for CAD content (Bentley ProjectWise). Retrospectively, it was felt that the major advantage of using such a server was the elimination of redundancy and potential error sources. It allowed all team members to access up-to-date project files at all times. The owner also appreciated the exchange server, as it provided continuous access to all project data and allowed for better monitoring of progress.

Collision detection

Multi-disciplinary collaboration was enhanced by relocating team members from all companies to a shared office environment. This enabled more efficient decision-making through direct communication. It also facilitated the use of

Navisworks for interference collision detection. Structural elements, HVAC, and piping were coordinated in the model, rather than on-site. The on-site 'first-come first-served' approach for occupying installation space typically leads to unnecessary costs. Required voids can be left for assembly teams to fill during later stages, as shown in Figure 9-1-5. These voids were analyzed in detail during the design process to guarantee not only enough space for physical installations but also for the additional space needed for placement of equipment. Initial early iterations allowed for some overlapping elements to permit rough layouts, but these tolerances were reduced to zero in the final sessions.

It is estimated that 3,000 to 4,000 interferences were identified and resolved, dealing with architectural components, structural steel, HVAC, mechanical equipment, lighting, cable trays, busses, ducts, and industrial waste. Decision-makers from all relevant teams met for collision detection sessions and relied on virtual models, such as shown in Figure 9-1-5. Problems were analyzed and the responsible partners committed to making the necessary changes. Records of these decisions were distributed to all parties.

It is striking that, even when collisions where discovered during the design phase, the solutions often had a very patched-on-site character. The sheer complexity of these numerous systems did not allow for optimizing solutions. Additionally, some systems, such as the coolant piping for the manufacturing lines, had to be integrated at a later stage, when most systems were already in place. Time for optimizing these solutions was not available, leading to C- and U-shaped route-arounds. These solutions not only required more work to implement, but depending on the pipes' usage, the resulting pressure loss may have lead to further disadvantages, such as increased energy consumption throughout the building's lifecycle. Thus, even further improvements are envisioned.

FIGURE 9-1-5 Avoided collisions.
3,000 to 4,000 interferences were detected and resolved. Despite finding problems at an earlier stage, optimal solutions could not always be found.

Project management

Apart from the BIM technologies mentioned above, the project management team also used a building block of the lean concept known as the "Kaizen Event." One such event is shown in Figure 9-1-6. These meetings were scheduled as needed. Kaizen events are short and focused workflow reengineering sessions attended by the owner, the designer, and other project participants.

During these four-hour events, which were conducted by Ghafari and attended by external BIM experts from other parties, the participants used techniques such as team empowerment, brainstorming, and problem solving. They rapidly mapped the existing processes, identified shortcomings, proposed improvements to the existing workflows by eliminating potential waste, and solicited buy-ins from all parties for implementing the new practices. Kaizen events not only provided tangible and immediate results but also fostered motivation for ongoing continuous improvements in the design and construction process.

The primary outcome of the Kaizen events in this project was eliminating the need for 2D drawing reviews and submittals. This accelerated the delivery of shop drawings and enabled the project team to meet the steel mill delivery deadlines and the MEP orders. In addition, Ghafari developed and maintained a 'lessons learned' document for the project duration and continuously solicited feedback from participants, owners, and other team members.

The project team did not use 4D technology for the purpose of construction site-management or for the coordination of equipment assemblies, however, GM and Ghafari piloted this approach in subsequent projects.

FIGURE 9-1-6
Team meeting
Decision makers from all parties during a 'Kaizen–style' collision detection meeting.

9.1.4 The Construction Process

During construction, the main benefits of BIM were an extraordinarily high degree of prefabrication and preassembly and a well-organized construction site. Fabricators assembled their systems into configurations that minimized onsite requirements. They also relied on Just-In-Time (JIT) delivery of materials and equipment, which was an outcome realized by complete coordination prior to the start of construction.

After many years of doing pilot projects with Ghafari, GM now demands the highest level application of prefabrication and pre-assembly methods for its projects. In this project, however, not all participants had experience in prefabrication and pre-assembly. The HVAC and steel subcontractors were the only parties experienced in this area, but even they had never worked to the extent of the Flint project and with such a complex array of trade subcontractors. Some other subcontractors, such as the piping contractors, first learned about this process during the project.

The detailed collision-free 3D model and the need for no field changes made prefabrication and pre-assembly possible and resulted in a more precise materials ordering process and time savings through the use of JIT delivery. This reduced waste and its associated costs. The various field teams could install their work based on the BIM model and rely on the proposed results.

Because of this well-planned process, the steel structure erection was complete 35 days early and without any changes during construction. Using the prefabrication and pre-assembly approach, the HVAC and MEP systems were installed without any onsite rework and in minimum time. Verification of the virtual model's consistency with the as-built is shown in Figure 9-1-7.

In addition, since field changes and the movement of people and materials onsite were minimal, site safety was improved. The morale of field teams was high, and they had greater trust and pride in their work.

9.1.5 Plant Operation

The facility commenced operation in September 2006. It is too early to speculate about the possible positive effects that BIM will have on the facility's operation. It is clear, however, that such a detailed representation of structural and technical elements will be a considerable benefit, not only for facility management but in the event of future expansion or alterations.

9.1.6 Conclusions and Lessons Learned

The overall success of the project was mainly due to good communication procedures and interfaces. All teams were working on fully-coordinated models,

FIGURE 9-1-7
Overlay of the model with the built project demonstrates the 'as-built' nature of the model.

allowing conflicts to be eliminated early on. No one wasted resources implementing unnecessary changes. Once all of the team members agreed to the use of BIM, Ghafari as the lead firm provided a unifying platform for all contractors. It was not an out-of-the-box solution, and some tweaking and know-how was necessary to enable frictionless procedures. In particular, collision detection workflows were based on best practices already used by Ghafari.

The 3D digital workflow enhanced the transition of information to project team members. It provided an environment in which they were able to interact and collaborate with each other more effectively and make real-time, integrated, and accurate decisions on a timely basis. Kaizen events were also useful in resolving problems and developing more effective processes.

Having eliminated many of the pitfalls in the early stages, the project was handled without big drawbacks. Time and money was saved despite the organizational overhead demanded by these communication protocols.

Most of the benefits extended seamlessly from design to construction, and many of BIM's advantages could be readily seen during construction, such as a quiet and well-organized building site that reflected JIT delivery, high-levels of prefabrication, and hardly any material waste onsite. With only a few crews working at a time, safety for the various teams on the jobsite was kept high. Throughout the construction process, almost no changes were required. Material was saved during manufacturing due to the precise quantity listing.

As a result of automated collision detection, the design team estimated that 3-5 % of the overall cost was saved and the 3D digital workflow contributed an additional 2-4 % in cost and time savings. Thus, despite a tough time schedule, work was finished with 5 weeks (12.5 %) to spare, without compromising the expected quality. General Motors is vague in enumerating an exact savings, so as to maintain a competitive advantage. However, the client, designer and contractors unambiguously reported a high return on investment, leading to a win-win situation.

Lessons Learned

A project benefits significantly when all subcontractors develop and work from their own 3D digital models. In a partially 3D team, the limitations of those not working in 3D will stand out.

Continuous model exchange of detailing or posting to a server allows team members to enter corrections while designing, not afterwards. This saves time.

This project shows what careful planning and strategic scheduling can accomplish to reduce the overlapping time of design and construction. Compared to traditional DBB projects, the Flint Extension demonstrated that through the use of a super-fast-tracked DB effort, project duration can be reduced 50 percent. This very fast execution included the following:

- A lean approach was adopted to closely integrate the design and construction processes to ensure smooth information flows from one to the other. The processes eliminated non-value adding steps, and appropriate mechanisms were set up to ensure pull flows for information. In particular, engineering detailing activity was scheduled according to pull from the construction activities. The 3D BIM technology was critical in supporting these aspects of lean production system design and process flow control, eliminating the wasteful processes that are required when using 2D CAD.
- Using full 3D modeling, many systems and groups of systems could be prefabricated and assembled offsite, reducing onsite work to placement and connection. Offsite work was more efficient and allowed for the application of automation tooling; offsite work did not encounter issues of congestion, temporary storage, material waste resulting from uncertainty regarding takeoff quantities, and weather and other onsite issues did not add waste.

There are significant benefits of co-locating all detailing teams, with modeling support for those just getting started. This approach prompted the sharing of information and provided a flexible learning curve.

Mechanical equipment routing, while greatly improved and collision-free, would have benefited from more and better conceptualizations as well as from layout planning prior to detailing. Such planning would eliminate most U- and C-route-arounds, which can reduce flow and pressure losses and adds extra hardware.

Fabricating 'to the model' can lead to an accurate as-built model for the owner.

Acknowledgments for GM Flint Case Study

This case study was originally written by Thomas Grasl and Hamed Kashani for Professor Chuck Eastman in a course offered in the College of Architecture Ph.D. Program at the Georgia Institute of Technology, Winter 2006 and has been adapted for use in this book. The authors thank Samir Emdanat of Ghafari for providing much of the material in this case study. The authors also thank General Motors for endorsing publication of this work.

9.2 UNITED STATES COAST GUARD BIM IMPLEMENTATION

BIM for scenario planning and facility assessment

This case study consists of three projects that highlight the use of BIM for scenario planning at a project and enterprise level and for facility asset management. The projects demonstrate the United States Coast Guard's (USCG) effort to implement BIM to support tactical and strategic business missions using web-based services and open standards enabled by BIM and accessible to a wide range of users.

9.2.1 Introduction

The United States Coast Guard (USCG) plans, designs, builds, and manages a portfolio of 8,000 owned or leased buildings and nationwide land holdings. For any given project, the USCG may be the owner, tenant, or design team. These multiple perspectives give them many potential opportunities to apply BIM and re-engineer the processes of their Civil Engineering Division.

In 2001, the USCG determined that BIM was a foundation technology for their Shore Facility Capital Asset Management (SFCAM) Roadmap. This decision was enabled by the Logistic Geospatial Information Center (LoGIC) under the direction of Paul Herold. David Hammond led the USCG SFCAM Roadmap effort, and members of the team included AECinfosystems, Inc.;

Onuma, Inc.; MACTEC; Standing Stone Consulting; and Tradewinds, a change management group.

The Roadmap (See Figure 9-2-1) is an enterprise focused on converging data and knowledge across multiple sectors—the various functional units within the USCG—to facilitate better decision-making for strategic asset planning and missions. Integral to this vision is the notion that process changes that support the capture of data and knowledge throughout a project lifecycle and across projects will produce a more efficient and sustainable facility delivery and management process workflow. For the USCG, the lifecycle of a business decision spans a very wide range, from very early planning, design, and construction to facility management and disposal. Additionally, the Roadmap identifies the need to manage information about an organization including its functions and operations, which are often more valuable than the physical assets. Consequently, any effort to implement BIM or related technologies must support the integration of a wide variety of data and decision-making processes very early during planning and align those decisions with key business drivers.

Figure 9-2-1 shows an example of the USCG Roadmap and the various milestones they have established with respect to the implementation of BIM. The following sections describe three projects that utilize BIM in the context of this Roadmap, for early design and facility operations and management. In these projects, BIM is the central enabling technology. Other technologies

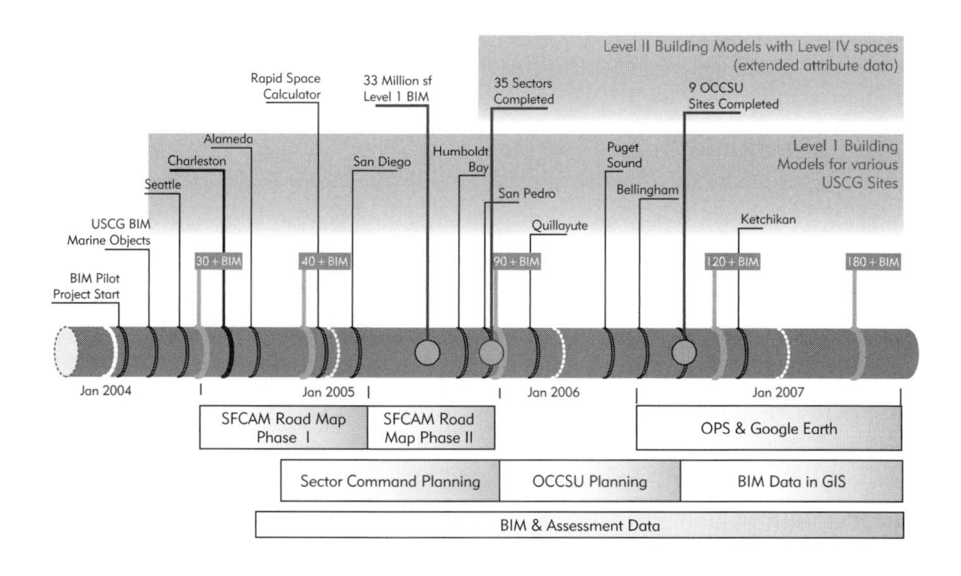

FIGURE 9-2-1 The United States Coast Guard's Roadmap for implementing BIM and their interoperability goals.

such as Web portals, GIS, and databases were integrated with BIM to fulfill specific project requirements and meet the goals established by the USCG. The projects also highlight the process changes that were necessary to successfully implement BIM and achieve an integrated process and data model.

Table 9-2-1 provides an overview of the participants and goals of the three projects.

9.2.2 BIM for Facility Assessment and Planning

The USCG continually assesses the condition of existing facilities, their mission dependency, and current space utilization. These metrics are used to analyze enterprise planning, including whether to renovate or maintain existing facilities or build new ones. In the traditional process, a team would collect and create floor plans of existing facilities and repeat this effort every few years. The analysis effort would occur in parallel with separate documents that typically

Table 9-2-1 Overview of the USCG and its partner service providers. projects.

Project name	Integrated Support Command (ISC) Alameda project	Sector Command Centers	Off Cycle Crew Support Units (OCCSU)
Location	Alameda site	35 sites	9 sites
Project Description	Consolidate and integrate facility assessment information into a single data repository	Perform conceptual design of 35 sector command facilities and define requirements for those facilities	Perform multiple what-if scenarios for future projects related to installations of new boats, including operational, staffing, cost, and security analysis
Project duration	3 Months Model of Alameda Island was created earlier by AECinfosystems, Inc.,Onuma, Inc.	6 months	5 months
BIM application	Facility assessment and asset management	Conceptual planning and requirement definition	Rapid conceptual planning and analysis
Key Participants	USCG Headquarters -SFCAM Logistics Geospatial Integration Center USCG Headquarter, Civil Engineering CEU Oakland FD & CC Pacific Onuma, Inc. AECinfosystems, Inc. NexDSS, MACTEC Standing Stone Consulting, Inc.		

are not directly associated or linked to the floor plan or facility documentation. For the USCG, this type of task was ripe for leveraging intelligent objects in BIM applications to optimize data entry, knowledge capture, and data reporting, as shown in their facility assessment Roadmap in Figure 9-2-2. This Roadmap communicates the ideal assessment workflow utilizing BIM to capture and store assessment information.

Requirements for a Facility Assessment System

A critical requirement for implementing a BIM-enabled assessment process is to ensure that the facility objects are modeled once—either during the design and construction process or post-construction—and assessment teams can associate those components with the following types of assessment information:

- **Facility Condition Index** represents the condition of various parts of the building, roof, walls, windows, and equipment etc. Each building component or aggregate of components, i.e., a building system, is associated with a numerical index typically assigned based on field surveys and ranging from 0 (failure) to 100 (all of its design life remaining).
- **Mission Dependency Index** represents parts of the facilities that are critical to business or mission operations, with a value of 100 representing a system with highest priority and 0 with lowest priority relative to a mission.
- **Space Utilization Index** represents the compliance of the actual space to USCG standards. A 1.0 value indicates a space complying exactly with USCG space standards, and values between .95 (slightly less than allowable) and 1.15 (slightly greater than allowable) are reasonable.

Although data will change over time—based on physical and operational changes—the system should support ongoing manual updates, and automatic updates. Some of the assessment data can be more readily updated or calculated using intelligent rules or associations with other enterprise data systems. Furthermore, the staff equipped with the knowledge and expertise to update this data is not typically trained in 2D or 3D CAD systems. The system must allow such staff to enter the data without extensive training. Finally, the system needs to support reporting capabilities for communicating information and analyses to a broad constituency, ranging from design professionals to laymen.

Description of the Facility Assessment System

The USCG contracted Onuma, Inc. and AECinfosystems, Inc. to develop and implement a BIM based assessment system. These service providers were

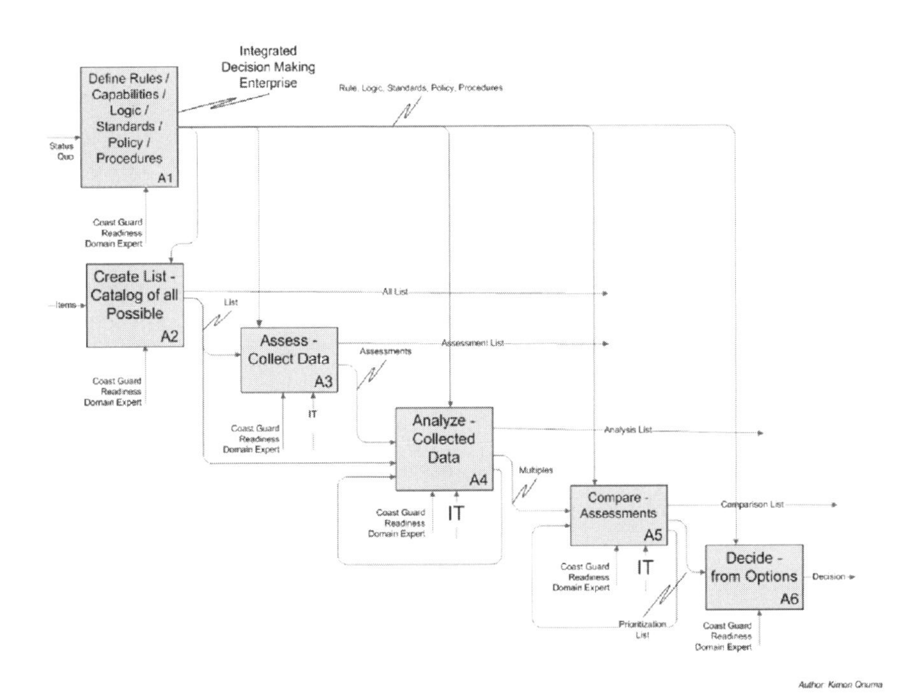

FIGURE 9-2-2A Roadmap for facility assessment showing the ideal information and process workflow to support facility assessment based on a building information model database.

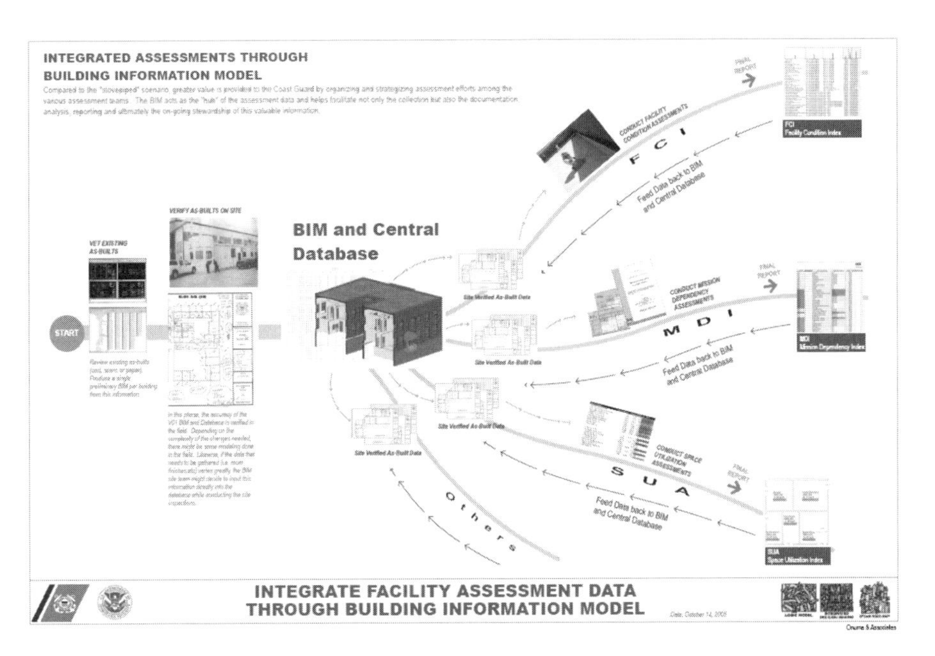

FIGURE 9-2-2B Assessment process as outlined in the Roadmap.

proficient with a variety of BIM tools as well as other enabling technologies, such as databases and Web-based tools. Additionally, Onuma and AECinfosystems had the expertise to model the facilities and to work with and train the USCG staff for widespread implementation. Both companies were familiar with the USCG organization and were involved in the Roadmap efforts. One challenge for the service providers was working within the task-centric or project-oriented deliverables mandated by USCG contracts.

Consequently, the development of various assessment tools and implementation and training of new assessment work processes spanned several projects. The service providers understood that integrating these disparate processes was a central goal governing the project's success and successfully developed and customized an assessment system over several projects and contracts.

Off-the-shelf BIM tools have built-in features to define custom objects and properties (see Chapter 2); but they lack the ability to integrate objects within enterprise systems such as central databases, and they cannot access component information via the Web. To address these shortcomings, Onuma, Inc. customized the ONUMA Planning System™ (OPS), a Web-enabled application that links Building Information Modeling (BIM) design tools with a central database built on open source environments (MySQL™ and Apache™).

For the assessment project, the critical components of OPS were the links between the BIM design tool (ArchiCAD®), a central model repository, and the Web portal for entering and viewing assessment data. The system is based on open standards including IFCs and XML to support interoperability and integration; and it includes interfaces to applications that are vital to the information or process workflows, such as Microsoft© Excel for data entry and reporting and Google™ Earth for earth-based visualizations.

FIGURE 9-2-3 The different levels of detail in a building information model.

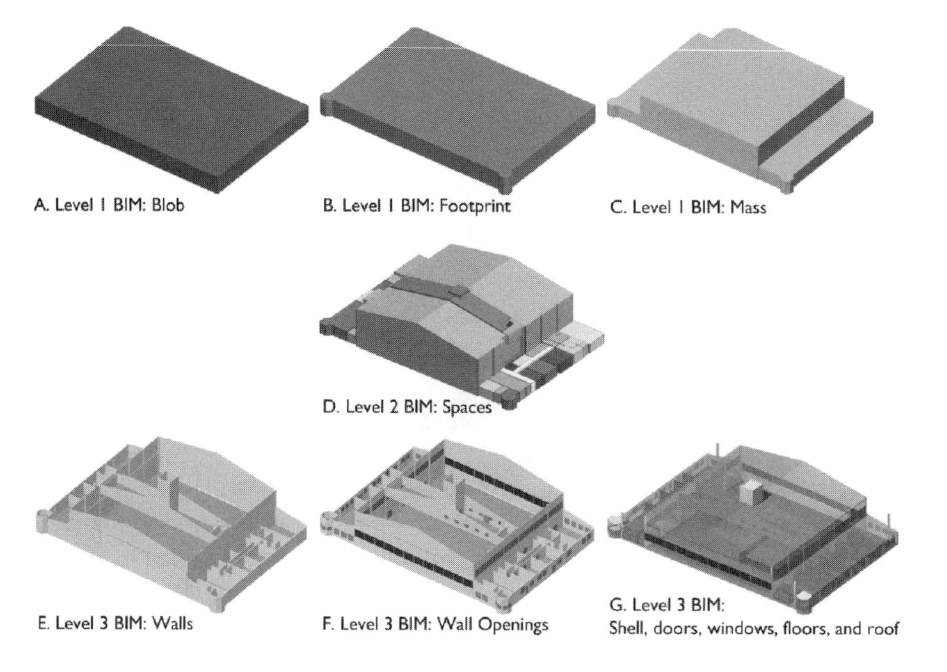

The data for facility information on the ISC project came from multiple teams and sources, including: 1) existing 2D drawings, 2) existing ArchiCAD building models, 3) assessment data from multiple assessment teams, 4) data from field assessors, and 5) new data created in ArchiCAD. Most critical was the as-built data. As-built room (zone objects in ArchiCAD and space objects in OPS) and building objects established the project information work breakdown structure and standards. Each of these objects was assigned a unique identifier (ID); and when combined with the building ID, this creates a unique identifier reference. Any data associated with the project must be associated with one of these core building objects via the ID.

After a team creates a Level 2 or Level 3 building model, the team exports it to a database which acts as the central data repository and model server. The database is populated with various building objects, each with its own unique identifier and associated object parameters (such as variable dimensions) and properties (such as the assessment index values shown in Figure 9-2-4). Once the building data is in the database, teams can access the model data via a Web interface and edit or update the object parameters and properties as shown in Figure 9-2-4. Many of the properties require that the user select from a predefined list of values to ensure validity and to mitigate potential data entry errors. For smaller projects, manual data entry is feasible. For example, a user can open a space setting, use pull down menus to see how the space is utilized, and manually select its mission readiness index. The user can repeat this process for doors, windows, walls, and assessment objects. In many cases, a user must adjust multiple properties for each object.

On large projects, entering data in a Microsoft Excel spreadsheet can greatly expedite data entry and the updating processes. The teams can import data from the Microsoft Excel file into the OPS database only if the building object contains a valid room ID and object property values. The OPS workflow requires that teams conform to the standards, and it enforces consistency. Any view of the building assessment data links to this database. Thus, if a team opens the building model in ArchiCAD after additional assessment data updates, the updates are available within the ArchiCAD environment.

Different methodologies for entering data in the OPS system support scalability and flexibility. An average building requires that teams enter or edit 10,000 data points or object property values, and each of these data points requires ongoing management. For a site encompassing ten facilities, the complexity of managing the assessment workflow process and managing the data dramatically increases. Thus, OPS provides multiple ways for teams to enter data, manually or automatically, via the web, spreadsheets, or BIM environments; and OPS manages the data in a central BIM-based model server.

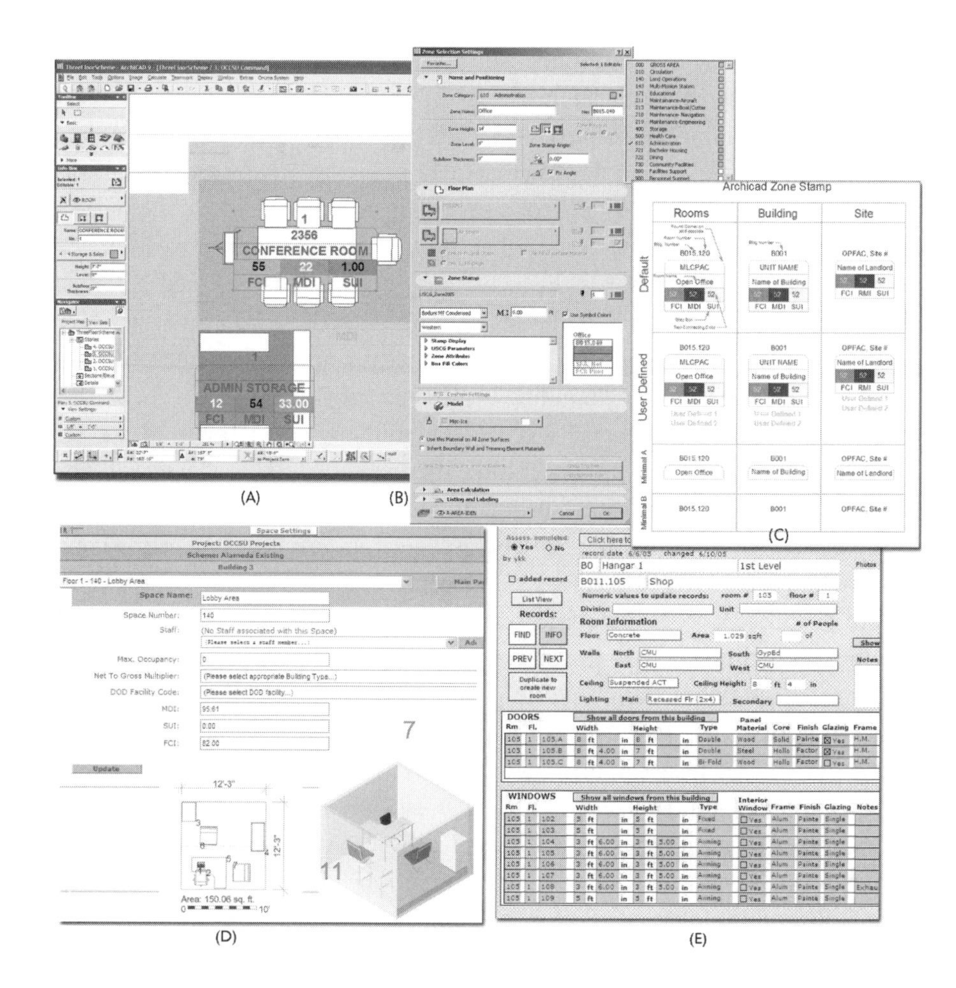

FIGURE 9-2-4
Interfaces to input room information in OPS.
A) View of room (zone) in BIM authoring tool with data from OPS; B) A custom zone-object interface in the ArchiCAD® environment;
C) User-defined room stamps to display data from OPS in ArchiCAD;
D) Room (space) visual UI in web-based OPS environment; and
E) room data view in OPS.

Use of the Planning System on the ISC Project

The USCG is faced with documenting and assessing over 33 million square feet of facilities. To test its Roadmap goals, the team (See Table 9-2-1 for list of team members) implemented the assessment system for the Integrated Support Command (ISC Alameda), located on Coast Guard Island in Alameda, CA. The ISC site contains thirty-five facilities totalling 700,000. This medium-scale implementation of OPS involved:

- Documenting 35 facilities at the Level 3 BIM detail, consisting of spaces, walls, doors, and windows
- Attaching assessment data and metrics to facility objects
- Managing updates to the assessment data

Figure 2-10

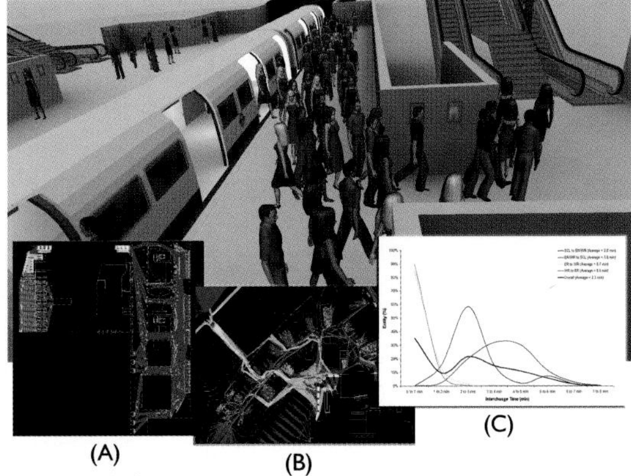

Figure 4-9

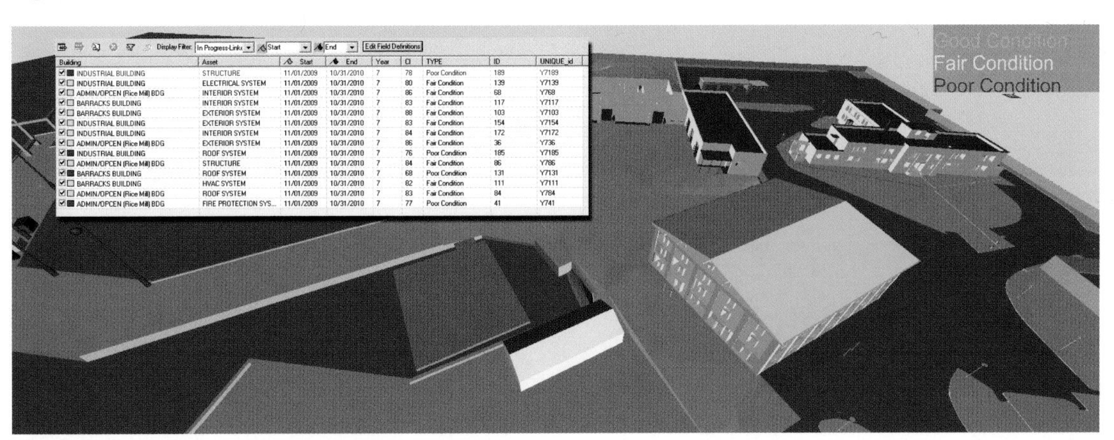

Building	Asset	Start	End	Year	CI	TYPE	ID	UNIQUE_id
INDUSTRIAL BUILDING	STRUCTURE	11/01/2009	10/31/2010	7	78	Poor Condition	189	Y7189
INDUSTRIAL BUILDING	ELECTRICAL SYSTEM	11/01/2009	10/31/2010	7	80	Fair Condition	139	Y7139
ADMIN/OPCEN (Rice Mill) BDG	INTERIOR SYSTEM	11/01/2009	10/31/2010	7	86	Fair Condition	68	Y768
BARRACKS BUILDING	INTERIOR SYSTEM	11/01/2009	10/31/2010	7	83	Fair Condition	117	Y7117
BARRACKS BUILDING	EXTERIOR SYSTEM	11/01/2009	10/31/2010	7	88	Fair Condition	103	Y7103
INDUSTRIAL BUILDING	EXTERIOR SYSTEM	11/01/2009	10/31/2010	7	83	Fair Condition	154	Y7154
INDUSTRIAL BUILDING	INTERIOR SYSTEM	11/01/2009	10/31/2010	7	84	Fair Condition	172	Y7172
ADMIN/OPCEN (Rice Mill) BDG	EXTERIOR SYSTEM	11/01/2009	10/31/2010	7	86	Fair Condition	36	Y736
INDUSTRIAL BUILDING	ROOF SYSTEM	11/01/2009	10/31/2010	7	76	Poor Condition	185	Y7185
ADMIN/OPCEN (Rice Mill) BDG	STRUCTURE	11/01/2009	10/31/2010	7	84	Fair Condition	86	Y786
BARRACKS BUILDING	ROOF SYSTEM	11/01/2009	10/31/2010	7	68	Poor Condition	131	Y7131
BARRACKS BUILDING	HVAC SYSTEM	11/01/2009	10/31/2010	7	82	Fair Condition	111	Y7111
ADMIN/OPCEN (Rice Mill) BDG	ROOF SYSTEM	11/01/2009	10/31/2010	7	83	Fair Condition	84	Y784
ADMIN/OPCEN (Rice Mill) BDG	FIRE PROTECTION SYS...	11/01/2009	10/31/2010	7	77	Poor Condition	41	Y741

Figure 4-11

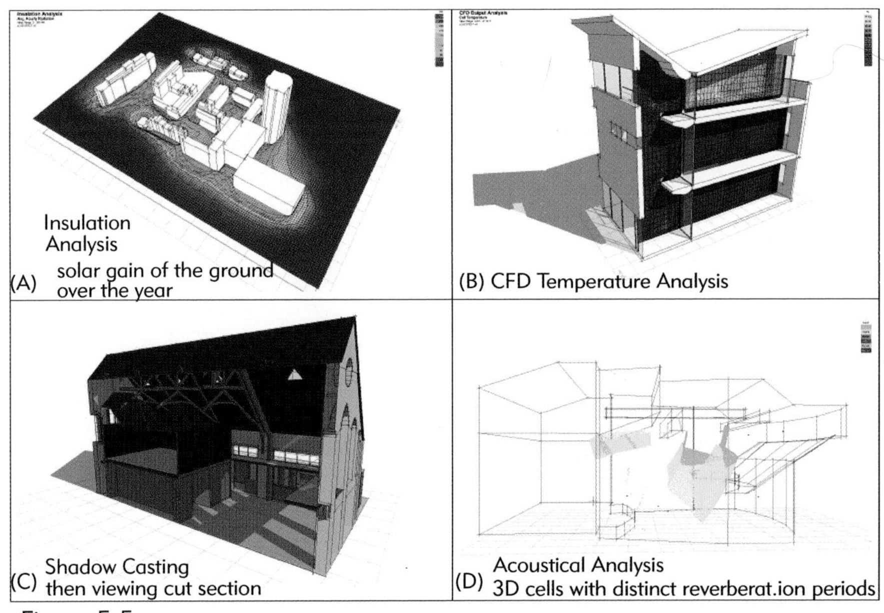

Figure 5-5

Figure 5-11

Figure 6-10

Figure 6-15

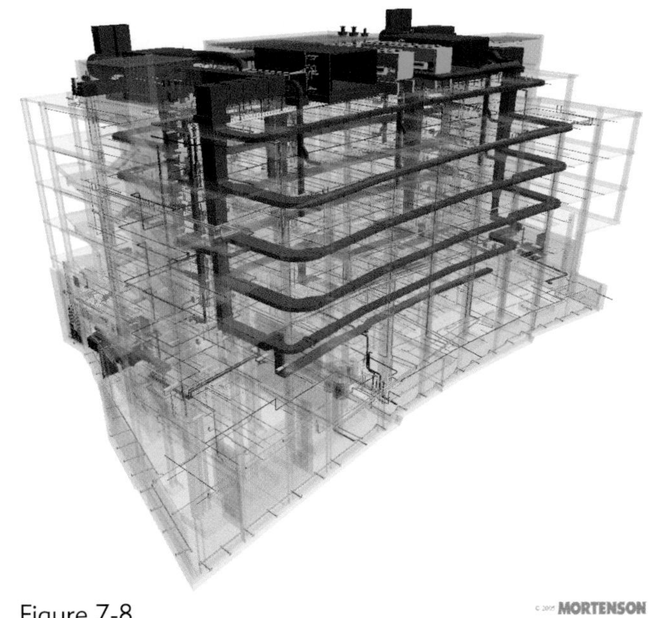

Figure 7-8

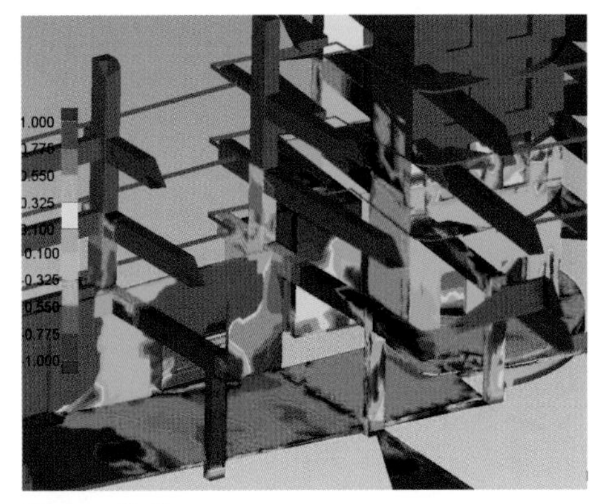

Figure 8-3

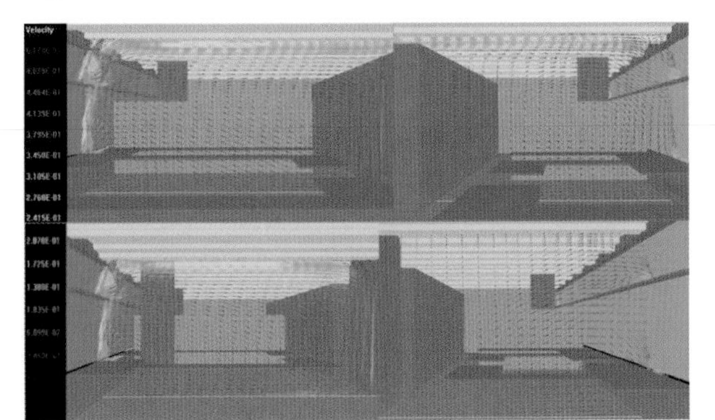

Figure 9-5-5

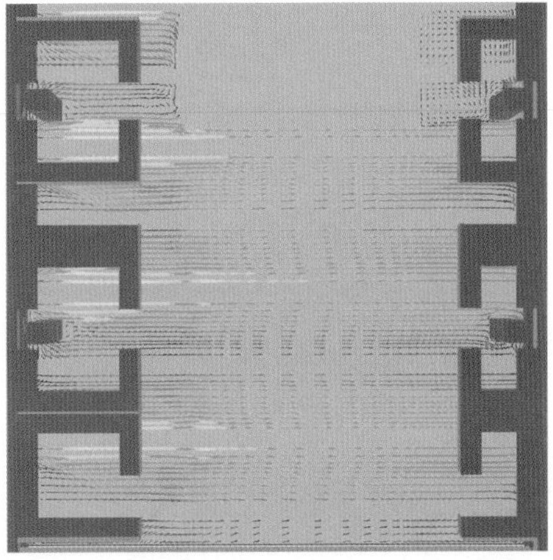

Figure 9-5-12

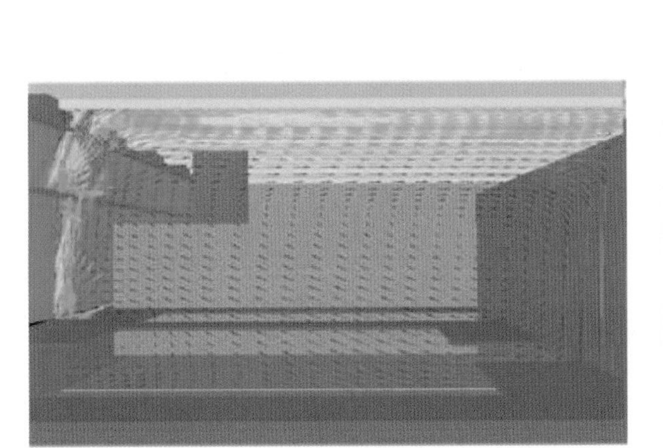

Figure 9-5-14

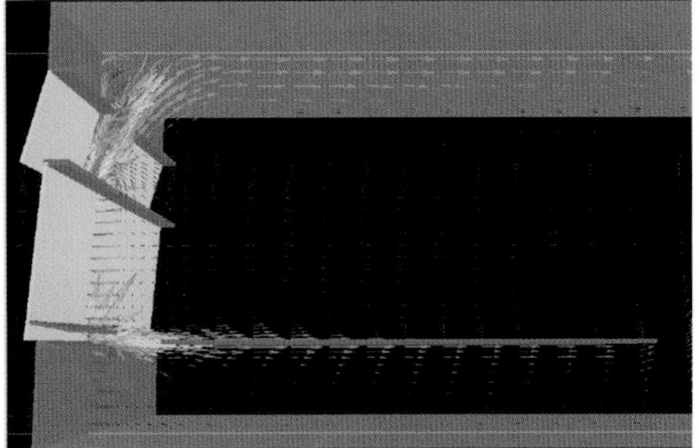

Figure 9-5-17

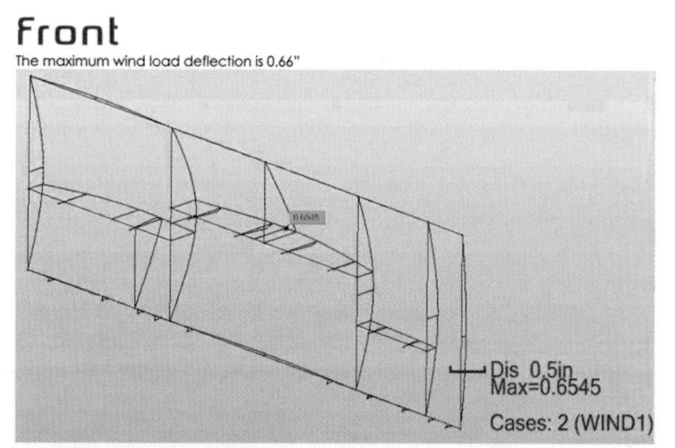

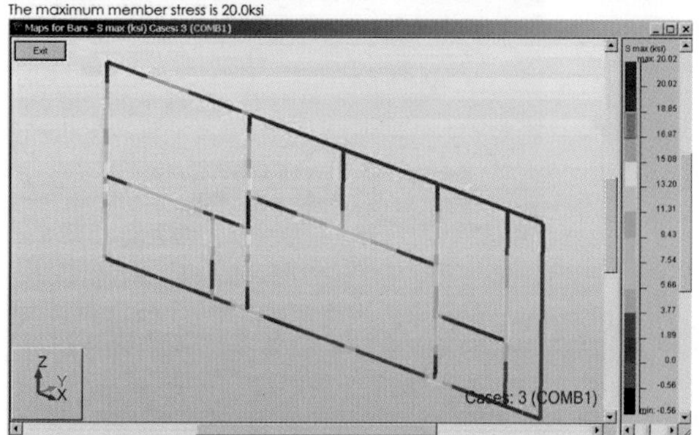

Figure 9-6-8

Figure 9-7-10

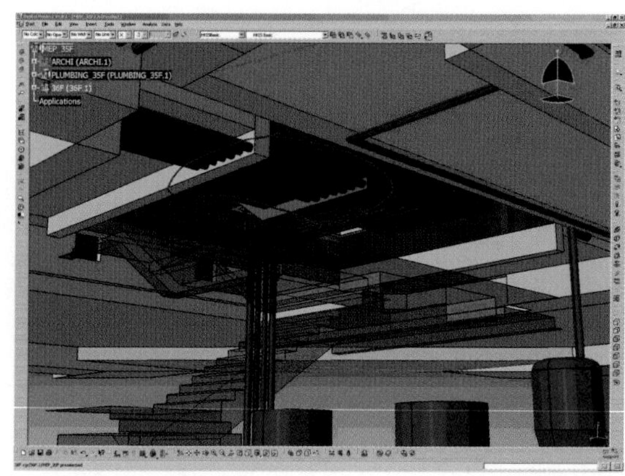

Figure 9-7-12B

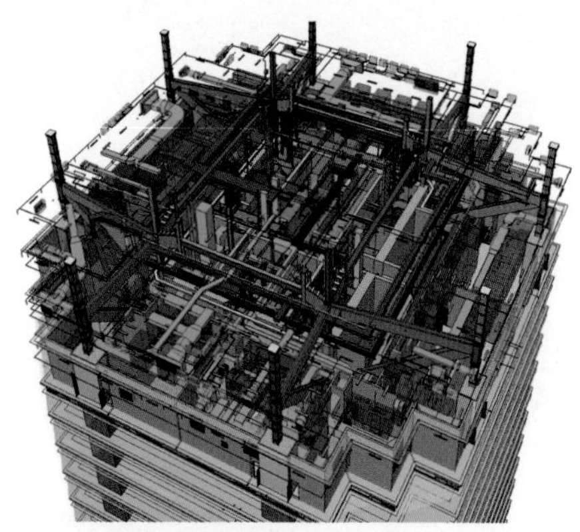

Figure 9-7-12A

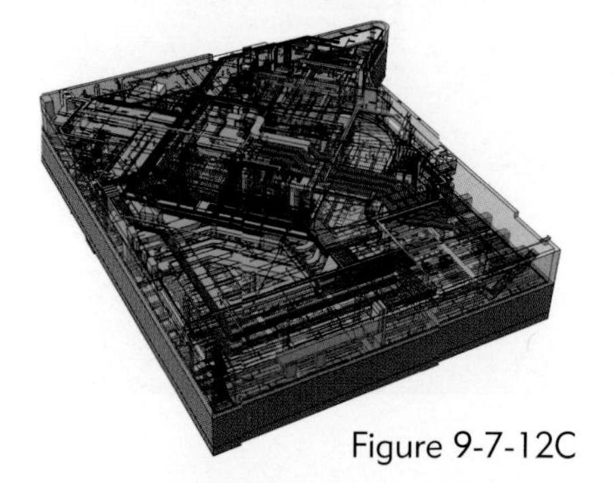

Figure 9-7-12C

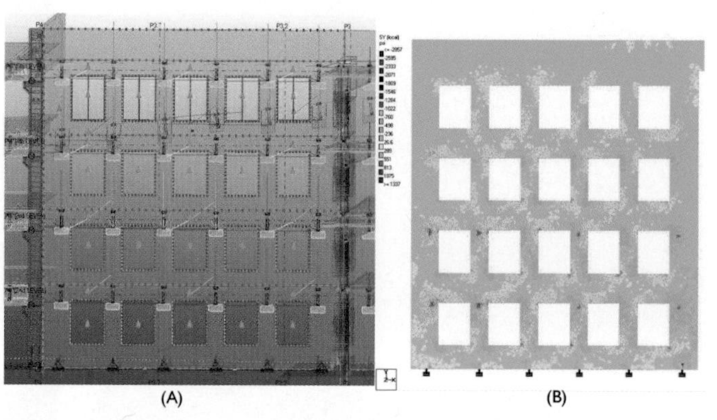

Figure 9-8-2

The as-built documentation consisted primarily of paper drawings or 2D CAD files. The as-built data was created in ArchiCAD using a combination of custom room and building objects, custom USCG objects, and built-in ArchiCAD objects. These included:

- Site plan with features
- Buildings
- Interior walls
- Doors and windows
- Furniture and equipment
- Assessment data associated with each object

Figure 9-2-5A shows a sample building model and the level of detail for that model. Figure 9-2-5B shows a Web-based data view of a room object with a list of associated doors, windows, and parameters and properties for all objects as well as the room's assessment status.

The assessment information came from multiple teams, many not trained to use ArchiCAD. The OPS' Web interface provided a simple and easy way to enter assessment data while situated in the office or field. For example, capturing knowledge about hazardous materials is critical to each assessment. The team in the field records the types of hazards (lead paint, asbestos, etc.), the degree of the condition, its quantity, location, height above the finish floor, and

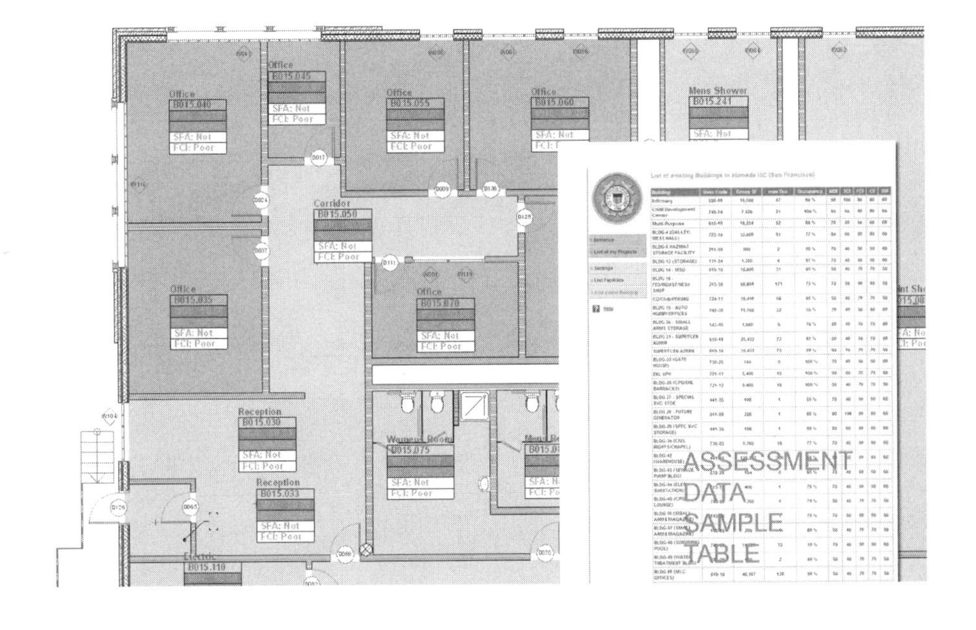

FIGURE 9-2-5
Plan view showing assessment values assigned to each room and an integrated assessment data table combining data from multiple facilities.

HAZMAT SCHEDULE										
ID	2D Symbol	Condition Status	Degree of Condition	Quantity	Zone	Zone #	Height AFF	Length (A)	Width (B)	Area
10		Lead Paint	Minimal	1			4'	6'-8"	6'-8"	44.61
11		Asbestos	Medium	1			4'	5'-1"	4'-1"	20.74
11		Asbestos	Medium	1			4'	7'-1 1/2"	4'-5"	31.56
12		Item-3	Entirety	1			6'-10"	2'-5 1/2"	2'-1"	5.11
13		Lead Paint	Minimal	1	Room	127	3'			
14		Item-4	Minimal	1	Room	127	3'			
15		Lead Paint	Low	1	Room	127	7'			
16		Item-5	Low	1	Room	127	7'			
17		Asbestos	Low	1	Room	127	3'			
18		Asbestos	Low	1			3'			
19		Asbestos	Low	1			3'			

(A) (B) (C)

FIGURE 9-2-6 A) Schedule to view hazardous material data from all hazardous material objects in a building. B) Plan view showing the hazardous object symbols. C) 3D views of the hazardous material object.

other data points directly in the model. They can create a new hazard object, add it to the model, and enter object parameters via pull-down menus for quick capture with the added benefit of data consistency across team members.

Upon return to the office, the team can generate a report in schedule form that summarizes information captured in the field or related to a specific room (Figure 9-2-5B).

One of the teams used specialized software called Vertex® (an engineering management system) to calculate the condition index based on the density and severity of observable defects (Mactec 2007). Using Microsoft Excel's import tool, the teams were able to automatically populate building objects with condition index values exported from Vertex.

By employing these methods to enter asset data and combining the information into an integrated model, the team reduced the updating effort by 98% compared to the traditional manual-based data entry and updating of assessment information. Table 9-2-2 shows a comparison between the manual method of updating information and the BIM-based method, either at the creation of asset information, post-construction, or during regular assessment updates. The time required to edit a single data point is reduced from two to .04 seconds. This amount represents time savings derived from several sources, including automatic updates from Vertex, Microsoft Excel data, and the reduction in manual errors. The overall time savings is significant when spread across four projects.

Lessons Learned from Small-Scale Implementation of a BIM-Based Assessment System

A BIM tool alone is insufficient to support enterprise level demands for knowledge capture and sharing, but it is integral to the process. OPS allowed the

Table 9-2-2 Comparison of the manual and BIM-based assessment processes for the USCG, based on Alameda project data.

	Manual BIM	Database and BIM-Based
Typical Building Size (SF)	20,000	20,000
Data Points Per Bldg	10,000	10,000
Data Points Per SF	.5	.5
USCG Total SF	33,000,000	33,000,000
USCG Data Points	16,500,000	16,500,000
Time to Edit One Data Point (sec)	2	.04
Total Time in Hours	9,166	183 hours
4 Iterations Per Project	36,666	733 hours

USCG to integrate building information model data with enterprise data from other analysis tools, such as Vertex in a usable and integrative manner that supports critical work processes, reduces the effort required to update data, and improves quality and accuracy.

9.2.3 BIM for Scenario Planning

The USCG must rapidly adopt and respond to ever-changing missions. In response to the 9/11 terrorist attacks, budgetary constraints, altered mission objectives and other factors. It established requirements for integrating the Group Commander and port operations into a single Sector Command Center. This presented a unique challenge. Not only did they need to define a new facility type and develop official standards for it, but they had to rapidly plan 35 of these centers in strategic locations as quickly as possible.

Requirements for BIM-based Scenario Planning System

The USCG needed a system that provided the structure and template for a Sector Command Center that would enable teams to rapidly design them while conforming to specified space standards. At the start of the project, however, these standards did not exist, or were incomplete; nor did the templates for defining them. Thus, the system needed to include functionality for consistently defining space standards and features for application to new facility designs.

Description of the Scenario Planning System

The USCG contracted Onuma, Inc. and AECinfosystems, Inc. to develop and implement a scenario-based planning system using OPS. The goal was to create

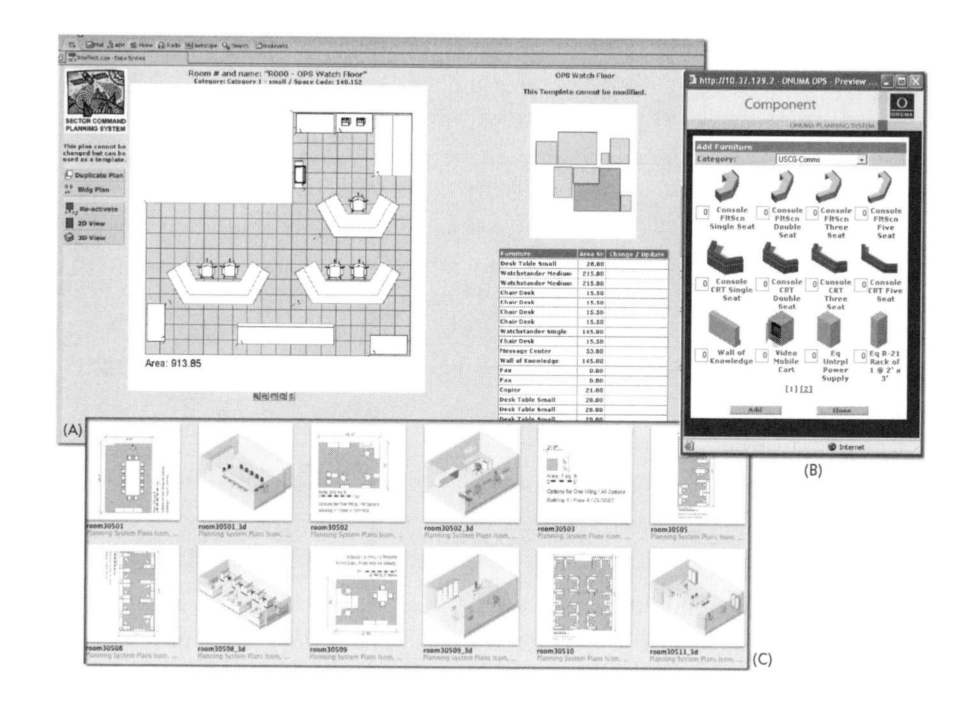

FIGURE 9-2-7
A) A sample room showing a template layout and a list of furniture for the room; B) A USCG furniture library; and C) Samples of room templates in 2D and 3D views.

a system that supported scenario planning without any training or understanding of BIM, yet the system had to be integrated with a building model. The way this was accomplished was by using a Web-based interface that worked with BIM through OPS.

Figure 9-2-7, 9-2-8, and 9-2-9 shows various parts of the Sector Command Center planning tool:

1. **Project management** (Figure 9-2-8A) is a Web portal to access current projects and scenarios (schemes) and to create new projects. A specific Sector Command Center, for example, might have three or four schemes or design options. Users can import the functional requirements. Functional Requirements are data rich space and/or room models that support additional information to support rapid decision processes and extraction of data. Additional information may include metric data such as Mission Dependency Index, space use requirements, equipment, level of security, costing, adjacencies, etc.
2. **Building planner** (Figure 9-2-8A, B) provides the user with a high-level layout of a single floor of a Sector Command Center consisting of room objects based on room templates (See Figure 9-2-8A). A command center could be a stand-alone facility or part of an existing or new facility.

3. **2D Room layout** (Figure 9-2-7B) is a detailed view of a room or room template with the ability to modify and edit the layout if the user has edit privileges. The room view also includes a list of the associated furniture and room properties and parameters.
4. **3D Room layout** (Figure 9-2-7C) 3D view of a room for visualization purposes.
5. **Export utilities** include exporters to ArchiCAD, AutoCAD or in IFC format.
6. **Reporting tools** provide the user with multiple ways to view a scenario and related analyses (similar to those in Figures 9-2-4 A, B, and C).

Implementation of the Scenario Planning System

OPS provided the USCG with a tool for generating and comparing Sector Command Center designs. Through charrettes, the teams used the system to design and evaluate Sector Command Centers in various locations. At the start of the project, the team had a set of initial room templates and programmatic requirements based on previous projects. Throughout the process, these template

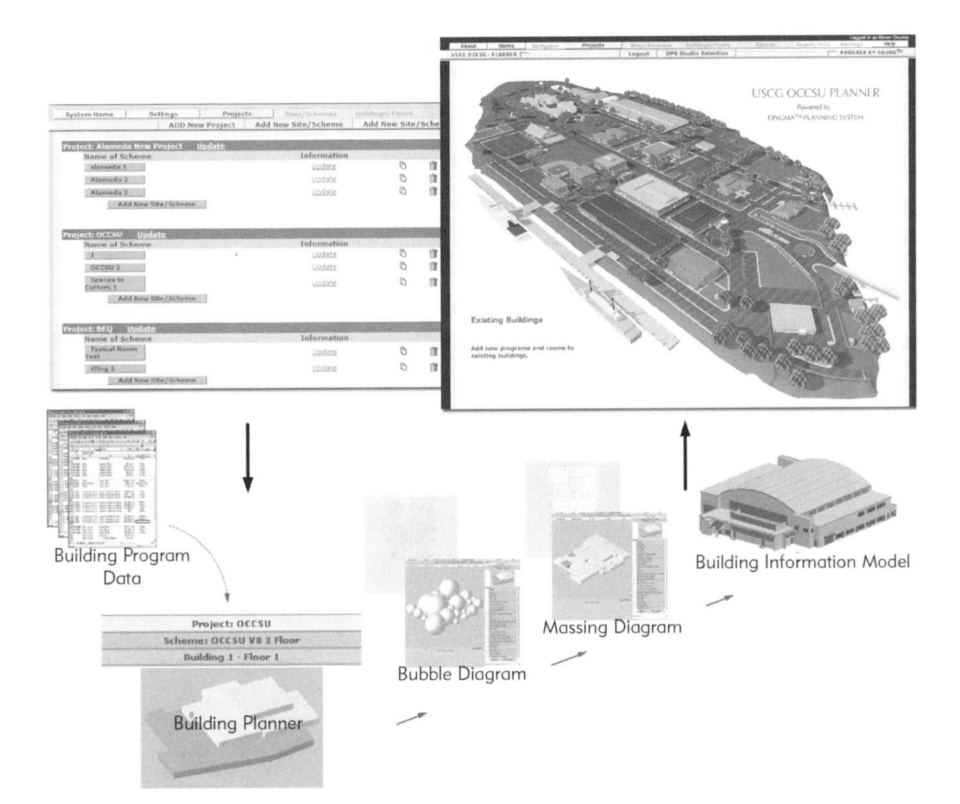

FIGURE 9-2-8 The workflow of the OCCSU Planner system. A) the project management portal to create schemes, B) the building planner to design a scheme using building program data, C) refinement of the scheme from masses, relationships between spaces in a bubble diagram, to more detailed massing and then a building information model; and D) publishing of the model and viewing the scheme in the project site.

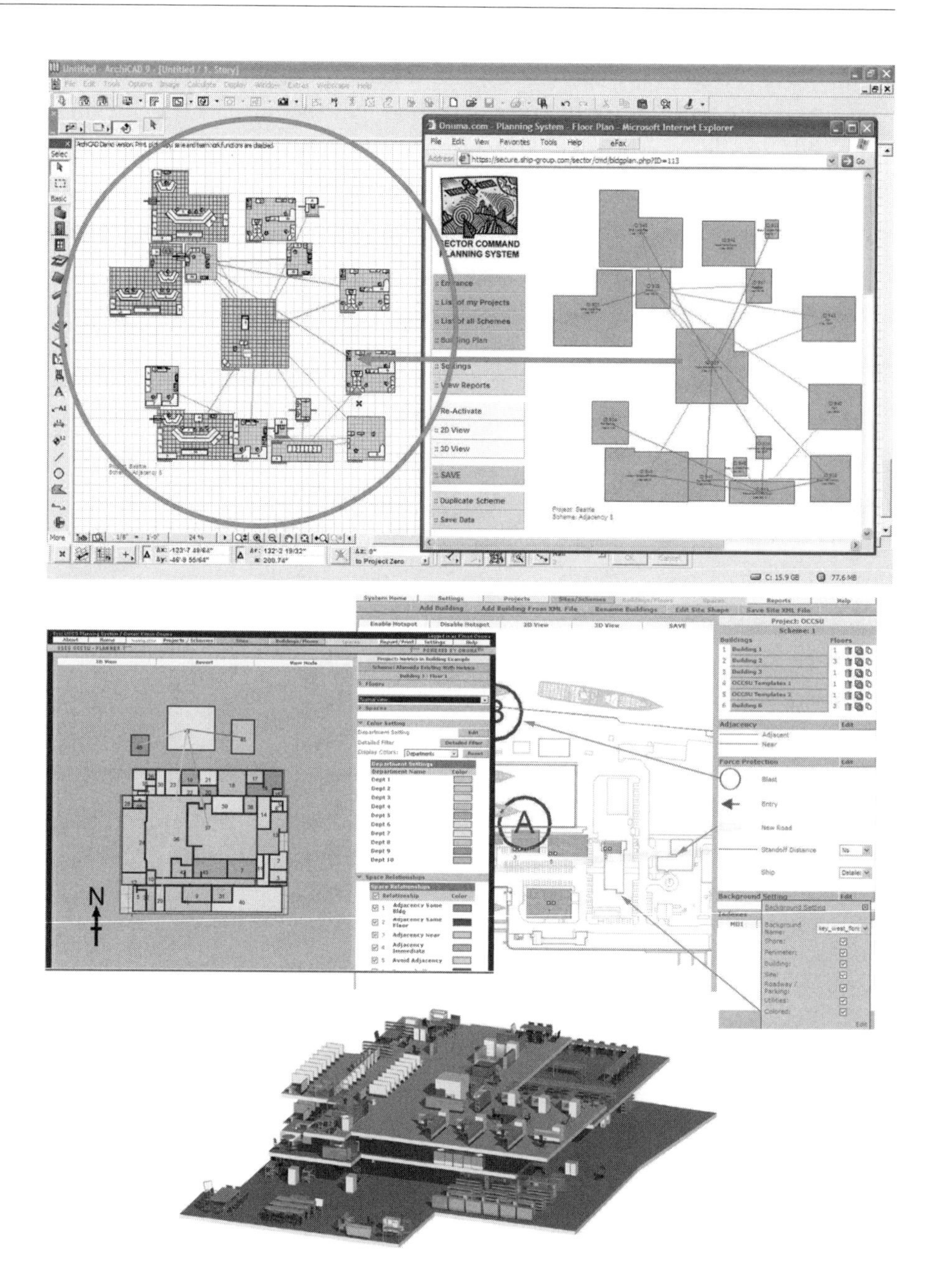

FIGURE 9-2-9 Views of building model data created and developed in the OCCSU Planner system.
2D layout views in A) Archicad and B) web-based OPS system. Views of the intelligent relationships between spaces and other objects such as C) *adjacency* and *avoid* relationships between spaces; and D) a 2D bomb blast object for security analysis. E) View of the building model detail produced from the scenario planning system including floors and furniture.

designs evolved and the requirements changed. The management features of OPS allowed them to build template rooms and record new requirements. As a new Sector Command Center was designed, they often discovered missing template rooms and other necessary requirement changes. This self-documenting

Table 9-2-3 Metrics for implementing BIM-based scenario planning for the design of Sector Command Centers.

	Traditional Design	BIM-based design with OPS
Number of Projects	35	35
Average size each	5,000 sf	5,000 sf
Average time to complete design for one project using traditional process	10 Months	1 Month
Total time to complete 35 projects using traditional processes	350 Months	6 Months
Estimated savings for design only		344 Months

process was a direct byproduct of the use of a data-rich and visually interactive environment.

Over a period of six months, the team successfully generated conceptual designs and related requirements for 35 sector command facilities. Traditionally, each project would take 10 months as documented in Table 9-2-3. With OPS, the conceptual design process was reduced to one month. With OPS, the team could analyze each design in terms of functional requirements, cost, and conformance with USCG standards. Similar to the ICS project, the use of templates forced consistency. When the teams discovered a need for a new room template or modification to an existing one, the system automatically captured these requirements.

At the conclusion of concept design for each facility, the team produced programmatic requirements for the sector command facility, 2D and 3D conceptual layouts, and reports documenting the equipment and furniture lists. This detail at the conceptual stage provided the USCG with much richer information to bid and procure a facility in a less time.

Lessons Learned from the Implementation of a BIM-Based Scenario Planning System

The parametric template-based approach to scenario planning dramatically reduced the overall time required to design each Sector Command Center. It allowed the USCG to continually build new prototype requirements and develop standards, even as they were designing and as new mission requirements evolved. OPS provided the benefits of building information modeling via an easy-to-use interface for early conceptual design, and it provided the teams with rapid feedback and real-time analysis of conceptual designs in visual and interactive formats. The rapid analysis capability allowed decisions to be made faster and eliminate options that would not support the needs of the client.

Quotes from USCG Personnel discussing benefits of SCCP:

"The main benefit from using a system like this goes beyond the actual process of designing and coming to consensus with a large group of stakeholders. The process of using an integrated system like this in itself creates an engine that supplies answers and collects knowledge that can be used in many different ways beyond the immediate need to define the requirements for a project. "

"The Sector Command Center Planning Tool was the genesis of the Web-enabled, rapid planning, 3-D system. Planners all over the Coast Guard used this tool to generate 35 detailed reports in a matter of six months with little to no training on the intuitive SCC Tool. We have found that this has categorically improved communication in the early planning stages between our engineers, architects and our customers."

VK Holtzman-Bell, Captain US Coast Guard

9.2.4 OCCSU System

The previous two projects focused on facility asset management and conceptual design. The OCCSU project addresses the feasibility phase of a project when an owner like the USCG must make critical and costly decisions to define a project and align the business drivers (demand) with available facilities. The business drivers for the OCCSU project are new cutters (ships) that will be completed over the next 10–12 years. These boats represent a new business model and a new project type for the USCG. The delivery date and location is unknown, but the USCG must plan for their deployment and be ready to procure the work for design and construction as delivery dates and locations are finalized.

Requirements for the OCCSU System

All variables in the OCCSU project are dynamic including timing, function, organization, and location. Each new boat might require 100 people on standby support and crew rotation. Scenarios such as how to rotate crews and support the boats, would have a direct impact on what facilities would be needed. This would entail facilities for housing standby support and crews and the maintenance of shores-side facilities and command centers. Thus, scenario planning would entail visualization of the entire site, multiple facilities, and individual buildings across multiple floors.

The template-based approach to designing the Sector Command Centers was the starting point for the OCCSU project; and the assessment models developed for the ICS project provided some of the as-built information. Two critical pieces were required to support the what-if scenarios: 1) links to a geospatial database or GIS system, where most existing information resided and 2) richer analysis tools to support analysis of the scenarios. The most important types of analyses were related to security.

Implementation of the OCCSU System

There were about ten sites that the USCG chose to evaluate for the OCCSU project. Some sites were part of their previous efforts to produce as-built models of their facilities, such as for the ICS project. These facilities were already in OPS and existed in ArchiCAD. Other sites only had 2D site building drawings and some buildings had no information at all. For example, Figure 9-2-9 shows a 3D view of one of the project sites consisting of various levels of details for each facility. This model also serves as an interface to a project and its various scenarios (schemes).

Each project can have multiple schemes, and each scheme represents a specific project location and associated requirements. The workflow is similar to that described for the Sector Command Center planning system. The team starts with a template or defines a new one. They view the space in 2D or 3D and define and create one or multiple floors. The addition of new furniture or equipment is only from a USCG approved furniture database.

Each space can have *adjacency* or *avoid* relationships with other spaces, and the team can view these in the 2D layout view, as shown in Figure 9-2-9. As a scenario is created, the team can view reports that show cost (based on square footage historical data) or building capacity. (A LEED analysis checklist was added to OPS after OCCSU). At any time they can export the scenario to IFC or ArchiCAD (via XML) to view the facility in 3D and add walls, doors, and windows. For security analyses, they can add a *blast object* from a database, which represents different types and sizes of blasts (shown in Figure 9-2-9B).

Lessons Learned

The OCCSU planning system demonstrates how owners can use BIM to support enterprise scenario planning to define and evaluate projects before they are financed or funded. OPS provides the USCG planning teams with an integrated and shared operational picture of various scenarios to support real time decision making. The decision-making shifts to a much earlier part of the design process and allows the USCG planning teams to predict the potential outcomes of various 'what-if' scenarios. This type of scenario based planning is different from today's reactive, linear, and time consuming approach to facility design and construction. These process changes yield significant benefits, yet require significant cultural and contractual changes.

Today's contractual methods are designed to support the single project approach. To extend the success of the OCCSU Planning System enterprise-wide within the Coast Guard, the contracts and relationships with consultants will need to change and evolve to support multiple projects, non-linear processes, and integrative approaches. Today, BIM is commonly used as a tool for architects to design a project and create construction drawings within a well-defined deliverable based process. When BIM is integrated with metric data and supported at an enterprise level, it becomes a visual decision tool for strategic planning. It also feeds the architectural design tasks with useful and more complete information from the client. Issues related to contracting services to add this metric data, often from multiple services and across projects, still exist and need to be addressed.

For the USCG, and other owners, the investment in the site and building data can payoff tremendously if the data is integrated, consistent, and accessible. That is, the USCG needs to ensure that any service provider can have access to the data, such as templates, and leverage enterprise data. Since the OCCSU system forces normalization, all services providers should benefit from access to as-built data and integration of site and other operational data.

Quotes Summarizing Benefits of OCCSU

The Onuma Planning System shifted the reality of true scenario based planning to an enterprise level, enabling us to link shore facilities and infrastructure to mission execution and strategic Coast Guard wide outcomes. This better enabled us to allocate infrastructure resources to the most important mission execution and outcomes.

The integration of BIM, geospatial data, real property data and mission requirements supports the need of a common operational picture for the USCG. This common operational picture can be real time tactical information as well as longer term strategic information, which is enabled by the Onuma Planning System.

David Hammond, Chief,
SFCAM Division Commandant, (CG-434) US Coast Guard

9.2.5 Conclusions and Lessons Learned

These projects demonstrate the dramatic cost and time savings associated with BIM-based planning and facility asset management. Much of the savings are attributed to standardizing work processes and capturing knowledge digitally, rather than through labor intensive manual processes. In implementing these BIM-based systems, the key lessons learned were:

- BIM-based work processes require significant cultural changes. These range from simple work process changes related to digital data entry to working with templates instead of freeform design. To prepare and plan for the impact of these changes, the USCG understood that wide-scale implementation would require realistic roadmaps, and they opted to perform small-scale implementations, such as the ISC project, before implementing large-scale efforts, such as the OCCSU project.
- It is imperative to make BIM accessible to a wide range of users. Broad access to the model for viewing and editing as well as creating scenarios was paramount to the success of OPS on these projects.
- Parametric object-based tools provide the building blocks for capturing knowledge and project requirements that inevitably change over time.
- BIM promotes standards and work processes designed on BIM-based OPS, which forced data normalization, reduced errors, and increased the value and quality of as-built information.

The value of the information from multiple projects such as the ISC Project, Sector Command Center Planning System and OCCSU Planning System exponentially increases as more integration happens. Unexpected connections between data in these projects, such as the value of the as-built assessment data in the ISC to the 'what-if' scenario planning in OCCSU demonstrate a return on the investment in an integrated, consistent, and standard enterprise building model.

Acknowledgments for USCG Case Study

The authors are indebted to Kimon Onuma of Onuma, Inc. and Dianne Davis of AECinfosystems, Inc. for their assistance in providing the information contained in this case study. Additional participants that were instrumental include: from USCG Headquarters Captain Jay Manik and David Hammond, Lieutenant Commander Robert Bevins, (SFCAM) Lieutenant Commander Scott Gesele, and William Logan, and Hassan Zaidi (Civil Engineering), Commander James Dempsey from CEU Oakland; and from FD&CC Pacific Captain Virginia Holtzmanbell, William Scherer, and Jeffrey Brockus; Paul Herold and team from the Logistics Geospatial Integration Center; Jim Watson and team from NexDSS (Mactec); Ian Thompson and team from Standing Stone Consulting, Inc.

All figures were provided courtesy of Onuma, Inc. and AECinfoystems, Inc.

9.3 CAMINO MEDICAL GROUP MOUNTAIN VIEW MEDICAL OFFICE BUILDING COMPLEX

BIM for design and construction coordination and integration with lean processes

9.3.1 Description of Project

The El Camino Medical Office Building and Parking Garage are located in Mountain View, CA in the San Francisco Bay Area. The medical center is intended for short-term (one-day) medical care and surgery. It is operated by the Camino Medical Group (CMG), which has 6 clinics on the Bay Peninsula and functions as a division of the Palo Alto Medical Foundation (PAMF). Sutter Health is its parent organization. The project manager, John Holms, played an important role in promoting the creation and use of 3D models in conjunction with lean construction techniques. Figure 9-3-1 is a rendering of the facility, based on the architect's design. Note the large covered garage in front of the office building, which is required to accommodate the daily influx of staff and patients.

As of November, 2006, the project was near complete, with an expected turn-over to the owner planned for April 2007. This date is about 6 months earlier than a traditional design-bid-build delivery model and represents a significant benefit to the owner and the project team. This case study describes how this was achieved. Table 9-3-1 provides an overview of the scope of the project and Table 9-3-2 lists the key milestones for the project.

9.3.2 General Description of Project Management Techniques

The project team evolved from the traditional model (shown in Figure 9-3-2) to the integrated model (shown in Figure 9-3-3) to allow for the effective use of a 3D model and lean construction techniques.

FIGURE 9-3-1 Visualization of El Camino Medical Office Building and Parking Garage.
Image provided courtesy of DPR Construction, Inc.

Table 9-3-1 The scope of the project.

Site & Infrastructure development	420,790 sf
Parking Structure	1,110 stalls, 420,000 sf
Medical Office Building	110 Providers, 250,000 sf
Urgent Care Center	6,000 sf
Outpatient Surgery Center	5 Suites, 20,000 sf
Pharmacy	6,000 sf
Laboratory and Diagnostics Radiology Center	30,000 sf
Construction Costs	$94.5 Million

Table 9-3-2 Milestones for the project.

Start of design	October 2003
DPR engaged	April 2004
Schematic Design complete	November 2004
Detail Design complete	March 2005
Construction Design (9 packages)	Nov. 2004 – Nov. 2005
Start of Construction	February 2005
Completion Date	April 2007

While a typical design-bid-build project would prevent subcontractors from detailing the design until after the project was bid, an integrated organization (Figure 9-3-3) allowed the general contractor and subcontractors to add constructability knowledge to the design as it developed. The collaboration was enhanced by weekly meetings held in a large trailer room at the project site. This space was also used by detailers working under the subcontractors, to allow for quicker resolution of conflicts. The goals established by the owner are summarized in Table 9-3-3.

The primary goal of the owner was to reduce overall project duration so that the clinic could be operational as soon as possible. To achieve this goal, construction was started before detailed design was complete. Each design assist mechanical and electrical subcontractor was asked to provide a 3D model of their portion of the design to be merged with the 3D models provided by the architect and structural engineer. The general contractor managed the technology and required that the model be maintained on its servers at the job site. Various software tools were used to design in 2D or 3D, as shown in Table 9-3-4. For members of the

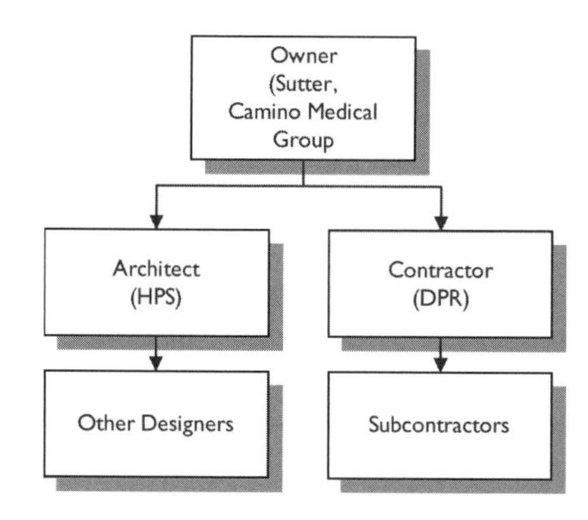

FIGURE 9-3-2 Traditional project team used at the start of the project.

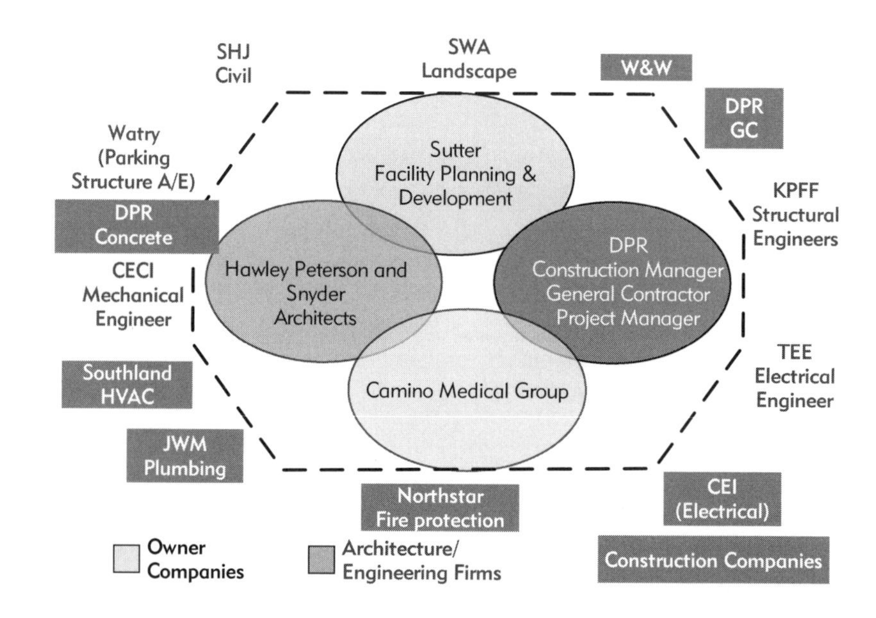

FIGURE 9-3-3 An integrated project team evolved early in the project, allowing for improved integration, an overlap of design and construction, a reduction in redundant effort, and savings in time.
Adapted from image provided courtesy of DPR Construction, Inc.

team using 2D tools, they were required to either convert their designs to 3D or hire a consultant for this task. When selections of designers and subcontractors were made, emphasis was placed on their experience using 3D design and detailing tools. In some cases, however, no designer or subcontractor with 3D experience that also had an excellent reputation was available. In these cases, it was necessary to convert 2D designs to 3D models by a third party. The project was a learning experience for all involved, regardless of prior background.

The use of a virtual model facilitated and demanded close collaboration among the project team; and it resulted in the anticipated as well as some unexpected benefits. Among the anticipated benefits were:

Table 9-3-3 Goals for the integrated project team, established by owner's project manager; additional specific goals are shown in Table 9-3-4 below.

Lean Approach	Goals of Sutter Health Lean Project Delivery
Subcontractors to be brought in early to participate in detail design	Better collaboration to improve constructability, reduce errors in field, save time and cost
Target costing	Better collaboration to reduce costs Optimize the project, not the pieces
Applying the Last Planner System™	Projects are networks of commitments
Weekly work planning	Tightly couple learning with action
Work streams	Increase relatedness among all project participants
Build virtually in 3D before constructing	Better collaboration using an accurate and complete 3D model; Optimize the project, not the pieces Tightly couple learning with action

Table 9-3-4 Design and Modeling responsibilities for each member of the project team

Team Member	Scope of Work	Software Used
Architect	3D architectural model & documentation	Architectural Desktop (ADT)
Parking Garage Designer	2D design & documentation 3D model built from 2D AutoCAD-built by consultant after design completed	2D AutoCAD 3D AutoCAD
Structural Engineer	2D design & documentation 3D structural Model of 2D Design	Revit Structures
Mechanical Engineer	2D design & documentation	2D AutoCAD
Electrical Engineer	2D design & documentation	2D AutoCAD
HVAC Contractor	3D shop drawings	CadDuct
Plumbing Contractor	3D shop drawings	Piping Designer 3D
Electrical Contractor	3D shop drawings	3D AutoCAD
Fire Protection Contractor	3D shop drawings	SprinkCAD

- **For the client:** greater understanding of how the facility would serve patients, doctors, and nurses
- **For the designers:** how design decisions would impact the facility, influence its constructability and cost, and greatly reduce field conflicts and clashes

- **For the contractor and subcontractors:** how the facility could be efficiently constructed and phased; how materials and offsite prefabricated assemblies could be installed

Some of the realized benefits included:

- All important design decisions were made during the design phase
- No misunderstanding among project team
- No conflict of systems in the field
- Hospital administration, staff, and doctors could see and understand the spaces
- Inform the community about proposed design
- Better coordination of subcontractors onsite
- Plan logistics and sequence installation of prefabricated assemblies

To overlap design with construction, the subcontractors started detailing before the A/E firms completed their designs. This workflow is shown in Figure 9-3-4. This reduced the detail design effort normally required of A/E firms, which typically delays the start of detail design by subcontractors; and it allowed construction and prefab assembly to begin earlier. The planning of these detailing efforts proved to be critical. If the subcontractors began their work too early, then further changes requested by the client would inevitably lead to considerable rework and extra workloads. On the other hand, a late start would cause delays in construction and fabrication. To deal with this, an overall

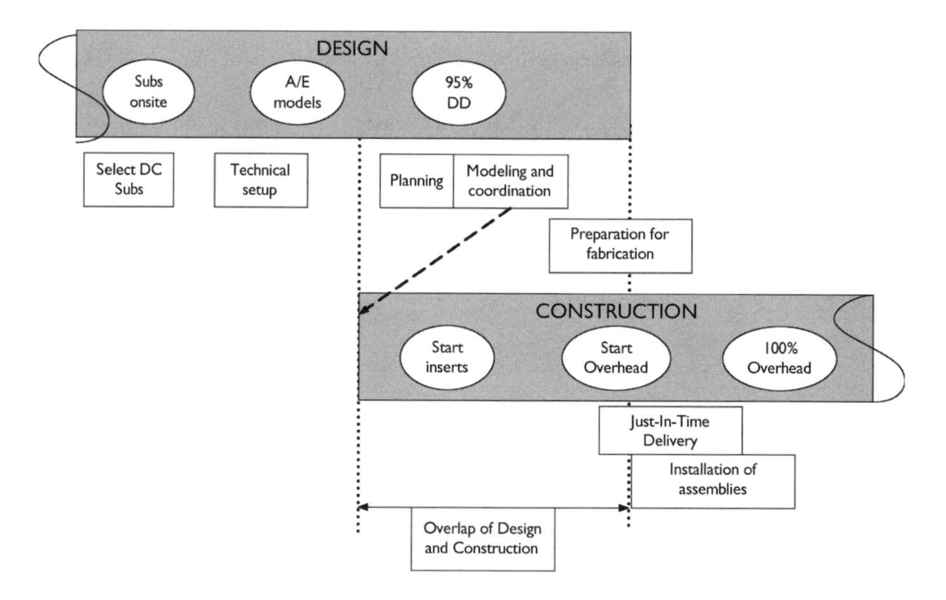

FIGURE 9-3-4 Workflow of virtual design and construction (not to time scale) showing how the overlap of design and construction was organized. Note the early selection and participation of subcontractors in design assist roles. Foundation work began midway through the modeling and coordination effort.

a) 3D virtual assembly of facility

b) Picture of actual onsite assembly

FIGURE 9-3-5 Comparison of a) virtual 3D model to b) actual onsite assembly.
Adapted from image provided courtesy of DPR Construction, Inc.

schedule was developed to show when each sector of the project would be constructed and when detailed design would be needed for each type of work within that sector, such as slabs, columns, sprinklers, HVAC, plumbing, electrical, etc. The 3D CAD model showed exactly how each object interacted with others and allowed for close collaboration among the various trades. Figure 9-3-5 shows how closely the 3D CAD model mirrored actual field assembly.

The schedule was carefully reviewed, and it guided the entire project. It was used to 'pull' the schedule for material deliveries, indicate when detailed design would be needed, and ensure sufficient lead time for field construction, prefabrication, and installation. The schedule also tracked the delivery of resources to keep construction continuous and on track. By and large, this planning system worked well. The only delays were caused by changes requested by the client and permit delays from the county or Hospital Regulatory Agency. In addition, not *all* subcontractors were involved early enough to collaborate during the detailed design phase. This necessitated some field changes, such as for the fixtures, furniture, and equipment (FF&E) subcontractor. Changes to the equipment requirements and locations caused associated changes in electrical, piping, and HVAC. Thus, despite very significant efforts to smooth the construction process and avoid delays in the field caused by design errors (clashes), there were some problems that required additional detailing work and some field delays. Nevertheless, the use of a shared 3D model linked to lean construction techniques led to reduced project cost and time.

In the following sections, we present the role of each participant, their individual contributions and provide detailed solutions to the problems encountered.

9.3.3 Subcontractors

The responsibilities of the GC included typical tasks, such as selecting subcontractors, planning and supervising the construction, coordinating the rest of the project team and outside agencies responsible for permits. Additionally for this project, they were responsible for setting up the computer environment for virtual design and coordinating the use of 3D modeling for planning, visualization, and clash detection. Working with the subcontractors, they also assisted in the ordering and installation of materials and fabricated assemblies.

In selecting the design assist and design build subcontractors, the GC placed greater emphasis on the need for prior experience with 3D modeling systems. The instructions to bidders are shown in the box below.

After each bidder was interviewed and pre-qualified, the most experienced bidder was selected. In almost all cases, they found an adequate number of

Camino Medical Campus 3D Object Modeling Requirements

Approach

Sutter Health has asked DPR to lead the effort to coordinate detailed construction designs. Each Design Assist and Design Build contractor will be required to create a 3D object model of their design. These models will be brought together into a 3D CAD integration software application where they can be viewed and analyzed. Designers will be asked to revise their designs to resolve conflicts. Each contractor will be responsible for the accuracy of their design, including any necessary revisions. After the changes have been made, the 3D model will be analyzed again in an iterative process. Contractor should plan to coordinate work with other trades, as they would normally, but also for the additional coordination effort based on 3D object models. Along with coordination, the 3D object models will be linked to schedule activities so that the sequence can be analyzed using 4D software. Installation sequences will be changed to create the best possible schedule.

Object Modeling

Detailed design will be done in CAD 3D object modeling software, which can create objects having the following characteristics:

- Each object in the model has a unique name or object identifier, and can have attributes attached to it, i.e., geometry info, weight, cost, schedule, procurement info, etc.
- These objects are either 3D solids or true parametric objects. They are not 2D polygons that are combined to create a 3D object, as is the case in a surface model.
- The object model is assembled from libraries of predefined parts representing a constructible component or assembly of components.
- Object properties should reside in a database which can be queried for attributes of the various objects.
- All views of the model are depictions of the object properties stored in the database so that a change made to one or more object parameters is reflected in every plan, section and elevation view.

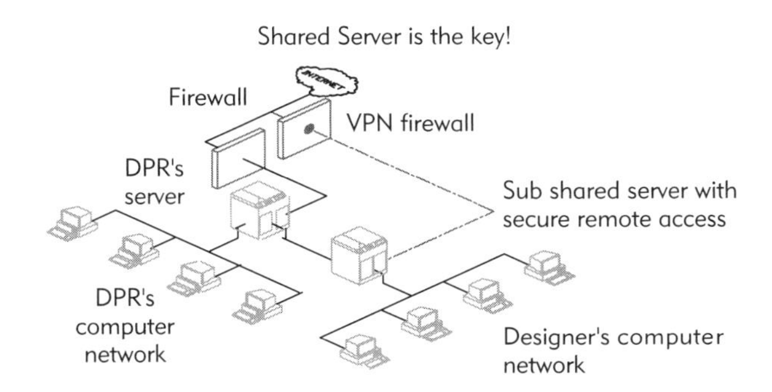

FIGURE 9-3-6
Server and network setup for the project. Note: internet access was through a firewall. *Image provided courtesy of DPR Construction, Inc.*

subcontractors with the required experience. In a few areas, it was necessary to accept less than the desired capability in 3D design. This resulted in a few problems later on, when the 2D designs caused problems (clashes) in the field. Among the most important lessons learned for this project was the importance of having ***all*** subcontractors input their designs into the shared 3D model.

Since a variety of 2D and 3D modeling systems were used, Navisworks JetStream software was used to merge the models and detect clashes. This technical setup is shown in Figure 9-3-6. The network was maintained at the job site with internet access through a firewall. The subcontractors shared the work for all detailing. This ensured excellent collaboration among the team and reduced the time needed to resolve problems.

So that subcontractors could work together easily, a set of standards was established for merging their individual efforts. These included:

- File naming
- Common scales
- Common reference point for all trades (0,0,0 point set by architect)
- Identification and colors used for objects
- Title blocks and view ports
- File formats compatible with Navisworks JetStream software

To establish a budget for the 3D modeling effort, each team member was assigned a scope of work. The budget was used to calculate the actual cost of 3D modeling, beyond what it would have cost to produce only 2D drawings. The extra costs were caused by using a combination of 2D and 3D tools. Additional effort was required to model objects that might otherwise not be shown in drawings, such as small pipes, pipe supports, and sheet metal supports, etc. The primary reason for the higher cost of detailing the piping was that design changes required rework. While this did add cost to the project, early detailing

FIGURE 9-3-7
Picture of detailers in the shared workspace onsite.
Image provided courtesy of DPR Construction, Inc.

facilitated field progress and reduced the overall project duration. This is a good example of how added cost for one subcontractor led to a reduced cost for the total project. This did not give piping subcontractors a disadvantage because all of the subcontractors involved were able to benefit from the overall project savings (see below for contract details).

Extra costs were as listed in Table 9-3-5.

9.3.4 Owner

Sutter Health and its subsidiaries are major owners of hospitals and medical office buildings in California and elsewhere. As such, they have considerable experience in the permitting, financing, and design and construction processes. In recent years, the pace of construction has increased significantly due to a recently enacted law in the state of California that requires seismic upgrading of all medical facilities. The pressure to upgrade and replace older facilities has resulted in increases in the cost of work and longer waiting periods for obtaining permits and undergoing reviews by the Office of State-wide Health Planning and Development (OSHPD) and other local regulatory agencies. Healthcare developers in general have been particularly eager to find more efficient approaches to the design and construction process. At the time of the project's completion, the affiliates of Sutter Health had not attempted to implement the

Table 9-3-5 The extra costs of modeling in 3D, as opposed to using 2D drawings, for each project member. The extra costs were considerably lower than the savings in time and labor hours.

Company/Function	Modeling scope	Extra cost for 3D effort	Percent of total modeling cost
HPS/Architecture	Architectural design	$80,000	19.3%
KPF	Structural design; foundations and steel frame	$20,000	4.8%
Watry Design	Parking structure shell	$15,000	3.6%
Southland / Sheet Metal Detailing	Add sheet metal supports and seismic bracing to 3D duct model from Mechanical Eng. Design; 3D detailing and coordination	$50,000	12.0%
McClenahan / Piping Detailing	Add pipe, pipe supports, and seismic bracing to coordinate with other trades; 3D detailing and coordination	$200,000[1]	48.2%
Northstar	Fire sprinkler shop drawings and coordination	unable to report cost	0.0%
Cupertino	Electrical shop drawings and coordination	$50,000	12.0%
Total Modeling Cost		$415,000	100.0%
Total construction cost		$94,500,000	0.44%

3D model for use in facility management. As a result, design changes were only entered on as-built drawings and not to the 3D model.

The owner's goals are summarized in Table 9-3-6. The last item in Table 9-3-6 was supported in writing through the use of a unique clause in the contract for specifying how the cost savings achieved by any individual team member would be shared collectively by the team, including the owner.

This clause was an important factor for facilitating a successful collaboration, therefore it is described in further detail in the section that follows, which is a description of the *Incentive Fee Plan* developed by the owner specifically for this project.

The first section defines the goals and underlying principles of the plan. The remaining sections define the three financial pools created by this plan and how they were to be measured and distributed to the project participants.

Calculation of the amounts represented by the Cost of Work and Incentive Pools is too complex to describe here. In summary, these pools depended on the quality, safety, and timeliness of work performed. Performance was defined by a list of factors, with points assigned to each. Periodic evaluations were performed by the OAC team throughout the project. The owner required that each participant submit a final report summarizing their performance (following an evaluation), which served as the basis for distributing payments from the two pools.

Table 9-3-6 The owner's goals and how they were implemented.

Owner Goal	Achieved by..
Reduce the time for design and construction	increasing the overlap of these two functions, with earlier involvement of subcontractors for design assistance.
Reduce the need for Office of State-wide Health Planning and Development (OSHPD) reviews	minimizing design changes after the start of construction,
Reduce the cost of design and construction	using lean techniques; and eliminate design errors and field delays through the use of clash detection.
Increase field productivity	increasing off-site prefabrication; and improve collaborative planning with the support of a shared3D model.
Increase project safety	reducing the number of workers in the field; and rely on better task visualizations to anticipate potential safety problems.
Improve collaboration among the project team	including incentive terms in the contract to ensure that the cost savingsachieved would be shared among all participants.

Owner's Incentive Fee Plan

Underlying Principles

1. Maximize the creation of value from the owner's perspective while minimizing waste.
2. Increase relatedness throughout the design and construction process. Approach each problem with an attitude of inquiry by first asking "who might I ask to help solve this problem" and then focusing jointly on exploring, rather than independently coming up with the solution.
3. Increase relatedness among the members of the design and construction team. Strangers cannot be expected to deeply collaborate and achieve higher levels of performance.
4. Pursue coordination of work on the project by recognizing that a project is a network of commitments. Work is to be coordinated through requests, promises, and 'pull' scheduling. The goal of project management and planning is to articulate and activate this network of commitments.
5. Constantly seek to maximize value at the project level, not at the individual or enterprise level, by asking "how can I create coherence between the goals of individual team members and the project as a whole?"
6. Approach each action with a commitment toward continuous improvement; if the team does not 'learn as it goes,' the project will not benefit from these opportunities, for which the project has paid.

This description is taken from the Owner's contract documents with some minor modifications in wording to clarify the text.

Creation of an Incentive Pool

The Incentive Pool has two funding sources: (1) savings in the cost of work; and (2) unexpended contingency, both in the design and construction phases. The anticipated distribution is shown in Figure 9-3-8 and described below.

Cost of the Work Savings Pool

Sutter Health anticipates that major trade contractors – such as Southland Industries, J. W. McClenahan Co., Cupertino Electric, and Northstar Fire Protection – will be performing work under subcontracts to compensate for the Cost of Work plus fixed fee, each with a guaranteed maximum price. Similarly, the self-performed trade work of DPR's will also be performed on a Cost Plus Fixed Fee and guaranteed-max-price (GMP) basis. With full implementation of Lean Project Delivery and the Last Planner System™ to control production, Sutter Health anticipates that reliable workflow and a better understanding of the satisfaction conditions will result in decreased labor and material costs and reduced management expenses. As a result, if the actual Cost of Work – including contractual fee – is less than the GMP for these identified trades, then 25% of the savings will be transferred to the Trade Contractor Incentive Pool (TCIP), 25% to the Owner-Architect-Contractor Incentive Pool(OAC), and the remaining 50% to the Owner Project Contingency (OPC). Distribution to the participating trades – from the TCIP – is to be based on the relative size of each contract, as compared to the overall value of the GMP trade contracts, to be reviewed and approved by the owner.

Contingency Preservation Pools

The project budgets and design and construction contracts consider that the Design Contingency (DC) and Construction Phase Contingency (CPC) are to be earmarked in the project budget and available throughout the construction process to address cost overruns, either resulting from design errors or omissions (DC) or extra work not otherwise entitling CM/GC to a change order (CPC). Preservation of the DC and the CPC should be a by-product of well performed lean project delivery. As a result, to the extent that the agreed upon DC and CPC are preserved, 50% of the savings – together with the 25% savings from the Cost of Work – are to be transferred to the OAC Incentive Pool, with the remainder going to the OPC. The OAC Incentive Pool is to be divided equally: with 50% paid to the Architect (with 40% paid to their sub-consultants); and 50% paid to the General Contractor, (with 40% to their subcontractors). Distributions to sub-consultants and subcontractors are to be reviewed and approved by the owner.

Over time and with experience, responses to this contract agreement proved to be mixed. As a practical matter, the provisions for calculating each of the pools were quite complex and difficult to implement in a timely manner. Because of this complexity, individual participants sometimes had difficulty

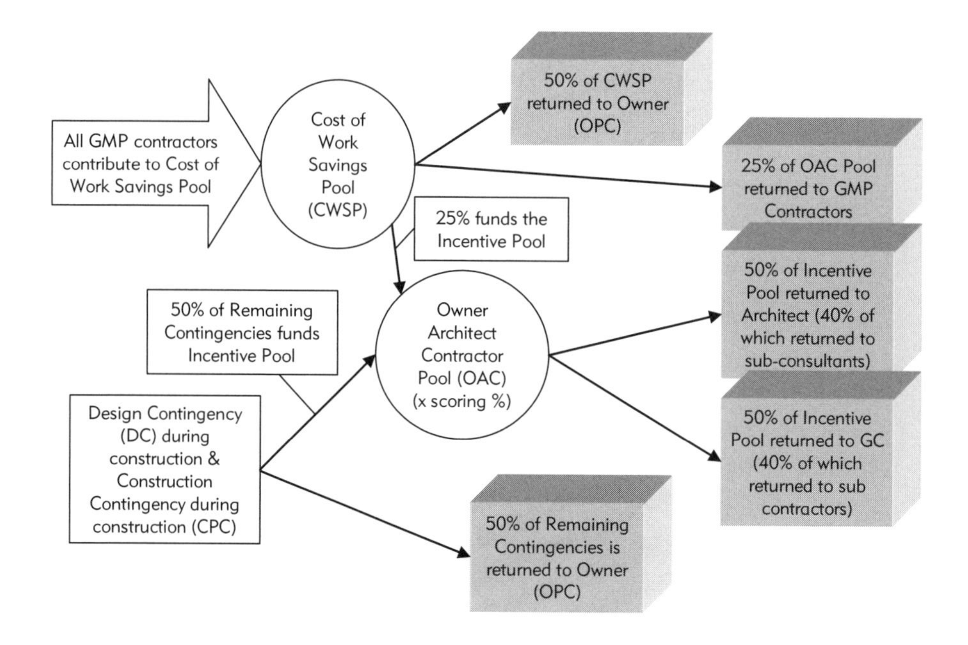

FIGURE 9-3-8 Diagram showing how the Cost of Work savings pool and contingency pools were distributed to the owner and project team.

evaluating how their actions would influence the final results. At the time of this case study, the project was not yet complete and final allocations of these pools were not made. The owner is experimenting with other contract approaches for subsequent projects.

9.3.5 General Contractor (GC)

DPR was the GC and also self-performed some of the work, specifically the concrete and drywall. Their major responsibilities were to pre-qualify and select mechanical and electrical contractors; to coordinate the detailed design and construction process; to obtain necessary permits and inspections; and to ensure that cost, time, and safety goals were achieved or improved upon. The key tools for achieving these results were the use of lean construction methods; early participation of key subcontractors for design review and detailing of systems; extensive offsite fabrication; and use of an accurate 3D model for planning, clash detection, and correction.

Productivity of the crew increased, on average, between 10% and 30% for the following reasons:

- Components fit without the need for field modifications, with no Requests for Information (RFI) and only five field changes for MEP trades with bulk of work.
- Less rework (only 41 out of 25,000 trade work hours).

- Just-in-time delivery meant less cluttered site in the way and fewer hours moving material from one place to another.
- Eliminated the rush to "get in first" to avoid clashes with other preceding trades.
- Conflicts were limited to material storage and handling, not component installations.
- Smaller crews were required to install prefabricated components in the field (approximate 1/3 reduction).
- Less-skilled workers could be productive (both offsite and onsite).
- A safer job-site resulted in:
  - Reduced field hours and smaller crews.
  - Less improvisation and better planning.

Because of these benefits, the HVAC sub contractor (Southland) anticipated a Cost of Work savings of $250K to $340K on a total contract amount of $8.4M, or about 3% to 4% of the total contract value.

A critical requirement for the GC was to coordinate the design process with a deep understanding of the interdependencies among various contractors. To determine the last responsible moment for making a decision, these dependencies had to be understood clearly. For example, the structural fabricator required information, such as weight and location, of mechanical equipment prior to sizing the roof members. Until these commitments were made, the structural members could not be sized. Due to the increase in dependency among the team, work had to be done in smaller batches to facilitate collaboration. This is an important deviation from the traditional process, where each design type is done independently and coordinated only when the work is near complete.

Another important revision in this work process was that more work had to be done by designers and subcontractors earlier than typically happens in the design-bid-build process. Building an accurate 3D model involves all trades and requires early participation by subcontractors, who would normally be involved after the start of construction. On this project, most of the major subcontractors were involved early and contributed to the 3D model, which aided in the elimination of clashes and the coordination of planning sequences, etc. Their involvement also eliminated the usual need for re-drawing layouts during design and then again during layout planning. The designers and subcontractors, however, did not all use 3D design tools. This led to deficiencies in the model and field problems that needed to be addressed. For example, the designer of the parking facility used 2D AutoCAD and – after the design was complete – converted to 3D using Architectural Desktop. Because of the fast track nature of the project, they did not have time (and the owner did not want to pay for the extra cost) of portraying the sloping second-level deck. Instead, flat slabs were shown for the garage and the model was inaccurate. This introduced problems for the piping subcontractor during construction. Again, this illustrates the importance of building an accurate and complete model for design prior to construction.

9.3.6 Architect

Hawley Peterson and Snyder (HPS) served as the Project Architect and used Architectural Desktop from Autodesk for modeling, visualization, and drawing. They worked under a conventional contract until the General Contractor (DPR) and key subcontractors were selected. At this point, the owner implemented a new agreement based on a guaranteed maximum price (GMP) together with the Incentive Fee Plan excerpted above. HPS was responsible for producing a 3D object-based model containing the building's architectural components. The structural engineer (KPFF) was a subcontractor of HPS and used Revit Structure and ETABS 3D to model and design the structural components. Other participants, including engineering consultants and subcontractors, were responsible for their respective building systems, such as electrical, fresh and waste water piping, fire safety piping and sprinklers, HVAC systems, etc. Weekly meetings were held at the job-site to review progress and analyze and correct perceived hard clashes using the 3D model. As the project neared the construction phase, detailing of piping and other systems began with the work being broken-up into packages that were coordinated with the construction and installation schedule for the field.

The level of detail needed for the model's definition was determined early in the project. The following items were ruled out of the model's scope:

- Connections between structural members
- Furniture
- Wall mounted equipment, fixtures, etc.
- Trim, handrails, other misc. architectural elements

After some experience using the model, however, additional objects were added as needed, such as footings, to coordinate the subgrade utilities. As construction progressed, it became clear that some objects lacked necessary details – such as structural connections, handrails, and rainwater piping – and resulted in unanticipated clashes in the field. These required extra time and cost to correct. In fact, lack of detail in the model was the only source of field errors, making this a remarkable accomplishment for a project in which some participants had no prior experience using a collaborative 3D model.

9.3.7 Electrical Subcontractor

Cupertino Electric had some prior experience with 3D design detailing and used Pipe Designer 3D for the layout of rigid conduit runs and FabPro for the design of prefabricated assemblies. They fed their 3D models into JetStream software to perform clash detection. They were responsible for modeling feeders, the telecommunication cable tray and conduit, homerun conduits, branch power and lighting system j-hangers, distribution equipment, and light fixtures. Along with other subcontractors, they found that for clash detection, 3D modeling was far more accurate than overlaying 2D drawings on a light table. It also provided a much improved basis for collaboration. The extra cost of producing 2D drawings from the 3D model – for use in the field – was estimated at $50,000. This expense was more than compensated for by the more productive use of field crew time, as a result of 3D task visualization and a reduction in clashes. The 3D model's improved accuracy also allowed for the effective use of prefabricated wiring assemblies and contributed to a reduction in waste materials. Smaller crews in the field led to a better safety record for all of the subcontractors that used prefabricated assemblies. The only problems encountered stemmed from starting the 3D detailing before the design was sufficiently complete, which caused the need for additional effort when changes were requested by the client, such as changing the location of equipment or using equipment with different electrical requirements.

9.3.8 Lessons Learned

Lessons learned by the design-construction team:

- Begin planning with the 3D model at 50% design development; start detailing only after 95% design development (to avoid rework).
- Make sure *all* trades use a 3D model for design, so that the merged model is as accurate as possible; designing in 2D and then building a 3D model is a poor substitute for designing directly in 3D.
- Prepare and make a weekly design, coordination, and construction schedule that links activities together based on install dates for the week ahead; make sure "responsible individuals" from each trade attend every

week, and that they are able to make commitments for their respective groups.

- Do not change coordination sequencing during the course of work, i.e., make commitments and stick to them.

9.3.9 Conclusion

This was a very successful project that started with a clear vision from the owner and excellent support from most of the project team. A genuine collaboration effort made it possible to overcome any lack of experience with 3D models and lean production planning. Not every participant had the skills, resources, and experience to fully participate in the project's goals. As in most projects, it took some time for the team to learn how to collaborate effectively. There were some delays caused by inadequate detail and errors in the 3D model. The architect was experienced in using 3D modeling, but this was not the case for some of the other designers and engineers. Converting 2D models to 3D – only after the design was complete – was the cause of most problems. There were owner-caused design changes that led to significant amounts of additional detailing effort, particularly for piping and electrical work. There were also 'normal' problems, caused by OSHPD and city permit delays. Despite all of these issues, the project was forecast to complete six months sooner than had originally been anticipated and under budget. Labor productivity was 15% to 30% above the industry standard. Crew sizes were smaller. Prefabricated assemblies were more extensive and more accurate than could have been achieved without the use of an accurate 3D model. Almost all project participants felt that the overall collaboration effort and lean work processes were successful. The owner plans to continue to develop this process on future projects.

Acknowledgments for El Camino Sutter Health Case Study

The author is indebted to Dean Reed of DPR for his assistance in providing access to the project team for this case study. The access included personal interviews, presentation materials (some of which are used in this case study), and field inspections.

9.4 BEIJING NATIONAL AQUATICS CENTER

A unique building made possible by BIM for conceptual design and structural engineering.

9.4.1 Introduction

The Beijing National Swimming Centre is being built for the 2008 Olympic Games. Often referred to as the 'Water Cube' (see Figure 9-4-1), the design was selected in an international design competition. The winning entry was submitted by the China State Construction and Engineering Corporation (CSCEC) together with Arup Consulting Engineers and PTW Architects. This case study addresses issues relevant to building information modeling (BIM), including new BIM innovations, the advantages and disadvantages encountered by the team, how the team operated, and the lessons learned.

The Water Cube design concept is all about water. Water is deeply expressed in the building in its bubbly state. A remarkable duality is experienced in the building, where the detailing and micro-detailing of the building's components achieve simplicity of form and monolithic monumentality. Water is also an important symbol for the people of Beijing, because it is a valuable resource and a treasure for such an inland city, and it is a sign of luxury in a person's life. The building represents a paradise in the heart of the city that brings happiness and joy and all kinds of fantasies associated with the presence of water. The project associates water structurally and conceptually as a recurring theme together with the square, the primal shape of the house in Chinese tradition

FIGURE 9-4-1
A rendering of the Water Cube pavilion.
Image courtesy of Arup, PTW and CSCEC.

and mythology. The project expresses the personal experience of water-based leisure and sports and attempts to provide an appropriate airy and misty atmosphere in an intentional setting that merges molecular science, architecture, and phenomenology.

The building is located on the Olympic Green in Beijing, China, the home of the 2008 Olympic Games. It is situated across from the main stadium (designed by Herzog & De Meuron), creating a duality between water in the pavilion and fire in the main stadium (Figure 9-4-2).

The building encompasses a gross floor area of 90,000 square meters (177 × 177 meters wide and 30 meters high). It contains five pools, including one—with a wave machine and rides—that is six times the size of an Olympic pool. There is also an organically shaped restaurant carved out of the bubble structure. The project holds seating for 17,000 spectators (6,000 permanent and 11,000 temporary). Events featured during the Olympic Games will be swimming, diving, water polo, and synchronized swimming. After the games, the pavilion will serve as one of Beijing's premier recreation centers.

To create the bubbly appearance, the building's structure is covered in 100,000 square meters of a modified copolymer called ethylene-tetra-fluoroethylene (ETFE), a tough recyclable material weighing just 1% of an equivalent sized glass panel, turning the building into an insulated greenhouse. Ninety percent of the solar energy that permeates the building gets trapped within the structural zone and is used to heat the pools and interior areas.

FIGURE 9-4-2 The Water Cube pavilion is located opposite the main stadium on the Olympic Green in Beijing, China. *Image courtesy of Arup, PTW and CSCEC.*

Table 9-4-1 The project participants.

Owner and operator:	People's Government of Beijing Municipality, Beijing State-owned Assets Management Co. Ltd.
Team leader:	China State Construction Engineering Corporation (CSCEC) (www.cscec.com.cn/english)
Design partners:	Arup Consulting Engineers (www.arup.com), PTW Architects (www.ptw.com.au), and China State Construction International Design
Engineers:	Arup Consulting Engineers and China State Construction International Design
Contractor:	CSCEC
Schematic design:	Arup and CSCEC Shenzhen Design Institute

Sustainability issues were a special concern in this project. The aim was to deliver a range of sustainable features while focusing on the building's entire lifecycle and identifying its scope for continual improvement. Water conservation and the design of efficient water conduit systems were especially important, given the high utilization levels, presence of pollutants, evapo-transpiration, and other unreliable climatic factors. Arup, through its SPeAR sustainability tool, proposed the reuse and recycling of 80% of the water harvested from roof catchments, pool backwash systems, and overland flows by incorporating water sensitive urban design principles. Reliance on the district's local water and sewage systems was thus reduced.

Arup is a global firm of designers, engineers, planners and business consultants providing a diverse range of professional services to clients around the world. The firm has over 7,000 staff in 75 offices in more than 33 countries.

Sydney-based firm PTW Architects (formerly known as Peddle Thorp and Walker) was established in 1889. PTW Architects currently employs over 150 people. It maintains offices in Sydney as well as Beijing, Hanoi, and Shanghai. The firm works with staff around the world on various projects across the disciplines of architecture, interior design, master planning, urban design and planning, and computer aided design, in addition to supporting staff administration, accountancy, IT, graphic design, marketing, and communications.

Established in 1982, CSCEC (China State Construction Engineering Corporation) is considered the largest construction enterprise and international contractor in China. It has been listed as one of the world's top 225 international contractors by ENR since 1984, ranking 16th in 2002. By the end of 2002, CSCEC had a total workforce of 122,500 people.

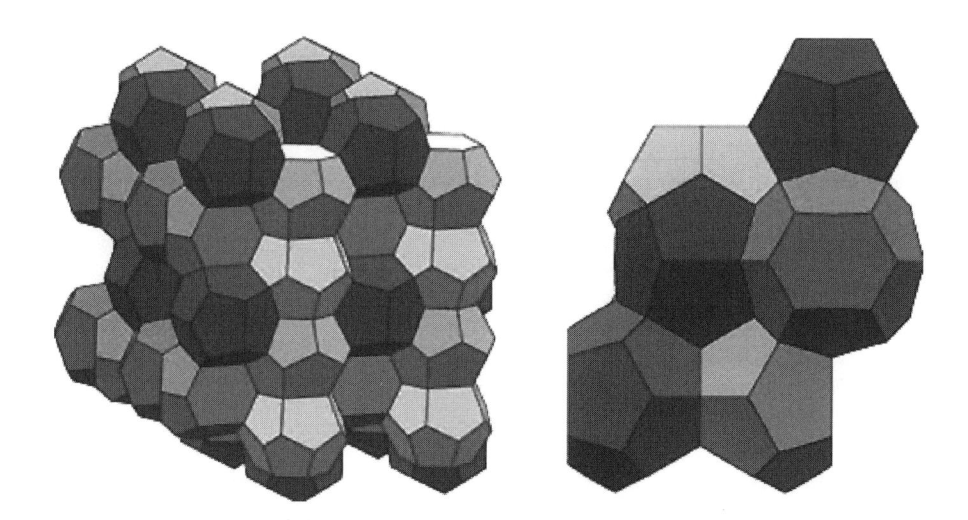

FIGURE 9-4-3 Weaire and Phelan's proposal for partitioning 3D space.
A) A cluster of repetitive units and B) the repetitive module of two dodecahedra and six 14-sided cells. *Images provided courtesy of Prof. Ken Brakke, Susquehanna University.*

9.4.2 Structural Concept

The Water Cube's structure is based on a unique lightweight construction technique developed by PTW and CSCEC with Arup and derived from the structure of water bubbles in the state of aggregation found in foam. Behind the seemingly random appearance of the facade is a strict geometry typically found in natural systems like crystals, cells, and other molecular structures. This arrangement, formulated by Weaire and Phelan (Weaire and Phelan 1994) and shown in Figure 9-4-3, is the most efficient subdivision of 3D space with equally sized cells. The transparency and apparent randomness is equally transposed into the inner and outer building-skins. A total number of 3,000 pneumatic ETFE cushions restrained in aluminum extrusions are supported by a lightweight steel structure. The cushions are inflated with low-pressure air to provide insulation and resist wind loads.

Unlike traditional stadium structures with gigantic columns and beams, cables and spans—to which the facade system is applied—the architectural space, structure and facade of the building are one and the same element. The wall cavity is 3.6 meters deep, and the cavity that forms the roof is 7.2 meters deep. The entire building contains more than 22,000 steel beam members and 12,000 nodes, with approximately 6,500 tons of steel. This presented a challenge of optimization, where every beam was supposed to be as small and as light as possible. As the roof usually consumes most of its strength in holding itself up, the need for optimization was extremely important, as self-weight is critical in long span roof structures.

9.4.3 BIM in Design and Construction

The project was designed in three distinct stages: preparation for the design competition including preparation of a physical model using rapid prototyping, design development, and preparation of tender documents. Structural analysis, design, and optimization continued throughout these process stages.

The pertinent BIM-related issues in this project were conceptual design, structural optimization, rapid prototyping, interoperability, and drawing production. The complexity and originality of the building meant that the design and the models representing it were constantly evolving and changing. By preparing and applying a variety of sophisticated add-on tools to standard BIM systems, the project team was able to achieve better optimization (structurally and functionally), better analysis schemes, and to meet and sometimes beat short project schedules. The following paragraphs discuss some of the BIM-related design innovations in the Water Cube project.

Competition Submission

Geometry was still being defined at the competition stage, so the focus was on writing scripts (using Microstation VBA) that would *skin* a wire-frame to provide a representative 3D solid model and help evaluate a potential wireframe model for other purposes. The 3D model created by the scripts consisted of members and node spheres of the same size, while some simple rules defined how to handle members of various lengths.

Full model exports to other modeling packages were only done during the competition stages for the purposes of visualization and rapid prototyping for the formal presentation. Arup developed a virtual reality walkthrough in a package called Arup Realtime, which uses game engine technology. This package allows any 3D model or data from other sources, such as 2D drawings, sketches, and photographs, to be imported through a script into the virtual environment. The software allows modes of navigation via a keyboard, mouse, or joystick and a level of interaction that allows for customization by the anticipated end user. This was a very time consuming process, involving exports via IGES format into Rhino and then into 3D Studio Max for generating AVI files, and it resulted in file sizes in excess of 1.2 GB. The data transfers are summarized in Figure 9-4-4.

Rapid prototyping was used to physically model the building's complex structure with the goal of conveying the idea to the design competition's jury, as was stated in the competition rules. It also provided a way for Arup to better understand the underlying concepts and elements. An STL file* was created from the 3D MicroStation model, and a physical model was built using that file through a stereo lithography (SLA) process, in which liquid epoxy resin is

*STL (Standard Tessellation Language) is a file format defined for the stereo lithography CAD software created by 3D Systems of Valencia, CA, USA. STL files describe only the surface geometry of a 3D object.

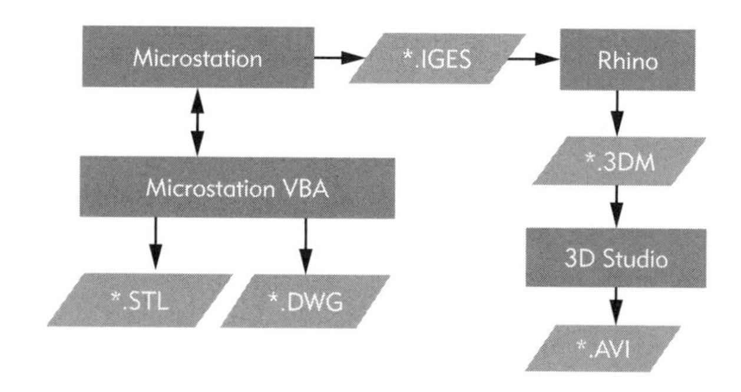

FIGURE 9-4-4
Data transfers at the competition stage.

solidified by a computer-controlled UV laser beam according to the instructions contained in the STL file. A robust semi-transparent 3D plastic model was produced; it is shown in Figure 9-4-5.

Before Arup made progress with the design and VBA scripting, the Arup geometry was exported as a wire-frame model through DXF into the structural analysis package Strand 7.0, but only after CSCEC confirmed that this geometry worked mathematically with a script using the Weaire and Phelan solution (Figure 9-4-3).

Design Development

In the design development stage, Arup used Strand 7.0 finite element analysis (FEA) software for structural analysis and optimization, going back and forth from the Microstation model. They developed VBA scripts to export the analyzed and optimized model to AutoCAD drawing files (DWG format), Microstation TriForma drawings (DGN), and Microsoft Excel spreadsheets (XLS) for data extraction.

FIGURE 9-4-5
Rapid prototype model of the Water Cube pavilion.
Image courtesy of Arup, PTW and CSCEC.

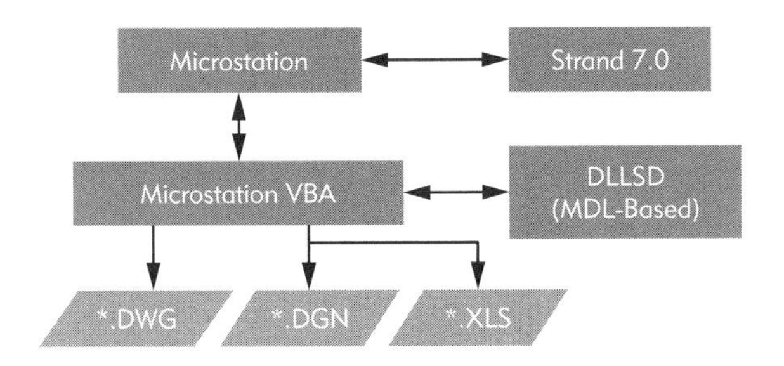

FIGURE 9-4-6
Data exchange during design development.

The analysis data exported to Microstation included end points, section types, and coordinates and dimensions for section types (Figure 9-4-6). The scripts imported data files and then recreated wireframe models for integrity checks and final production purposes, with annotations of element number and section type. VBA scripts, called precompiled DLLSD (Microstation's MDL-based 3D modeling routines), were used to create primitive 3D objects. All members were constructed in an element-aligned local coordinate system before being translated back into the global coordinate system.

Tender Documentation

The application of true structural elements was achieved by selecting specific section elements from the 3D wire-frame model and importing them from the analysis model. Part of the VBA script applied sectional information, beam reference, and property numbers to the wire-frame as invisible attributes, thus allowing search within Microstation for the required structural sections.

Following the information in the beam schedule, which had 37 different section properties, each section type was found via the *Select by Attribute* function, which displayed all stick elements within the model that required a specific section-type attribute. Each stick was skinned with a Triforma structural element, giving a fully defined and structurally correct model.

Documentation and extraction of the relevant information was made easier using the Bentley Structural software package (a sample drawing is shown in Figure 9-4-7). Huge libraries containing sections of steel members and their sizes were developed by the software. In addition, non-standard sections—in the case of fabricated rectangular, circular, and square hollow structural sections (RHS, CHS and SHS sections)—were added to text files and used within the structural model. Data extraction made these sections available in material list reports, providing total length, weight, grades, and quantities for each section. The section files were set up as empty files containing every vertical, horizontal, and detail section location defined within each file. In this case, 112 sections were needed to appropriately document the entire project. Each section had a unique ID.

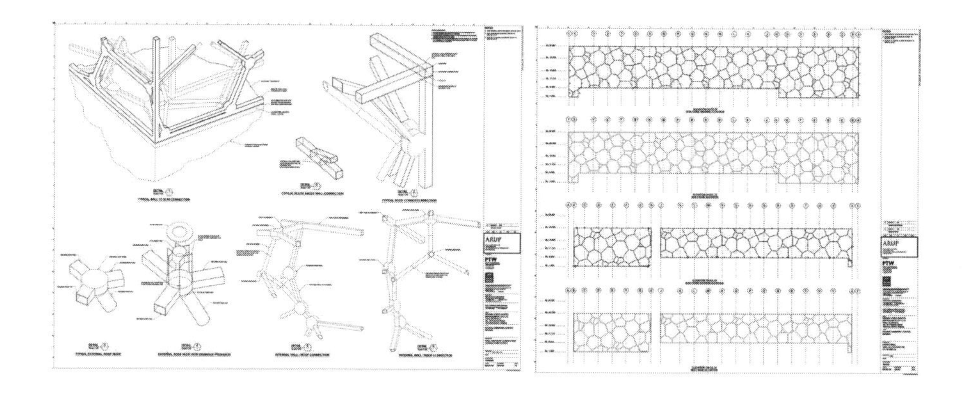

FIGURE 9-4-7 A tender drawing produced from the 3D model.
Image courtesy of Arup, PTW and CSCEC.

One of the many benefits of the Bentley Structural software package was reduced risk of human error, due to the extensive automation encountered in drawing production. Another important aspect was the ability to rebuild the model several times from scratch as the model and the structure were modified. This quick and accurate method led to the generation of a completely new 3D model from the structural analysis model. Eventually, extraction of all documentation drawings, plans, sections, elevations, and details was easier to manage and update, as all 65 drawings were updated over one weekend. All the drawing sheets were set up in advance, each with a relevant pre-loaded reference file. As soon as a section file was created, it automatically referenced into a drawing sheet, thus creating the documents.

Microstation Structural Triforma has a function called *Family and Parts,* meaning there are several parts which include beams, columns, and purlins under the title *Steel.* This allowed for the *unification* of structural objects in the data extraction file and enabled all CHS, RHS, and SHS members to treat nodes as bounding elements, giving a true and accurate detail of the connections.

Fabrication and Construction

Arup proposed prefabrication to ease construction and limit expensive onsite welding. This idea, however, was rejected by the client in China, who preferred to use a large labor workforce and onsite welding. Figure 9-4-8 and Figure 9-4-9 show the Water Cube pavilion under construction. Approximately 12,000 spherical nodes and 22,000 tube and box sections were individually fixed onsite. There were approximately 3,000 workers onsite including more than 100 welders.

Tube and box section preparation was done manually at a fabrication plant from the cut length spreadsheets that Arup provided in its tender documents. This process is unique to China. In other countries, Arup's technique would be to hand over the information in a format for continued use down the supply

FIGURE 9-4-8 Construction of the Water Cube pavilion. *Image courtesy of Ben McMillan and Arup.*

FIGURE 9-4-9 Interior view of the Water Cube pavilion showing the steel structure nearing completion. *Image courtesy Ben McMillan and Arup.*

chain, such as a Tekla Structures model in Steel Detailing Neutral Format (SDNF), to cut down fabrication time and enable the use of CNC machinery.

A very large number of part drawings were produced in this non-standard steel job because of the variety of complex connections. The number of part drawings Arup produced exceeded 15,000 drawing views (see Figure 9-4-7). These were required for fabrication on the shop floor, including all small items such as plates and stiffeners, which needed to be attached manually.

Project Management & Scheduling

Recognizing the short timelines and the complexity of their role, Arup relied on internal specialist project managers to act as the focal points for all project communication. This enabled the client to build a close working relationship with its project managers and removed any ambiguity about information flows or responsibility. The fast track design was delivered from competition stage through to a full scheme design package in 10 weeks. The whole process, from model to documentation, required a seven month design and documentation period, which was split into three modeling blocks: competition, preliminary design, and tender documentation. Structural design and optimization continued through this seven month period.

The team mobilized and coordinated more than 80 Arup engineers and specialists spread across 12 disciplines in four countries. Although elements of design were being shared between Arup's Sydney, Beijing, Hong Kong, and London offices, coordination was centered in Sydney.

Arup's involvement on other high profile projects in Beijing, including the Stadium, CCTV Tower, the international airport and interchange, also gave the team an advantage. Two years prior to commencement of this project, Arup completed the Shenzhen Aquatic Centre. The team was therefore familiar with certain elements of the Chinese review process, such as the types of questions they would be asked and who would review the work. This experience assisted greatly in a smooth transition.

Structural Optimization

The greatest challenges in the Water Cube project were associated with structural design and fabrication. The first was to optimize the overall weight of the steel structure. The second was to fulfill the seismic design requirements stipulated by the city of Beijing. Whether to use compact sections for roof design that would behave plastically under seismic loads or stiffened slender sections that would behave elastically under such loads was a key issue.

To overcome these and other challenges, Arup developed and implemented a structural optimization program to perform all structural analysis, design, and design optimization. Modeling of the complex structure was the fundamental aspect, thus allowing continuous on-demand refinements and adjustments. The process of nonstop alteration of designs to reach maximum optimization in addition to correctly documenting the building with sophisticated components was a significant factor in developing the optimization program.

Optimization of the building involved determining member sizes for 22,000 beams to meet the requirements of 13 Chinese steel code strength equations at five points on each beam for 190 load combinations. Thus, there are 22,000 design variables and $22{,}000 \times 13 \times 5 \times 190 = 271.7$ million design constraints.

Gradient-based optimization methods were not possible due to the large number of constraints. Software was written from scratch to determine minimum sets of member sizes for 22,000 beams. A damped constraint satisfaction method was adopted to find three sets of well-graded discrete member cross section choices. Methods used to assess each set included simulated annealing, plastic versus elastic stiffener design, and engineering judgment.

The optimization code was written in Visual Basic 6.0. The VB software controlled the entire optimization process. Structural analysis was performed using Strand7 Finite Element Software, and its API was used to interface with the VB code.

9.4.4 Lessons Learned

The Water Cube pavilion is a unique project. The ongoing processes of information exchange, interoperability, optimization, and building information modeling added great value to the project. The positive impacts of BIM were enhanced modeling time and speed resulting in shorter schedules and a very short design response cycle time. It proved possible to rebuild the 3D fabrication model based on weekly structural analyses and sometimes even daily, updating all related information in a seamless process as the design was optimized. Issues of sustainability, building performance, fire protection, and safety were effectively resolved. Up-to-date tender documentation was easily extracted from the full design model; and reductions in human error were a significant plus.

It was only in the early competition stage that full model exports to other modeling packages were executed, for the purposes of animation and presentation. It was then realized that this process, which involved exports via IGES format into Rhino and then Studio Max, was very time consuming; and it created large files that were difficult to handle. In the post-competition stage, where construction documentation was required, Arup had to decide how to coordinate the analytical and fabrication model information sets. This was done by linking the analytical software to the CAD software via a script. This was significant, as within this structure any member change would ripple throughout the structure affecting all elements around the base. Due to the sheer number of potential changes during the design stage, the script exported an entirely new model every day for re-documentation purposes.

Arup developed the conversion program to run from inside its CAD package to ensure that all elements and procedures operated in the structure optimization and analysis program were properly imported and translated into the detailed design model. This conversion program had numerous advantages, the most important of which were:

- **Full design model.** All plans, sections, elevations, details, and final documentation drawings were easily extracted from the fully coordinated 3D model, which contained all necessary information and design elements.
- **Reduced modeling time with greater speed.** The entire steel structure and building was modeled in only 25 minutes. If modeled manually, it would have taken weeks or months.
- **Improved visualization.** With the physical model in hand, 3D wireframe models were developed and saved to 3D AutoCAD, allowing the client and contractors to easily visualize the project's details with corresponding embedded information.

The main disadvantage in this project was the lack of complete and integrated coordination between the design development and the construction phases. The complex computerized design, analysis, and optimization processes were followed by a manually coordinated fabrication and construction process, which is unexpected for a project of such complexity. The use of labor-intensive construction methods and onsite welding prevented a streamlined fabrication approach, in which automated CNC machinery would have been used.

Better use could have been made of the model-based coordination between the architect and engineering firm. A key lesson learned was the importance of architectural priorities in projects of this sort, as bulky cross sections, thick walls and roofs, and focusing on building performance issues alone can all adversely affect the aesthetic, functional, and spatial qualities of a project.

Previous collaboration between all team members and their expertise and experience with Olympic projects helped the team succeed. Arup's wide ranging involvement in Beijing projects, including earlier ones related to the 2008 Olympic Games such as the National Stadium, Convention Center, and airport terminals, created a clear understanding and familiarity of the workflow and processes for dealing with authorities, design codes, and relevant matters. Many of these projects were in collaboration with CSCEC, thus creating a mutual understanding between the two firms. PTW is highly regarded for its major civic projects and large-scale sports facilities, including involvement in the International Athletics Centre and the Aquatic Centre for the 2000 Sydney Games. Its strong and existing relationships with the Chinese partners played a pivotal and significant role in the development process of the Water Cube. Arup and PTW had also collaborated previously on a number of projects including other venues for the 2000 Sydney Olympic Games, which established an easier mutual partnership. These factors eased collaboration during the project's design and execution and had a positive impact on the design duration.

One of Arup's important observations about the project was related to the implemented software. They pointed out the importance of Bentley's software. Arup Senior 3D Modeler Stuart Bull explained: "Bentley Structural's capabilities, such as automatic drawing extraction, dramatically reduced the time needed to produce 2D documentation. Since we didn't have to worry about that part of our workload, we could focus on the 3D model. Using a MicroStation VBA routine to automatically model the structure saved us quite a bit of time. Also, being able to save files in other formats let us quickly issue drawings to clients and consultants in the formats they needed. Yet, we didn't have to give up the enhanced capabilities that Bentley solutions offered us. That was quite important. If we had been using any other software package, it's unlikely that we could have produced such complicated geometry and documentation and integrated it with structural analysis, especially in the time frame available."

Another observation made by Bull concerned Arup's software innovation and its developed scripts, which facilitated data exchange between the integrated model and analysis packages: "The ability to use the VBA scripts to create our geometry, which gave us the link from the engineering and analysis model to our working 3D CAD model, was very important. In terms of a consultant's model, we provided everything we could downstream to our client, detailer/fabricator, and facade contractor. It could have been more BIM-enabled if the architect had contributed to the model and we passed the model back and forth; but in terms of what a BIM model should be, we probably weren't that close. It must be remembered that this project was done nearly three years ago and still is a cutting edge 3D analysis/CAD project that most newer programs like Revit would still not be able to cope with."

It was clear, however, that coordination of the model's information exchange between Arup and PTW was lacking. More effort was needed to achieve a fully-integrated building model at the early stages of information exchange. More importantly, at the later stages of fabrication and construction, most of the fabrication information produced by the complex model was not fully utilized. Manual execution of construction processes, onsite welding, and the preference for labor-intensive fabrication over the use of automatic CNC machinery all reduced the value of the exported model information. The Excel spreadsheet information was used to manually fabricate cross sections and operate them, instead of using the 3D information for direct operation of the fabricated products.

9.4.5 Conclusion

The use of BIM in this project related to structural optimization, information exchange software, and interoperability. Innovations included rapid prototyping, visualization, and treatment of building performance issues, such as sustainability and fire protection. The factor that added most value, however, was the constant modeling of the evolving and changing complex structure achieved through structural optimization. Many useful outputs were achieved, such as optimized structural members with low steel tonnage and high strength, easy data extraction from the full design model, enhanced modeling time and speed, automatic referencing, rapid preparation of updated tender documentation, easy model rebuilding, and reduced human error in drawing production. All of these helped reduce ambiguity, leading to optimized design and construction results. This type of project would not have been possible without sophisticated computer tools for structural analysis and fabrication detailing.

Acknowledgments for Beijing Aquatics Center Case Study

This case study was originally prepared by Sherif Morad Abdelmohsen for Professor Chuck Eastman in a course offered in the College of Architecture Ph.D. Program at the Georgia Institute of Technology, Winter 2006. It has been adapted for use in this book. The authors wish to acknowledge and thank the members of the Arup, PTW, and CSCEC teams, whose contributions were essential to this case study. The insight offered by Stuart Bull of Arup was particularly helpful.

9.5 SAN FRANCISCO FEDERAL BUILDING

A study in energy efficient design with support from BIM

9.5.1 Introduction

The San Francisco Federal Building (SFFB) is located near the city's Civic Center on a 3-acre site across the street from the Ninth Circuit Court of Appeals Federal Courthouse Building, built in 1905. Architectural firm Morphosis served as the design architect for the SFFB. The building reflects the U.S. General Services Administration's (GSA) commitment to design excellence and sustainable architecture. The design goals established by the GSA included a commitment to sustainable living and a reduction in the consumption of resources, by minimizing waste materials and creating healthy and productive working conditions. The building enhances the surrounding area by introducing new amenities and public spaces.

Figure 9-5-1 is a digital simulation of the building in its San Francisco setting. This study describes the salient aspects of the building's design and engineering and the software and procedures used during design and construction. It is an early example of Building Information Modeling. It was designed in the 2001-2004 timeframe and built shortly after. The character of the building's mechanical aesthetic can be seen in Figure 9-5-2.

9.5.2 Description of Building Program

The Federal Building consists of an 18-story (605,000 SF) office tower, a four-story annex building (118,511 SF) projecting from the tower, and a 47-space parking structure (19,244 SF) at the lowest level. In addition, there is a cafeteria (3,196 SF).

FIGURE 9-5-1 Simulated picture of SF Federal Building.
Image provided courtesy of Morphosis.

FIGURE 9-5-2 Facade of the SF Federal Building, as viewed from the childcare play yard. *Image provided courtesy of Morphosis.*

9.5.3 Participants

The building complex is divided into several components, with the main component being an 18-story tower, placed along the site's Northwest boundary that includes offices for the US. Department of Health and Human Services and the Department of Agriculture, as well as conference rooms, fitness facilities, and a child care center. All will be open for use by the general public (see Figure 9-5-3).

Perpendicular to the tower is a four-story annex with spaces for the Department of Labor and the Social Security Administration, with certain areas and services open to the public. A plaza space will be built for hosting concerts or an informal market that, in addition, features a cafeteria. This tower houses 1,700 workers.

9.5.4 Fresh Air and Natural Ventilation

"With only five percent of the world's population, the United States consumes 25 percent of the world's energy. The ongoing challenges of resource management

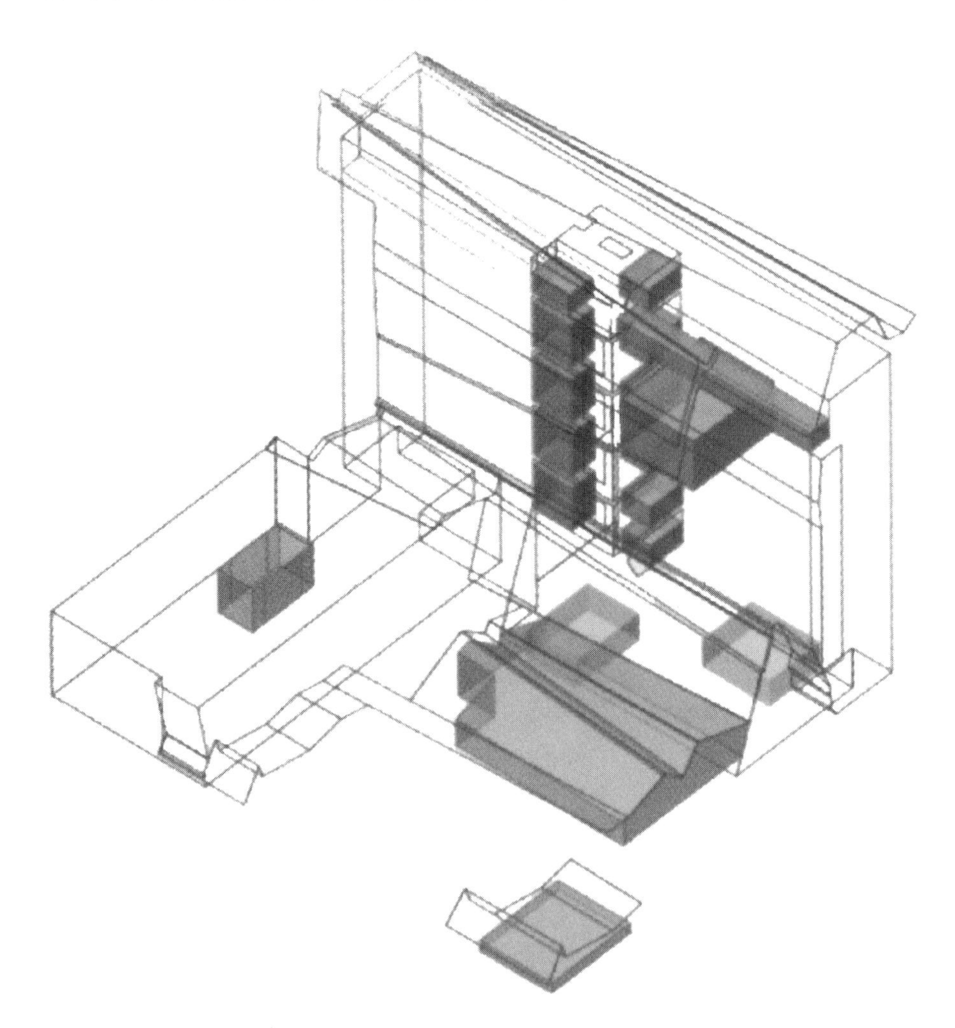

FIGURE 9-5-3 Diagram showing public spaces included in the design (volumes identified in gray). *Image provided courtesy of Morphosis.*

demand that architects confront the ethics of ecology and the politics of energy. The design for the San Francisco Federal Building recognizes these challenges and poses a solution."—Thom Mayne, Principal, Morphosis.

The design team envisioned a solution capable of substantially reducing resource consumption primarily through the use of natural ventilation. Such a proposal was possible due to the site's prevailing Northwest winds and average temperature range of 44° to 78° Fahrenheit (7° to 18° Celsius). Morphosis designed a narrow but tall structure capable of admitting plenty of natural light. They also proposed a unique cooling system based on the interaction of three design features: operable windows, thermal mass, and cross ventilation.

The cooling system is managed by an automated system that senses the pressure of the prevailing winds and reacts by opening and closing windows to permit or restrict airflow into the building. Users maintain some level of control

Table 9-5-1 List of participants in the design and construction of SF Federal Building.

Design Functions	Participant
Project	San Francisco Federal Building
Client	General Services Administration
Architect	Morphosis, Santa Monica
Executive Architect	Smith Group, San Francisco
Structural, Mechanical, Electrical, and Plumbing Engineers	Ove Arup and Partners
Natural Ventilation Modeling	NaturalWorks, Lawrence Berkeley National Laboratory
Landscape Architect	Richard Haag Associates inc. with J.J.R
Civil Engineer	Brian Kangas Folk Company
Geotechnical	Geomatrix
Lighting Consultant	Horton Lees Brogden Lighting Design, inc
Curtain Wall	Curtain Wall Design & Consulting, inc.
Signage	Kate Keating Associates
Collaborative Artists	James Turrell, Ed Ruscha, Rupert Garcia, Hung Liu, Raymond Sanders, William Wiley
Blast Consultant	Hinman Consulting Engineers
Building Code Consultant	Rolf Jensen & Associates
Acoustics	Thornburn Associates
Vertical Transportation	Hesselberg, Keesee & Associates, Inc.
Cost Estimator	Davis Langdon Adamson
Construction Functions	**Participant**
General Contractor	Dick Corporation-Morganti Joint Venture
Construction Manager	Hunt Construction
Steel detailers	B.D.S. Steel Detailers
Miscellaneous metals	T. & M. Manufacturing
Cladding	Permasteelisa Cladding Technologies

by manually adjusting the operable windows. The automated system evaluates building performance by taking into account users' interactions, automated window settings, and the thermal mass of the concrete columns and slabs.

The building's Southeast facade – the side that faces the afternoon sun and absorbs more heat – has a perforated metal sun-screen to mitigate heat gain. The building's Northwest facade has a series of translucent sunshades to inhibit the glare caused by the curtain wall's glazed surfaces. These energy control concepts were developed collaboratively by Morphosis and Arup.

Given the complexity of the building's design, it was important to develop an accurate digital building model that could be used for simulation analyses. This required the input of environmental data, including wind pressure on both sides of the building and outdoor and indoor air temperatures.

For both the computational fluid dynamics analysis and the definition of the Building Automated System (BAS), each floor of the building was divided into two symmetrical sides, with each area divided into five slices. The dark gray square in the center of Figure 9-5-4 is an elevator/service core that partitions the building.

Figure 9-5-5 shows a side-view of the velocity field (m/s) in the NW and SW wings, with all operable windows open. In the upper figure, air is able to flow above the meeting rooms. In the lower figure, the obstructive effects of the service core causes air to re-circulate around the building's core.

Designs such as this one benefit greatly from the use of computer-simulated engineering analyses. To validate the effects of the proposed design, GSA asked

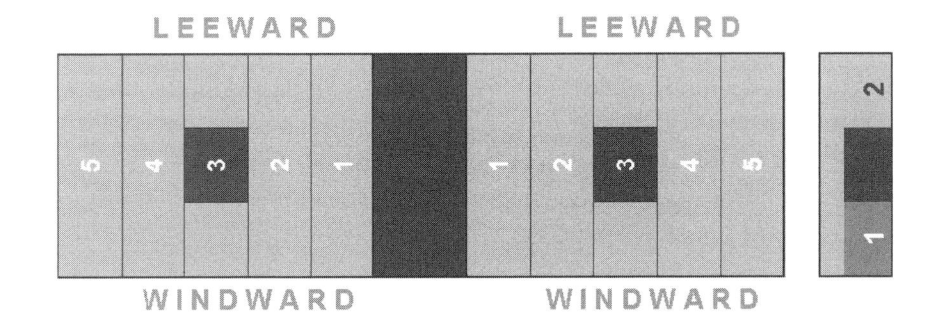

FIGURE 9-5-4 Plan view showing how each floor was modeled for thermal analysis (Carrilho da Graca et al. 2004).

FIGURE 9-5-5 Output from the thermal simulation model: an elevation view of air flow for the building's two symmetrical sides, with the NW facade shown on top and the SW below (see color insert for full color figure).

Image provided courtesy of NaturalWorks.

NaturalWorks (http://www.natural-works.com/naturalworks/hvacdesign.php) to serve as advisors to the project and carry out substantiating simulations. NaturalWorks undertook the analysis of the cross ventilation system. This required definition of the window geometry, opening areas, and specification of the Building Automated System (BAS) control rules. The model predicted air flow patterns, maximum air flow velocities, air exchange rates, and temperature profiles for different environmental conditions. The level of detail provided by these digital simulations assessed and verified the proposed design. This level of expert analysis provided the design team and client with a high level of confidence in the system's performance before it was built. This building was intended as a prototype for showing how – by harnessing natural ventilation – buildings could be designed with better working environments and energy savings.

Controls for the operable windows represented a crucial aspect of the design. EnergyPlus was coupled with a routine to calculate the effects of different control strategies and to design the modes for the building's systems. A test run of EnergyPlus is shown in Figure 9-5-6.

Cross ventilation (CV) is a common feature of many naturally ventilated buildings, with air flowing through windows, open doorways, and large internal apertures and across the building's rooms and corridors. Displacement ventilation (DV) is characterized by the supply of cool air at low elevation levels and the extraction of warm air at high elevation levels.

Simplified airflow models were developed using a combination of scaled model experiments, computational fluid dynamics (CFD), and scaling analysis. This combined approach provided insight into the mechanisms and system

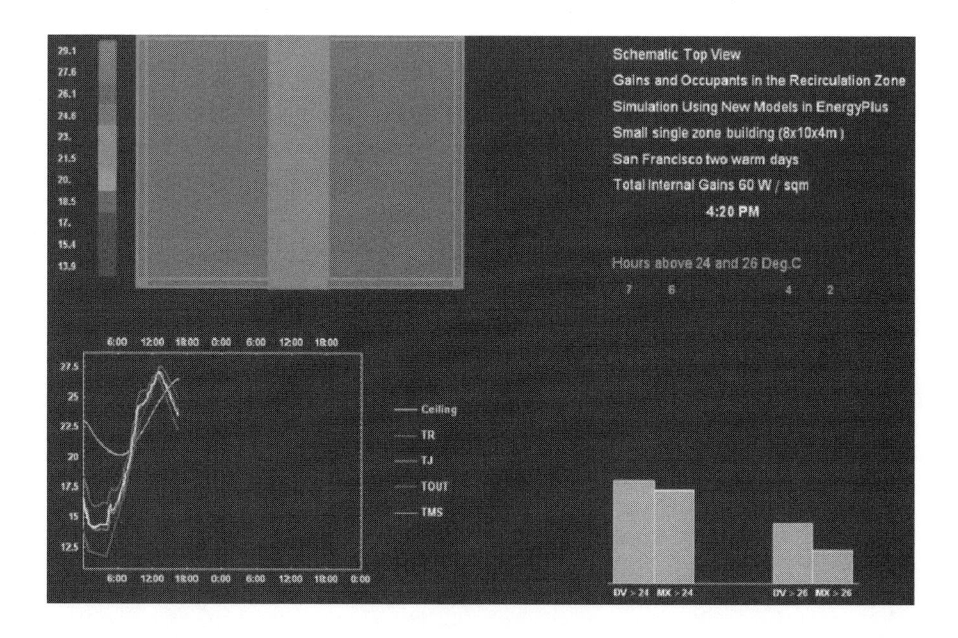

FIGURE 9-5-6 Results of an EnergyPlus simulation of a single zone office building using the new UCSD displacement ventilation model. *Image provided courtesy of Lawrence Berkeley Laboratories.*

EnergyPlus and Energy Plus Extensions Description

EnergyPlus was developed as a collaborative effort between Lawrence Berkeley National Laboratory's Simulation Research Group, led by Fred Winkelmann of the Environmental Energy Technologies Division, the University of Illinois at Urbana-Champaign, and the U.S. Army Construction Engineering Research Laboratory, with assistance from other research organizations. The recently developed EnergyPlus Displacement (DV) and Cross-Ventilation (CV) models were implemented by NaturalWorks. The new models resulted from research funded by the California Energy Commission (CEC), done at the University of California at San Diego (UCSD). The project was part of a large research effort in Low Energy Cooling systems, coordinated by Lawrence Berkeley National Laboratories (Dr. Philip Haves).

parameters needed to control the airflow pattern and vertical temperature variations (Linden and Carrilho da Graca 2002).

9.5.5 Uses of BIM Tools

The use of three-dimensional parametric modeling and interoperability fell into four basic areas; architectural design, architectural detailing, structural design, and CFD simulation. Each operation was performed by a different company. At the time that these aspects were developed, software was not yet available to support the integration needed for collaborative work by Morphosis and Arup, their engineering advisors.

The use of BIM tools, however, did allow the designers to identify and solve problems at an earlier stage in the detailing process than would normally occur, allowing them to correct certain problems prior to construction. This helped keep the project on budget despite its complex design.

9.5.6 Use of Building Modeling Tools for Architectural Development

Morphosis is a longtime user of FormZ for early stage design. Likewise, SFFB's architectural model and renderings were developed in this application*. Later in the project, the collaborative team developed a more detailed model, and

*In the future, Morphosis intends to use Bentley's Generative Components (GC), which is capable of both generating and visually showing the design components as well as abstract relationships between them. (An example of GC can be found in Chapter 2, Figure 2-7.) This capability allows GC to capture and communicate the logic behind the geometry.

Morphosis turned to Bentley Triforma to detail the steel members and describe the complex geometry. As the design developed, steel members were added or modified in both the Triforma and FormZ models. While it was possible to reliably exchange geometry between the two systems, it was not possible to exchange objects and attributes. When the architects desired to design from a purely geometric point-of-view, they reverted back to FormZ. Snapshots of the Triforma model are shown in Figures 9-5-7 and 9-5-9.

The Triforma model evolved to become the project's primary data repository. Three-dimensional models from other contractors and the data that these contained were imported into the Triforma model. The steel frame that covers the Southwest facade was developed by BDS using Tekla XSteel (now called Tekla Structures) (see Figure 9-5-8). Concurrently, Permasteelisa handled the detailing of the panels in AutoCAD (see Figure 9-5-10).

At this point, the project was being modeled using three different tools, with the architectural model constructed in Triforma, the steel structure modeled in Tekla XSteel, and the definition of metal facade panels in AutoCAD.

Data consistency problems may arise when using complex geometry, such as that developed for SFFB. The use of 2D drawings for the metal sunscreen cladding by Permasteelisa required close coordination, particularly for the warped surfaces. An example of an alignment review is shown in Figures 9-5-9. and 9-5-10. Aligned surfaces take on a dappled pattern as a result of round-off errors in visibility calculation. Morphosis provided the geometry control points, which facilitated the

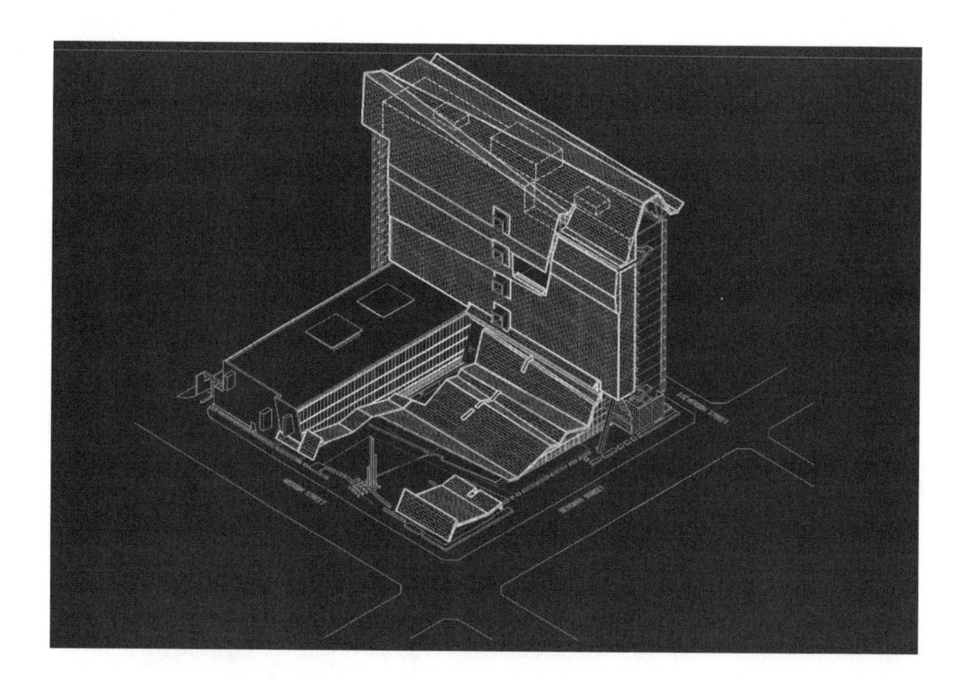

FIGURE 9-5-7 Triforma model with the metal skin highlighted.
Image provided courtesy of Morphosis.

FIGURE 9-5-8
Tekla XSteel model of the steel structure.
Image provided courtesy of BDS.

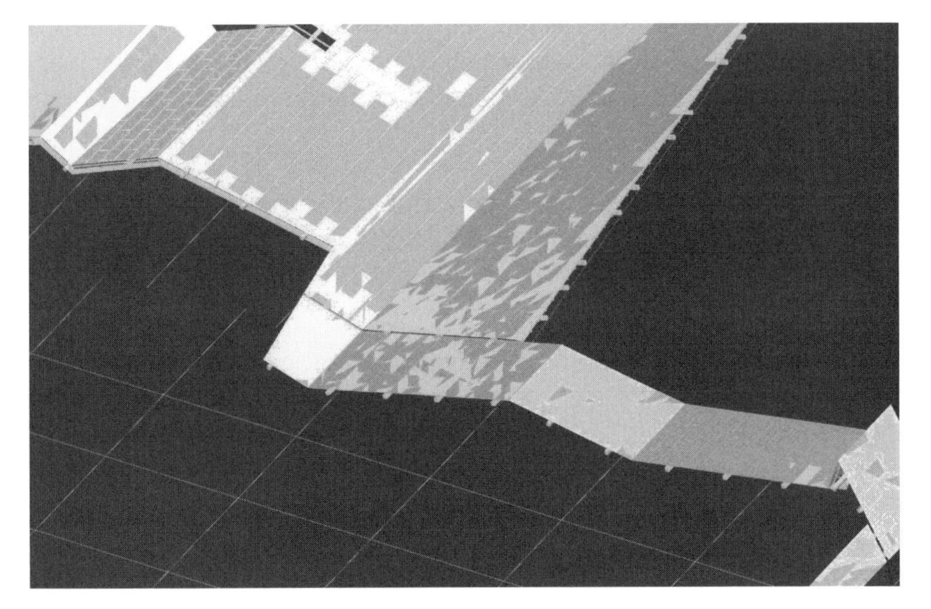

FIGURE 9-5-9
A detail of the metal skin panels shown in Figure 9-5-7, showing the verification of geometric consistency between the Triforma and the AutoCAD models.
Image provided courtesy of Morphosis and BDS.

proper alignment of each model. Two-dimensional drawings were generated from the Triforma model to provide definition of the concrete structure.

The exchange of information between Morphosis (using Microstation Triforma) and BDS (using Tekla XSteel) was extremely successful. No problems

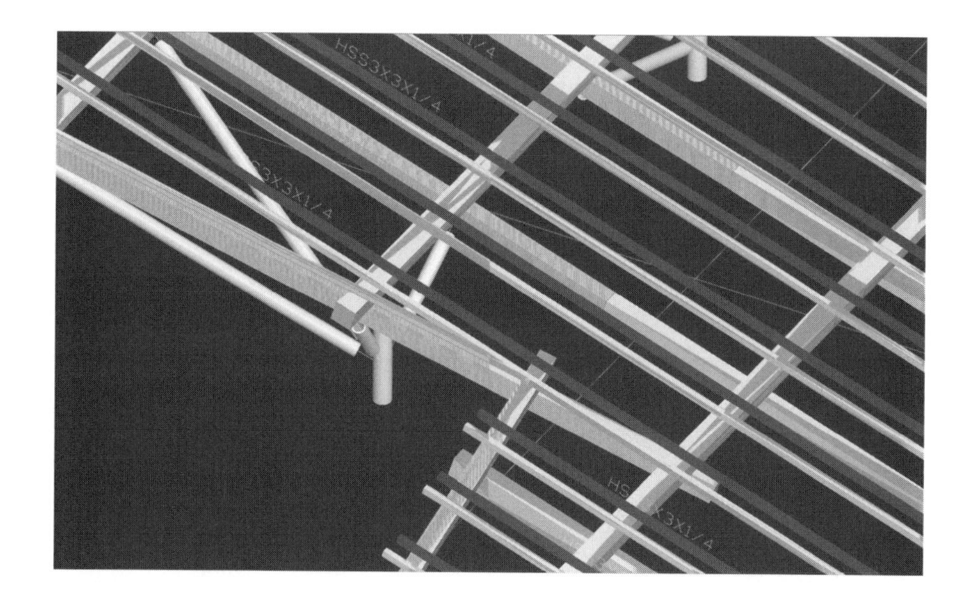

FIGURE 9-5-10 Verification of geometric consistency for the steel structure using the Triforma, XSteel, and AutoCAD models.
Image provided courtesy of Morphosis, BDS, Permasteelisa.

were reported. Integration between the two applications using the Structural Design Neutral File (a precursor to CIS/2) data exchange format was very accurate.

By combining all models in Triforma, Morphosis was able to check for visual conflicts and alignment, allowing them to verify the accuracy of information coming from the detailers and fix errors before approving the information to generate shop drawings.

Exchanges among the team doing the CFD simulations were generated through the use of 2D drawings. Information was restricted to the building's representative vertical sections. Analyzing air flow within the building required abstraction of building geometry that could not be generated automatically.

9.5.7 How the Design Was Affected by Natural Ventilation

Because this is a federal building, GSA requires there to be a 'hard,' i.e., impenetrable, perimeter at the base. The outside air intakes for the building had to be kept inaccessible to the public, and the structure had to be designed for blast-resistance. Despite the appeal of a building capable of opening up in response to internal and external environmental conditions, security concerns mandated that the lower and more vulnerable floors remain completely sealed. This required full air conditioning for the first five floors, while floors 6 through 18 could take advantage of natural ventilation.

According to the California Energy Code, interiors are only considered naturally ventilated if every unit is within 20 feet of an operable element. The structural core of this particular type of building must be approximately 20 feet from the outer skin. These two requirements necessitated the use of two natural ventilation zones running along the Northwest and Southeast facades, while service areas, conference rooms, and private offices requiring continual mechanical ventilation were to be located within 20 feet of the exterior. An enclosed spine contains the mechanical ventilation equipment, so that it does not interfere with the conditions of the areas that feature natural ventilation. The spine's height was restricted to 9.2 feet (2.8 m), leaving a minimum gap of 1.0 foot (0.3 m) between the lower surface of the upper level slab, to provide space for air flow generated by the natural ventilation system.

Since the building's concept was to replace mechanical ventilation with natural ventilation wherever possible, the surfaces of the concrete structure were left exposed to take advantage of their thermal mass properties. Concrete either traps or releases thermal energy, depending on environmental conditions, thereby acting as a temperature equalizer.

The Building Automated System (BAS) actively captures solar energy for use during colder periods to warm the concrete slabs, if the release of stored energy is not desired. Night cooling of the concrete ceiling slab is performed during warmer months, when previous day temperatures reach 75°F (24°C) or higher between the hours of 11 AM and 4 PM. When night cooling is indicated by the temperature control routine, the ventilation system uses the maximum allowed opening until slab temperatures fall below 66°F (19°C) or until early morning the following day (7 AM).

At the national level, GSA strives to limit annual energy consumption to 55,000 BTU per sf. Consumption for the San Francisco Federal Building's naturally ventilated floors is estimated to be less than 25,000 BTU per sf. If these projections are accurate, the sustainable design will translate to a taxpayer savings of approximately $250,000 annually (about $2 per sf per year).

In the future, the design team would like to integrate weather prediction data into the control system, basing the decision to initiate night cooling on anticipated weather patterns for the following day, by combining this data with that of interior heat accumulation during the previous day.

The SFFB's system is an example of integrating structural and mechanical engineering to achieve natural ventilation. The structural model of the concrete employs a non-standard up-stand beam, where each beam sits above the slab it supports, rather than below it. A photo of an early full-sized model is shown in Figure 9-5-11. These up-stand beams allow for smooth air flow, with no obstructions on the underside of each slab during overnight pre-cooling periods to reduce the thermal mass load. A simulation of this airflow pattern is shown in Figure 9-5-12. This structural system also allows for better penetration of sunlight into the interior. Due to the 13 foot ceilings, it is estimated that

FIGURE 9-5-11 Physical model of the slab showing the sinusoidal lower surface.
Image provided courtesy of Arup.

approximately 85% of the office will be illuminated using natural light. Lastly, this structural system allowed for the possibility of creating an under floor plenum for routing communication and electrical lines. And the plenum space provided additional under-floor air distribution for those offices that could not permit natural ventilation.

The lower plane of each concrete slab, which remains exposed to the interior level below it, was designed using a sinusoidal shape to increase the area of exposed concrete and improve its thermal mass exchange properties, as compared to traditional flat slabs. The sinusoidal waves are oriented parallel to the prevailing wind direction and help conduct cross-ventilation across the building's entire width (McConahey et al. 2002).

Another design consideration that supported natural cross-ventilation had to do with the location of the furniture in the offices located near the core of the building. Workstations near the building's perimeter were specified with a height of 4'-4", to allow natural light to reach the interior core. Their surfaces were specified with perforated metal to allow maximum air flow over and through them.

Definition of the operable windows was also affected by natural ventilation. CFD simulations were performed to test the viability of this concept. Windows were to be operated manually but also by the Building Automated

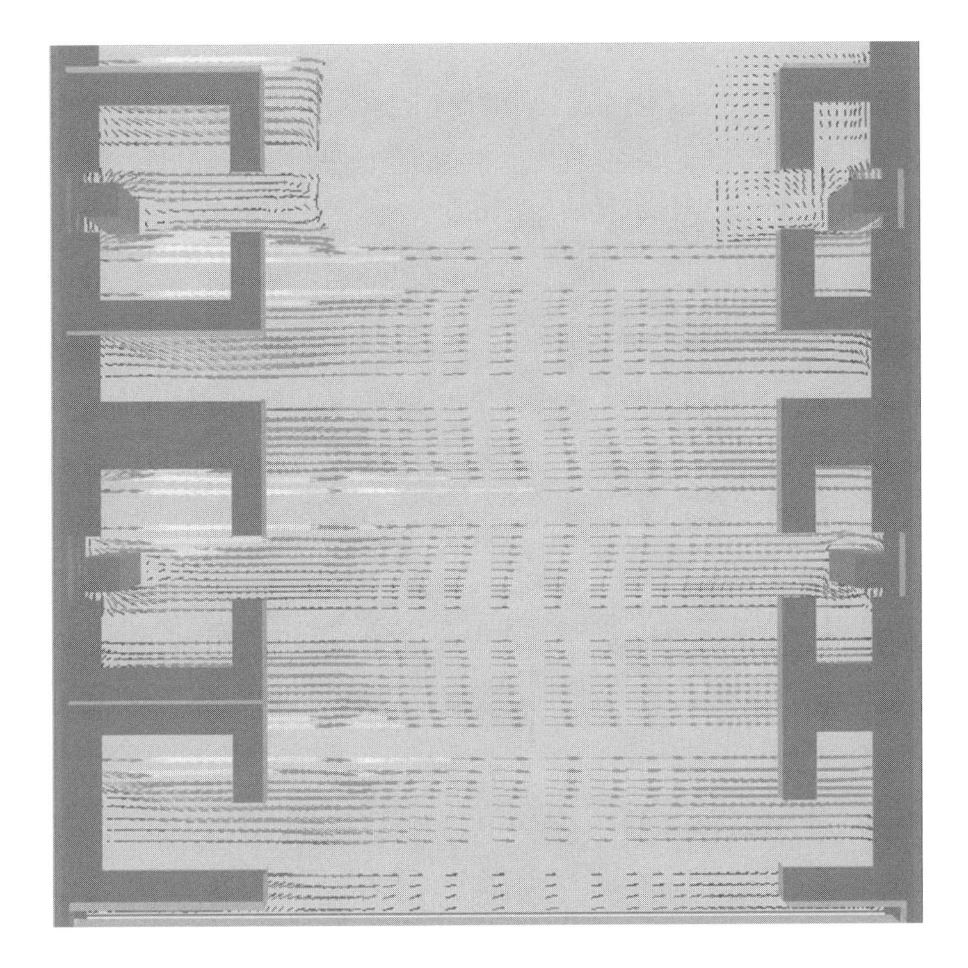

FIGURE 9-5-12 CFD simulation, showing the behavior of the air flow when in contact with the sinusoidal surface (see color insert for full color figure).
Image provided courtesy of Natural Works.

System (BAS). Two strips of BAS-controlled windows were located just below the user-controlled window (Trickle vent) and 7.2' (2.2 m) above it.

The simulations showed that the row of operable windows located at 6.2' (1.9 m) did not work optimally (see Figure 9-5-14) and required modification to improve performance. Because the window panel only allowed for a limited 1.0' (0.3 m) opening, incoming airflow circulated toward the interior instead of following a vertical path near the window panel. A deflector was designed to alter the direction of air flow. Four different deflector positions with 30° tilts were tested before the final configuration (Linden and Carrilho da Graca 2002) (see Figure 9-5-15) was determined. A fine grained graphic simulation of airflow was undertaken (Figure 9-5-16) and the overall effect remodeled (Figure 9-5-17). Comparison of Figures 9-5-14 and 9-5-17 shows a decreased flow of air traveling vertically up the windows and an increase in circulation toward the room's interior.

FIGURE 9-5-13 Artistic rendering showing the interior of the building; note the perforated surfaces in the furniture. *Image provided courtesy of Morphosis.*

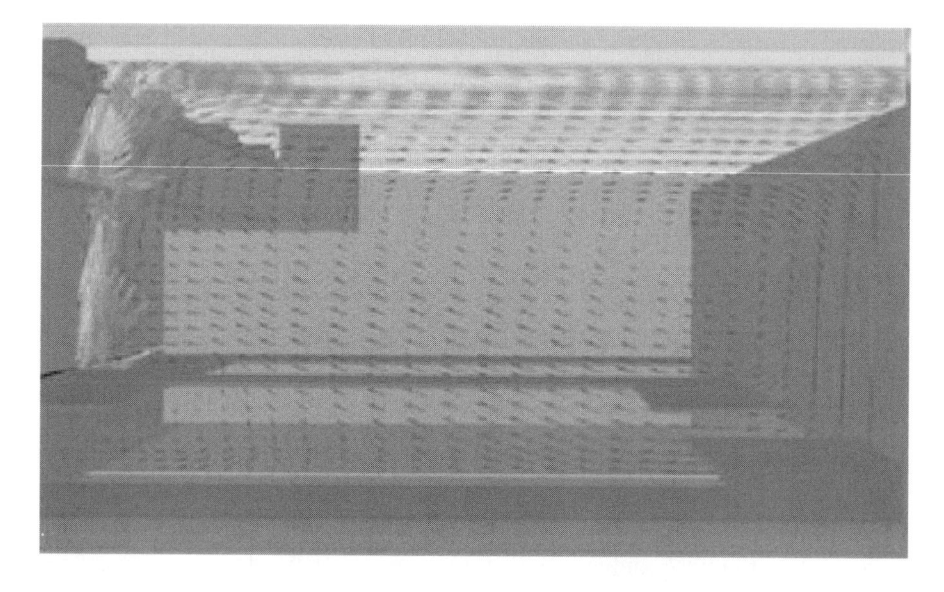

FIGURE 9-5-14 Airflow simulation without deflector in place. (See color insert for full color figure.) *Image provided courtesy of Natural Works.*

All in all, the San Francisco Federal building was able to incorporate many healthy living amenities and urban features for the surrounding neighborhood. As seen in an earlier application in the Caltrans Building in Southern California, the SFFB also incorporated skip-floor elevators, providing office workers with opportunities for cardiovascular exercise. Workers are able to maintain some

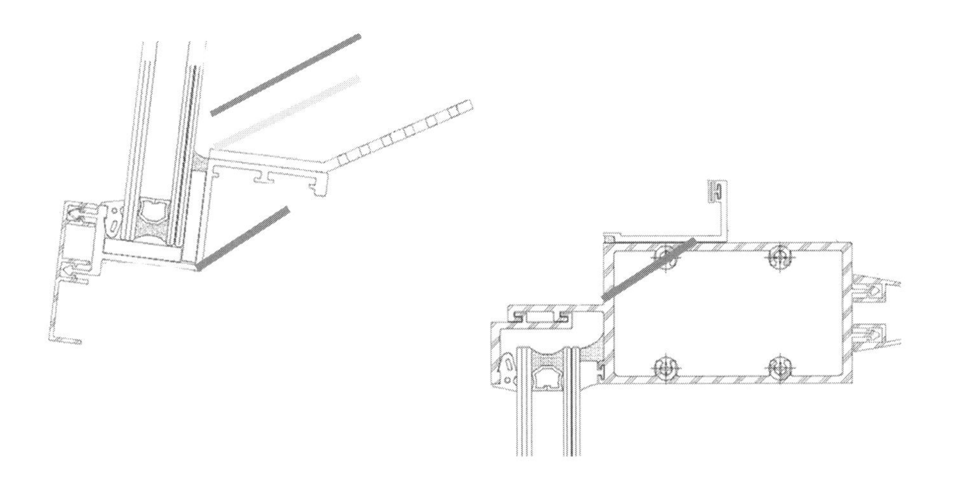

FIGURE 9-5-15 Analysis of possible positions for the deflector.
Image provided courtesy of Morphosis.

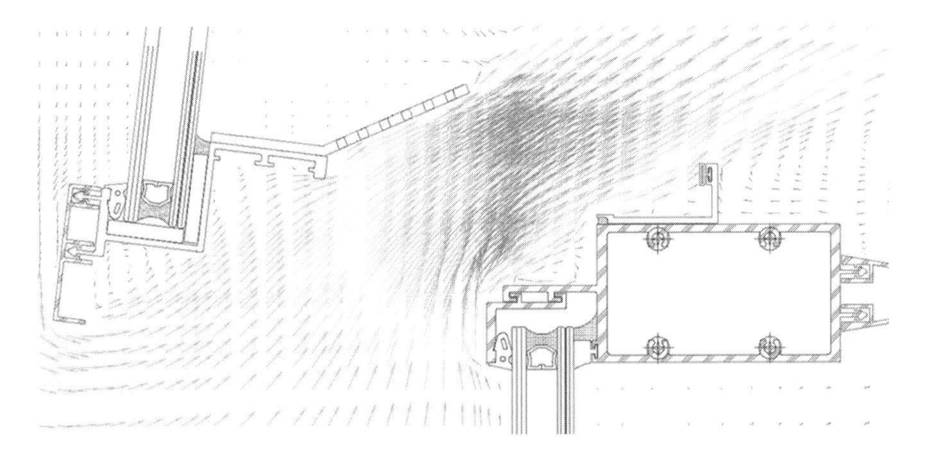

FIGURE 9-5-16 Analysis showing airflow around the selected deflector.
Image provided courtesy of Morphosis and Natural Works.

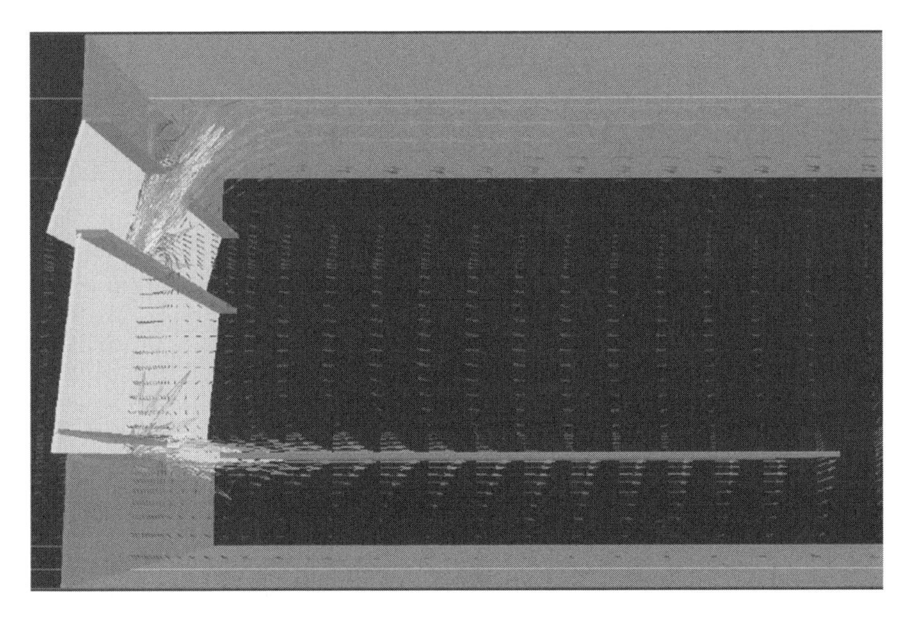

FIGURE 9-5-17 Airflow simulation with the deflector in place.
(See color insert for full color figure.)
Image provided courtesy of Natural Works.

control over their office environment, by adjusting windows to accept breezes from the San Francisco Bay while remaining protected from glare and direct sun. An outdoor sky-garden on the eleventh floor also provides excellent panoramic views. The park-like ground level plaza is for everyone's use; the facility includes an auditorium for local meetings, a daycare center, and a cafeteria for workers. It provides an attractive new stopping place for this section of the City.

9.5.8 Conclusions and Lessons Learned

The San Francisco Federal Building offers several lessons in the application of BIM technologies.

The use of Triforma gave Morphosis the ability to continuously check for geometric consistency between the conceptual development model and the various fabrication models. In this way, it allowed them a great deal of control over the geometric definition of their design and its corresponding details in fabrication models. These visual consistency checks helped them identify and solve problems early in the detailing process. Had these errors not been detected, conflicts would have arisen during the fabrication or installation processes, resulting in unforeseen expenses and delays.

The novel cross-ventilation system – with no mechanical ventilation for offices located near the tower's perimeter – motivated GSA to invite LBNL, UCSD, and NaturalWorks to serve as technical advisors on the project and validate Morphosis's design proposal. The use of computational fluid dynamics (CFD) allowed the design team to refine aspects of the system, such as the small wind deflectors located above the operable windows.

Information contained in the architectural models was not structured properly for CFD modeling, requiring the testing team to create an abstraction of the building based on drawings, in order to perform the simulations for testing building performance under varied conditions, including those generated automatically by the BAS. In the future, these simulations might also become useful in post occupancy evaluations of the building, to verify the accuracy of anticipated airflow patterns.

It is important to note that the contractor relied on 2D drawings extracted from the Triforma model for coordination and that some of the fabricators also worked in 2D. This early stage BIM effort is useful in documenting the kinds of problems designers were attempting to solve using the limited capabilities available between 2001 and 2004. It is interesting to speculate how the design and construction processes might have been even more efficient using current-generation BIM tools with more advanced interoperability capabilities. Still, further work must be done to extend the use of BIM tools to all team members involved in the construction industry and to extend the associated workflow interoperability to facilitate their use.

Acknowledgments for SF Federal Building Case Study

This case study was written by Hugo A. Sheward for Professor Chuck Eastman in a course offered by the Design Computing Department at Georgia Institute of Technology, Winter 2007. It has been adapted for use in this book. The authors wish to acknowledge the assistance of Tim Christ and Martin Doscher from Morphosis in the preparation of this study.

9.6 100 11TH AVENUE, NEW YORK CITY

BIM to facilitate design, analysis and prefabrication of a complex curtain wall system.

9.6.1 Introduction

This case study identifies innovative approaches in the implementation of Building Information Modeling (BIM), with a special focus on the ways in which BIM facilitates design, communication, and analysis for curtain wall design. The building, located in Manhattan near the Westside Highway and 19th Street, is a 21-story residential condominium with overall dimensions of 150 ft. L × 75 ft. W × 235 ft. H. Figure 9-6-1 shows a rendering of the building and a hidden-line view of the curtain wall.

The site is located in an area of Chelsea that does not have the same physical qualities and amenities as other surrounding areas, such as Greenwich Village, where the Perry Street condominium project by Richard Meier has a nearby park, high-end retail, and a subway stop. The Highline project, however, is catalyzing a significant transformation of this area, with a series of projects planned by various renowned architects.

Table 9-6-1 The project team

Ownership Group:	Cape Advisors + Alf Naman Real Estate
Design Architect:	Ateliers Jean Nouvel
Architect of Record:	Beyer Blinder Belle
Construction Manager & General Contractor:	Gotham Construction
Facade Consultant:	Front Inc.
Structural Engineer:	DeSimone Consulting Engineers
Acoustic Consultant:	Cerami & Associates
Curtain Wall Fabrication Team:	CCAFT, SGT, KGE

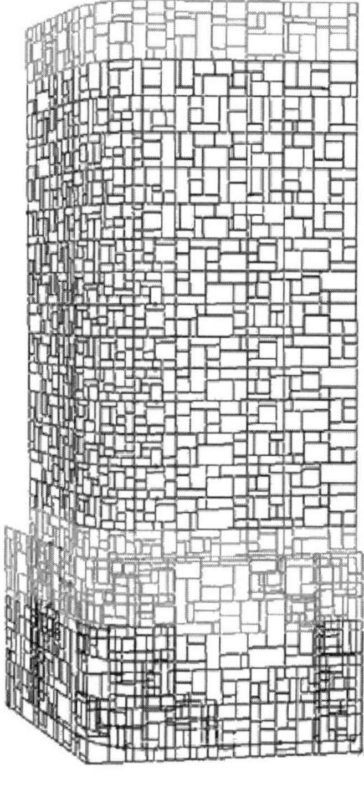

FIGURE 9-6-1 100 11th Ave. in Manhattan, a condominium project by architect Ateliers Jean Nouvel.

This building provides a river-view of the downtown financial district and the New Jersey shore. The curtain wall facade is therefore an iconic element of the design. A key concept of the curtain wall is that it is not a load-bearing enclosure but hangs from the structure of the building. There are also technical and environmental issues relevant to a glass curtain wall, such as water impermeability and insulation, which constrain the morphology of its surface. In this case, housing ordinances affecting new buildings along the Westside Highway prescribe a sound transmission class (STC) of 42 for the facade. In this project, the view is the driving element for the curving of the facade and the configuration of its surface. Another key issue is the faceted surface, which consists of a series of planes tilted along multiple angles and axes.

Front Inc. was brought in early in the process, at the concept design stage, to assist the ownership group in evaluating the feasibility of two preliminary proposals prepared by the design architect. The proposal selected for the overall exterior surface has two systems. Glass covers 40% of the surface area, which complies with the New York State energy code provision, which imposes a 50% maximum.

The remaining 60% of the surface area is clad in custom black brick with random window openings, within which each window is set at a tilted angle. The curtain wall system changes at street level to a hybrid system of a faceted curtain wall and a series of cavities with large planters that affect the loading system.

The average cost for the overall exterior surface is calculated by weighing the cost of each system typology by its proportion of the total surface area. The glass and steel facade system, covering 40% of the building, was procured at approximately 2.5 times the cost of standard uniform curtain wall systems, while the brick and window facade was purchased at approximately 1.0 times the same benchmark. Thus the average cost was:

$$2.5 \times 40\% + 1.0 \times 60\% = 1.6\times \text{ the standard uniform curtain wall cost.}$$

Relative to the cost typically incurred by condominium developers for facades, this system is expensive. The construction cost is estimated at approximately 25% the overall hard construction cost of the entire building. A typical range is 12% to 15% percent. Although 25% is high, it is still less than cladding systems designed by architects such as Gehry Partners, Asymptote, or Sejima, Nishizawa & Associates, which have costs in the range of 25% to 40% of the total construction cost for some building projects. It is clear that for specific projects, special facade systems have generated cultural, environmental, marketing, and other values.

9.6.2 BIM Process: Innovation and Challenges

Building Information Modeling systems are being adopted as a powerful tool in the AEC community. Their many features include a robust parametric modeler, analysis tools, spreadsheet functionality, API, and file export functionality. In the case of Front Inc.'s use of Digital Project (DP) for this building, the major innovation was in the implementation of modeling tools early in the design process. They were able to update and store all product information for the curtain wall design and exchange this data with the rest of the project team.

Parametric Modeling

Marc Simmons and his team at Front Inc. worked as consultants for the ownership group for approximately six months during the concept phase of the project. The design architect, Ateliers Jean Nouvel, and Front Inc. collaborated in the conceptualization of the facade system. The challenge in this phase was to identify rational ways to systematize the design without compromising the aesthetic value of the design concept.

Material Selection. The choice was made to use steel over aluminum for the framing system, because aluminum would require connections to be detailed as moment connections with large fasteners and exposed bolts. In addition, with the complex pattern proposed by Jean Nouvel, linear continuity for the load path would be difficult; however, a welded steel frame could be variegated to provide this type of *Mondrianesque* irregular pattern and still carry the loads as one network of welded steel members.

Parametric Panels. As can be seen in Figure 9-6-2, Ateliers Jean Nouvel provided a breakdown of the facade system as a composition of glass panels with four directions of rotation: tilting up, down, left, and right; four glass variations; and angles of rotation varying through 0, 2, 3, 4, and 5 degrees off vertical. Front Inc.'s first step was to create an Excel spreadsheet (Figure 9-6-3) for organizing these parameters along with the glass panel dimensions. The Excel

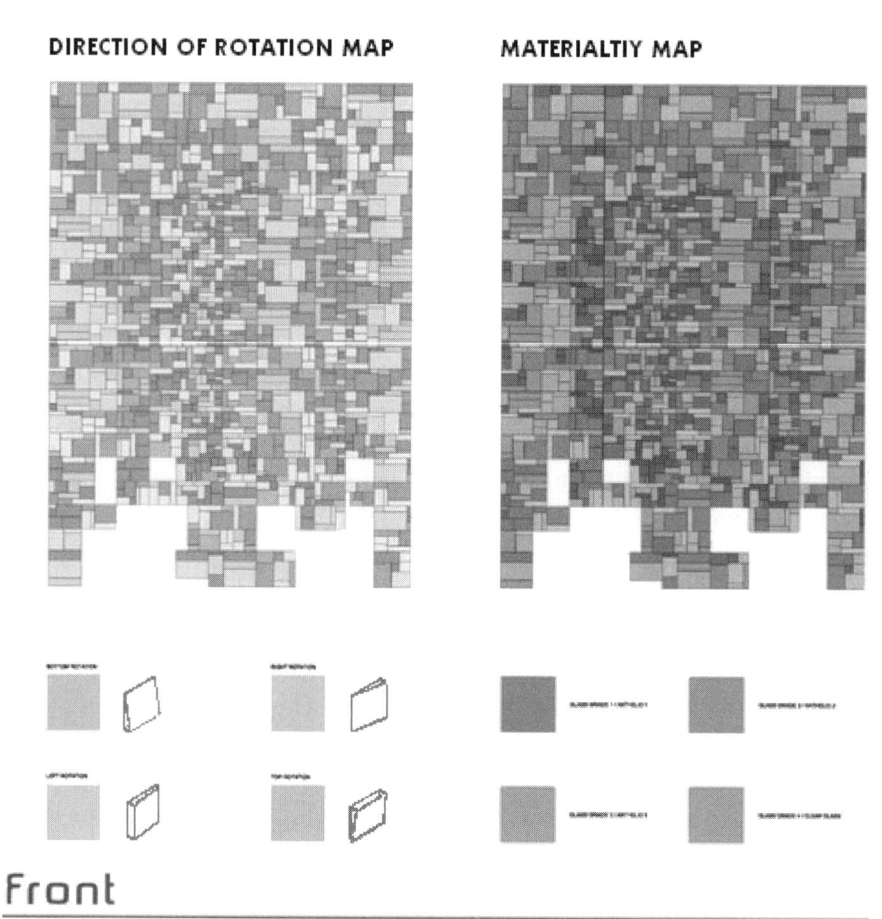

FIGURE 9-6-2 Design architect's drawings produced from a FormZ model. Colors (shown in grayscale) denote angle and direction of rotation and glass material type.

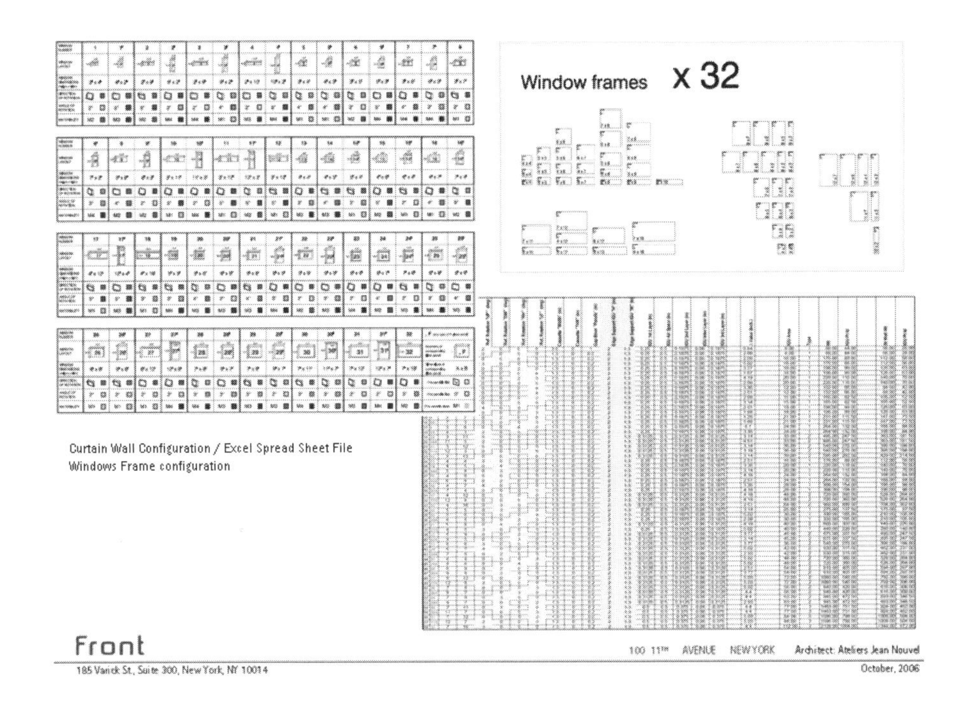

FIGURE 9-6-3 Master spreadsheet defining the variations in the glass panes. The schedule (top left) details all of the different permutations of size, angle and coloring used for the panes of glass. The different frame sizes needed are listed on the top right sheet, and the Excel worksheet that drives the schedule and its graphics is shown at bottom right.

file would be referenced as a design table in Digital Project to associate all parametric variations.

Mega-Panel Assembly. Front Inc. proposed a basic grid system that could be concealed within the pattern of the facade by keeping the width of the steel members constrained to 3 in., as agreed upon with Ateliers Jean Nouvel as a reasonable dimension allowing for both structural efficiency and integration of panel mechanics. This allowed for the subdivision of the system into mega-panels for framing glass sub-panels of varying dimensions, tilt, and materiality. In Digital Project, the mega-panels were part-body assemblies with a basic parametric wire-frame, as shown in Figure 9-6-4.

The mega-panels were associated with a design table and created as a *Powercopy* that could be initiated using values from the spreadsheet. A Powercopy is a type of feature template in Digital Project, where the elements are accessible for editing. In the case of each mega-panel, the overall dimensions conform to each room within the building and become a picture frame. The mega-panel dimensions vary from 11 ft. × 18 ft. to 20 ft. × 37 ft., and they affect the dimensions and number of component sub-panels.

Curtain Wall Assembly. The overall facade system is made of steel with a regulated grid composition of mega-panels and a randomized subdivision of glass panes held in aluminum cassettes. In Digital Project, a design table organizes

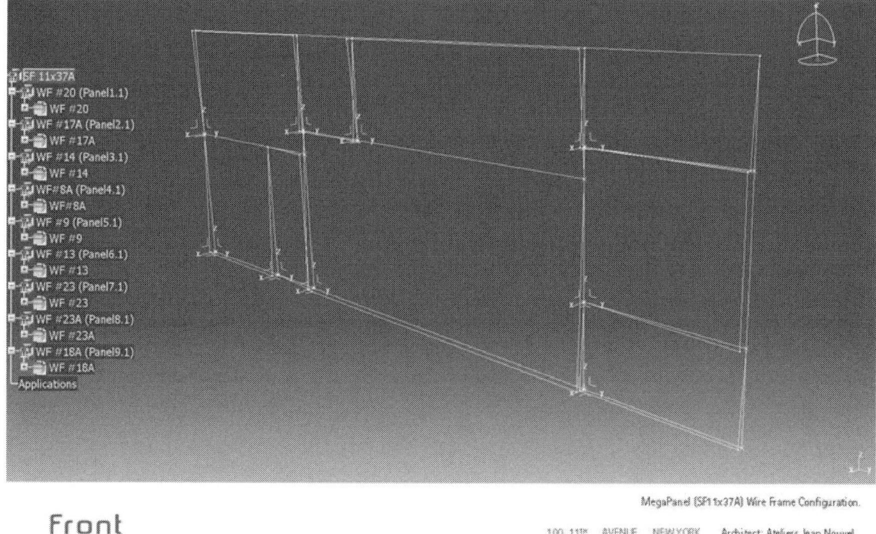

FIGURE 9-6-4 Original parametric Powercopy of a mega-panel in Digital Project.

the variations of the 1,351 individual glass panes that make up the facade, by dimension, configuration, and location in the grid. The frames of the mega-panels have members 1½ in. wide by 6 in. deep on two sides and the top. The bottom member is 3 in. wide by 6 in. deep. The profiles of the mullions that subdivide the mega-panels vary in cross-section (3 × 3 in., 3 × 4 in., 3 × 5 in., and 3 × 6 in.) to account for variations in loading. Steel profile protrusions at the intersections of the mullions vary in length to provide the specified angular tilt of each sub-panel. The triangular gaps between the glass panes and resulting mullions are closed with steel plates at the head and sill of each pane. The extensions are welded and sanded smooth to maintain visual continuity between the mullion and the cassettes and to provide a place for thermal and acoustical seals. These details are shown in Figure 9-6-6 and Figure 9-6-7.

The mega-panels are supported by floor slabs and connect through a steel spreader beam. The 4 in. × 10 in. beam is engineered to minimize deflection to 1/8 in. The beam has two dead-load connections to the concrete slab and one wind-load connection. The mega-panel has multiple connections to the spreader beam, which uniformly distributes the loads. Therefore, the beam handles the deflection between the two systems. In fabrication, each mega-panel would be pre-assembled and the beam connected to the panel onsite.

Fabrication Activities

Fabrication Team. For the bidding phase, Front Inc. did preliminary engineering of the steel and produced a set of drawings for pricing. All bidders came back with bids much higher than the cap price stipulated by the ownership group. Front Inc. believed that their skills and BIM technology would enable them to deliver the facade at a reasonable cost and asked permission from the owners to form a team to galvanize resources that could deliver the project. There was an opportunity for great profit, but the financial risk was also very high. Marc Simmons, the principal of Front Inc., was prepared to work as design consultant on the facade contractor side, executing all aspects of the design and incorporating them into the master Digital Project model. Private investors joined forces with China Construction America (CCA), a subsidiary of China State Construction (the largest construction company in China), to establish a new facade contractor named CCAFT. This new company had the financial strength to provide bonding for the project, including the ability to subcontract fabrication to China-based fabricators, such as SGT and KGE; to subcontract design work to Front Inc.; and to subcontract installation work to Island Industries in New York.

Visual Mock-up. A 15 ft. × 42 ft. visual mock-up was fabricated by SGT and reviewed in Shenzhen with the ownership and architects in January 2007. The mock-up consisted of two curved corner mega-panels and two flat mega-panels, as shown in Figure 9-6-5. For the visual mock-up, the steel, aluminum,

FIGURE 9-6-5 Photo of a 15 ft. × 42 ft. visual mock-up, fabricated by SGT in Shenzhen, China.

and glass were generated as a 3D solid model in Digital Project, with the design table information driving the associative parametric sub-assemblies in the wire-frame armature. From this 3D model, SGT was able to: extract fabrication geometry and prepare piece drawings according to their own format and language requirements using Digital Project; and view and interrogate the model, which was exact and fully engineered in all regards. At the time the visual mock-up was fabricated, the extruded aluminum profiles were not yet available; however, SGT was able to use the same parametrically-generated geometry to create CNC cutting patterns for a break-formed aluminum sheet, to replicate the geometry of the actual aluminum caps. In the performance mock-up, this geometry would be used to describe the milling and cutting paths for delivering the same effect with extruded aluminum profiles. Having the digital model available with such specificity at this stage was critical and enabled the creation of an exacting visual mock-up, despite it being out of the normal fabrication sequence.

Performance Mock-up. A 32 ft. × 55 ft. performance mock-up was to be fabricated and delivered in July 2007 for testing at ATI in Pennsylvania. This mock-up consisted of two curved corner mega-panels and two flat mega-panels, as shown in Figure 9-6-6. The mock-up mega-panels were fully detailed to include locations for the cranes to hook and transport the pre-assembled panels, to keep them level as they are hoisted, and to minimize any risk of glass breakage. The mock-up was also intended to provide valuable information on

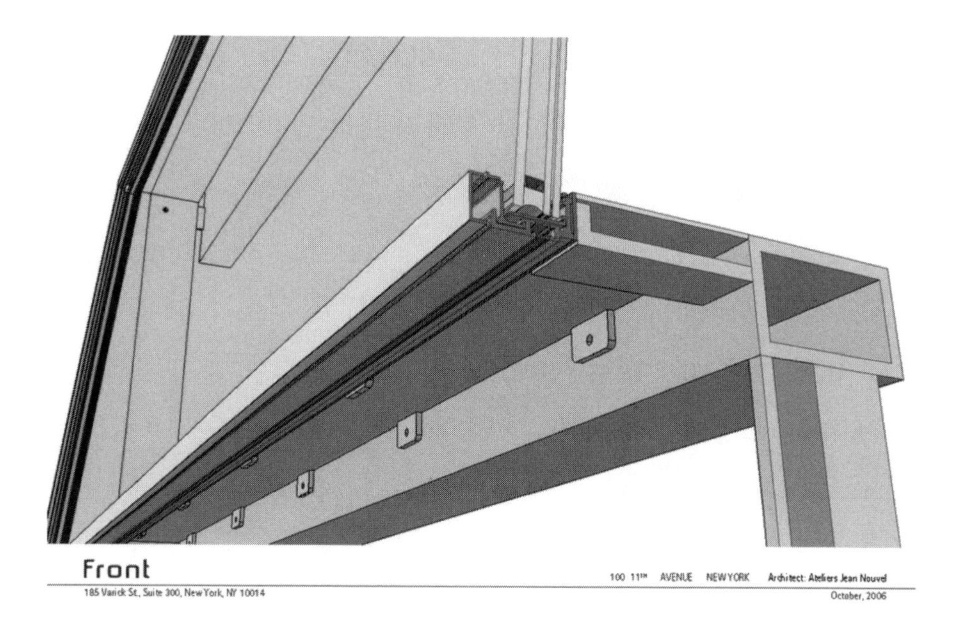

FIGURE 9-6-6
Cross-section view of mullions, extensions, and a glass panel.

the process of exchanging product model data with SGT, their ability to meet design and engineering specifications and to work within the shipping and delivery time constraints. It was also a critical learning and proving stage for the installers, who would be responsible for taking ownership of the mega-panels and installing them correctly in the testing rig. Production and delivery of the mega-panels was planned to begin after review of the mock-up, in a sequence matching the construction schedule. The first installation onsite is scheduled for January 2008.

9.6.3 Information Exchange and Interoperability

One of the most important aspects of BIM is the facilitation of information storage and exchange to enhance communication and collaboration in the design and construction process. Contractually, Front Inc. was responsible for engineering the design of the steel and overall system geometry. Although the glass gasket, aluminum, and silicone were not part of their contract, Front Inc. modeled all of this information. The model included high-level detail, such as the location of air-filled cavities and the meeting points of rain gaskets, to provide absolute continuity. The model also included different thicknesses of glass, edge bevels, and the thickness of the PVB (Polyvinylbutyral) interlayer laminated between the glass panes. Front Inc.'s model became the data repository for the entire curtain wall system.

Front Inc. used several different software programs, including Rhino, AutoCAD, SolidWorks, and CATIA. They found that CATIA and Digital Project (DP) have additional parametric capabilities, as compared to other BIM tools. In-house engineering design and analysis for this project was done using DP (for geometry and other information) and Robot and Strand (for structural analysis). For example, the mega-panel wire-frame was exported from DP in an IGES format and brought into the structural analysis tool (Robot) to test deflection. Figure 9-6-8 shows how the mega-panels were analyzed for strength and deflection in Robot.

Profile cross-section details for the framing system were developed in AutoCAD, and the 2D information was imported into DP. This was done in 2D, because the architect of record required 2D drawing files as well as due to the limited number of skilled DP users. The facade was rendered in Flamingo and Viz.

At the start of the project, Front Inc. received a 3D model with a single-polygon representation of tilted glass panels produced in Rhino from Ateliers Jean Nouvel. Communication was done principally with renderings and other similar information to aid in the visualization of the design proposal and its implementation in the new model produced in Digital Project. The DP model

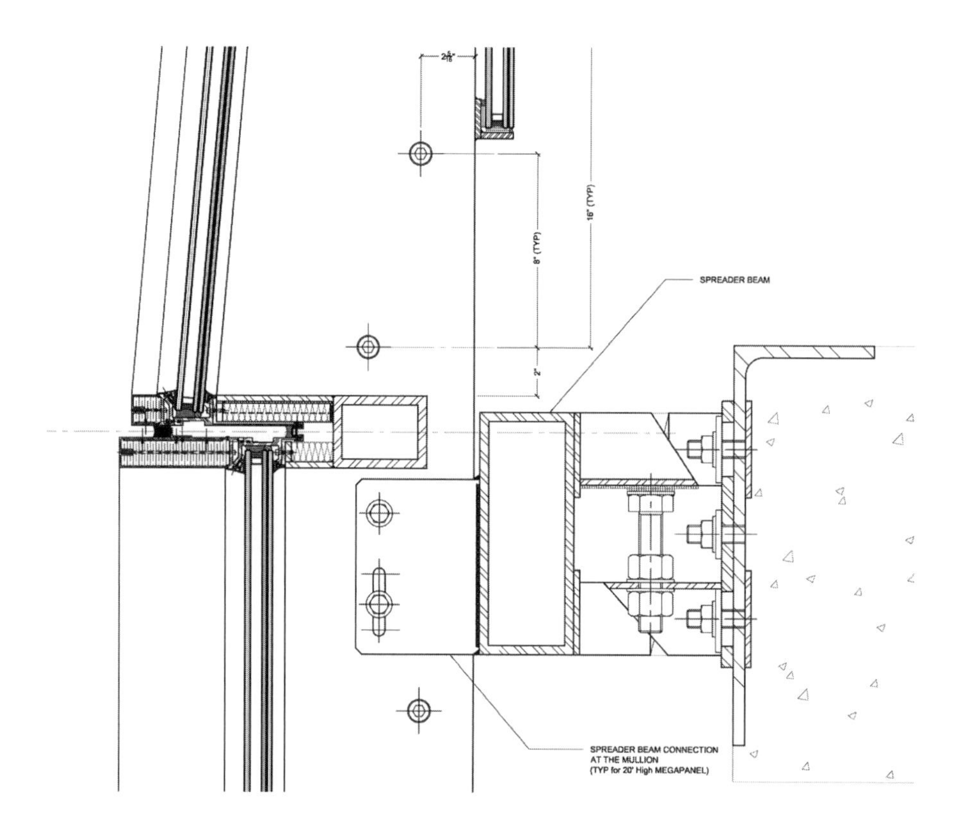

FIGURE 9-6-7
Slab edge detail.

was the facade contractor's (represented by Front Inc.) expression of the designer's intent, as defined in the Rhino model.

The facade had to be coordinated with the building itself, and that was done between the façade contractor and the architect of record, Beyer, Blinder, Belle (BBB), who produced the construction documents for the entire building. No construction documents were prepared for the steel and glass facade, allowing the DP model and contractor-side drawings to serve this function. For the facade, the approach was design-build, so Front Inc. and BBB shared shop-drawing information. A great deal of coordination with many iteration cycles was needed. Front Inc. produced documents and passed them to BBB as PDF files. They verified the coordination with their documents and sent information back to Front Inc. such as for slab edge profiles in AutoCAD.

Front Inc. provided the fabricator in China with the 3D model of the final geometry and all other product model information in a CATIA file format. From this model, the fabricator extracted all of the drawings necessary for fabrication.

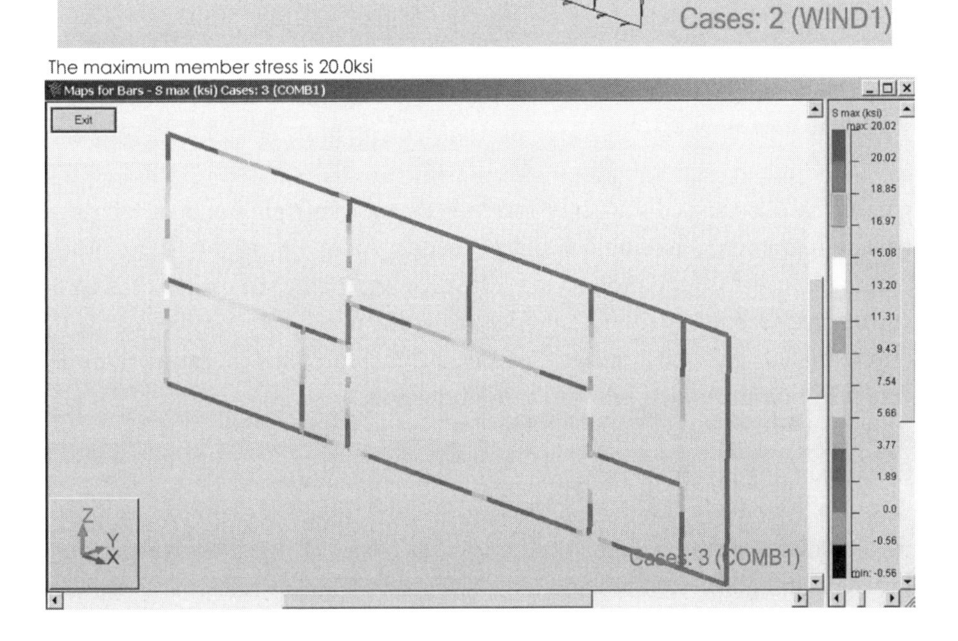

FIGURE 9-6-8 **Structural analysis of a mega-panel in Robot.** (See color insert for full color figure.)

9.6.4 Lessons Learned

Parametric Curtain Wall Catalog

The most valuable result of the full use of a building model was the ability to use parameters associated with a spreadsheet and predicate rules in the production of feature templates that could be re-used. The parametric assembly was used to generate multiple solutions for various conditions. By using this model, Front Inc. could produce a curtain wall system that was very different from the design proposal provided by Jean Nouvel but utilizing the same

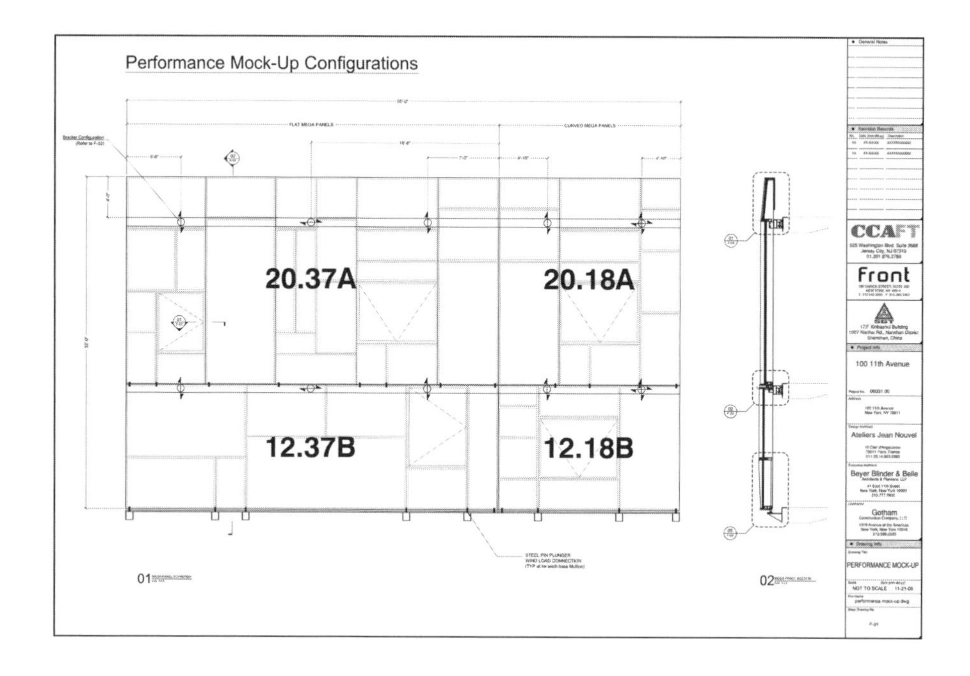

FIGURE 9-6-9 Shop drawing of a mega-panel.

parametric assembly. Front Inc. foresees that an emerging catalog of these types of parametric assemblies could expedite their work in the future. Marc Simmons described the process as, "Effectively mapping the essential DNA of customized facade assemblies and deploying them to create families of highly differentiated mass-customized facades. This represents a future entirely dependent on integrated practice and fabrication."

Changing the Business Model

Dassault Systemes' CATIA is an integrated tool designed for companies in aerospace engineering, where the person responsible for design is also responsible for engineering, sales, and marketing. The use of Digital Project, as a product of the consortium of CATIA and Gehry Technologies, could have similar ramifications in architecture, engineering, and construction in a design-build paradigm.

As a result of this experience, Front Inc. has broadened its services beyond consulting for owners, architects, and contractors. The company is working with architect Point B Design on a gallery project in Philadelphia, where they have designed, engineered, and are in charge of fabrication and delivery of the facade system. They are now engaged in similar deployment strategy for facades designed by Neil Denari Architects in New York City and Tod Williams and Billie Tsien in Amagansett, New York.

The Need for a BIM Skill Set

Front Inc. had a team of architects and engineers with various backgrounds. Although the majority of the team had received in-house training in Digital Project, Marc Simmons, the company principal, explained that one individual in particular was truly capable of using this software to its full potential at that time. Thanks to his training as an aeronautical engineer, this individual exemplified the new type of role, where the designer is also responsible for detailing and performance.

Front Inc. now looks for a different skill-set in the pool of applicants eager to join the firm, with greater emphasis placed on the use of BIM technology. Recent graduates from Rensselaer Polytechnic Institute, California Polytechnic State University, and Georgia Institute of Technology seem to meet the expected level of skill and knowledge of building information modeling tools.

9.6.5 Conclusion

In this project, a complex aesthetic concept for a curtain wall system was resolved in a building information model with a set of regulating lines in order to structure a hierarchical nesting of parameters bound to design tables. Visually, the effect of the regulating grid is neutralized, because of the complexity of the subdivision of the system. Therefore the legibility of the vertical and horizontal lines is reduced, and the variegated texture of the facade is emphasized.

Because this innovative curtain wall system entailed a financial risk, the ownership group looked at the contract early and asked Front Inc. to find solutions that would enable the design proposal to be realized with the lowest possible cost risk.

Building information modeling is impacting the way architects design and how buildings are constructed. Because the model becomes the central database of information, the analysis of design ideas and building performance, cost estimation, and construction scheduling can be done with greater accuracy. By using the same model, architects, engineers, contractors, and manufacturers can communicate and implement changes to the design quickly. Communication, however, is limited by two main factors that affect the BIM process. At the basic level, not all members of the AEC team may be using the same or compatible software; therefore information is lost in the exchange of model files. In addition, for BIM tools to be used to their full capability, additional expertise and skills are required of architects and designers, beyond the limits of commonly accepted skills. In this case study, these problems were avoided because of the unique skill-set of the Front Inc. team and their management of all information contained within the curtain wall design. At a more complex level, communication during the design and construction process is fragmented by other structures in place, having to do with the definition of professional roles and responsibilities within the AEC team. In this example, a different model of services was developed.

Acknowledgments for 100 11th Ave. NYC Case Study

This case study was prepared by Paola Sanguinetti for Professor Chuck Eastman in a course offered in the College of Architecture Ph.D. Program at the Georgia Institute of Technology, Winter 2006. It has been adapted for use in this book. The authors wish to acknowledge and thank Marc Simmons and Dario Caravati of Front Inc. for providing the bulk of the information. The images are courtesy of Front Inc. (Marc Simmons, Dario Caravati and Philip Khalil) and Ateliers Jean Nouvel.

9.7 ONE ISLAND EAST PROJECT

Owner/developer BIM application to support design management, tendering, coordination, and construction planning

9.7.1 Introduction

This case study documents the implementation of BIM to manage the functional and financial relationships between design, construction, and facility management on a large, complex project by an owner-developer. The owner identified the potential of BIM to manage information more efficiently and save time and cost over the project lifecycle. This case study discusses how the owner initiated the BIM effort on the One Island East (OIE) project after design using 2D tools had already started, worked closely with their design/construction team and technology team, and integrated BIM into the design and construction of the project. The study discusses training of the project team in the Digital Project BIM software to support design, clash detection and correction, quantity takeoff and tendering, coordination of the team, and 4D planning.

Swire Properties is one of the top developers and industry leaders involved in the construction industry's transformation in Hong Kong. One of their projects, OIE, is a large commercial office building with seventy floors that is currently under construction in Hong Kong. Table 9-7-1 provides an overview of the basic project information. Figure 9-7-1 shows a computer rendering of the planned office building. As an owner, Swire's organization is responsible for managing the design and construction and the leasing and operations of the facility. In addition to this facility, Swire manages hundreds of facilities or projects at any given time; and was seeking better tools to both oversee and coordinate the design and construction process and potentially link the building

Table 9-7-1 Summary of One Island East Project Information.

Location	Hong Kong, China
Project Name	One Island East (OIE)
Contract Type	Competitive tendering
Construction Cost	$300 million (approximately)
Project Scale	70 Floors with 2 basement levels Total floor area: 141,000 m^2 (1,517,711 sf) Typical floor area: 2,270 m^2 (24,434 sf)
Schedule	Period of construction: 24 months Expected completion: March 2008
Current Stage	Under construction as of January 2007
Owner	Swire Properties Limited
Architect	Wong & Ouyang (HK) Limited
Quantity Surveyor	Levett & Bailey Quantity Surveyor Limited
Contractor	Gammon Construction Limited
BIM Consultant	Gehry Technologies
Functions	Office and Commercial Facilities
Structure	Reinforced Concrete
Exterior	Aluminum Curtain Wall

information to their facility management systems. In particular, Swire was looking for a building management tool capable of managing a very large project or several such projects with the capability to link design information, cost and schedule data, construction process management, quality assurance and facilities management.

Furthermore, as part of their commitment to improving the quality of their buildings, Swire recognized the potential of BIM to increase the quality control and efficiency of their buildings over the entire facility lifecycle.

9.7.2 Pre-Tender Stage BIM Implementation Process

The OIE project was in schematic design phase when Swire was researching BIM systems. They attended a presentation by Gehry Technologies (GT) together with the Hong Kong Polytechnic University held in early 2004 where

FIGURE 9-7-1 Computer rendering of the proposed building.

the Swire team saw a demonstration of Digital Project software. After serious consideration, Swire Properties adopted Digital Project (DP) as their company-wide BIM management tool in February 2005. The DP system was designed to support large, complex projects and manage relationships to various information sources such as cost, construction, and facility management.

At the time of DP adoption the OIE project team was performing schematic design with traditional 2D drawings. The four key project organizations were already on the project:

- The design consultant team consisted of the architect, the structural engineer, the mechanical and electrical engineer (M&E), and the quantity surveyor. Wong & Ouyang Hong Kong Limited led the design.
- Structural design was provided by Ove Arup & Partners Hong Kong Limited.
- Meinhardt (M&E) Limited was responsible for the M&E engineering design.
- Levett & Bailey Quantity Surveyor Limited was responsible for all aspects of cost control.

Initially, these four companies, which comprised the design consultant team, communicated with each other using 2D drawings but subsequently

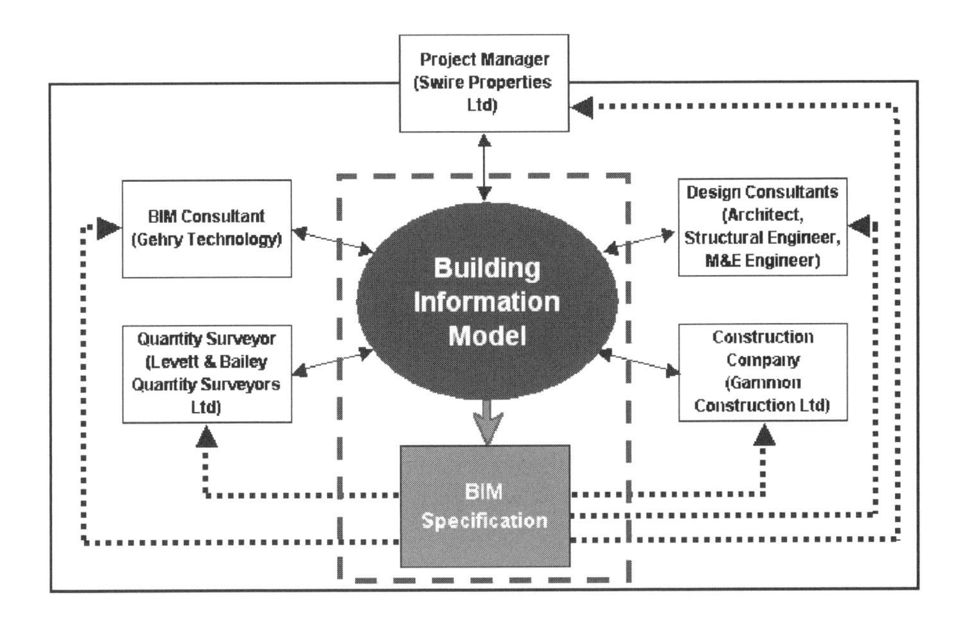

FIGURE 9-7-2 Integration of BIM within the project team.

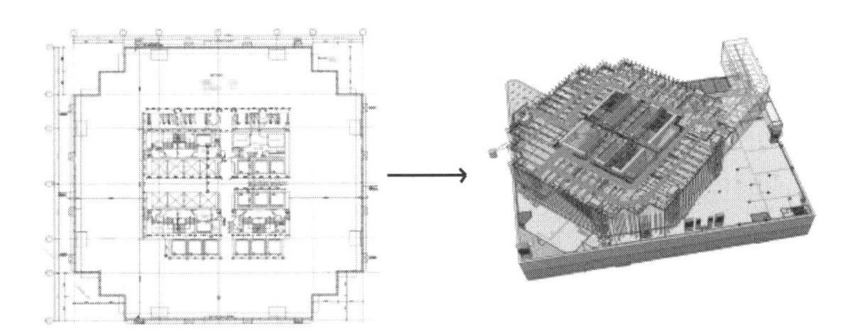

FIGURE 9-7-3 Translation from 2D to 3D.

developed a full 3D process using DP's BIM software as shown Figure 9-7-2. To expedite the process, GT provided BIM consultancy services and led the BIM implementation and training.

The building information model was initiated by Gehry Technologies, who then trained the design consultant team for three weeks. The actual modeling team consisted of architects, structural engineers, MEP engineers, and quantity surveyors. After becoming proficient in the use of DP, the design consultant team took over the BIM process and completed the building model. GT and the design consultant team cooperated closely in this effort.

Creation and Coordination of the Building Information Model

Almost all coordination issues were managed using BIM. The BIM consultant team, the OIE design team, and the project manager worked in one room for the first year. The design team consisted of four architects, four structural engineers, six MEP engineers, two quantity surveyors, one project manager, one MEP project manager, and four GT BIM consultants. Everyone involved worked either directly or indirectly on coordinating the input of information into the building model. Project team members also communicated with each other through a portal site maintained by Swire Properties, which became the main communication platform for the BIM process. Project information that was not contained directly in the BIM was delivered and shared through the portal. The BIM implementation team met weekly for the first year to identify and resolve errors, clashes, and other design problems using the BIM.

In practice, many unofficial coordination meetings also took place, because the team was located in the same workspace. A number of clashes and errors were identified and managed before tendering and construction. Figure 9-7-4 shows how the clashes between elements of different disciplines could be identified easily in the 3D model. Clash detection was mainly achieved using functions contained within the DP software, which is able to identify geometric clashes

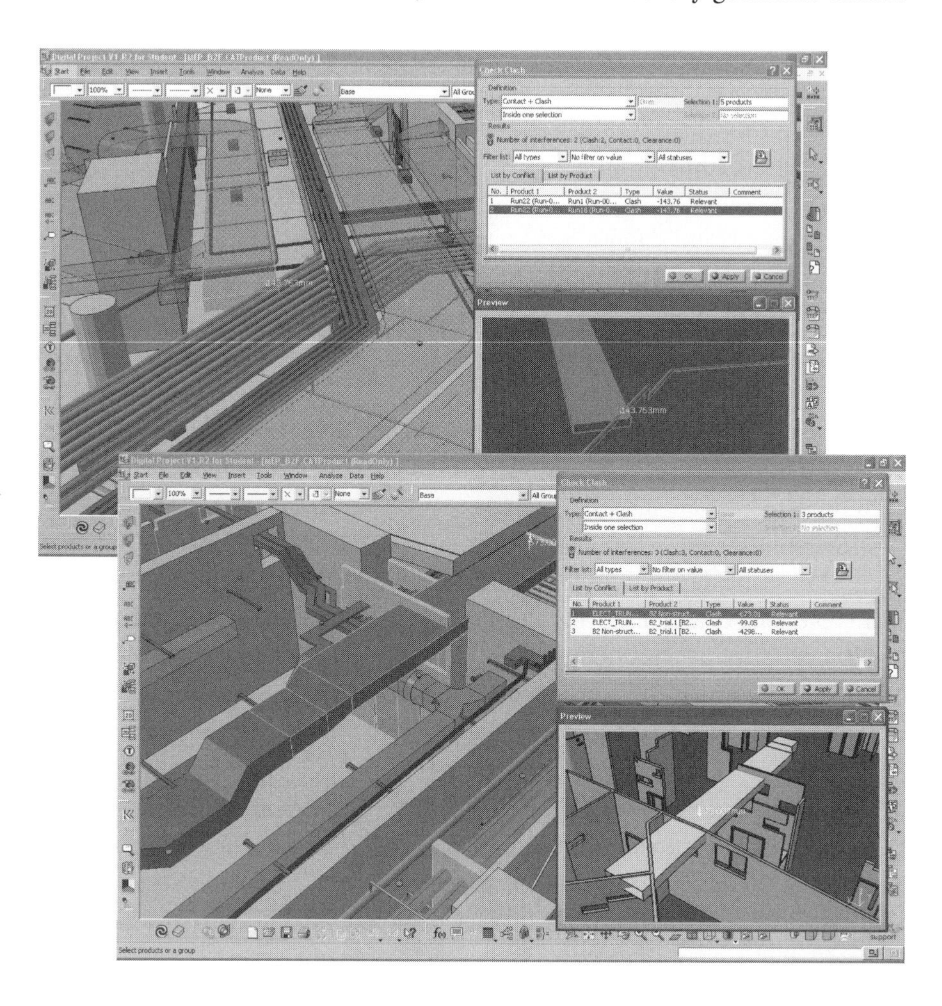

FIGURE 9-7-4
Examples of automated clash detections.

and automatically generate a list. Double-clicking on an item in the list takes the user to the virtual geometric location of the clash. The structure is then corrected and re-designed by the appropriate project team member. The user can specify the tolerance of the clash-check, and this clash tolerance can be designated as the standard.

Traditionally, clashes are identified manually by design consultants using overlaid drawings on a light table. Correspondingly, in the past, a great deal of clash detection and management was left to constructors. Through BIM, over 2,000 clashes and errors were found prior to tendering and construction, which means that a substantial cost savings was achieved, compared to the incomplete design information inherent in a traditional 2D process.

Organization and Structure of the Building Information Model

This section describes the way information was structured using Digital Project.

A typical building information model can consist of hundreds or even thousands of parts (referred to in Chapter 2 as an object or building component). A part can be a wall, columns of a floor, an escalator or an HVAC run. Another type of file, called the *Product Structure* is used to organize these parts within an hierarchical structure (also called a tree structure). This approach has several powerful implications:

- One can open the entire building model or just one branch of it. An example would be a floor of a building or a building service system, such as HVAC or Drainage.
- Large teams can work on the same DP master-model concurrently, because each file is an autonomous entity. Modelers can use a tool called CVS (Concurrent Versioning System) to manage file permissions for modelers.
- It is possible to load a part from different Product Structures. For example, one can create a tree to load the building floor-by-floor and at the same time create another tree organized by function (Figure 9-7-5).
- Parts files can be interlinked and therefore influence each other's geometry in a parametric way.

A detailed example of a typical project tree-structure is shown in Figure 9-7-6.

A typical BIM model is structured in a way that has the main building parts, such as PODIUM and TOWERS, e.g., POD, T01, at the highest level. Each main building part is then organized by floor and also contains branches for parts that are not floor specific. Examples of the latter are the branches for

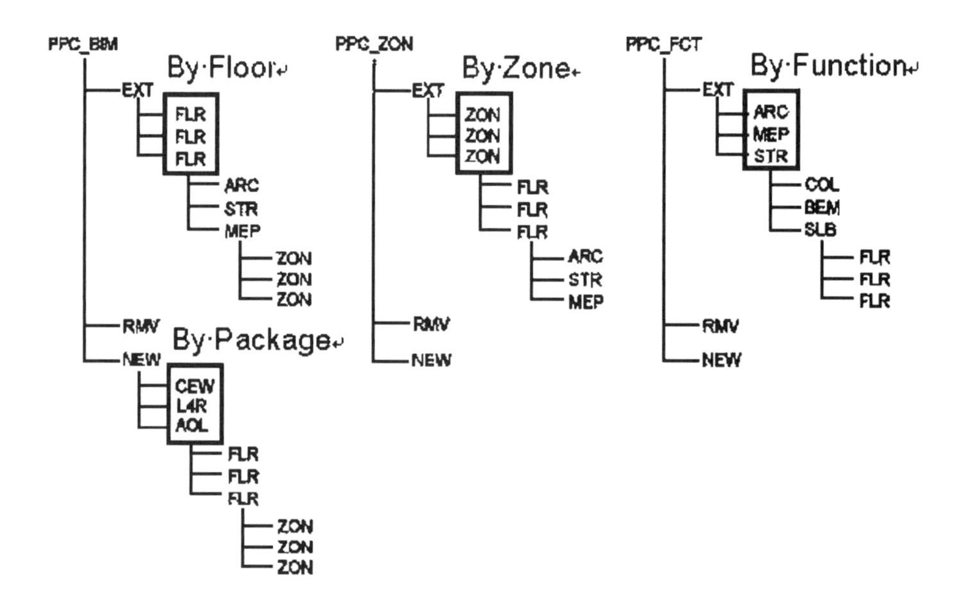

FIGURE 9-7-5 BIM database structure. This figure shows that files can be organized in multiple ways: by floor, by HVAC zone, by function and by construction package. The same part file can be referenced in multiple product structures.

the main driver files, lift cars, and section drawing files. Driver files contain information for details, such as column centers, floor-level planes (e.g., Finishing and Structural Floor Levels, SFL & FFL respectively), and zoning cuts (for terminating columns and beams that extend beyond each zone). The parametric nature of DP allows one to change a floor level and automatically update all of its related geometries. Driver elements are published and thus made available as linkable objects. Each floor contains five sub-branches: (1) a branch with floor specific-drawing files, such as plans; (2) a branch containing driver files called DRV; and (3-5) three sub-branches that contain the actual geometry. There is a branch for architectural elements: walls, finishes, balustrades, landscape elements, sanitary, escalators and lifts; a branch for structural elements: core walls, columns, slabs, beams, stairs and ramps; and a branch for MEP elements: HVAC systems, drainage pipes, etc.

The current OIE BIM model is approximately five gigabytes in size. It opens easily on a laptop, and there are no data management problems. Information is organized properly, and only material that is being worked on is opened and shown in full detail so that it can be updated. Items surrounding the item being worked on are shown grayed out and in less detail using *Computer Graphic Representation* (CGR). This reduces data size but can still provide clash detection and measurement information.

Tendering

DP's BIM tool measured many of the quantities automatically; however, the quantity of reinforcing bar in the structure was calculated manually

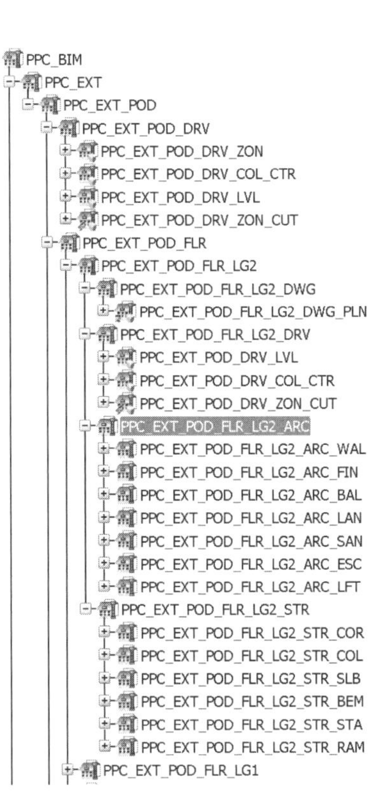

FIGURE 9-7-6
BIM Data Structure for the Podium showing the tree structure for this part of the building.

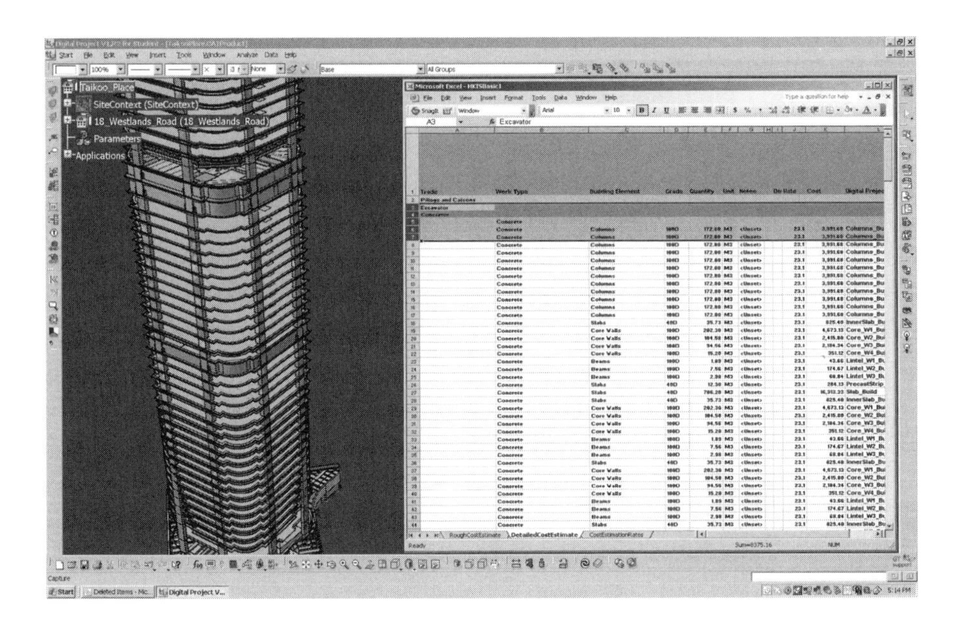

FIGURE 9-7-7
Automated quantity takeoff for the generation of the Bill of Quantities in Excel Format.

using the ratio of rebar to concrete. DP could have provided this capability if the rebar had been modeled explicitly. All elements modeled in the BIM process included their own attributes, such as size, weight, area, volume, etc. Such quantitative information was tabulated automatically and imported into the Excel spreadsheet shown in Figure 9-7-7. This approach increased accuracy

FIGURE 9-7-8 Site progress as of January 2007.

and reduced the effort and time normally required for manual quantity takeoff. Furthermore, the quantities were linked to the BIM, which enabled them to automatically update when the design changed. The BIM model was provided to all of the tenderers, enabling them to confirm the bill of quantities firsthand using the model. This improved the tender process and resulted in lower cost estimates and reduced risk.

Tendering for the OIE project was greatly improved by precise quantity extractions from the BIM. Construction companies that bid on the OIE project saved time and money because they did not need to measure the quantities manually. The increase in accuracy of quantity takeoffs also helped the quality of construction and reduced the contractors' risk. BIM was used to track the cost implications of contemplated changes onsite. Designers and engineers could easily coordinate with each other by checking all elements using 3D BIM (Figure 9-7-12), which has traditionally been a difficult process using 2D drawings.

9.7.3 Post-Tender Stage BIM Implementation Process

At the time of writing, nearly one-half of the 24-month construction program had passed, and the structure had progressed on schedule (Figure 9-7-9). After tendering, the contractor's construction BIM model was updated by Gammon Construction Limited. A substantial amount of additional information regarding construction objects, such as formwork and other temporary structures, was added by the contractors. Of course, the model is in accordance with the pre-tender design intent. Gammon has realized the added value of using BIM technology in construction.

By utilizing the BIM elements developed by the design consultant and contractor project teams, advanced construction process modeling was also carried out to further verify the construction methodology. Gammon Construction Limited called upon the expertise of the Hong Kong Polytechnic University's Construction Virtual Prototyping Laboratory, led by Professor Heng Li, to produce detailed visualizations of the construction sequencing. Links were created between DP's BIM elements and the detailed Gammon Primavera construction schedule. This enabled the visualization of construction sequence which was a helpful tool for the contractor, called 4D CAD (see Figure 9-7-10). Visualizations of the sequence of erection of building elements could be created easily, according to the Primavera early or late start sequence. In this way, Gammon was able to visualize and analyze various scenarios (see Figure 9-7-10); and spatial/safety issues could be identified prior to construction (see Figure 9-7-11). For example, the sequence of formwork erection for a typical floor was checked and re-checked to ensure that a construction cycle of four floors per day could

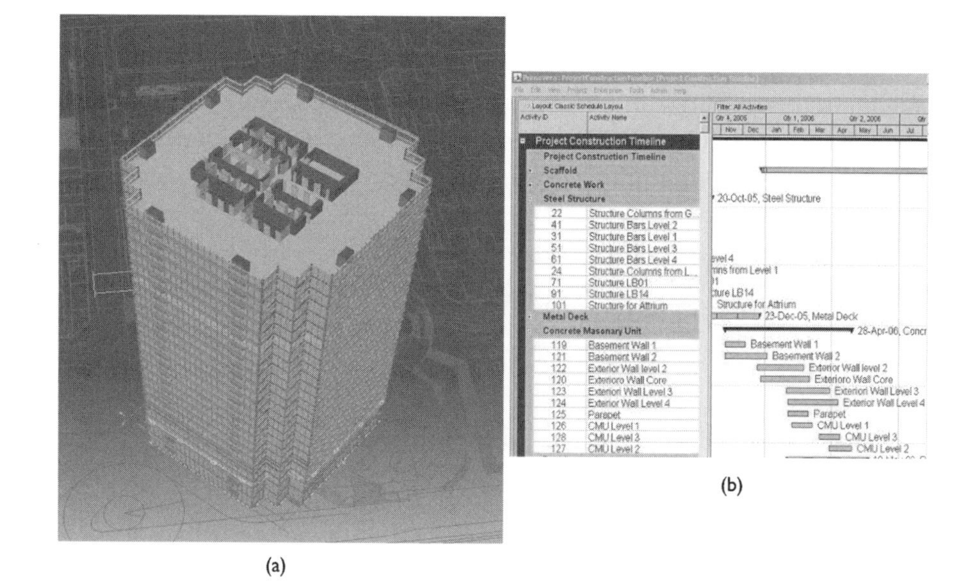

FIGURE 9-7-9 Schedule integration and visualization. *Image provided courtesy of Gammon Construction Ltd. and Professor Heng Li at Hong Kong Polytechnic University.*

FIGURE 9-7-10 Illustration of the construction sequence (see color insert for full color figure). *Image provided courtesy of Gammon Construction Ltd. and Professor Heng Li at Hong Kong Polytechnic University.*

FIGURE 9-7-11 Illustration of clash detection.
Courtesy of Gammon Construction Ltd. And Professor Heng Li at Hong Kong Polytechnic University.

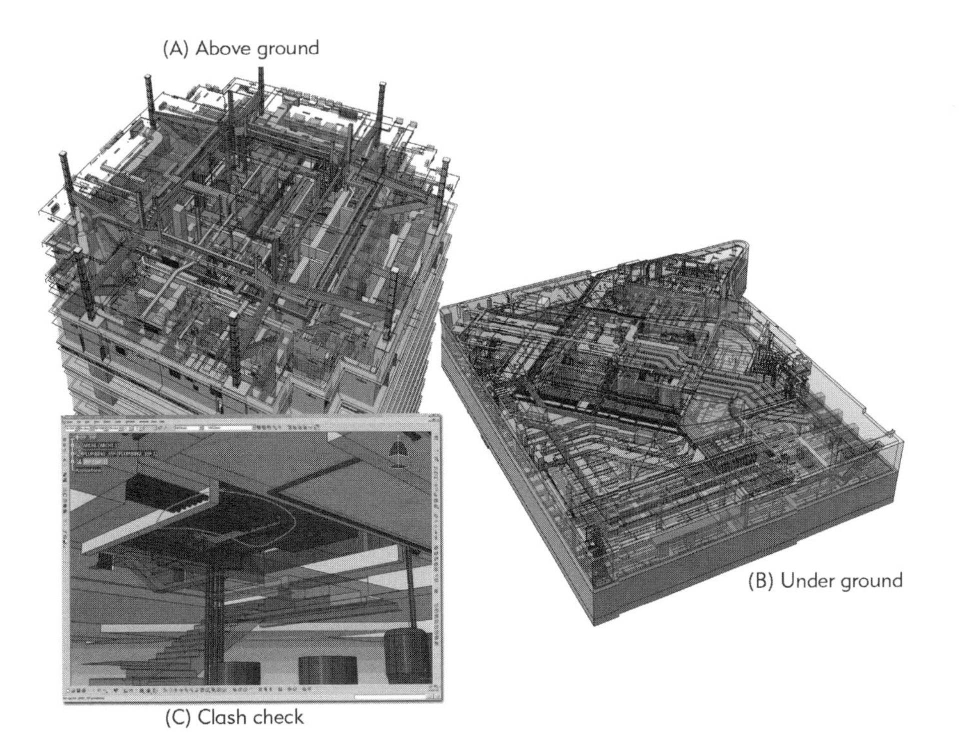

FIGURE 9-7-12 Three-dimensional coordination of all project elements. (See color inserts for full colored figures.)
A) Structural and MEP elements at a typical floor lebel;
B) Piping in basement area under ground level;
C) MEP elements in the basement levels of the building.

be maintained. The construction methodology for the difficult outrigger floors was carefully examined to ensure safety and practicality. This saved time and money in the field.

9.7.4 Conclusions and Lessons Learned

One of the main challenges for the project team was transitioning from 2D to 3D. Swire Properties mitigated potential cultural issues by requiring the use of 3D and the DP system on the project. Furthermore, Swire hired Gehry Technologies as a consultant to provide the adequate training and support during the first year of the BIM effort. In the future, Swire Properties hopes to achieve even greater value when the BIM technology and working methods are implemented from the very beginning of the process.

A second critical lesson is to select a BIM application and system appropriate for the type of projects and business goals of an organization. Swire carefully reviewed and assessed technologies based on their long-term organizational needs. Since Swire builds large, complex facilities, the selection of DP was critical both in supporting their long-term investment as well as ensuring that their project team would not face daily issues with managing model size or complexity.

The post-tender stage is not yet complete, so it is impossible to quantify all of the benefits and problems associated with BIM implementation in this phase. At this point it is possible to say that there have been significant savings in cost and time resulting from the reduction in design errors found by clash detection. The careful 4D analysis performed by Prof. Heng Li's group provided assurance that a fast and safe schedule could be achieved. Quantitative information, such as the number of change orders, safety records, budget performance, schedule performance, etc. will ultimately help measure the value added by the use of BIM.

Acknowledgments for One Island East Project

This case study was originally written by Sung Joon Suk and Martin Riese for Professor Chuck Eastman in a course offered by the Design Computing Department at the Georgia Institute of Technology, Winter 2006 and has been adapted for use in this book. The authors wish to acknowledge the assistance of Swire Properties and Gehry Technology in the preparation of this study.

9.8 PENN NATIONAL PARKING STRUCTURE

Use of BIM by a precast concrete subcontractor for detailed design and prefabrication

9.8.1 Introduction

In 2005, the first entirely precast concrete parking structure engineered and designed by a precast producer using parametric 3D modeling was built at the Charles Town Racetrack in West Virginia. The precaster, however, used the 3D model only for internal processes. Communication between the architects, engineers, and precast fabricator was carried out using traditional 2D drawings. The precaster produced 2D erection drawings and shop drawings of the pieces (called *shop tickets*) for review.

Soon after completion of the Charles Town Racetrack, the same team—owner, architect, engineer, contractor and precaster—found themselves working together on a similar but larger project, the Penn National Parking Structure in Grantville, Pennsylvania. This time, the model was used more extensively for design collaboration and construction. The second project is the subject of this case study.

The Penn National Parking Structure is 303 ft. wide, 497 ft. long, and provides 773,498 sq. ft. on five levels (slab-on-grade and four levels above grade). It is a total precast structure, with cast-in-place concrete toppings on the decks, aluminum and glass curtain walls enclosing the stair towers and decorative column covers, and other details built of studs and EIFS boards.

Design of the building began in July 2005, and implementation of the precast model commenced in June 2006. Construction was scheduled to begin during the first quarter of 2007. The owner employed the construction manager under a guaranteed maximum price contract, and the precaster contracted with the construction manager under a fixed price contract for design, construction, and erection. Figure 9-8-1 shows a general view of the building, as it was designed.

Because the parking deck is open to the environment, it did not require ventilation and heating/cooling analyses or coordination with window and door suppliers, plumbing fixtures, or mechanical systems. For these reasons, the case study focuses primarily on the experience of the precast fabricator, who used BIM to interact with architects and structural engineers. For the precast fabricator, the innovation consisted of using a parametric 3D model throughout the process. In the company's earlier projects using BIM, the model had served exclusively for sales and automated preparation of general arrangement and shop drawings. Here its use was extended to structural analysis and design, design coordination with architects, submittals for structural design approvals,

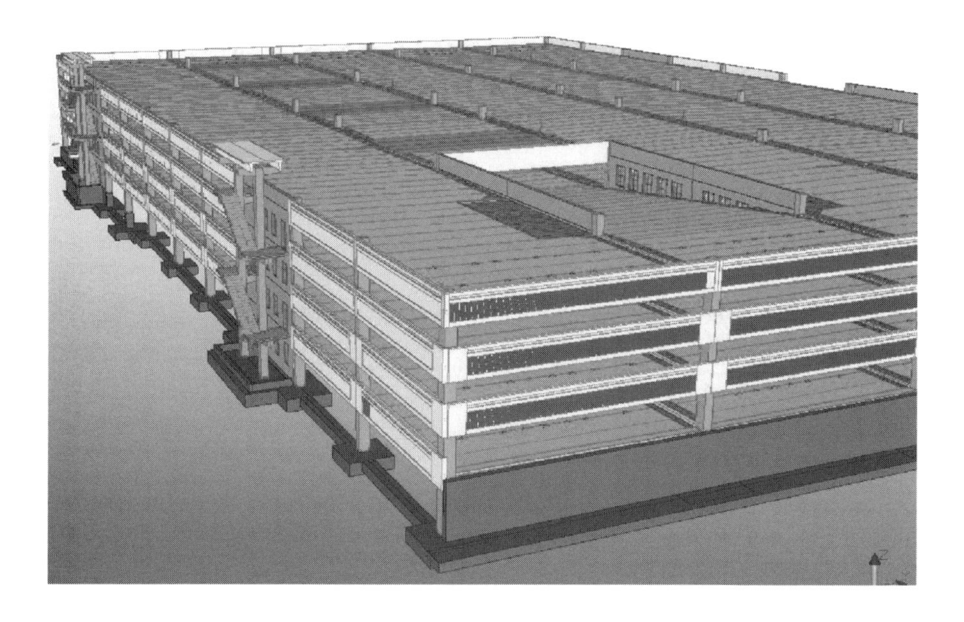

FIGURE 9-8-1 General view of the Penn National Parking Structure.

Table 9-8-1 The project participants:

Owner:	Penn National Gaming
Architect:	Urban Design Group Inc, Atlanta, GA
Structural Engineer:	Gregory P. Luth Associates (GPLA), Santa Clara, CA
Precast Contractor:	High Concrete Structures Inc., Denver, PA
Construction Managers:	Keating Building Corp., Philadelphia, PA

quantity takeoffs, and production planning. The software used by both the precaster and the structural engineer was Tekla Structures Version 12.0.

9.8.2 Conceptual Design

The building was first modeled by the precaster after a bid was requested by the construction manager. The precaster was given the freedom to propose a structural system, including the internal layout of columns and spans to provide a constructible and cost effective structure. In the first iteration, all of the structural elements were modeled, showing only their external geometry, and no connections were detailed between them. This phase took only nine hours, partly due to the team's experience with parking structures.

The precast modeler explained: "After this conceptual model (nine hours) we e-mailed the Web viewer and PDF files of a typical plan and the exterior elevations to the architect, owner, and engineer of record. The following morning we had a conference call with all parties, reviewed the concept, and discussed the foundations (more of a cast-in-place interface) and our concerns/recommendations for framing at stair towers and fire requirements, as the garage adjoined the casino. After this call we worked on the model for another six hours to incorporate the changes discussed."

For each design iteration, a detailed list of the precast components, with accurate measures of their concrete volume, was extracted and imported into Excel to generate the company's price estimate for the building.

Using the model as a medium of communication, they conveyed the concept to the owners, coupled with a very quick turnaround time to respond to their requests for modifications. This proved to be a powerful tool and was instrumental in the company winning the contract for the building. The precaster was able to rapidly model and estimate the costs of various alternatives, which allowed the construction manager and the owner to make decisions with confidence. For example, the decision was made to build precast stairs throughout, unlike the steel with concrete-fill stairs used in the earlier Charles Town project.

9.8.3 Design Development and Analysis

Once the contract phase was complete, the building was handed over to the company's engineering department. In the handover meeting, the modelers and estimators projected the model onto a wall-sized screen and manipulated it to explain the project, instead of the accepted practice of providing reams of 2D drawings, notes, and calculations to the engineers. The handover proved to be extremely efficient. The engineers were able to quickly focus their questions, because the presentation was clear and concise. More importantly, the scope of the project was defined with precision. Lack of clarity about exactly what would work was the responsibility of the precast company and was included in the contract, as opposed to various tasks being performed by others, which was a source of conflicts with general contractors in previous projects.

The sales model was then developed into a design and production model. The engineer of record was responsible for overall structural stability, which included concern for interfaces between the precast building, its cast-in-place foundations, and a neighboring structural steel building. Approximately one year earlier, a number of precast concrete companies, with whom the engineer had collaborated, began using the 3D modeling software used by structural steel detailers. The engineer's practice had decided to adopt the same software. Therefore, they were able to receive the precaster's model directly.

The results of the engineer's lateral analyses were provided to the precaster for detailed design of the individual pieces and connections. Structural analyses of the pieces were performed in three ways: (1) using the modeling software's interface to the STAAD PRO finite-element (FEM) analysis package; (2) using Microsoft Excel design tables prepared in-house; and (3) using LEAP Software Inc.'s PRESTO and AXSYS applications. Figure 9-8-2 shows a model of the stack of four lite-wall pieces and the analysis software. The typical turnaround time for a single analysis run using the FEM package was 40 minutes and included: definition of loads, exporting to the FEM package, execution of the analysis, and importing the results back to the model. Engineering the design of the individual pieces to carry the necessary loads was performed directly within the model using LEAP software's PRESTO and AXSYS plug-in applications.

Modeling the precast structure revealed potential misunderstandings or inconsistencies in the set of architectural design drawings but also expedited their resolution. For example, the modeler could not determine the correct location of the EIFS board column cover to the precast column, shown in Figure 9-8-3, because it appeared differently in the elevation compared to the detail. Issues such as this were clarified rapidly by using model images over telephone conferences. Similarly, the modeler encountered situations not fully defined in the architectural details. For example, a spandrel adjacent to a shear wall was not detailed in the architectural drawings, because there were no sections cut there. These issues were resolved quickly, because the modeler could demonstrate to the architects exactly which details lacked definition.

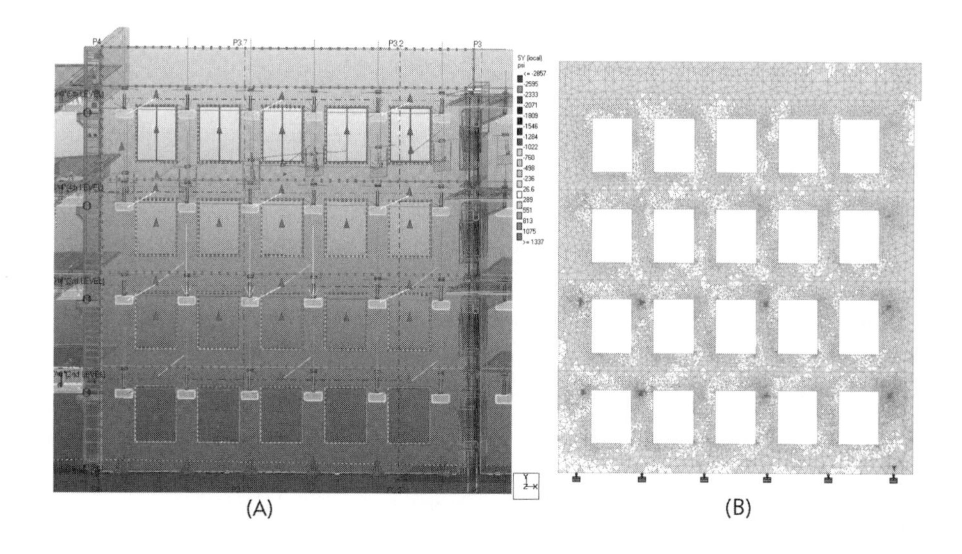

FIGURE 9-8-2
A) A stack of lite-wall precast pieces in the Tekla Structures model with loads defined, and B) the same section in the STAAD PRO finite element analysis. (See color insert for full color figure.)

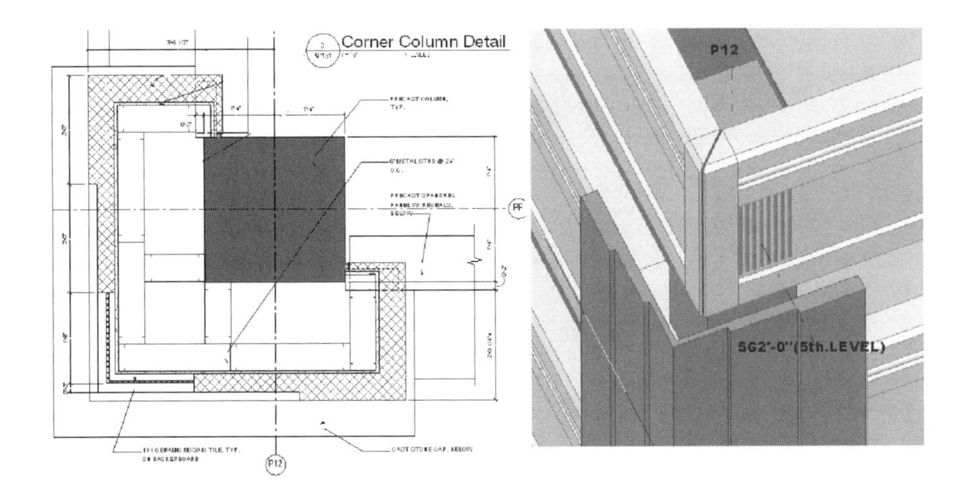

FIGURE 9-8-3
A corner column detail (left) and the column cover in the model (right).
The location of the edge of the cover relative to the primary axis appeared different in the detail compared to the elevation drawings and was resolved by modeling it in the precast model.
Image courtesy of Urban Design Group, Inc.

Another issue identified by the architects involved a situation where double tee bearings would have obstructed egress through a door in the stair tower. The precaster changed the bearing type from a corbel to a pocket, solving the issue. The principal architect responded to the BIM process by saying, "What I found so appealing about the model is the way in which the connections could all be seen clearly and had the ability to avoid future conflict."

9.8.4 Detailing for Fabrication

The submission of drawings for review, both structural and architectural, is usually a lengthy process and is often performed in batches due to the slow pace of shop drawing generation. The Penn National building has 1,300 individual pieces of precast concrete, with 50 unique designs. Structural review was performed directly using the model. It was straightforward, because the structural engineer used the same BIM software (Tekla Structures), obviating the need for translations. The precaster was still required to submit drawings through the channel specified in the contract (the construction manager) to meet legal requirements. This highlights the need for new contract terms that would enable the realization of BIM's benefits without perpetuating practices that were appropriate for communication using 2D drawings alone.

As in most precast buildings with at least some architectural detail, the precaster was required to produce a mock-up piece, showing the salient features that would appear in the real pieces and invite the architect to view it. The mock-up piece was modeled to float above the building in the real model and manufactured using information drawn directly from the model itself. Both the virtual and the real mock-up pieces are shown in Figure 9-8-4.

FIGURE 9-8-4
Mock-up of a precast piece, showing the depth surface treatments on the facade, in the model and as it was produced.

A) Panel in model

B) Reinforcing in model

C) Mock-up of precast piece

The construction manager expressed satisfaction with respect to the coordination of the interface between the parking structure and the main building it is attached to. The structural steel frame of the main building was also modeled, and the visual clarity of both models allowed the construction manager's firm to feel confident that the two structures had been coordinated properly. In an earlier project, the manager commissioned the construction of a physical model to understand the interfaces between parking, retail, and hotel areas of a building. In this case, a physical model was not needed.

Once the model was fully detailed, the construction sequence was determined with the erection crew superintendent, by consulting the model. Setting the sequence is a complex task, because pieces key into one another in different ways that restrict the way they are to be maneuvered into place, and also because the partly erected structure must be checked for stability at every step. The visualization proved sufficiently powerful to enable the erection superintendent to consider using two cranes instead of the usual one. In addition to ensuring that the two cranes would not interfere with one another, the fabricator was able to address other issues involved in scheduling the two cranes. Erection crews commonly experience delays when design or drafting errors result in pieces not fitting together. In such situations, repairs, adaptations to connectors or even replacement pieces are needed, resulting in downtime for the crane and crew. The Charles Town project mentioned earlier, however, was erected without a single error in the pieces, which increased the superintendents' confidence in simultaneously scheduling two cranes and crews.

The ability to reliably plan an erection schedule is not only important for crews but for production itself. In this project, erection sequencing data was used to schedule production. Phasing data was first used to create stepped bills of material for purchase and/or fabrication of rebar, embedded hardware, and other components. This has a number of benefits that made the process leaner: by reducing work-in-progress, raw material, and parts inventory. It also meant that ordering was controlled and the waste of excess parts typically ordered in previous jobs was avoided. Finally, fabrication of the pieces themselves in closer relation to the erection sequence reduced the number of pieces in the storage yard.

The most easily measured (but not necessarily the largest) benefit in this project was a reduction in engineering and drafting hours. The precast company spent approximately 52% of the hours budgeted for an equivalent 2D CAD design process. Some of this benefit can be attributed to prior experience with the Charles Town project, including the availability of the building model and the custom connections and parts that had been prepared for it.

9.8.5 Conclusions and Lessons Learned

Representatives of the precaster reported some specific lessons learned in this project. First, visual coordination using a model helped prevent constructability errors in shop drawings from reaching the production plant. This was underlined during modeling with some of the company's standard connections. For example, the modeling of a standard girder-column haunch detail, which had been repeated for years in CAD drawings, revealed a spatial conflict between a steel plate and a set of rebars. Production personnel simply bent the rebars to accommodate the plate. The error was fixed in the new BIM connection detail.

Another visualization benefit experienced earlier in the Charles Town project was emphasized: the power of the model to communicate the design intent to the owner. In these two projects, the owner raised no complaints or issues at the end of construction, and this was attributed in part to the explanatory power of the model, according to the precasters.

In this case study, the degree of BIM adoption across various parties was not uniform. Along with concomitant technology issues, this led the participants to feel that the technology's full potential had not been realized. The principal architect expressed this succinctly:

> "The model being shared outside of High Concrete Structures, Inc. for design verification, coordination, and for the resolution of clash conflicts has been worthwhile. However, the use of the model has not appeared to go beyond this. What is missing is the ability to have other team members actively interface by using the model (outside of the two structural engineering groups). We as the architects could only view a slimmed-down version and would not be in anyway capable of interlacing our other non-precast components into the model. Other projects we are doing are moving into a full BIM format."

Other participants suggested changes made possible by BIM that would improve their processes in the future:

- The engineer is responsible for the integrity of the structure as a whole and envisions preparing a conceptual design model in Tekla Structures, then passing it to the subcontractors of all structural components (steel, precast and cast-in-place concrete) for detailing. In this way, the engineer has a higher degree of confidence that everything will fit, obviating the need for additional review and possibly rework for all parties at the submittal review stage. Ideally, they would then use an intelligent software tool to check whether the subcontractors' detailed designs conform to their original design intent by performing comparisons directly within the model.
- Coordination with the cast-in-place (CIP) concrete and electrical subcontractors. The precaster was responsible for coordinating drainage slopes, electrical block-outs in tees for conduit and lighting, and anchor bolts for light standards on the roof. Coordination with CIP is needed for base plates and embeds, such as *Lenton* splice sleeves (a pivot which is grouted after assembly and needs ¼ in. tolerance) and dowel sleeves. In this project, they produced PDF documents of model views, but they could easily provide a full model to the CIP detailers.
- Standard connections and details in the modeling software could improve productivity well beyond what was achieved.
- Using model viewers in plant production shops and also for erection could improve productivity of those activities.
- The precast company is in the process of implementing a corporate-wide management information system using SAP. Exporting of bills of

material directly to the corporate SAP system is planned, but they were entered manually for this project.

A specific demand made by several of the precast company's quality control personnel provided an interesting anecdote that underscores the extent to which BIM adoption required changes in their organizational culture. Some QC inspectors, who check that drawings submitted to the plant for production are correct and complete, found fault with the fact that a series of embedded plate assemblies in a precast beam were each shown on the BIM shop drawing with identification codes annotated separately. Standard company practice required that such plates be labeled collectively, with a single text ID code, to ensure identical fabrication and installation. The demand that shop drawings look exactly like 2D CAD drawings resulted in the modelers having to erase the accurate associative annotations placed automatically by the system and replace them with manually applied and more error-prone dummy text notes.

While this case study demonstrates that significant progress has been made, it is important to note that the use of BIM is still in its infancy. The structural engineer emphasized that the full potential of BIM for design development and coordination will only be realized when all three of the following conditions are met:

All of the designers and detailers use BIM tools. If any one of them uses 2D drawings, then the whole team's workflow remains as a 2D workflow, with long cycle times for submittal and review and likely coordination problems.

The richness of data exchange between BIM tools is significantly improved. Exchange using IFCs (see Chapter 3) is still inadequate for transferring anything but 3D geometry, due to discrepancies between the ways that each software vendor implements their export and import routines.

The project management and participants are committed to a more collaborative and lean design process, in which information is shared openly and cycle-times are reduced.

Acknowledgments for Penn National Case Study

The authors greatly appreciate the help of the following professionals, whose assistance in compiling this case study was invaluable: Dave Bosch, Bill Whary, Karen Laptas, and Dave Foley at High Concrete Structures Inc.; Gregory Luth and Michael Loomis at GPLA; Dave Hofmann at Keating Building Corp.; and Don Buenger and Elroy Sutherland at Urban Design Group. Image 9-8-3 courtesy of Urban Design Group, Inc. all other images courtesy of High Concrete Structures.

9.9 HILLWOOD COMMERCIAL PROJECT

BIM for Conceptual Cost Estimating

9.9.1 Introduction

This case study demonstrates the potential for building information models to support conceptual estimating early and often in a project and during the conceptual design and development phase. The use of parametric models by the design-builder, the Beck Group, showcases the benefits of providing informed design options to an owner early in the process and enabling both the owner and the design-build team to explore more options and, ultimately, to provide better overall design, in terms of programmatic and cost requirements.

The project is located in the Victory area of downtown Dallas, TX on an old railroad yard that is currently under remediation in preparation for the Victory Park Development Project (www.victorypark.com), which includes an office-retail facility and several other buildings (See Figure 9-9-1). The owner and developer, Hillwood Development, plans to lease the retail and office space. The project was initiated in August 2006, with a lump-sum fee for design services provided by the Beck Group. As of March 2007, the project was in the schematic design phase, and lease discussions were in progress. This case study focuses on the two-month period of schematic design, when the conceptual estimating took place.

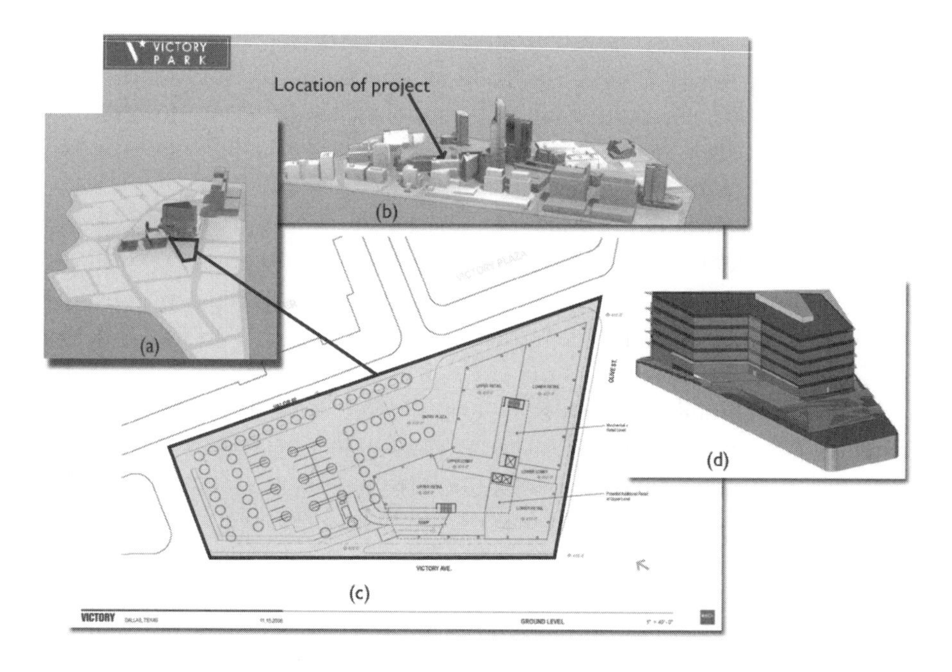

FIGURE 9-9-1 The project site shown in A) 3D project site rendering, B) 3D rendering of Victory project, C) 2D plan view in AutoCAD, and D) 3D conceptual rendering of building.

Table 9-9-1 Overview of project details.

Developer	Hillwood
Architect	Beck Group
Lot size	1.6 acres
Office Space	115,439 sf
Retail Space	22,712 sf
Parking Spaces	81
Floors	6

The project is a six-story 135,000 sf office/retail building on a 1.6 acre lot. The site involves unique constraints for vehicles and pedestrians, requiring accessibility changes due to the topography (See Figure 9-9-1). The Beck Group is predominantly a commercial builder with many repeat clients, and their services included conceptual estimating use of their propriety software.

9.9.2 The Conceptual Estimating Process

The Beck Group provides conceptual estimating as part of its standard services for architectural design. Based in Dallas, TX, the firm is a leading-edge design-builder on the forefront of using parametric-based CAD to support their design and building processes. In 2000, the Beck Group acquired intellectual property from the Parametric Technology Corporation (PTC) to better provide custom design services. With a small full-time team devoted to the customization and enhancement of this technology, they were able to combine their expertise in design-build with that of PTC's technology. Their initial efforts focused on supporting the quick exploration of different design options.

On this project, the Beck Group began the digital modeling effort as soon as the owner signed-off on the conceptual design and prior to the schematic design phase. The Beck Group's architectural design team developed a conceptual cost model and estimate while exploring multiple design revisions and evaluating the costs associated with each of the alternative features. The iterative process involved exploration of design alternatives, calculation of cost estimates, and presentation to the client. The participants were predominantly architects from the Beck Group and the construction project manager, who provided input and guidance on constructability issues and estimating and preconstruction services. Since the Beck Group is a design-build firm, they can rely on internal knowledge. Members of both teams benefited from the multidisciplinary collaboration.

9.9.3 Overview of BIM Technology to Support Conceptual Estimating

The central tool employed is DProfiler, a BIM-based solution capable of generating accurate cost estimates from a digital design model. DProfiler is a 3D parametric BIM tool that allowed the team to quickly generate a design based on specific features, parametric variables, and custom design with parametric components. The major enhancement within DProfiler, compared to other building information modeling tools, is its association with cost information.

As designers build a digital model using components from a building component library, it is possible to view real-time cost information. Each component is associated with cost items from a database. The DProfiler software package is integrated with RSMeans cost data, which includes 18,000 assemblies and more than 180,000 line-items. RS Means is a cost construction database provided by Reed Construction Data (see www.rsmeans.com for more information). This association allowed the team to quickly calculate specific features and design alternatives, allowing the owner and designer to also work in real-time.

Beck's experience using DProfiler for estimating, compared to traditional manual-based estimating, has resulted in a 92% time reduction in producing an estimate with the digital estimating process; and an estimate with a 1% delta from the manual estimate on similar projects. Consequently, the design team can achieve the same results in far less time, with potentially more accuracy and the ability to explore more scenarios.

9.9.4 Overview of the BIM Estimating Process

Once the design team had developed an initial concept, the dedicated modeler used DProfiler to create a parametric building model with links to cost items (Figure 9-9-2). A critical first step involved entering the project information including the project zip code. This allowed the team to account for regional cost factors (See Figure 9-9-3). The modeler then selected the *Building Type* that most closely resembled the project. The *Building Type* sets the project's default assumptions based on pre-defined parameters within the DProfiler database. The *Building Type* is basically a roadmap that links additional building components, for example, it used a template for an office building of 4-6 stories that included cast-in-place concrete structures and slab assemblies commonly found in this building type. Alternatively, an *8-24-story Apartment Tower with a Steel Frame* would include steel member components. These templates were

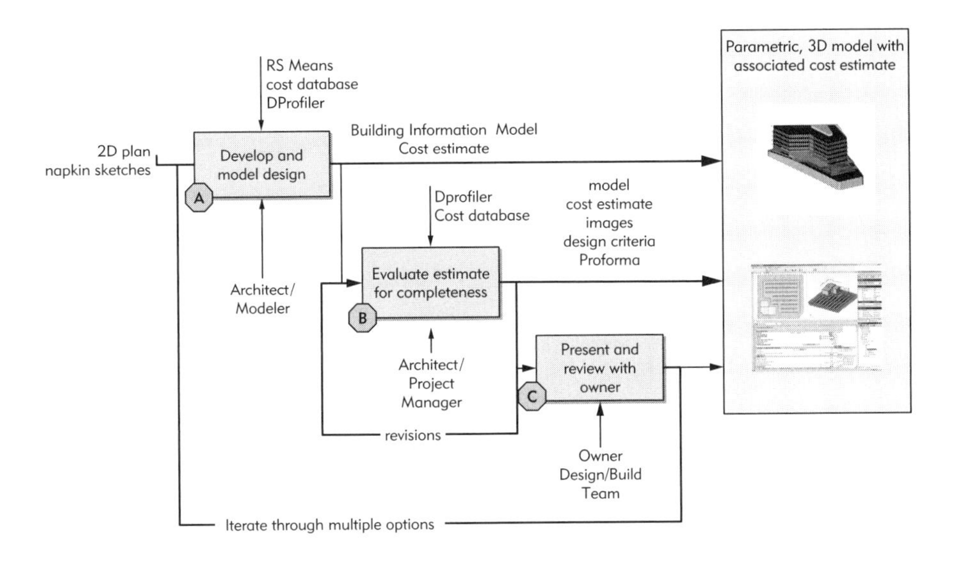

FIGURE 9-9-2 Conceptual estimating workflow using DProfiler.
It included: A) development and modeling of a design scenario using parametric building components and/or project templates; B) evaluation-based estimating using cost information associated with building components from a cost database, such as RS Means, with insight from an experienced designer and project manager; and C) presentation and review of the estimated design option, involving the owner and the design-build team. This entire process can be performed for multiple design scenarios.

created based on the Beck Group's input and experience in building similar projects, as were their other 23 project templates. Organizations can develop and modify such templates based on the types of facilities they design and build.

The modeler then lays out the project site and building based on an initial concept and using building components from the selected template. The modeler can import 2D plans and use them as an underlay to expedite the initial process. As the modeler creates the site or building mass, the summary data is updated in real-time, as shown in Figure 9-9-3.

Each of these components or assemblies is associated with cost items in the database. As the modeler adds detail to the model, the cost estimate updates in real-time and the modeler can view the estimated information, as shown in Figure 9-9-3. In this case, items are associated with information in the RS Means cost database, but creating custom line items or assemblies is also a possibility.

These costs include line items with rules for extracting parameters from the model's components as well as for building components not represented geographically. For example, the slab assembly may include rules to account for a *fire extinguisher and cabinet* for every 2,500 sf of slab in the building.

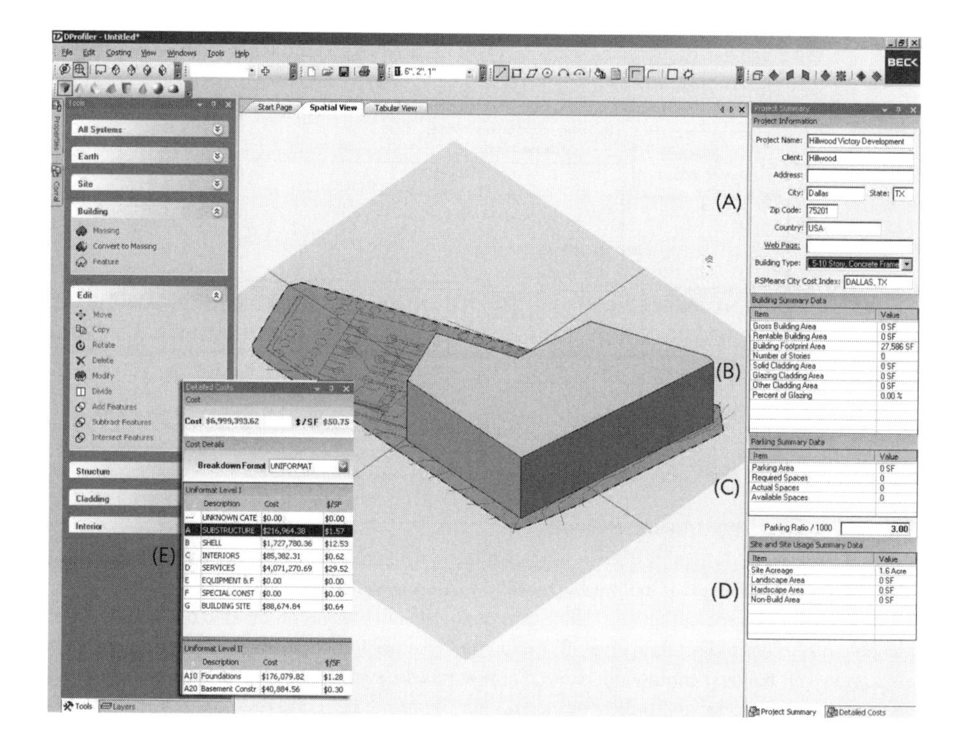

FIGURE 9-9-3 Snapshot of DProfiler showing (A) Project Information, (B) Building Summary Data, (C) Parking Summary Data, and (D) Site and Site Usage Summary Data; and (E) showing a real-time cost estimate for the modeled design. The cost estimate includes overall and cost-per-square-foot as well as UniFormat Level I and II cost breakdowns. Alternatively, cost can be organized using the 16 CSI (Construction Specification Institute) divisions.

Additionally, cost line items can be associated with model parameters and variables, such as overall project square footage. These types of variables can be used to calculate the costs associated with temporary services or other less tangible building items. Throughout the modeling process, the modeler can switch between 3D model views and a detailed cost view, as shown in Figure 9-9-4.

It is important to note that the design and construction team must work together to identify such items and determine how to best incorporate them into the estimating template and components or assemblies.

The modeling process may also involve the creation of a new component or assembly representing an uncommon or unique component. For example, the shading canopy components were created from scratch for this project. This involved creating the geometry and representing the component and its associated assembly item in the cost database.

As the model is created, real-time information associated with the cost database becomes available. This type of information provides the designer with estimated costs for the current design alternative and gives the ability to associate costs with specific design features.

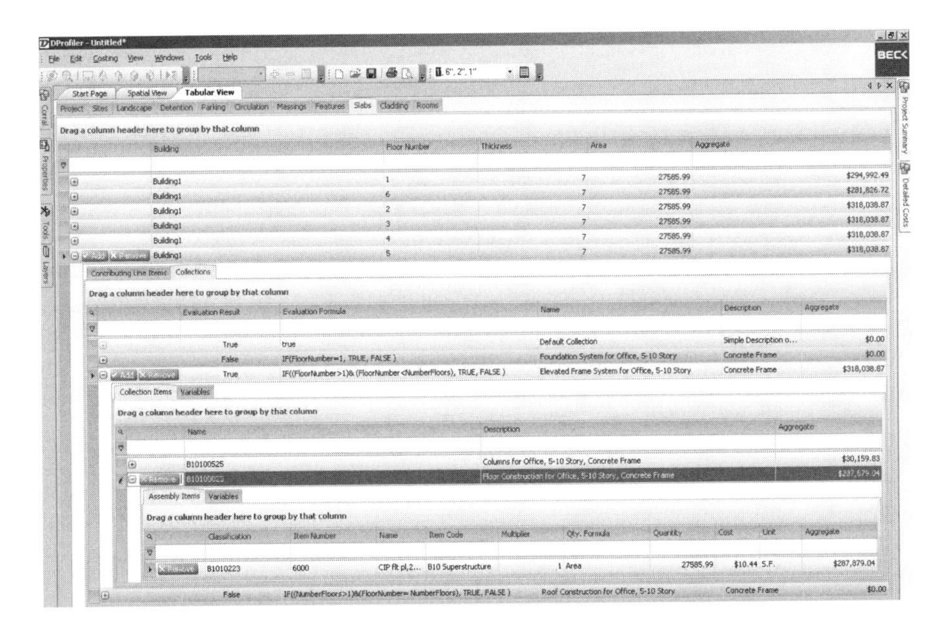

FIGURE 9-9-4 Snapshot of DProfiler's detailed cost estimate showing the hierarchy and information.

At the top of the tabular view are 11 model component tabs. Within these tabs, the Collections and Assemblies each contain individual Line Items from the cost database. In the example shown above, the slab component tab is selected and the 5th level (level five of the building) is expanded to include the Collections contained in that level. The Elevated Frame Assembly is also selected and expanded to reveal the Line Item contained within it.

9.9.5 Design Alternatives That Were Evaluated

The Beck design-build team also used DProfiler to run several *what-if* scenarios, once the initial design concept model and its associated estimate were found to be over budget and did not work within the owner's Proforma framework (see Figure 9-9-5). The team evaluated multiple cost options, such as changing floor-to-floor heights, adding and removing a floor to increase or decrease square footage, relocating the garage component from below to above-grade, and evaluating the current plan against a more rectilinear and potentially more efficient shape to determine the cost premium of the site constraints requiring them to utilize a less efficient plan and perimeter.

One design option included the use of a glazing frit film on the exterior window wall system, in lieu of constructing costly metal-panel eyebrow overhangs to cope with direct solar exposure from the South and West angles. Figures 9-9-6 A and B show these two options. The team combined different options (whenever possible) and reviewed this information with the owner.

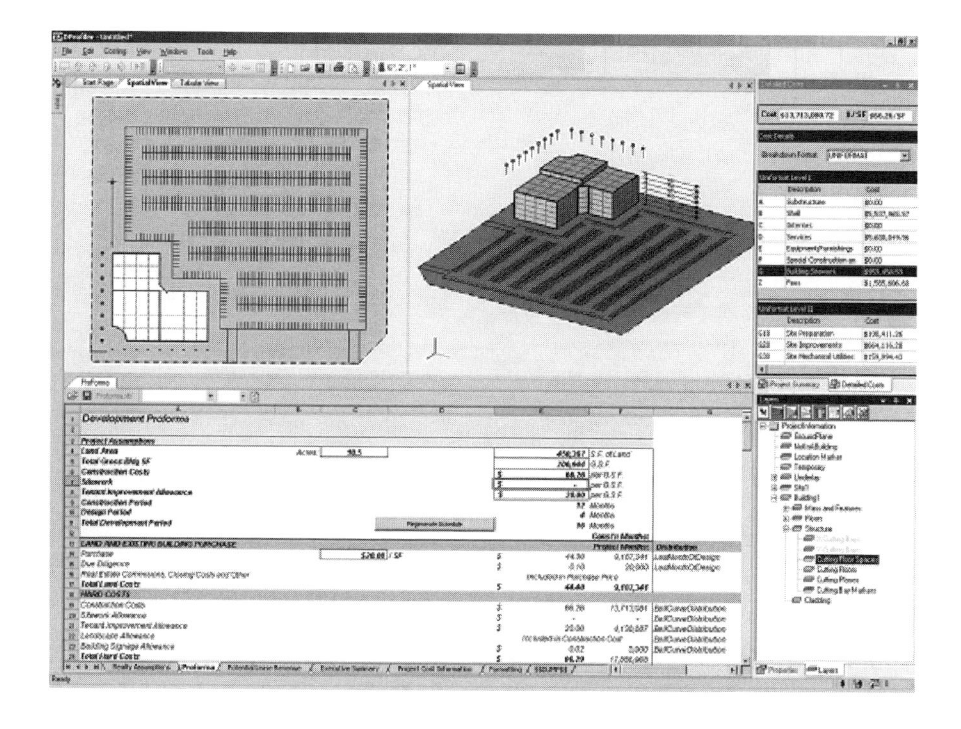

FIGURE 9-9-5 Screenshot of the DProfiler system, showing a 2D spatial view of the project, a 3D view, the related Proforma (left corner view), and cost details (along the right side).

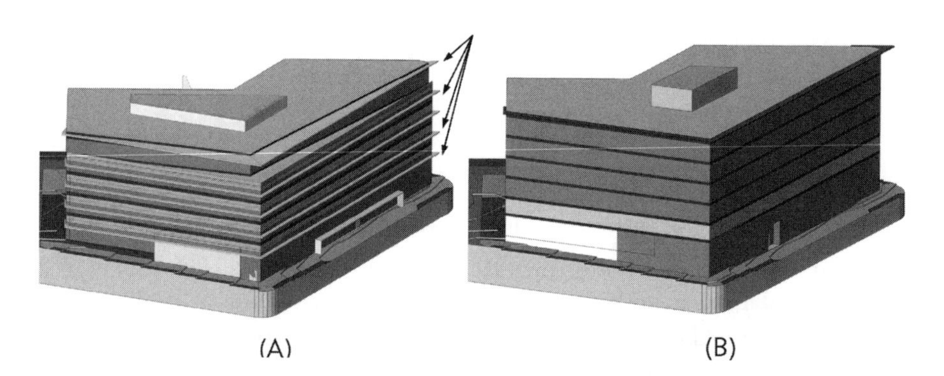

FIGURE 9-9-6 A) Snapshot of the model, showing a design option with eyebrow canopies for shading the direct sun. B) Snapshot of the model, showing a design option with canopies removed that utilizes glazing frit instead.
Glazing frit is a film applied to the interior of the glass and is not visible in the model.

The team was capable of inputting the design variables directly into the owner's financial Proforma (see Figure 9-9-5), such as for leasing square footage, but did not do so on this project. The potential advantage of linking the model directly with the Proforma is that real-time feedback then includes estimated building costs and operating income and expenses. While owners typically view this information as proprietary, the ability to rapidly evaluate design options based on real building parameters is invaluable. The DProfiler system provides capabilities to link model variables to spreadsheet variables and formulas in Microsoft Excel. In lieu of a direct link to the owner's Proforma, the Beck team used a modified Proforma based on the owner's input in order to evaluate the design options and observe how the increased or decreased square footage impacted the overall results. In response to these features, the owner said,

> "By modeling the design options with pricing impacts, we were able to identify the right products for this market. Potential tenants will have high demands, and we have to provide both the right architecture and the right lease rates to satisfy them."

9.9.7 Benefits Realized

Conceptual stage cost estimating yields numerous benefits to the owner and the design team. These benefits include:

Reduction in Preconstruction/Estimating Labor-hours to Produce an Estimate. The reduction in hours came from two sources. First, the typical estimating process requires participation from personnel in the pre-construction team, which includes estimators and experienced construction project managers. With DProfiler, an experienced member of the design team can produce an estimate with guidance and input from the project manager, thus reducing their time commitment and overall labor-hours required by the estimating team for the coordination effort. In a traditional setting, the same estimate would require more than one person to perform the same quantity takeoff and estimate in the same time frame. Second, DProfiler performs the tedious quantity takeoff process, greatly reducing the time required in a traditional approach.

Accurate Estimates in Real-time. The parametric model ensures takeoff accuracy by verifying that all items are counted and included, thus reducing the potential for estimating errors. Rapid feedback on cost allows the design team to focus on analyzing the financial impact of design changes, rather than evaluating the accuracy of the estimate.

Visual Representation of the Estimate. The cost estimate is represented graphically in the 3D model. This reduces potential errors and omissions caused by oversight from the cost estimator. For example, in traditional estimating the exterior wall cladding is quantified by calculating the square footage of the surface area. If an estimator fails to account for the entire wall area or otherwise omits a section, this area will not be calculated. With DProfiler, if a modeler is missing an area of exterior wall cladding, it becomes apparent in the 3D graphic.

In other words, the slab area is a parametric value, meaning that the model calculates its area based on its inherent geometry. The slab is also represented graphically, therefore making it possible to see the slab visually in its geometric form. Because the cost estimate is directly linked to physical components in the model, it is nearly impossible to have a component in the model without an associated cost or vice versa, to have a cost in the estimate not represented by a physical component in the model. For example, the modeler found an area of extra square footage within the building that was not as obvious in the 2D plan and building sections. In response to this, the owner said,

> "Using this software offered us the advantage of being able to get the information we needed about the total project cost and simultaneously be able to have visual documentation of what that cost represented."

9.9.8 Conclusion

Conceptual estimating early in the design process yields potentially significant benefits to the owner and design-build team. This case study shows how an organization, over several years, adapted their design process to take advantage of BIM technologies to better serve their clients and produce more cost effective designs that are better aligned with owner requirements. Achieving these benefits required:

- Experienced designers and project managers who could provide invaluable insight into the digital estimating process. There is a common misperception that these types of new technologies are intelligent solutions and powerful tools that allow younger and less experienced employees to be more productive. This case study demonstrates that use of the tool by skilled and knowledgeable employees with field experience, who understand the assemblies and complexities associated with constructing a building, is immensely valuable. In many cases, these tools require more intelligent input to yield efficiencies and quality in the output.
- Investing time and effort earlier in the process to properly train employees in using the software.
- Customizing the database to fit an organization's standard estimating processes. As organizations adopt these types of conceptual estimating tools, significant up-front investment is required to translate the estimating rules and methods to model variables, parameters, and model component properties or attributes. Over time, this up-front investment will decrease substantially for similar types of projects. This may involve developing a standard method to associate building components with

cost items using industry formats, such as UniFormat fields or a custom property that can be easily associated with items in the cost database.

- Cooperation of the owner or client to provide proforma templates or spreadsheets for evaluating and analyzing the estimated information. Linking these templates to the spreadsheets removes another step in the process and provides the owner with the tools to view the estimate, not just in terms of construction costs but in terms of its impact on operating expenses and income.

Acknowledgments for Hillwood Commercial Project

The authors are indebted to Brent Pilgrim, Stewart Carroll, and Betsy del Monte of the Beck Group for their assistance in providing the information for this case study. Additionally, we would like to thank Ken Reese, Executive Vice President with Hillwood for his comments and support for the case study. All figures used in this case study were provided courtesy of the Beck Group and Hillwood Development.

9.10 U. S. COURTHOUSE, JACKSON, MISSISSIPPI

BIM to enhance architectural design and delivery, as part of GSA's BIM Program

This case study reviews an ongoing project and early participant in the General Services Administration's (GSA) 3D and 4D BIM demonstration projects. It outlines a sequence of early applications of advanced IT and BIM technologies that served to advance this project and more generally to explore effective utilization of advanced information technologies on large federal projects.

The Jackson Mississippi Courthouse has gone through multiple phases of congressional review and funding. It will have taken approximately a decade from the start of initial planning through to occupancy. The planning, design, and construction of the Jackson Courthouse are managed by the GSA, the agency that develops facilities for various federal agencies. The building will be leased to other branches of the federal government; in this case the US Courts are the primary tenant. Another occupant of courthouses is the US Marshals Service, which provides security for prisoners and for the facility. Courthouses often include leases with additional federal occupants.

The Jackson Mississippi Courthouse is approximately 400,000 sf with six stories above grade and two below. It includes twelve district, bankruptcy, and magistrate courtrooms and a grand jury space. Federal courthouses have elaborate building programs that reflect best practices learned from over a century of experience. A main repository of this knowledge is the *US Courts Design Guide* (1997). A major factor of all federal courthouses is the requirement for three distinct circulation systems: for the public, for judges, juries and staff, and for the defendants.

The courthouse is located about a mile from the Mississippi State Capital building with both on axis with East Capital Street. The design was conceived to reflect this relationship and not upstage the center of state government. Civic buildings in downtown Jackson are often enclosed by a monumental green space. These 'greens,' established in 1822, are the only vestiges of Thomas Jefferson's "checkerboard plan" which alternated squares for urban development with squares for public use. The design is conceived around a central rotunda that is the inverse or reflection of the capital dome.

Over its long tenure, the Jackson Courthouse has been used as a test bed for several different BIM technologies. These have included:

- Use of virtual reality for user assessment (see Figure 4-7)
- Early use of BIM on a large project with emphasis on early integration of design teams
- As a platform to integrate cost estimating
- As a test bed for automated design guide checking.

Some of these trials are still ongoing. In this case study, an overview of the project is followed by an explanation of each of the IT and BIM applications according to their sequence of use in the project.

9.10.1 H3 Hardy Collaboration Architecture LLC and the Jackson Mississippi Courthouse

In 1999, H3 (H3 Hardy Collaboration Architecture LLC) received the planning commission for the Jackson Mississippi Federal Courthouse. In 2002, funds for the design phase were awarded to H3. At this time, the Estimated Total Project Cost (ETPC) was $68 to $78 million. The project was to follow the Design-Bid-Build (DBB) process. The cost of design and construction were included in the ETPC.

For subcontracting in August of 2002, the United States General Services Administration (GSA) held a "Small Business Subcontracting Networking Session" for prospective small business subcontractors and consultants involved in the courthouse's design phase Other major participants in the design and construction team are shown in Table 9-10-1. Jacobs Engineering Group was

FIGURE 9-10-1 Front facade and entrance to the Jackson Mississippi Courthouse. *Images courtesy of H3 Architects, rendering by Eric Schuldenfrei.*

Table 9-10-1 Fact sheet for the Jackson Mississippi Federal Courthouse

Hard Facts	
Project	Federal Courthouse
Location	Jackson Mississippi
Client	General Services Administration
Architect	H3 Hardy Collaboration Architecture, LLC
BIM Consultant	Ghafari Associates, LLC
Contract Type	Design-bid-build
Construction Manager	Jacobs Engineering Group, Inc.
General Contractor	Yates Construction Company
Structural Engineers	Walter P. Moore (Structural)
Mechanical Engineers	Cooke Douglass Farr Lemons (CDFL) (Mechanical/Plumbing)
Software Consultants	Contractors
Neutral Formats	IFC
BIM Areas	Virtual reality design review, 3D BIM Collaboration, integrated cost estimation, design guide review
Area	400,000 sf
Duration	seven years (est.)
Cost	$122 million

contracted for construction management; the Yates Construction Company was hired as the general contractor; Walter P. Moore was selected as the structural engineer; Watkins and O'Gwynn was selected as the electrical engineer; and Cooke Douglass Farr Lemons (CDFL) was selected as the mechanical/plumbing contractor.

The project timeline is shown in Figure 9-10-2, marking the times when various activities were undertaken: the virtual reality study by Arup, BIM software review and selection, training on Revit, and other activities that are reported in this case study.

9.10.2 Virtual Reality Use in Courtroom Design

As the GSA BIM program was just being initiated in late 2003, they launched a pilot project to explore the use of 3D visualization in courtroom design. Courtrooms are the primary public stage for many life-affecting dramas. They are also the center of action for federal judges, who are granted lifetime appointments. In most courthouses a judge is assigned to a courtroom where most of his or her cases are heard. Not only does the *US Courts Design Guide* include extensive criteria, but federal judges are often actively involved in the design and detailing of "their" courtrooms. A judge may voice preferences for room geometries and the placement of elements, such as the bench, counsel tables, jury box, and witness stand. In the past, it has been common practice for courtrooms to be mocked up at full scale in plywood, with locations for all of the major actors, fenestration, and lighting, to allow for the assessment of a courtroom before building it for real. These mock-ups cost in the range of $50,000 or more.

The GSA team retained Ove Arup and Associates to undertake multi-dimensional virtual reality simulation and modeling. A visualization model was developed by H3 staff and students from the Center for Integrated Facilities Engineering (CIFE) at Stanford University. The Computer Automated Virtual

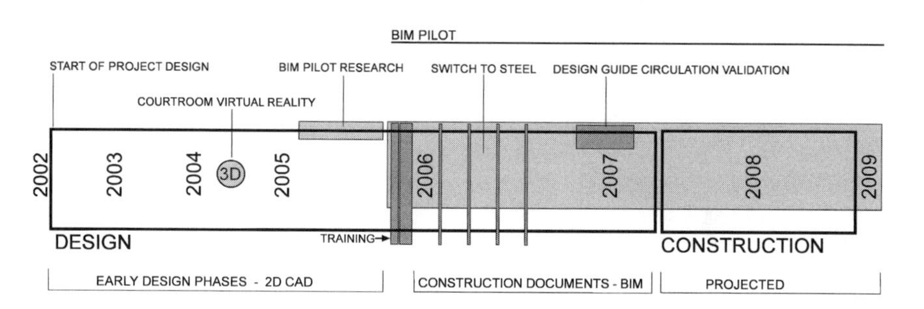

FIGURE 9-10-2 Time table for development and implementation of BIM Pilot program.

Environment (CAVE) at Walt Disney Imagineering was used to mock up a courtroom to stereoscopically reproduce a full size representation of the room in semi-wraparound fashion. The judges wished to visualize the project's sightlines and materials from different critical courtroom locations. Because the CAVE did not provide complete immersion, the sightline reviews were only partially successful. Through this exercise, judges' feedback resulted in several changes to the courtroom's design, including lowering the view-blocking rail on the top of the judge's bench. The CAVE sessions also resolved where the counsel tables would be located.

To validate the design's acoustic characteristics, the GSA relied on a 3D sound model created by Arup Acoustics, based on the courtroom's geometry and planned materials. By electronically transforming sounds in a manner that replicated the geometry and materials, they tested judges in a sound lab accurate simulation of speech from different parts of the courtroom in relation to various seating locations. These were also compared to the reverberation and intelligibility measures that had been documented in other existing courtrooms with good and poor sound characteristics. Several speech obscuring aspects were found and rectified. Revisions included changing some surface

FIGURE 9-10-3
The Jackson Mississippi Virtual Courtroom.
Images courtesy of H3.

shapes and adding sound-absorbing materials to reduce the reverberation time and improve speech intelligibility. Arup Acoustics applied metrics regarding the cost of the analysis using a 3D model, as opposed to the usual plywood mock up. http://www.arup.com/americas/project.cfm?pageid=6708

Working closely with the client and design team, Arup optimized the design to balance reverberation and clarity of speech. The SoundLab also provided an environment in which all the design and client team members could experience the courtroom's acoustics and make collaborative decisions quickly and efficiently. ARUP estimates that the whole process can be conducted for approximately 40% less than the cost of a physical mock-up.

The third simulation effort dealt with lighting. The goal was to assess the planned lighting in relation to materials and surfaces based on luminance, using Radiance, developed at Lawrence Berkeley Labs. Several hot spots that resulted in too high contrast ratios were identified, resulting in some material and color changes. Paul Marantz, the project's lighting designer and a partner of Fisher Marantz Stone of New York, noted some intrinsic limitations of the virtual-reality process for lighting analysis. These included contrast ratios that can be seen by the human eye but are not visible in a virtual environment, due to the intensity limitations of digital projection technology. Consequently, the mock-up failed to show some shadows and highlights in the 3D environment that people would normally see in a real room. But lighting isn't usually evaluated in the plywood courtroom mock-up, so Marantz felt that the lighting experience was valuable. "We were able to fix a half-dozen issues, none of which were fatal. It gave us the opportunity to improve the design," Paul Marantz said (Architectural Record 2005).

Supreme Court Justice Stephen Breyer noted that a courtroom is to promote a conversation between plaintiff and defendant. Key design issues that were essential to the judges that participated in the virtual mockup of the Jackson project involved room lighting, visual sight lines, room acoustics and how these enhanced communication among all parties. The judges felt virtual reality was a useful tool and several design issues were found and fixed as a result of the simulation study. Changes to the courtroom design were then integrated into the 2D early Design Development drawings.

9.10.3 Exploratory Application of the GSA BIM Program

In late 2003, the General Services Administration's (GSA) Public Buildings Service's (PBS) Office of the Chief Architect (OCA) established the National 3D/4D-BIM Program. This program was undertaken to advance the ability of US construction companies to provide better value for government projects.

Since the program's initiation in 2004, its achievements have been featured in a number of professional publications and have been positively received by the international community (e.g., International Alliance of Interoperability, the Workplace Network, and AIA TAP BIM Awards).

In 2005, GSA and H3 agreed to explore the use of BIM tools in the Jackson Mississippi Courthouse project. GSA offered H3 an opportunity to assess BIM technology and to review its merits and limitations. GSA provided consultation to advise the new users and offered guidelines for adopting new technologies:

- Develop a building model of the new Jackson Courthouse for use in the coordination of architectural and engineering disciplines.
- Develop Real Estate Assignment Plan square footage tabulations for each of the courthouse tenants. The accuracy of these figures is important as they establish the amount of rent paid to GSA by each tenant.
- Apply quantity surveying and estimating in conjunction with conventional construction estimates.
- Produce a set of consistent 2D Construction Documents.
- To serve later as a basis for integration with shop models and drawings.
- To be used in 4D construction planning and sequences.

While the courthouse project was in Design Development, H3 spent eight months carrying out a selection process to determine which software suite was to be used. They used a multi-stage question and interview process for four potential software vendors: Autodesk, Bentley, Graphisoft, and Gehry Technologies. H3 performed initial investigations to identify the four invited vendors. A proposal request and questionnaire was distributed to the selected vendors based on the GSA's building information modeling services scope document and questions formulated by H3. To aid them in this task, H3 hired Ghafari Associates as consultants. Responses to the questionnaire were used to narrow the selection to two vendors (Autodesk and Graphisoft), who were invited to participate in a day-long interview and modeling demonstration for the project team (H3 and primary consultants) and the GSA.

The final selection of Revit Building (for use by H3), Revit Structure (for use by Walter P. Moore), and Architectural Building Systems (for use by CDFL) was based on a combination of software evaluation, costs, and services provided in the vendor proposals, among other factors. In addition to the above selection, NavisWorks was evaluated and selected as a software partner to provide "collaboration and coordination" software to be used by all members of the design team.

One of the major risks involved in the adoption of BIM technologies was the concern that it simply would not work. H3 was concerned that the whole BIM concept of documentation and data extraction based on a 3D model would not meet their requirements. Another issue was the time conflict between the team's learning process and the project schedule of document production. To effectively produce documents based on the 3D model, their team would require a knowledge level that would take time to acquire.

Soon after the software adoption process, in April, 2006, a decision was made to switch from reinforced concrete to steel for the upper level structural system. The switch was based on analysis by the CM indicating that a steel structure would reduce the construction schedule by 5 months. This change affected the structural frame from the second floor framing up to the roof. The first floor frame and columns and the lower parking levels and foundations remained as structural in-situ concrete. The change shortened the contractor's General Conditions requirements for construction activities, resulting in significant savings. A directive from the GSA for this change led to a complete redesign by the structural engineer, Walter P. Moore.

9.10.4 BIM Design Integration

H3 originally managed the project's content and the structure of the design team's files and folders in the original 2D development phases. The adoption of BIM in the middle of the project led to the decision to utilize a hybrid 2D and 3D BIM approach. This was based on multiple factors, including:

- Existing project drawing data that did not have to be modeled in the building model. (i.e., standard details)
- Interim schedule constraints and requirements that conflicted with training and the ongoing learning curve required to become proficient with the new software.
- Existing contractual relationships with sub-consultants that were initially not part of the BIM pilot due to funding limits. Non-BIM sub-consultants worked with conventional 2D CAD and referenced 'background' drawing data extracted from the building model.
- Project team members who (for various reasons) did not adapt to the BIM software tools and continued working in conventional 2D software.
- The principal BIM participants were Architecture (H3), Mechanical and Plumbing (CDFL), and Structural (WPM) consultants.

The document production plan using the hybrid BIM approach is shown in Figure 9-10-4. It shows those drawing sets that are produced entirely in 3D BIM, those produced in 2D AutoCAD, and those that are a mixture of 3D massing and 2D detail infill. Information flow from the three principal BIM users—architects, structural designers, and MEP engineers—into derived drawings is outlined.

To achieve strong levels of design collaboration, a central coordination server using Bentley ProjectWise was set up for internet access at Jacobs Engineering Group, the project construction manager. All participants in the BIM process were given explicitly defined roles, including different levels of authorization to access the building model. In general, every participant has authorization to edit their part of the project and reference other parts. The original H3 site existed as the central drawing repository prior to the BIM pilot, and its structure became the BIM pilot's on the server. It provided the structure for the BIM data (original and extracted reference data) as well as all conventional 2D drawing data from consultants.

The higher level server was structured according to systems, as shown in Figure 9-10-5, allowing for version management. In the 2D conventional early

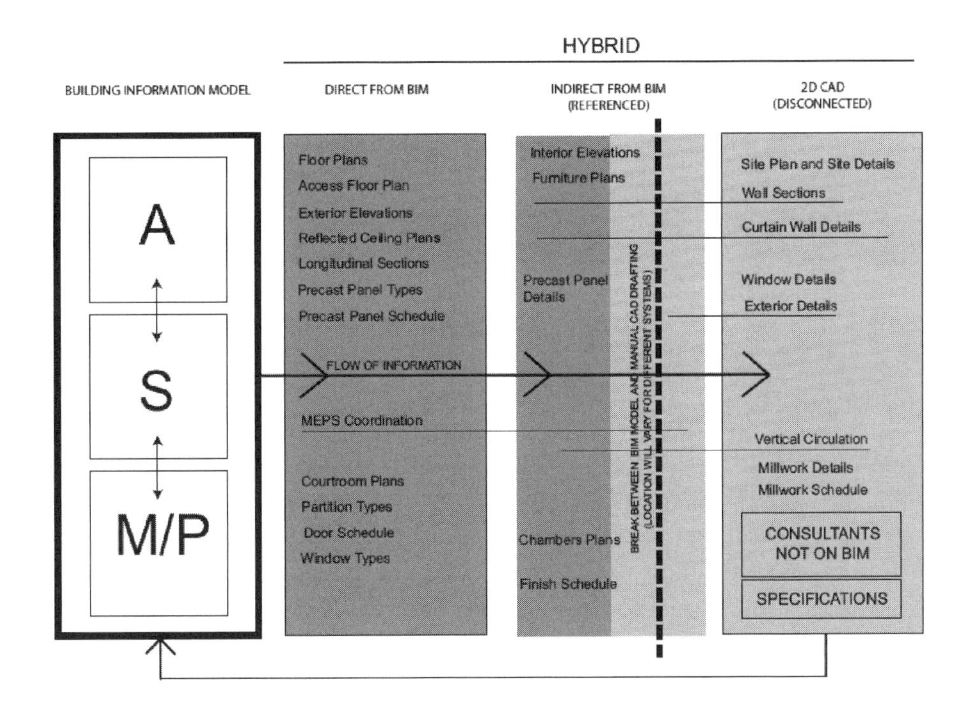

FIGURE 9-10-4 Drawing production responsibilities in the hybrid approach.

FIGURE 9-10-5 High level structure of project files on the FTP site. This site carried the project files for all 3D and 2D systems.

H3 Hardy Collaboration Architecture
PROPOSED FOLDER STRUCTURE FOR PROJECT FTP SITE
Organized by discipline then by type
Follows existing ProjectWise folder structure, organized by discipline

Root	Sub-Folder	File format / Description	Responsibility	Frequency
Active Models				
702.01 Architectural				
	3D Active Models	Revit and Navisworks model files	H3	*tbd*
	3D Reference Models	AutoCAD 3D exported from Revit	H3	*tbd*
	2D Extracted Drawings	AutoCAD 2D exported from Revit (for reference)	H3	*tbd*
	2D Original Drawings	AutoCAD created by manual input.	H3	*tbd*
	includes 2D Sheets	AutoCAD sheet files (not from Revit model)	H3	*tbd*
702.05 Structural				
	3D Active Models	Revit and Navisworks model files	WPM	*tbd*
	3D Reference Models	ADT 3D exported from Revit	WPM	*tbd*
	2D Extracted Drawings	AutoCAD 2D exported from Revit (for reference)	WPM	*tbd*
	2D Original Drawings	AutoCAD created by manual input.	WPM	*tbd*
	includes 2D Sheets	AutoCAD sheet files (not from Revit model)	WPM	*tbd*
702.06 Mechanical (HVAC)				
	3D Active Models	ABS native and Navisworks files	CDFL	*tbd*
	3D Reference Models	AutoCAD 3D exported from ABS (proxy)	CDFL	*tbd*
	2D Extracted Drawings	AutoCAD 2D exported from ABS (for reference)	CDFL	*tbd*
	2D Original Drawings	AutoCAD created by manual input.	CDFL	*tbd*
	includes 2D Sheets	AutoCAD sheet files	CDFL	*tbd*
702.07 Plumbing (Piping)				
	3D Active Models	ABS native and Navisworks files	CDFL	*tbd*
	3D Reference Models	AutoCAD 3D exported from ABS (proxy)	CDFL	*tbd*
	2D Extracted Drawings	AutoCAD 2D exported from ABS (for reference)	CDFL	*tbd*
	2D Original Drawings	AutoCAD created by manual input.	CDFL	*tbd*
	includes 2D Sheets	AutoCAD sheet files	CDFL	*tbd*
702.08 Fire Protection				
	3D Active Models	ABS native and Navisworks files	RJA	*tbd*
	3D Reference Models	AutoCAD 3D exported from ABS (proxy)	RJA	*tbd*
	2D Extracted Drawings	AutoCAD 2D exported from ABS (for reference)	RJA	*tbd*
	2D Original Drawings	AutoCAD created by manual input.	RJA	*tbd*
	includes 2D Sheets	AutoCAD sheet files	RJA	*tbd*
702.09 Electrical				
	3D Active Models	Revit and Navisworks model files	H3/Wo'G	*tbd*
	3D Reference Models	ADT 3D exported from Revit	H3/Wo'G	*tbd*
	2D Extracted Drawings	AutoCAD 2D exported from Revit	H3/Wo'G	*tbd*
	2D Original Drawings		Wo'G	*tbd*
	includes 2D Sheets		Wo'G	*tbd*
Etc.	(similar for additional consultant disciplines where applicable)			
Coordination Reviews	(posted in new folder "702.17 Coordination Review Models and Reports")			
	3D published models	Published Navisworks models / reports for review	H3 / Ghafari	*tbd*
Issued Documents	(posted in existing "Project Deliverables" folder within appropriate sub-folder)			
Review Sets				
	Architectural	PDF DWF NWD		
	Structural	PDF DWF NWD		
	...	PDF DWF NWD		
Construction Documents				
	Architectural	PDF DWF NWD		
	Structural	PDF DWF NWD		
	...			

phases of the project, design participants were responsible for specific 2D drawings. In BIM, however, drawings are reports extracted from the building model. Because of the cross linkage of objects within a parametric model, the architectural model in Revit is a single large file. In order to manage multiple users' concurrent access to parts of the model, Revit supports projects being divided into *worksets* (as opposed to plans, sections, and elevations). Worksets define volumes of space and the project information within it. In this case, the project was partitioned into floors and East and West wings (2–6), roof, exterior, rotunda, and parking.

The building model developed for this courthouse was one of the largest projects attempted on the Revit platform during this time period. There were three models: the architectural model, the structural model and the mechanical/plumbing model. Together, they included details of each courtroom's layout, the structural system and much of the ductwork and piping (see Figure 9-10-6). The architectural model is largest and includes some of the other elements. For some uses, the architectural model became too large to deal with in a single file, leading to interaction problems. Various partitioning strategies were attempted, including the one set up for the server. Through experience, they found that project files over 250 megabytes in size were to be avoided. While trying to maximize the benefits of BIM, H3's challenge was to decide what to model and what to detail using 2D sections (See Sec. 2.2.3). For instance, many repeated standard details were modeled in 2D to limit 3D complexity. Also, features like site plan and details were also excluded. These were new model management issues for the architects at H3.

The architects, structural engineers, and MEP engineers all worked through construction level models to produce construction documents using BIM. Bi-weekly online meetings were held using Webex conferencing, and NavisWorks software was used to review the model's current status and intervene where needed. Some adjustments were required for NavisWorks to manage the data with Revit's worksets. Many spatial conflicts were discovered and

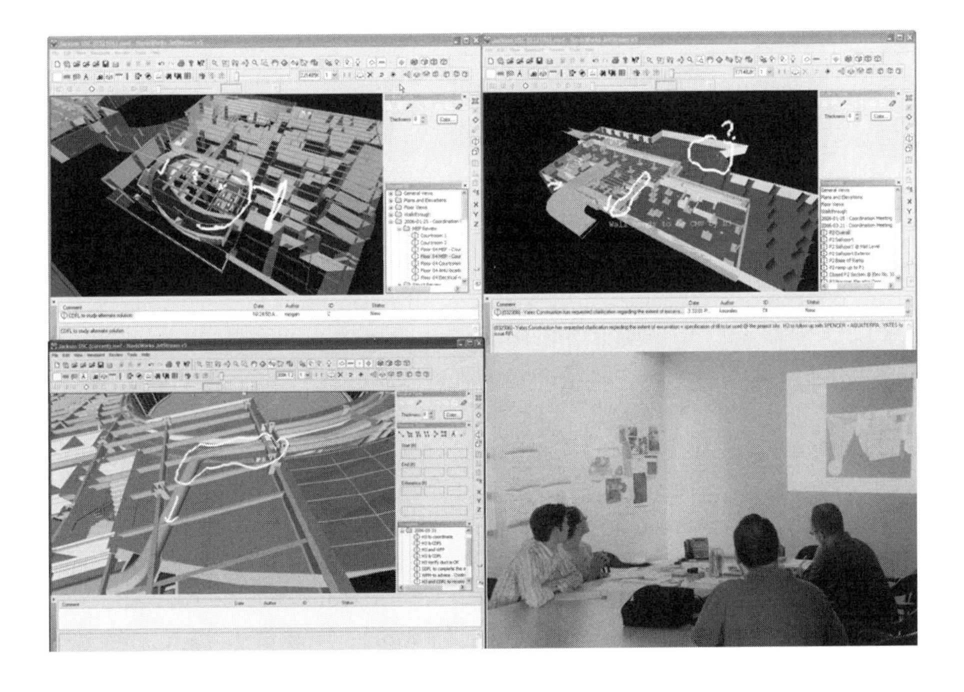

FIGURE 9-10-6 Multiple images of layout conflicts caught in the bi-weekly review meetings, using NavisWorks.
Images courtesy of H3.

corrected. Some of the problem reports generated during those meeting are shown in Figure 9-10-6. NavisWorks was able to deal with the full model.

9.10.5 Integrating Cost Estimation with BIM

Due to the length of time required of most federal projects and the inevitable inflation of construction materials and labor, cost estimation and controls are a necessary component of such work. As a result, the GSA was interested in supporting the use of BIM-integrated cost estimation. They arranged for an Atlanta-based software company, U.S. Cost, to create an integrated approach using this project as a test case.

On this project, the cost estimation team used a hybrid BIM estimating approach that included automatic takeoff from the building model and a direct link of the takeoff data into Excel, and then to U.S. Cost (see Section 6.6). To elaborate, the team used a plug-in developed by U.S. Cost to extract takeoff information at the component level. They associated cost item tags with assembly items via a component property. This takeoff data was then exported via ODBC to an Excel spreadsheet.

In this case, the estimates are for future material procurement. In order to cross-check these estimates, the plan was to generate multiple estimates. Figure 9-10-7 represents how different cost estimates could be used for comparison. The multiple estimates by different teams, based on different information structures (Masterformat, Uniformat) were to serve as the basis for comparison. Originally, the manual cost tracking was planned to consist of nine estimates at three stages of completion, by H3 (Architecture), Jacobs (Construction Manager), Yates (Construction Company), and at the end by GSA (Developer). The revised automated estimating plan was changed to consist of six estimates: three BIM-based estimates by U.S. Cost and three by Yates (reconciled by Jacobs). This was the plan.

Initially, the team attempted to generate estimating data for the whole project, using the architectural model. That model, however, was too large to be used. To reduce the size of the 280 megabyte project file, various attempts were made to split the model into different parts that could be handled in a productive way. Finally, the team ended up using only the structural model provided by Walter P. Moore and estimating quantities and costs of the steel and concrete structure. This 80 megabyte file worked acceptably with the US Cost plug-in.

The bidder's manual structural cost estimates were significantly larger then those derived automatically from the model. Upon inspection, the manual estimator provided extra material for waste, while the system optimized the cut-member lengths against standard shop lengths. Because connection

Jackson USCH

A. Current Estimating Requirements - 9 total estimates

	65%						95%						100%					
	Uniformat	CSI	Standard	BIM based	New	Update	Uniformat	CSI	Standard	BIM based	New	Update	Uniformat	CSI	Standard	Sub pricing	BIM based	New
H3	*	*	*		*		*	*	*			*		*	*			
Jacobs	*	*	*		*			*	*			*						
Yates		*	*		*			*	*		*			*	*	*		*
GSA														*	*			*

B. Proposed Estimating Requirements - 6 total estimates (4 using BIM automated quantity takeoffs)

	Week 15									
	Uniformat		CSI		Basis			Type		
	Detail Level	Summary Level	Detail Level	Summary Level	Quantification	BIM based	Subcontractor	New	Update	Reconciliation
H3	N/A	III	*	**		*		*		
Jacobs										*
Yates	N/A	III	*	**	*			*		
GSA										

	Week 30									
	Uniformat		CSI		Basis			Type		
	Detail Level	Summary Level	Detail Level	Summary Level	Quantification	BIM based	Subcontractor	New	Update	Reconciliation
H3	N/A	III	***	**		*			*	
Jacobs										*
Yates	N/A	III	***	**	*		*	*		
GSA										

	Week 45									
	Uniformat		CSI		Basis			Type		
	Detail Level	Summary Level	Detail Level	Summary Level	Quantification	BIM based	Subcontractor	New	Update	Reconciliation
H3	N/A	III	***	**						
Jacobs										*
Yates	N/A	III	***	**			*	*		
GSA					*	*		*		

* The level of the cost estimate in CSI Masterformat shall correspond with Uniformat Level V as defined in PBS 3440.5. See sample forms
Unit prices shall be broken down into labor, materials and equipment

** The summary of the cost estimate in CSI Materformat shall correspond with Uniformat Level III as defined in PBS 3440.5

*** The level of the cost estimate in CSI Masterformat shall correspond with Uniformat Level VI as defined in PBS 3440.5. See sample forms
Unit prices shall be broken down into labor, materials and equipment

FIGURE 9-10-7 The original cost estimating plan (A) and the revised one (B). The original plan involved nine estimates. The revised schedule relied on six – three BIM-based ones by U.S. Cost, to be cross-checked with three by Yates.

details were missing, these aspects of the estimate were also initially ignored. The structural material bill of material was defined in metric units, which introduced a further conversion factor. In addition, GSA generally uses the ASTM version of Uniformat that is slightly different from that provided by the Construction Specifications Institute (CSI).

US Cost offers automatic takeoff plug-ins for Revit and extraction of take-off information from an IFC model.

A complex situation triggered one of the innovations in this project. The GSA was required to have a 1% tolerance in the total gross area of the project in relation to the number established by the building program. Because this is a small and difficult target to maintain, the team decided that the best approach would be to extract these quantities from the BIM model. This way, they would have a better understanding of the floor areas in comparison with the program, and they could keep a 'real time' check on the areas and spaces within the project. This required careful setup of the model and its data. Relying on the numbers extracted directly from the model meant that no rooms could be left unlabeled or unaccounted for.

9.10.6 Design Guide Validation

When the Jackson Courthouse was at 65% construction documents (CDs), the architects were asked to allow the building model to be used to support a new GSA BIM initiative. It was stated at the beginning of this case study that extensive criteria were applied from the *U.S. Courts Design Guide* (1997). This Guide is maintained by the US Courts as a summary of best practices gained during more than a century of courthouse design and operation. Currently, manual checking of candidate designs against the design guide is both very time consuming and inevitably incomplete. Errors that are supposed to be caught at the concept design stage are often not identified until much later. Many of the guide's requirements are security driven; the U.S. Marshals Service spends man-months or more reviewing designs regarding security, communication, and emergency issues.

In October, 2006, the GSA contracted Georgia Tech to develop a prototype implementation of a Court Design Guide Validation Tool. One of the goals of the Design Guide Validation project was to facilitate the checking of a courthouse concept design to identify issues early for easy correction. The GSA Design Guide Validation is based on the availability of a neutral building model representation. The only such public domain representation is Industry Foundation Classes (IFC), which is discussed at some length in Chapter 3. Any BIM design tool qualified for use by GSA must be able to export an IFC building model that is structured to support such testing.

One of the fundamental aspects addressed in the existing design guide is circulation layout, dealing with the three separate and disjoint circulation routes: one for the public, another for the judge, jury and court staff, and the last for prisoners and defendants. The guide also addresses ease of use and directness of access. For the first phase of the validation project, the Georgia Tech team* extracted over 300 rules from the guide that dealt with passage between spaces. Since these rules were written in terms of space types—there were twelve courtrooms and over 300 separate spaces in the Jackson Courthouse—and because there are multiple paths between the same two spaces, the set of rules expanded to almost 50,000 different circulation paths to be examined.

In order to apply these rules, a building model must contain appropriate content and structure. The structure is straightforward, listed in Table 9-10-2.

All spaces must be defined consistently with the definition of IfcSpace to represent the volume and named in a manner consistent with the way the rules

*The Georgia Tech GSA Design Guide Team was led by Chuck Eastman and includes Yeon-Suk Jeong, post doctoral research scientist, and Ph.D. students Jaemin Lee, Jin Kook Lee, Hugo Sheward, and Sherif AdbulMohsen.

Table 9-10-2 Required entities to be carried in the GSA BIM Circulation Validation View.

Modeling	IFC Entity Name	Description
Space	IfcSpace	Spaces are bounded by walls or virtual walls. Space names should be defined on the basis of space naming conventions. (See Courthouse naming conventions).
Security Zone	IfcZone	Security zone should be one of "public,""restricted," and "secure." A zone for each security level is defined and associated with spaces having that security level.
Door	IfcDoor	There is no special requirement.
Stairs	IfcStair	A stair must be defined as an IFC stair entity. A stair should be contained in a space.
Ramp	IfcRamp	A ramp must be defined as an IFC ramp entity. A ramp should be contained in a space like stairs.
Elevator	IfcSpace	Currently, elevator objects are defined through a space name. For example, "judges elevator" or "prisoner elevator."
Wall	IfcWall, IfcCurtainWall	There is no special requirement.

are defined. The space names were extracted from the design guide rule-set, and all synonyms were merged to a single name. These names also had to be integrated with GSA STAR rental space categories. The development of a general set of standard space names consumed several months, for Georgia Tech, the US Courts, and the GSA.

Each space must also have a security type designation, which was originally defined using property-sets, an easy way to add data to an IFC object. In addition, stairs, ramps and elevators had to be defined, not just geometrically but as *IfcStair* or *IfcRamp* objects, and as an elevator space. This way, the software could identify them and access their attributes.

While these rules seem straightforward, they imposed conditions where none existed previously. Moreover, the size of the Jackson Courthouse model was such that it was difficult to operate. The IFC export capability could not function on the full model. Only through careful rebuilding of the courthouse model by GA Tech could the rules be effectively applied.

After developing the names, renaming all spaces, and rebuilding most of the Jackson Courthouse model so that it carried just the information needed for circulation tests, the model was reduced to about 28 megabytes. The path traversals reviewed by the plug-in take less than a minute to execute the 50,000 paths.

The test results showed that several minor aspects of the design guide were not followed. While these test results came along too late for effective correction, it showed that such testing could be done effectively and reliably. The rigor and care required in the definition of the model to support these tests was also a new experience.

9.10.7 Lessons Learned

Because the project is still in the Construction Documents Phase, it is too early to survey any metrics that could give us a comparison between this project and previous similar projects done by the GSA or H3. In addition, the project was used as a test-bed in ways that are not likely to be replicated.

While the virtual reality review of the courtrooms was not the direct result of BIM, it provides an example of effective client involvement facilitated by BIM (as described in Chapter 4). We expect this kind of virtual review of both spatial and environmental parameters to increase with the availability of building models. But as we move beyond raw simulation capabilities, the accuracy of those simulations become important. In this case, lighting and visual simulation limitations were noted and some strong benefits were reported from the acoustic simulation.

In adopting BIM partway through the project, H3 knew that this project was going to have a steep learning curve for their firm but they also expected that the effort would later increase their productivity on similar projects. The fact that some consultants and team members still relied on conventional 2D drawings led to a hybrid approach for document generation, requiring careful management of 2D and 3D versions.

The Jackson Courthouse design team, including H3, Walter P Moore and CDFL, tried to make maximum use of BIM capabilities and developed an impressively detailed model. But even with reasonably configured PCs, such an endeavor today requires careful methods of modeling to minimize model size, which the team only learned by experience as they proceeded. We normally think of design as a process that continuously adds detail to the base design. This project, however, involved removing data from the model, which became more critical as the quantity and level of detail of the information in the model increased and became more complex.

Another general challenge in BIM exemplified in this project involves integrating other software to make use of the building model data, such as cost estimating and automatic design guide checking packages. Each application has specialized data requirements that must be harmonized with the others and with careful definition of specific model constructs. This goes beyond getting the model or drawings to "look right" and imposes general information and structure requirements on the building model itself. These requirements are best defined early in the project, so that iterative rebuilding of the model is not required.

This project highlighted some of the challenges associated with being an early adopter and outlined issues that other firms should anticipate when undertaking new applications of BIM technology.

Acknowledgments

This case study was written by Eliel de la Cruz and Haldun Kececigil for Professor Chuck Eastman in a course offered by the Design Computing Ph.D. program in the College of Architecture at Georgia Institute of Technology, Winter 2006. It has been adapted for use in this book. The authors wish to acknowledge the assistance of H3 Architects and GSA District Four in the preparation of this study.

Glossary

Building Data Model
An object schema suitable for representing a building. A building data model may be used to represent schemas for file exchange, for XML-based web exchange, or to define a database schema for a repository. The main examples of building data models are IFC and CIS/2.

Building Element Model (BEM)
A digital representation of a building product that can be inserted and used in a BIM tool or other application that utilizes integrated product information.

Building Information Modeling (BIM)
We use BIM as a verb or an adjective phrase to describe tools, processes and technologies that are facilitated by digital, machine-readable documentation about a building, its performance, its planning, its construction and later its operation. Therefore BIM describes an activity, not an object. To describe the result of the modeling activity, we use the term 'building information model', or more simply 'building model' in full.

Building Model
A digital, machine-readable record of a building, its performance, its planning, its construction and later its operation. A Revit® model or a Digital Project™ model of a building are examples of building models. 'Building model' can be considered the next-generation replacement for 'construction drawings', or 'architectural drawings'. Downstream in the process, the term 'fabrication model' is already in common use as a replacement for 'shop drawings'.

Building model repository
A building model repository is a database system whose schema is based on a published object based format. It is different from existing project data management (PDM) systems and web-based project management systems in that the PDM systems are file based, and carry CAD and analysis package project files. Building model repositories are object-based, allowing query, transfer, updating and management of individual project objects from a potentially heterogeneous set of applications.

Building objects
Building objects are the things or parts that make up a building. Objects can be aggregated into higher level objects, such as 'Assemblies'; assemblies are also objects. More generally, an object is any unit of a building that has properties associated with it. Thus the spaces in a building are also objects. In parts of the text, *element* or *component* is used as a synonym for object.

BIM application
A specific use of a building information model to support a work process or work task by the project team.

BIM process
A process that relies on the information generated by a BIM design tool for analysis, fabrication detailing, cost estimation, scheduling, or other use.

BIM system
A software system that incorporates a BIM design tool and other applications that utilize the BIM data. The system may be connected through a local area network or the Internet.

BIM tool
A software application used to generate and manipulate building information models. The term can be further qualified to denote specific application areas. For example, 'BIM Design Tool' is often used to refer to tools used primarily for architectural design, such as Revit® Building, Bentley Architecture, Digital Project™ and ArchiCAD®.

Interoperability
The ability of BIM tools from multiple vendors to exchange building model data and operate on that data. Interoperability is a significant requirement for team collaboration.

Model server
See *Building model repository*.

Bibliography

ACM (1969–1982). Annual Design Automation Conferences, ACM New York.

AGC (2006). *The Contractors' Guide to Building Information Modeling,* Associated General Contractors of America. Arlington, VA.

AIA (1994). *The Architect's Handbook of Professional Practice* Washington, DC, AIA Document B162, American Institute of Architects.

AISC (2007). *AISC Design Guide,* 20 vols. 2007 Edition, AISC Chicago, Ill.

Akintoye, A., and E. Fitzgerald (2000). "A survey of current cost estimating practices in the UK," *Construction Management & Economics* 18(2): 161–172.

Alberti, L. B. (1987). *The Ten Books of Architecture*: The 1755 Leoni Edition. New York, (Paperback) Dover Press.

Anderl, R., and R. Mendgen (1996). "Modelling with constraints: theoretical foundation and application," *Computer-Aided Design* 28(3):155–168.

ANSI (1996). ANSI/BOMA Standard Method for Measuring Floor Area in Office Buildings ANSI. Z65.1.

Akinci, B., F. Boukamp, C. Gordon, D. Huber, C. Lyons, and K. Park (2006). "A formalism for utilization of sensor systems and integrated project models for active construction quality control." *Automation in Construction* 15(2): 124–138.

American Institute of Architects (2006). "Business of Architecture: 2006 AIA Firm Survey." Washington, D.C., AIA.

Arnold, J. A. (2007). "Tectonic Vision and Products." Tectonic Networks website. http://www.tectonicbim.com/

Ashcraft, H. W. J. (2006). "Building Information Modeling: A Great Idea in Conflict with Traditional Concepts of Insurance, Liability, and Professional Responsibility." Schinnerer's 45th Annual Meeting of Invited Attorneys.

Autodesk (2004). "Return on Investment with Autodesk Revit." 25 June 2007. Autodesk Website. Autodesk, Inc. http://images.autodesk.com/adsk/files/4301694_Revit_ROI_Calculator.zip

Autodesk (2007). "DWF: The Best File Format for Published Design." 25 June 2007. Autodesk DWF Community Website. 2007. Autodesk, Inc. Inc. http://usa.autodesk.com/adsk/servlet/item?siteID=123112&id=8675679.

Bader, F., D. Calvanese, D. McGuinness, D. Nardi, and P. Patel-Schneider (2003). *The Description Logic Handbook: Theory, Implementation, and Application,* Cambridge University Press, Cambridge.

Baer, A., C. Eastman and M. Henrion (1979). "Geometrical modeling: A survey." *Computer-Aided Design* 11(5): 253–272.

Ballard, G. (1999). "Improving Work Flow Reliability," Proceedings IGLC-7, UC Berkeley, CA.

Ballard, G. (2000). *The Last Planner™ System of Production Control,* PhD Dissertation, University of Birmingham, Birmingham, U.K.

Barton Malow (2006). General Motors Corporation, Global V6 Engine Facility. Barton Malow website. 27 June 07. http://www.bartonmalow.com/specialties/corporate/gmengine.htm.

Beard, J., M. Loulakis and E. Wundram (2005). *Design-Build: Planning Through Development,* McGraw-Hill Professional.

Bijl, A., and G. Shawcross (1975). "Housing site layout system," *Computer-Aided Design* 7(1): 2–10.

Birx, G. W. (2005). "BIM Evokes Revolutionary Changes to Architecture Practice at Ayers/Saint/Gross," *AIArchitect* 12/2005. 29 June 2007. http://www.aia.org/aiarchitect/thisweek05/tw1209/tw1209changeisnow.cfm.

Bjork, B.C. (1995). *Requirements and information structures for building product models.* Helsinki, VTT Technical Research Centre of Finland.

BLIS. (2002). "Building Lifecycle Interoperable Software." BLIS project website. 25 June 07. http://www.blis-project.org/index2.html.

Booch, G. (1993). *Object-Oriented Analysis and Design with Applications* (2nd Edition), Addison-Wesley.

Boryslawski, M., (2006). "Building Owners Driving BIM: The Letterman Digital Arts Center Story." *AECBytes.* Sept. 30 2006. 27 June 07. http://www.aecbytes.com/buildingthefuture/2006/LDAC_story.html

Bozdoc, M., (2004). *"The History of CAD."* MB Solutions website. 27 June 07. http://mbinfo.mbdesign.net/CAD-History.htm

Braid, I. C., (1973). *Designing with Volumes.* Cambridge UK, Cantab Press, Cambridge University.

Brucker, B. A., M. P. Case, E. W. East, B. K. Huston, S. D. Nachtigail, J. C. Shockley, S. C. Spangler and J. T. Wilson (2006). "Building Information Modeling (BIM): A Road map for Implementation to Support MILCON Transformation and Civil Works Proejcts within the U.S. Army Corps of Engineers," US Army Corps of Engineers, Engineer Research and Development Center.

CAD (1976,1978,1980). Proceedings of CAD76, CAD78, CAD80, IPC Press, London.

Caltech (1997) "Boeing 777: 100% digitally designed using 3D solids technology." 29 June 07. http://www.cds.caltech.edu/conferences/1997/vecs/tutorial/Examples/Cases/777.htm

Campbell, D. A., (2006). "Modeling Rules," *Architecture Week,* http://www.architectureweek.com/2006/1011/tools_1-1.html

Chan, A., D. Scott, and E. Lam (2002). "Framework of Success Criteria for Design/Build Projects," *Journal of Management in Engineering* 18(3):120–128.

CIFE and CURT (2007). "VDC/BIM Value survey result." CIFE, Stanford University. 10 March 07. CIFE/CURT. 25 June 07. http://cife.stanford.edu/VDCSurvey.pdf

CIS/2 (2007). "CIMSTEEL Integration Standards." Steel Construction Institute. 05 June 07. http://www.cis2.org/

Construction Clients Forum (1997). "60% of clients dissatisfied." *Building,* 8.

CORENET (2007). "Integrated Plan Checking Systems." CORENET. 14 Nov 2006. 05 June 07. http://www.corenet.gov.sg/

Court, P., C. Pasquire, A. Gibb and D. Bower (2006). "Design of a Lean and Agile Construction System for a Large and Complex Mechanical and Electrical Project," *Understanding and Managing the Construction Process: Theory and Practice, Proceedings of the 14th Conference of the International Group for Lean Construction,* R. Sacks and S. Bertelsen, eds., Catholic University of Chile, School of Engineering, Santiago, Chile, 243–254.

Crowley, A. (2003). "CIS/2 Interactive at NASCC," *New Steel Construction* 11:10.

CSIRO (2007). "Building & Construction." CSIRO. 27 June 07. http://www.csiro.au/science/ps2oz.html

Dakan, M., (2006) "GSA's BIM Pilot Program Shows Success." *Cadalyst.* 19 July 2006. 27 June 07. http://aec.cadalyst.com/aec/article/articleDetail.jsp?id=359335

Day, M. (2002). "Intelligent Architectural Modeling." *AECMagazine* September 2002. 27 June 07. http://www.caddigest.com/subjects/aec/select/Intelligent_modeling_day.htm

Debella, D. and R. Ries (2006). "Construction Delivery Systems: A Comparative Analysis of Their Performance within School Districts." *Journal of Construction Engineering and Management* 132(11): 1131–1138.

Demkin, J. (2001). *Architect's Handbook of Professional Practice.* New York, John Wiley and Sons.

Department of Defense (2000). "*Guidebook for Performance-Based Services Acquisition (PBSA).*" Defense Acquisition University Press, Washington, D.C.

Ding, L., R. Drogemuller, M. Rosenman, D. Marchant and J. Gero (2006). "Automating Code Checking for Building Designs – DesignCheck. Clients Driving Innovation: Moving Ideas into Practice". K. Brown, K. Hampson and P. Brandon. Sydney, Australia, CRC for Construction Innovation for Icon.Net Pty Ltd.

Dolan, T. G., (2006). "First Cost vs. Life-Cycle Costs: Don't get caught in the trap of saving now to pay later." *School Planning and Management.* January 2006.

Douglas Steel (2007). "General Motors V-6 Engine Plant." Douglas Steel Fabricating Corporation. 27 June 2007. http://douglassteel.com/gmv6.htm

Dunwell, S. (2007). "Linking Front and Back Offices: The ERP Vendor's Perspective." Presentation at buildingSMART 2007: The New Business Model for Design, Construction and Facilities Management, RIBA, London, UK.

Eastman, C. M. (1975). "The use of computers instead of drawings in building design." *Journal of the American Institute of Architects* March: 46–50.

Eastman, C. M. (1992). "Modeling of buildings: evolution and concepts." *Automation in Construction* 1: 99–109.

Eastman, C. M. (1999). *Building Product Models: computer environments supporting design and construction*. Boca Raton, FL, USA, CRC Press.

Eastman, C. M., I. His and C. Potts (1998). Coordination in Multi-Organization Creative Design Projects. Design Computing Research Report. Atlanta, GA, College of Architecture, Georgia Institute of technology.

Eastman, C. M., R. Sacks and G. Lee (2001). Software Specification for a Precast Concrete Design and Engineering Software Platform. Atlanta, GA, USA, Georgia Institute of Technology.

Eastman, C. M., R. Sacks and G. Lee (2002). "Strategies for Realizing the Benefits of 3D Integrated Modeling of Buildings for the AEC Industry" ISARC - 19th International Symposium on Automation and Robotics in Construction, Washington DC, NIST.

Ergen, E., B. Akinci and R. Sacks (2007). "Tracking and Locating Components in a Precast Storage Yard Utilizing Radio Frequency Identification Technology and GPS." *Automation in Construction* 16: 354–367.

Evey, L. (2006). Conversation between P. Teicholz and Lee Evey, President, Design-Build Institute of America.

FIATECH (2007). "*aecXML.*" FIATECH. 27 June 07. http://www.fiatech.org/projects/idim/aecxml.htm

FMI/CMAA (2005, 2006). "FMI/CMAA Annual Survey of Owners." FMI/CMAA websites.

Folkestad, J. E. and D. Sandlin (2003). "Digital construction: Utilizing three dimensional (3D) computer models to improve constructability." Colorado

State University. 27 June 07. www.cm.cahs.colostate.edu/Faculty_and_Staff/folkestad/PDF/Folkestad%20&%20Sandlin%20IEMS% 202005.pdf

Frampton, K. and J. Cava (1996). *Studies in Tectonic Culture: The Poetics of Construction in Nineteenth and Twentieth Century Architecture.* Cambridge, MA, The MIT Press.

Gaddie, S. (2003). "Enterprise programme management: Connecting strategic planning to project delivery." *Journal of Facilities Management* 2(2):177–191.

Gallaher, M. P., A. C. O'Connor, J. John, L. Dettbarn and L. T. Gilday (2004). Cost Analysis of Inadequate Interoperability in the U.S. Capital Facilities Industry. Gaithersburg, Maryland., National Institute of Standards and Technology, U.S. Department of Commerce Technology Administration.

Geertsema, C., G. E. Gibson, Jr., D. Ryan-Rose (2003). Emerging Trends of the Owner-Contractor Relationship for Capital Facility Projects from the Contractor's Perspective. Center for Construction Industry Studies Report. University of Texas at Austin. Report No. 8. http://www.ce.utexas.edu/org/ccis/a_ccis_report_32.pdf

Gielingh, W. (1988). "General AEC Reference Model (GARM)" Conceptual Modeling of Buildings, CIB W74+W78 Seminar, Lund, Sweden, CIB Publication 126.

Goldstein, H. (2001). "4D: Science Fiction or Virtual Reality" *Engineering News Record.* April 16, 2001.

Gonchar, J. (2007). "To architects, building information modeling is still primarily a visualization tool." *Architectural Record.* 07 27 2007. http://archrecord.construction.com/features/digital/archives/0607dignews-2.asp

Green Building Studio (2004). "AEC Design Practice Survey Identifies Opportunities to Accelerate 3D-CAD/BIM Adoption and Green Building Design." 25 June 2007. http://www.greenbuildingstudio.com/gbsinc/pressrelease.aspx?id=24

Green Building Studio (2007). "Green Building XML Schema" from http://www.gbxml.org/index.htm

Grilo, A., and R. Jardim-Goncalves (2005). "Analysis on the development of e-platforms in the AEC sector," *International Journal of Internet and Enterprise Management* 3(2), 187–18.

Grohoski, C. (2006). "Virtual Reality Isn't Just For Gamers Anymore." *Webwire.* 27 June 2006. 29 June 2007. http://www.webwire.com/ViewPressRel.asp?aId=15849

GSA (2006). 02 - GSA BIM Guide For Spatial Program Validation Version 0.90. Washington DC, United States General Services Administration: 61.

GSA (2006a). "3D-4D Building Information Modeling." GSA. 27 June 2007. http://www.gsa.gov/bim

GSA (2007). "GSA Performance-Based Acquisition." GSA. 27 June 07. http://www.gsa.gov/Portal/gsa/ep/channelView.do?pageTypeId=8203&channelPage=%252Fep%252Fchannel%252FgsaOverview.jsp&channelId=13077

GT-COA (2007). "GSA Project - BIM Enabled Design Guide Automation." GT Design Computing. 27 June 2007 http://dcom.arch.gatech.edu/gsa/

Hänninen, R. (2007). "Performance Metrics for Demand and Supply in the Building Lifecycle Process." Presentation available online at: www.uni-weimar.de/icccbe/late/icccbe-x_050.pdf.

Haymaker, J., and M. Fischer (2001). "Challenges and Benefits of 4D Modeling on the Walt Disney Concert Hall Project." Working Paper Nr. 64, CIFE, 17 pages.

Hendrickson, C. (2003). *Project Management for Construction: Fundamental Concepts for Owners, Engineers, Architects and Builders* Version 2.1. 27 June 07 http://www.ce.cmu.edu/pmbook

Hodges, C., and W. W. Elvey (2005). "Making the Business Case For Sustainability: It's Not Just About Getting Points." *Facilities Manager* 21 (4):50–53.

Hopp, W. J., and M. L. Spearman (1996). *Factory Physics*. Chicago, IRWIN.

Hospitals, V. (2007). "Dept. of Veterans Affairs." from http://www.va.gov/facmgt/cost-estimating/.

Houbaux, P. (2005). The SABLE Project: Towards Unification of IFC based Product Model Servers, IAI-Iberia Industry Day, Madrid.

Howell, G. A., (1999). "What Is Lean Construction – 1999" Seventh Annual Conference of the International Group for Lean Construction, IGLC-7, Berkeley, CA.

IAI (2007). "BuildingSMART and Interoperability." Industry Alliance for Interoperability. 18 June 07 http://www.iai-na.org/bsmart/index.php

IAI (2007a). "IFC/ifcXML Specifications." http://www.iai-international.org/Model/IFC(ifcXML)Specs.html.

Ibbs, C. W., Y. H. Kwak, T. Ng, and A. M. Odabasi, (2003). "Project delivery systems and project change: Quantitative analysis." *Journal of Construction Engineering and Management* 129(4): 382–387.

ICC (2007). "SMARTcodes™." International Code Council. 30 June 07. http://www.iccsafe.org/SMARTcodes/

Jackson, E., (2000). Prototype Practice for the Building Industry: Operation and Maintenance Support Information. Linking the Construction Industry: Electronic Operation and Maintenance Manuals: Workshop Summary (2000), Federal Facilities Council.

Jackson, S. (2002). "Project cost overruns and risk management." Proceedings of the 18th Annual ARCOM Conference, Glasgow.

J.E. Dunn (2007). "Building Information Modeling Specialist" from http://www.kcrevit.com/Job_Postings_Main.htm.

Johnston, G. B. (2006). "Drafting culture : a social history of Architectural graphic standards." Ph.D. Thesis, Emory University, Atlanta.

Kalay, Y. (1989). *Modeling Objects and Environments*. New York, NY, John Wiley & Sons.

Khemlani, L. (2004). "The IFC Building Model: A Look Under the Hood." 30 March 2004. *AECbytes*. 15 June 07. http://www.aecbytes.com/feature/2004/IFCmodel.html

Khemlani, L. (2005). "CORENET e-PlanCheck: Singapore's Automated Code Checking System." 26 October 2005. *AECbytes*. 30 June 07. http://www.aecbytes.com/buildingthefuture/2005/CORENETePlanCheck.html

Khemlani, L. (2006). "AIA CBSP Symposium on BIM for Building Envelope Design and Performance." 15 Nov 2006. *AECbytes*. 28 June 07. http://www.aecbytes.com/buildingthefuture/2006/AIA-CBSP_BIM.html

Khemlani, L. (2006). "Use of BIM by Facility Owners: An "Expositions" Meeting." 16 May 2006. *AECbytes*. 27 June 07. http://www.aecbytes.com/buildingthefuture/2006/Expotitions_meeting.html

Khemlani, L. (2005). "Multi-Disciplinary BIM at Work at GHAFARI Associates." 21 Nov 05. *AECbytes*. 27 June 07. http://www.aecbytes.com/buildingthefuture/2006/BIM_Awards.html

Khemlani, L. (2006). "2006 2nd Annual BIM Awards Part 1." 12 July 2006. *AECbytes*. 28 June 07. http://www.aecbytes.com/buildingthefuture/2006/BIM_Awards.html

Kieran, S., and J. Timberlake (2003). *Refabricating Architecture: How Manufacturing Methodologies are Poised to Transform Building Construction*. New York, McGraw-Hill Professional.

Koerckel, A., and G. Ballard (2005). "Return on Investment in Construction Innovation – A Lean Construction Case Study." Proceedings of the 14th Conference of the International Group for Lean Construction, Sydney Australia.

Konchar, M., and V. Sanvido (1998). "Comparison of U.S. Project Delivery Systems." *Journal of Construction Engineering and Management* 124(6): 435–444.

Koo, B., and M. Fischer (2000). "Feasibility Study of 4D CAD in Commercial Construction." *Journal of Construction Engineering and Management*, 126 (5): 251–260.

Koskela, L. (1992). *Application of the New Production Philosophy to Construction*, Center for Integrated Facility Engineering, Department of Civil Engineering, Stanford University.

Kunz, J., and M. Fischer (2007). Virtual Design and Construction: Themes, Case Studies and Implementation Suggestions, Center for Integrated Facility Engineering, Stanford University.

Laurenzo, R. (2005). "Leaning on lean solutions," *Aerospace America,* June 2005: 32–36.

Lipman, R. R. (2004). "Mobile 3D visualization for steel structures." *Automation in Construction* 13(1): 119–125.

Liston, K., J. Kunz and M. Fischer (2000). "Requirements and benefits of interactive information workspaces in construction," Proceedings of the Eight International Conference on: Computing in Civil and Building Engineering (ICCCBE-VIII), Renate Fruchter, Feniosky Pena-Mora, and W.M. Kim Roddis (eds), Stanford, CA, 2:1277–1284.

Liston, K., M. Fischer and T. Winograd (2001). "Focused Sharing of Information for Multidisciplinary Decision Making by Project Teams." *ITCon (Electronic Journal of Information Technology in Construction)* 6: 69–81.

Ma, Z., Q. Shen and J. Zhang (2005). "Application of 4D for dynamic site layout and management of construction projects." *Automation in Construction* 14(3): 369–381.

Majumdar, T., M. Fischer and B. R. Schwegler (2006). "Conceptual Design Review with a Virtual Reality Mock-Up Model," Joint International Conference on Computing and Decision Making in Civil and Building Engineering. Montréal, Canada.

McDuffie, T. (2007). "BIM: Transforming a Traditional Practice Model into a Technology-Enabled Integrated Practice Model." *AIA Cornerstone.* http://www.aia.org/nwsltr_pa.cfm?pagename=pa_a_200610_bim

McKinney, K., J. Kim, M. Fischer and C. Howard (1996). "Interactive 4D-CAD." Proceedings of the Third Congress on Computing in Civil Engineering, Jorge Vanegas and Paul Chinowsky (Eds.), ASCE, Anaheim, CA, 383–389.

McKinney, K., and M. Fischer (1998). "Generating, evaluating and visualizing construction schedules with 4D-CAD tools." *Automation in Construction* 7(6): 433–447.

McNair, S., and M. Flynn (2006). "Managing an ageing workforce in construction." Leicester, Department for Work and Pensions Centre for Research into the Older Workforce.

Mitchell, J., and H. Schevers (2005) "Building Information Modelling for FM at Sydney Opera House." CRC Construction Innovation, CRC-CI Project 2005-001-C Report No: 2005-001-C-4.

Moore, G. A. (1991). *Crossing the Chasm,* New York, NY, Harper Business.

Morgan, M. H. (1960). *Vitruvius - The Ten Books of Architecture,* New York, Dover Press. See esp. Book 1, Chapter 2, Section 2.

Munroe, C. (2007) "Construction Cost Estimating." American Society of Professional Estimators. 27 June 2007. http://www.aspenational.com/CONSTRUCTION%20Cost%20ESTIMATING.pdf

NASA (1978–1980). Conference on Engineering and Scientific Data Management, Hampton, VA, NASA.

National Research Council (1998). *Stewardship of Federal Facilities: A Proactive Strategy for Managing the Nation's Public Assets.* National Academy Press, Washington, D.C.

Neuberg, F. (2007). "MAX BGL - Progress is Built on Ideas." buildingSMART: The New Business Model for Design, Construction and Facilities Management, RIBA, London, UK.

NIBS (2007). "National BIM Standard." National Institute of Building Science. 27 June 07. http://www.facilityinformationcouncil.org/bim/index.php

NIST and Fiatech (2006). "Capital Facilities Information Handover Guide, Part 1." NISTIR 7259. Department of Commerce. 27 June 07. http://www.fire.nist.gov/bfrlpubs/build06/PDF/b06016.pdf

Oberlender, G. and S Trost (2001). "Predicting Accuracy Of Early Cost Estimates Based On Estimate Quality." *Journal of Construction Engineering and Management* 127(3): 173–182.

O'Connor, J. T., and S. C. Dodd (1999). "Capital Facility Delivery with Enterprise Resource Planning Systems." Center for Construction Industry Studies Report. University of Texas at Austin. Report No. 16.

OmniClass (2007). "OmniClass: A Strategy for Classifying the Built Environment," OmniClass. 30 June 07. http://www.omniclass.org/

Onuma, K. G. and D. Davis (2006). "Integrated Facility Planning using BIM Web Portals." 9 Oct 2006.from Federal Facilities Council. 30 June 07. http://www.onuma.com/services/FederalFacilitiesCouncil.php

Pasquire, C., R. Soar and A. Gibb (2006). "Beyond Pre-Fabrication - The Potential of Next Generation Technologies to Make a Step Change in Construction Manufacturing" *Understanding and Managing the Construction Process: Theory and Practice, Proceedings of the 14th Conference of the International Group for Lean Construction* R. Sacks and S. Bertelsen, eds., Catholic University of Chile, School of Engineering, Santiago, Chile, 243–254.

Pixley, D., J. Holm, K. Howard and G. Zettel (2006). "A Look at Sutter's Lean Program." LCI Congress, Lean Construction Institute.

PCI (2004). *Design Handbook of Precast and Prestressed Concrete.* Skokie, IL, Precast/Prestressed Concrete Institute.

Post, N. M. (2002). "Movie of Job that Defies Description Is Worth More Than A Million Words." *Engineering News Record* 8 April 2002.

Ramsey, G. and H. Sleeper (2000). *Architectural Graphic Standards.* New York, John Wiley & Sons.

Requicha, A. (1980). "Representations of rigid solids: theory, methods and systems." *ACM Comput. Surv* 12(4): 437–466.

Robbins, E. (1994). Why Architects Draw. Cambridge MA, MIT Press.

Roe, A. (2002). "Building Digitally Provides Schedule, Cost Efficiencies: 4D CAD is expensive but becomes more widely available." *Engineering News Record* 25 February 2002.

Roe, A. (2006). "The Fourth Dimension is Time" *Steel,* Australia, 15.

Romm, J. R. (1994). *Lean and Clean Management: How to Boost Profits and Productivity by Reducing Pollution,* Kodansha International.

Roodman, D. M., and N. Lenssen (1995). *A Building Revolution: How Ecology and Health Concerns Are Transforming Construction,* Worldwatch Institute.

Rundell, R. (2006). "1-2-3 Revit: BIM and Cost Estimating, Part 1: How BIM can support cost estimating." *Cadalyst* 7 August 2006.

Sacks, R. (2004). "Evaluation of the economic impact of computer-integration in precast concrete construction." *Journal of Computing in Civil Engineering* 18(4): 301–312.

Sacks, R., and R. Barak (2007). "Impact of three-dimensional parametric modeling of buildings on productivity in structural engineering practice." *Automation in Construction* (2007), doi:10.1016/j.autcon.2007.08.003

Sacks, R., C. M. Eastman and G. Lee (2003). "Process Improvements in Precast Concrete Construction Using Top-Down Parametric 3-D Computer-Modeling." *Journal of the Precast/Prestressed Concrete Institute* 48(3): 46–55.

Sanvido, V., and M. Konchar (1999). *Selecting Project Delivery Systems, Comparing Design-Build, Design-Bid-Build, and Construction Management at Risk,* Project Delivery Institute, State College, PA.

Sawyer, T. (2006). "Early Adopters Find the Best Models Are Digital Virtuosos." *Engineering News Record* 02 October 2006.

Sawyer, T., and T. Grogan (2002). "Finding the Bottom Line Gets A Gradual Lift From Technology." *Engineering News Record* 12 August 2002.

Schenck, D. A., and P. R. Wilson (1994). *Information Modeling the EXPRESS Way.* New York, Oxford University Press.

Schodek, D., M. Bechthold, K. Griggs, K. M. Kao and M. Steinberg (2005). *Digital Design and Manufacturing, CAD/CAM Applicatons in Architecture and Design.* New York, John Wiley and Sons.

Schwegler, B., M. Fischer and K. Liston (2000). "New Information Technology Tools Enable Productivity Improvements." North American Steel Construction Conference, Las Vegas, AISC.

Sheppard, L. M. (2004). "Virtual Building for Construction Projects" *IEEE Computer Graphics and Applications* January 6–12.

Smoot, B. (2007). Building Acquisition and Ownership Costs. CIB Workshop, CIB.

Stephens, S. (2007). "Crowding the Marquee." *Architectural Record* January 2007. 30 June 07. http://archrecord.construction.com/practice/firmCulture/0701crowding-1.asp

Still, K. (2000). *Crowd Dynamics*. PhD Thesis. University of Warwick, Department of Mathematics.

Sullivan, C. C. (2007). "Integrated BIM and Design Review for Safer, Better Buildings: How project teams using collaborative design reduce risk, creating better health and safety in projects." *McGraw-Hill Construction Continuing Education*. 27 June 07. http://construction.com/CE/articles/0706navis-1.asp

Teicholz, E. (2004). "Bridging the AEC/FM Technology Gap." *IFMA Facility Management Journal* Apr/Mar 2004.

Thomas, H. R., C. Korte, V. E. Sanvido and M. K. Parfitt (1999). "Conceptual Model for Measuring Productivity of Design and Engineering." *Journal of Architectural Engineering* 5(1): 1–7.

Thomson, D., and R. Miner (2006). "Building Information Modeling - BIM: Contractual Risks are Changing with Technology." *The Construction Law Briefing Paper*. 27 June 07. http://www.aepronet.org/ge/no35.html

TNO (2007). "Knowledge for Business." from http://www.ifcbrowser.com/.

Tournan, A. (2003). "Calculation of Contingency in Construction Projects." *IEEE Transacations on Engineering Management* 50(2): 135–140.

Tulacz, G. J. (2006). "The Top Owners." *Engineering News Record* 27 Nov 2006.

U.S. (2004, 26-May-2005). "2002 Economic Census Industry Series Reports Construction." 2002 Economic Census from. http://www.census.gov/econ/census02/guide/INDRPT23.HTM.

U.S. Department of Energy (DOE) (1997). "Cost Estimating Guide." DOE G 430.1-1, Washington, D.C.: Chapter 11.vUnited States. Department of Energy (1997). "Cost Estimating Guide." DOE G 430.1-1, Washington, D.C.: Chapter 11.

U.S. Department of Energy (DOE) and I. P. Public Technology (1996). "The Sustainable Building Technical Manual: Green Building Design, Construction, and Operations." Washington, D.C., U.S. Department of Energy: 34.

USDOT - Federal Highway Administration (2006). "Design-Build Effectiveness Study." USDOT Final Report. 27 June 07. http://www.fhwa.dot.gov/reports/designbuild/designbuild.htm

Warne, T., and J. Beard (2005). *Project Delivery Systems: Owner's Manual,* American Council of Engineering Companies, Washington, D.C.

Weaire, D., and R. Phelan (1994). "A counterexample to Kelvin's conjecture on minimal surfaces." *Phil. Mag. Lett.* 69:107–110.

Womack, J. P., and D. T. Jones (2003). *Lean Thinking: Banish Waste and Create Wealth in Your Corporation.* New York, Simon & Schuster.

Yaski, Y. (1981). *A consistent database for an integrated CAAD system.* PhD Thesis, Carnegie Mellon University, Pittsburgh PA.

ORGANIZATION WEB SITES

AISC (2007). American Institute of Steel Construction, http://www.aisc.org/.

CIFE (2007). Center for Integrated Facility Engineering, http://www.stanford.edu/group/CIFE/.

CURT (2007). The Construction User's Roundtable, http://www.curt.org/.

DBIA (2007). Design Build Institute of America, http://www.dbia.org/index.html.

Eurostep (2007). Information Solutions for a Global Age, http://www.eurostep.com/.

IAI (2007). International Alliance for Interoperability http://www.iai-international.org/index.html.

LCI (2004). Lean Construction Institute http://www.leanconstruction.org/.

NIBS (2007). National Institute of Building Sciences, http://www.nibs.org.

OGC (2007). Geography Markup Language http://www.opengeospatial.org/standards/gml

COMPANY AND SOFTWARE INDEX

(continued)

Company	Products	Website	Page Reference
COADE Engineering Software	CADWorx® CEASAR II®	www.coade.com/	61
Common Point, Inc.	Common Point 4D ConstructSim OpSim	www.commonpointinc.com	129, 231–2
Computers and Structures Inc.	SAP, ETABS,	http://www.csiberkeley.com/	61, 372, 438
CSC	3d+	3dplus.cscworld.com/	269t, 270t
Dassault Systemes	CATIA	www.3ds.com/home/	54, 60, 323t, 413–4, 416
Data Design System UK Limited	DDS IFC Viewer	www.dds-bsp.co.uk/IFCViewer .html	79
U.S. Department of Energy	Energy Plus	http://www.eere.energy.gov/buildings/energyplus/	170t, 324t, 394, 394f, 395
Design Data	SDS/2	www.dsndata.com	29, 40, 269t, 270, 323t, 328t, 333t
Digital Building Solutions	BIMContent Manager	www.digitalbuildingsolutions.com	195t
eRENA	ViCROWD	www.erena.kth.se	129t
Enterprixe	Model Server	www.enterprixe.com	127
EPM Technology	EDMserver	www.epmtech.jotne.com	127
Esteco	modeFRONTIER	www.esteco.com	180
Eurostep	ModelServer for IFC	www.eurostep.com	88, 127
Exactal Pty Ltd.	CostX™ CostX Viewer™	www.exactal.com	120–1, 122f
Flomerics	Flovent	http://www.flomerics.com/flovent/	170t
Flow Science	Micro Flow	www.flow3d.com	170t
FormFonts	Form Fonts 3D	www.formfonts.com	193, 195t
Gehry Technologies	Digital Project	www.gehrytechnologies.com	34t, 52–4, 60, 77, 127, 163, 177, 179t, 229t,269t, 273, 407–13, 410f, 415–8, 420, 422–4, 427,430, 467–8
Google	3D warehouse SketchUp	www.sketchup.com	193 53, 58–9, 158, 160, 164t, 193, 194t, 195t, 219
Graphisoft ‖	ArchiCAD® ArchiGlazing, Ductwork ArchiFM	www.graphisoft.com	29, 59, 127, 218, 220, 229, 269t, 273, 276, 455
GT STRUDL	GTSTRUDL	www.gtstrudl.gatech.edu/	170t

Company	Products	Website	Page Reference
IES	Simulex Apache	www.iesve.com	129t, 158, 162, 163t, 164t, 170t
IMSI/Design	TurboCAD	www.turbocad.com	235
Innovaya	Visual Simulation 3.0 Visual Estimating 9.4	www.innovaya.com	220–1, 231t, 292
Intellicad	IntelliCAD MEP	www.intellicad.org	323t, 328t, 333t
Interspec	e-SPECS	www.e-specs.com	186
Leap Software	AXSYS PRESTO	www.leapsoft.com	434
Legion	Legion Studio	www.legion.com	1095, 129t
Lawrence Berkeley National Laboratory	DOE-2 Energy Plus Radiance	http://www.doe2.com/	170t, 324t, 394, 394f, 395
LKSoft	IDS_STEP Database	www.ida-step.net	88
MAP	CAD-Duct	www.cadduct.com	269t, 276,
NaturalWorks	Displacement Ventilation	ep.natural-works.com	392t, 393f, 394, 401–403f
NavisWorks, Ltd*	JetStream Clash Detective JetStream Freedom JetStream Roamer	www.navisworks.com	125, 135, 144, 218, 231, 254, 276, 324, 328, 332–4, 365, 455, 459–60
Nemetschek	Allplan Engineering Allplan Building Services	www.nemetschek.com	xii, 259, 269t, 272–3
Objects Online, Inc.		www.objectsonline.com	194t
Octaga	Modeler	www.octaga.com	79
ODEON Room Acoustics Software	ODEON	http://www.odeon.dk/	170
OnCenter	On Screen Takeoff®	www.oncenter.com	221
Onuma and Associ-ates, Inc.	Onuma Planning System™	www.onuma.com	127, 323, 344, 356
Primavera Systems	P6	www.primavera.com	58, 138, 224, 229–32, 427
Quickpen	PipeDesigner 3D DuctDedesigner	www.quickpen.com	184, 269, 276
RAM International	Structural System Advanse Connection Concept	http://www.ramint.com/	58, 170, 324, 328, 333
Renkus-Heinx Inc.	EASE	http://www.renkus-heinz.com/ease/ease_intro/intro_about_ease.html	173t
Revit City	Downloads	www.revitcity.com	194t

(continued)

Company	Products	Website	Page Reference
Rhinoceros®	McNeel North America	www.rhino3d.com/	37, 53, 77, 159, 164t, 323t, 379, 380f, 385,413–414
RISA Technologies	RISA 3D RISA Foundation RISA Floor RISA Tower	www.risatech.com/	58, 170t
Robobat	ROBOT Millenium	www.robobat.com	58, 170t, 324t, 413, 415f
Sage	Timberline Office	www.sagetimberlineoffice.com	220
SCADA Soft AG	WizCAD™	www.scada.ch	61
SoftTech	SoftTech V6 Manufacturer	www.softtechnz.com	269t, 273
Solar Energy Laboratory, U. Wisconsin	TRNSYS	http://sel.me.wisc.edu/trnsys	170
Solibri	Model Checker™	www.solibri.com	xiv, 79, 89, 122, 123f, 170t, 171, 218, 291, 296
SolidWorks Inc.	SolidWorks	www.solidworks.com/	54, 273, 323t, 413
SprinkCAD	SprinkCAD	www.sprinkcad.com	269t, 276, 323t, 361t
Square One research Pty Ltd	Ecotect ESP-r	www.ecotect.com http://www.esru.strath.ac.uk/	59, 158, 162–3, 162f, 163t, 164t, 166, 170t, 302
Strand7	STRAND	www.strand7.com	324t, 380, 381f, 413
StructureWorks	StructureWorks	www.structureworks.org	29, 54, 269t, 271
Synchro, Ltd.	Synchro 4D	www.synchroltd.com	231t
Tectonic Network	BIM Library manager	www.tectonicnetwork.com	190, 196, 196f, 311
Tekla	Structures	www.tekla.com	29, 40, 40f, 52, 61–62, 81, 1707t, 181f, 184t, 204, 265f, 269t, 270, 271, 272, 273, 303, 323t, 383, 396–397, 397f, 432, 434f, 435, 438
Trelligence	Affinity™	www.trelligence.com	158, 161, 164t, 171, 302
U.S. Cost	Success Estimator	www.uscost.com	120, 121, 220, 460, 461
Vico†	Constructor Estimator	www.vicosoftware.com	59, 220, 224, 229t, 231t, 298
Vizelia	FACILITY	www.vizelia.com	127, 129f
WolfRam research Inc.	Mathematica®	http://www.wolfram.com/	178, 180

||In March, 2007 Nemetschek, Inc. acquired Graphisoft. Throughout the book, however, all Graphisoft products are referred to as Graphisoft products.

*In June, 2007 Autodesk, Inc. announced its plans to acquire Navisworks. In the book, all Navisworks products are referred to as Navisworks products.

†In March, 2007 Vico, Inc. was spun out from the Construction Solutions Division of Graphisoft.

Index

(See Software Index for software)

F

G

H

I

J

K

L